Springer
Proceedings in Physics   26

Springer Proceedings in Physics

Managing Editor: H. K. V. Lotsch

Volume 1 *Fluctuations and Sensitivity in Nonequilibrium Systems*
Editors: W. Horsthemke and D. K. Kondepudi

Volume 2 *EXAFS and Near Edge Structure III*
Editors: K. O. Hodgson, B. Hedman, and J. E. Penner-Hahn

Volume 3 *Nonlinear Phenomena in Physics*   Editor: F. Claro

Volume 4 *Time-Resolved Vibrational Spectroscopy*
Editors: A. Laubereau and M. Stockburger

Volume 5 *Physics of Finely Divided Matter*   Editors: N. Boccara and M. Daoud

Volume 6 *Aerogels*   Editor: J. Fricke

Volume 7 *Nonlinear Optics: Materials and Devices*   Editors: C. Flytzanis and J. L. Oudar

Volume 8 *Optical Bistability III*   Editors: H. M. Gibbs, P. Mandel,
N. Peyghambarian, and S. D. Smith

Volume 9 *Ion Formation from Organic Solids (IFOS III)*   Editor: A. Benninghoven

Volume 10 *Atomic Transport and Defects in Metals by Neutron Scattering*
Editors: C. Janot, W. Petry, D. Richter, and T. Springer

Volume 11 *Biophysical Effects of Steady Magnetic Fields*
Editors: G. Maret, J. Kiepenheuer, and N. Boccara

Volume 12 *Quantum Optics IV*   Editors: J. D. Harvey and D. F. Walls

Volume 13 *The Physics and Fabrication of Microstructures and Microdevices*
Editors: M. J. Kelly and C. Weisbuch

Volume 14 *Magnetic Properties of Low-Dimensional Systems*
Editors: L. M. Falicov and J. L. Morán-López

Volume 15 *Gas Flow and Chemical Lasers*   Editor: S. Rosenwaks

Volume 16 *Photons and Continuum States of Atoms and Molecules*
Editors: N. K. Rahman, C. Guidotti, and M. Allegrini

Volume 17 *Quantum Aspects of Molecular Motions in Solids*
Editors: A. Heidemann, A. Magerl, M. Prager, D. Richter, and T. Springer

Volume 18 *Electro-optic and Photorefractive Materials*   Editor: P. Günter

Volume 19 *Lasers and Synergetics*   Editors: R. Graham and A. Wunderlin

Volume 20 *Primary Processes in Photobiology*   Editor: T. Kobayashi

Volume 21 *Physics of Amphiphilic Layers*   Editors: J. Meunier, D. Langevin,
and N. Boccara

Volume 22 *Semiconductor Interfaces: Formation and Properties*
Editors: G. Le Lay, J. Derrien, and N. Boccara

Volume 23 *Magnetic Excitations and Fluctuations II*   Editors: U. Balucani, S. W. Lovesey,
M. G. Rasetti, and V. Tognetti

Volume 24 *Recent Topics in Theoretical Physics*   Editor: H. Takayama

Volume 25 *Excitons in Confined Systems*   Editors: R. Del Sole, A. D'Andrea, and
A. Lapiccirella

Volume 26 *The Elementary Structure of Matter*   Editors: J.-M. Richard, E. Aslanides, and
N. Boccara

Volume 27 *Competing Interactions and Microstructures: Statics and Dynamics*
Editors: R. LeSar, A. Bishop, and R. Heffner

Volume 28 *Anderson Localization*   Editors: T. Ando and H. Fukuyama

The Elementary Structure of Matter

Proceedings of the Workshop,
Les Houches, France, March 24 – April 2, 1987

Editors: J.-M. Richard,
E. Aslanides, and N. Boccara

With 207 Figures

Springer-Verlag Berlin Heidelberg New York
London Paris Tokyo

Professor Jean-Marc Richard
Inst. des Sciences Nucléaires, 53, Avenue des Martyrs,
F-38026 Grenoble, France

Professor Elie Aslanides
C.P.P.M.-C.N.R.S., Luminy Case 907, F-13288 Marseille Cedex 9, France

Professor Nino Boccara
Centre de Physique, Université Scientifique et Médicale, Côte des Chavants,
F-74310 Les Houches, France

ISBN 3-540-19013-9 Springer-Verlag Berlin Heidelberg New York
ISBN 0-387-19013-9 Springer-Verlag New York Berlin Heidelberg

Library of Congress. Cataloging-in-Publication Data. The Elementary structure of matter. (Springer proceedings in physics; v.26) Preface in English and French. Includes index. 1. Nuclear structure – Congresses. I. Richard, J.-M. (Jean-Marc), 1947–. II. Aslanides, E. (Elie), 1942–. III. Boccara, Nino. IV. Series. QC793.3.S8E44 1988   539.7'4   88-4650

This work is subject to copyright. All rights are reserved, whether the whole or part of the material is concerned, specifically the rights of translation, reprinting, reuse of illustrations, recitation, broadcasting, reproduction on microfilms or in other ways, and storage in data banks. Duplication of this publication or parts thereof is only permitted under the provisions of the German Copyright Law of September 9, 1965, in its version of June 24, 1985, and a copyright fee must always be paid. Violations fall under the prosecution act of the German Copyright Law.

© Springer-Verlag Berlin Heidelberg 1988
Printed in Germany

The use of registered names, trademarks, etc. in this publication does not imply, even in the absence of a specific statement, that such names are exempt from the relevant protective laws and regulations and therefore free for general use.

Printing: Weihert-Druck GmbH, D-6100 Darmstadt
Binding: J. Schäffer GmbH & Co. KG., D-6718 Grünstadt
2154/3150-543210

Preface

The meeting entitled "The Elementary Structure of Matter" was held at Les Houches, France, from 24 March to 2 April, 1987. The aim was to bring together experimentalists and theoreticians to discuss some topics relevant to both nuclear and particle physics: the quark structure of nucleons and nuclei, hadron interactions, lepton scattering, the study of hadrons and strange nuclei, collisions of heavy ions, implications of results from astrophysics, new accelerators.

The meeting was held in excellent conditions, thanks to the collaboration of the members of the Conseil des Houches and the generous support of the Institut National de Physique Nucléaire et de Physique des Particules, the Institut de Recherche Fondamentale du Commissariat à l'Energie Atomique and the Ministère des Relations Extérieures. We very much appreciated the zeal and the competence of the staff of the Ecole des Houches, in particular Mme Nicole Leblanc and Mme Anny Glomot.

Our warmest thanks go to the speakers for the effort they took with their presentations and the writing of their contributions to the proceedings. We are grateful to all the participants who enlivened the discussions throughout the meeting, and also to Sonia Fleck, whose assistance with the preparation of the book is greatly appreciated. Finally, we should not forget the guides from Les Houches who reassured even the least daring during an excursion to the Vallée Blanche.

Marseille
Les Houches
Grenoble, September 1987

E. Aslanides
N. Boccara
J.M. Richard

Preface

La réunion consacrée à "La structure Elémentaire de la Matière" s'est déroulée aux Houches, du 24 mars au 2 avril 1987. Le but était de réunir des expérimentateurs et des théoriciens pour discuter de quelques problèmes touchant à la fois à la physique nucléaire et à la physique des particules: structure en quarks des hadrons et des noyaux, annihilation sur nucléons et sur noyaux, interactions hadroniques, diffusion de leptons, étude des hadrons et des noyaux étranges, collisions d'ions lourds, implications des résultats d'astrophysique, nouveaux accélérateurs.

La rencontre s'est déroulée dans d'excellentes conditions, grâce à la collaboration des membres du Conseil des Houches, au généreux support de l'Institut National de Physique Nucléaire et de Physique des Particules, de l'Institut de Recherche Fondamentale du Commissariat à l'Energie Atomique et du Ministère des Relations Extérieures, et grâce au dévouement et à la compétence du personnel de l'Ecole des Houches, en particuliers de Mmes Nicole Leblanc et Anny Glomot.

Les remerciements les plus chaleureux vont aux orateurs pour leurs efforts consacrés à l'exposé et à la rédaction des comptes-rendus, aux participants qui ont animé les discussions tout au long de la rencontre, ainsi qu'à Sonia Fleck qui a participé à l'édition de ce livre, sans oublier les guides des Houches, qui ont su rassurer même les moins téméraires lors d'une excursion dans la Vallée Blanche.

Marseille
Les Houches
Paris, September 1987

E. Aslanides
N. Boccara
J.M. Richard

Contents

Part I	Quarks and Hadrons

Non-perturbative QCD on the Lattice
By B.J. Pendleton (With 8 Figures) 2

From Lattice QCD to Nuclear Physics
By H.J. Pirner (With 7 Figures) 14

Frontiers of the Quark Model. By H.J. Lipkin 24

Heavy Multiquark States. By L. Heller (With 2 Figures) 35

Stable Multiquark States
By C. Gignoux, B. Silvestre-Brac, and J.M. Richard 42

Strong Decay of Baryons. By Fl. Stancu and P. Stassart 46

Resonating Group Method Applied to Hadrons
By P. Bicudo (With 4 Figures) 51

Quark Confinement and Nuclear Dynamics
By E.J. Moniz (With 10 Figures) 56

The Cheshire Cat Principle Applied to Hybrid Bag Models
By H.B. Nielsen and A. Wirzba (With 10 Figures) 72

Pion and Nucleon Structure: Low Energy Aspects
By W. Weise (With 6 Figures) 101

Chiral Symmetry and Light Mesons. By A. Le Yaouanc, L. Oliver,
O. Pène, and J.-C. Raynal (With 4 Figures) 115

Chiral Field Theories as Models for Hadron Substructure
By S.H. Kahana (With 3 Figures) 125

Strange Skyrmions. By M. Praszałowicz (With 1 Figure) 133

Diquarks in Exclusive Reactions. By P. Kroll (With 7 Figures) 138

Diquark Clustering in Baryons
By S. Fleck, B. Silvestre-Brac, C. Gignoux, and J.M. Richard 148

Hadron Wave Functions with Condensate Induced Running Masses
By M. Lavelle, E. Werner, and St. Głazek (With 2 Figures)   155

Vector Meson Interactions in the Effective Lagrangian
By B. Moussallam   160

Infrared Aspects of QCD. By H.M. Fried   164

Many-Body Techniques Applied to QCD and Aspects of Confinement
By D. Schütte   168

Hadronic Reactions at Large Momentum Transfers
By J. Soffer (With 9 Figures)   173

Polarized Parton Distributions and the Magnitude of Spin Effects at
Very High Energies. By P. Taxil (With 2 Figures)   184

Part II Annihilation

$N\bar{N}$ Annihilation into Two Mesons
By A.M. Green (With 2 Figures)   190

Charged Two-Meson Production from $N\bar{N}$ Annihilation at Rest
By G.A. Smith (With 4 Figures)   197

The S-, P-Wave Problem in $N\bar{N} \to \pi\pi$ at Rest
By G.A. Smith (With 2 Figures)   202

Spin-Dependent Observables in $p\bar{p}$ Elastic Scattering at Low Energy
By R. Bertini, H. Catz, A. Chaumeaux, B. Fabbro, J.-C. Faivre,
H. Fanet, J. Pain, F. Perrot, E. Vercellin, E. Boschitz, W. Gyles,
C.R. Ottermann, T. Tacik, E. Descroix, R. Harountunian,
J.Y. Grossiord, A. Guichard, J. Arvieux, J.M. Durand, J. Yonnet,
S. Mango, J. Konter, and B. Van den Brandt (With 2 Figures)   205

Antiproton-Nucleus Annihilation. By J. Cugnon (With 4 Figures) ..   211

Search for Unusual Behavior in $\bar{p}$-Nucleus Annihilation at Rest
By G.A. Smith (With 5 Figures)   219

Part III Structure Functions

Nucleons in Nuclei from Quasi-Elastic Electron Scattering
By A. Gérard (With 2 Figures)   226

New Results on the EMC Effect
By G. D'Agostini (With 13 Figures)   235

Nuclear Effects in Quark and Gluon Distributions – Experimental
Perspectives. By K. Rith (With 12 Figures)   245

The EMC Effect and Related Issues
By Hong Jung and G.A. Miller (With 9 Figures) 259

Part IV Strangeness

An Overview of Hypernuclear Physics
By R. Bertini (With 14 Figures) 272

Perspectives in Strange Particle Physics
By P.D. Barnes (With 17 Figures) 292

Hyperon-Hyperon Interaction and the H Particle
By J. Carbonell, B. Silvestre-Brac, and C. Gignoux (With 4 Figures) 311

Part V Relativistic Heavy Ions

A Review of Quark-Gluon Plasma and High Energy Heavy Ion
Collisions. By L. McLerran (With 4 Figures) 320

New Processes and Old Spin Physics. By M. Jacob (With 1 Figure) 333

Hot Strange Matter in Relativistic Nuclear Collisions
By J. Rafelski (With 4 Figures) 340

First Results from the CERN Light Ion Program
By G.W. London (With 9 Figures) 352

Perspectives on Heavy Ion Physics at CERN in the 1990s (or What
Can We Gain from a Lead Beam?)
By G.W. London (With 5 Figures) 361

Part VI Axions

The Emission of Isoenergetic Electron Positron Pairs from Very
Heavy Ion-Atom Collisions. By K.E. Stiebing (With 5 Figures) ... 372

Introduction to Axions. By Kyungsik Kang 378

Part VII Round Table on Future Medium Energy Accelerators

A LEAR-like Option for Brookhaven
By G.A. Smith (With 2 Figures) 404

The Scientific Program of a Multi-GeV cw Electron Accelerator
By A. Gérard 406

European Proposals for a B-Factory. By J. Duclos 408

Physics at Laboratoire National Saturne with MIMAS
By M. Roy-Stéphan 410

A Hadron Facility for Europe. By F. Bradamante (With 1 Figure) . 412

Physics at Super-LEAR. By P. Dalpiaz, R. Klapisch, P. Lefevre, M. Macri, L. Montanet, D. Möhl, A. Martin, J.M. Richard, H.J. Pirner, and L. Tecchio (With 4 Figures) 414

Round Table Discussion on Future Accelerators: RHIC
By L. McLerran 437

Part VIII Astrophysics

Supernova Theory and 1987a (Shelton)
By S.H. Kahana (With 4 Figures) 440

Particle Physics and Astrophysics. By E. Schatzman 449

Part IX Conclusions

Theoretical Perspective. By H.B. Nielsen (With 3 Figures) 454

Index of Contributors 467

Part I

Quarks and Hadrons

Non-perturbative QCD on the Lattice

B.J. Pendleton

Physics Department, University of Edinburgh,
The King's Buildings, Edinburgh EH9 3JZ, Scotland

1 Introduction

Lattice regularisation has become the major tool for non-perturbative studies of Quantum Chromodynamics (QCD). In this talk I shall review the lattice formulation of QCD and discuss recent progress in numerical simulations. It is not meant to be an exhaustive study; for a comprehensive review see [1]. I will discuss three topics: the approach to the continuum limit, the calculation of the hadron spectrum in the quenched approximation, and the phase transitions to quark-gluon matter at high temperature and density.

QCD purports to be the correct quantum field theory of the strong interactions of quarks and gluons. Even after many years of effort we still have no good analytic solution of QCD. The short distance (high energy) properties of the theory are well understood because of asymptotic freedom. This permits the calculation of a whole plethora of quantities whose agreement with experiment is well known. However, we are not able to make any first principles analytic calculations of such important quantities as the hadron spectrum, decay rates, or low energy scattering processes. Instead, we perform computer simulations of lattice regularised QCD from which we extract quantitative results for many interesting observables. This is not as easy as was first thought.

Monte Carlo simulations require enormous computer resources for many reasons. The lattice spacing in physical units must be small enough so that the results obtained are independent of the cutoff (continuum limit); the number of lattice sites must grow as the lattice spacing decreases in order to prevent the physical volume shrinking to zero. Furthermore, for fixed lattice spacing, we must ensure that the finite size of the box in physical units does not affect the results (thermodynamic limit). One must also contend with statistical errors. In order that these be small, we must make measurements of interesting quantities on a large sample of statistically independent gauge field configurations. As we approach the continuum limit of lattice QCD the correlation between successive Monte Carlo generated configurations grows rapidly. This is the problem of critical slowing down and it occurs in all physical systems with long range correlations.

The rest of this talk describes results on lattices of sizes up to 20^4.

2 The Continuum Limit

As is well known, the QCD Langrangian is

$$\mathcal{L} = -\frac{1}{2}\text{Tr} F_{\mu\nu}^2(x) + \bar{\psi}(x)\left(\gamma^\mu D_\mu + m\right)\psi(x) \tag{1}$$

where $F_{\mu\nu}(x) = -ig^{-1}[D_\mu(x), D_\nu(x)]$ and the covariant derivative is defined in terms of the gauge field by $D_\mu(x) = \partial_\mu + igA_\mu(x)$.

The transcription to the lattice is carried out by dividing (Euclidean) spacetime into a grid of points with spacing a. A gauge invariant precription for doing this was given long ago by Wilson. All the results presented here use the staggered fermion formulation which preserves on the lattice a continuous remnant of the full continuum chiral symmetry. The lattice action is

$$S[U, \bar{\psi}, \psi] = S_G[U] + S_F[U, \bar{\psi}, \psi], \tag{2}$$

where $S_G[U]$ is the pure gauge action

$$S_G[U] = -\frac{\beta}{3}\sum_{x,\mu<\nu} \text{Tr Re}\left\{U_\mu(x)U_\nu(x+\hat{\mu})U_\mu^\dagger(x+\hat{\nu})U_\nu^\dagger(x)\right\} \tag{3}$$

and $S_F[U]$ describes the interaction of the quarks and gluons

$$S_F[U, \bar{\psi}, \psi] = \sum_{x,y} \bar{\psi}(x) M[x, y; U] \psi(y). \tag{4}$$

The gauge fields $U_\mu(x) = e^{iagA_\mu(x)}$ are SU(3) matrices living on the links of the lattice and the ψ, $\bar{\psi}$ fields are 3 colour component Grassmann fields without Dirac indices. The fermion kernel M is

$$M[x, y; U] \equiv m\delta_{x,y} + \frac{1}{2}\sum_{\mu=1}^{4}\left\{U_\mu(x)\eta_\mu(x)\delta_{y,x+\hat{\mu}} - U_\mu^\dagger(y)\eta_\mu(y)\delta_{y,x-\hat{\mu}}\right\}. \tag{5}$$

The phase factors $\eta_\mu(x) \equiv \prod_{\nu=1}^{\mu-1}(-)^{x_\nu}$ are the staggered fermion equivalents of the Dirac gamma matrices. The quark mass is m and β is related to the lattice coupling constant g by $\beta = \frac{6}{g^2}$. Because of fermion doubling, the staggered fermion action describes four flavours of quarks in the continuum limit. Expectation values of operators Ω which depend on both gluon and quark fields are given by

$$\langle \Omega \rangle = \frac{1}{Z}\int [dU][d\bar{\psi}][d\psi] e^{-S_G[U]-S_F[U,\bar{\psi},\psi]}\,\Omega[U, \bar{\psi}, \psi]. \tag{6}$$

Asymptotic freedom tells us how to take the continuum limit of the lattice theory. For small g^2 the lattice spacing is given by

$$a = \frac{1}{\Lambda_L} \left(g^2 b_0\right)^{\frac{b_1}{2b_0^2}} e^{-\frac{1}{2b_0 g^2}}, \qquad (7)$$

where $b_0 = (1/16\pi^2)(11 - 2n_f/3)$, $b_1 = (1/16\pi^2)^2(102 - 38n_f/3)$, n_f is the number of quark flavours, and Λ_L is the scale parameter for lattice regularised QCD. Suppose we measure, as a function of g^2, a physical quantity such as the proton mass. The continuum limit is obtained when this mass, measured in units of the inverse lattice spacing, scales as

$$M_l = aM_p = \frac{M_p}{\Lambda_L} \left(g^2 b_0\right)^{\frac{b_1}{2b_0^2}} e^{-\frac{1}{2b_0 g^2}}, \qquad (8)$$

where M_l (M_p) is the proton mass in lattice (physical) units. This is known as asymptotic scaling and should occur for small enough g^2. Just how small g^2 must be is determined by explicit calculation. A less stringent condition is that results independent of lattice artifacts may be obtained at slightly larger couplings where M_p does not scale as in (8) but where ratios of physical quantities are independent of g^2. This is known as "general scaling" and is a non-trivial requirement. It does not occur in the strong coupling limit $g^2 \to \infty$.

The essential singularity at $g^2 = 0$ in (8) shows that ordinary weak coupling perturbation expansions cannot be used to calculate hadron masses. (In a finite volume, where there is an additional scale, some weak coupling spectrum calculations can be made [2].) One major advantage of the lattice formulation is that it allows a perturbation series in $1/g^2$ (strong coupling expansion) which has provided very useful information about confinement and hadron masses on the lattice [3]. Unfortunately, the extrapolation from the strong coupling régime to the continuum limit at $g^2 = 0$ is very difficult.

As an example of what can be calculated numerically, Fig. (1) shows the potential between heavy quark sources calculated from large Wilson loops on a

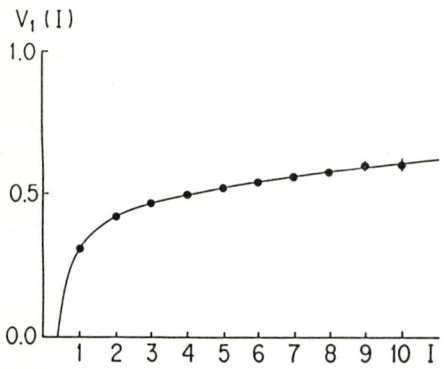

Figure 1: The quark-antiquark potential at $\beta = 6.8$. The data is from S. Itoh et al. [4]

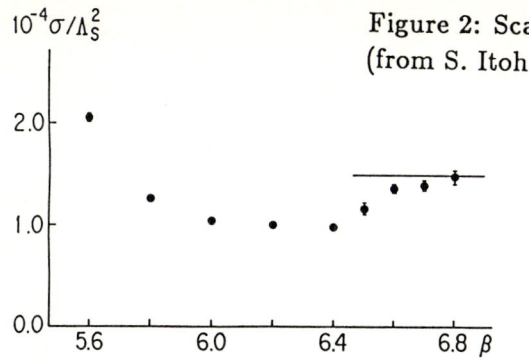

Figure 2: Scaling properties of the string tension σ (from S. Itoh et al. [4])

20^4 lattice. The linear potential at large distances turns into a Coulomb force at short distances in agreement with the predictions of asymptotic freedom.

Much work has gone into trying to determine the value of g^2 where asymptotic scaling begins. Figure (2) shows a plot of the string tension, normalised by Λ_L, against β for a 20^4 lattice. Clearly, asymptotic scaling holds at the 20% level for $\beta > 6$. However, the positive slope in what should be a constant quantity gives cause for concern. Calculations of the deconfining temperature for β values in the range $6.15 \to 6.45$ are also consistent with scaling behaviour [5].

3 Hadron Spectrum Calculations

Incorporating fermions into Monte Carlo calculations presents a problem. The fermion fields belong to a Grassmann algebra and cannot be stored in a computer as ordinary numbers. The usual procedure is to perform the integral over fermion fields in (6) to obtain a non-local effective action depending only on the gauge fields

$$S_{eff}[U] = S_G[U] - \ln\det(\mathcal{M}). \qquad (9)$$

For numerical simulations, the cure is as bad as the problem. We wish to generate a series of configurations, distributed as $e^{-S_{eff}}$, by a Markov process. This involves sweeping through the lattice making local changes to the U fields according to some prescribed process which includes comparing the action in the "old" and "new" configurations. The non-local nature of S_{eff} means that we have to calculate a determinant which depends on the state of the whole lattice each time we wish to change a single link variable $U_\mu(x)$.

The crudest aproximation is simply to leave out the determinant and to calculate expectation values of fermionic observables in a pure gauge field background. This is the quenched approximation used in most calculations of the hadron spectrum to date. It corresponds to omitting all diagrams with closed fermion loops.

To extract hadron masses, one measures the propagators of composite operators $\mathcal{O}$ having the appropriate quantum numbers to create/destroy the hadrons of interest. Examples are $\mathcal{O} = \bar{\psi}\gamma_5\psi$ for the pion and $\bar{\psi}\gamma_\mu\psi$ for the rho. Asymptotically, these decay as exponentials

$$\sum_{\mathbf{x}} \langle \mathcal{O}(\mathbf{x},t) \mathcal{O}^\dagger(0,0) \rangle \sim e^{-m_{hadron} t}. \qquad (10)$$

It is actually quite hard to extract the hadron spectrum from small lattices. The u and d quarks are extremely light giving rise to an approximate chiral symmetry which is broken spontaneously. The Goldstone pion is almost massless and can propagate easily to the extrema of the lattice causing large finite size effects. From chiral perturbation theory, one expects the pion mass to behave as

$$m_\pi^2 \propto m_q. \qquad (11)$$

Deviations form this relation are interpreted as finite size effects.

Figure (3) shows some recent results for the pion mass obtained by the Edinburgh group [6] on a 16^3 by 24 lattice at $\beta = 6.15$. Clearly, there is a large range of quark masses for which the relation (11) holds. However, the lightest mass

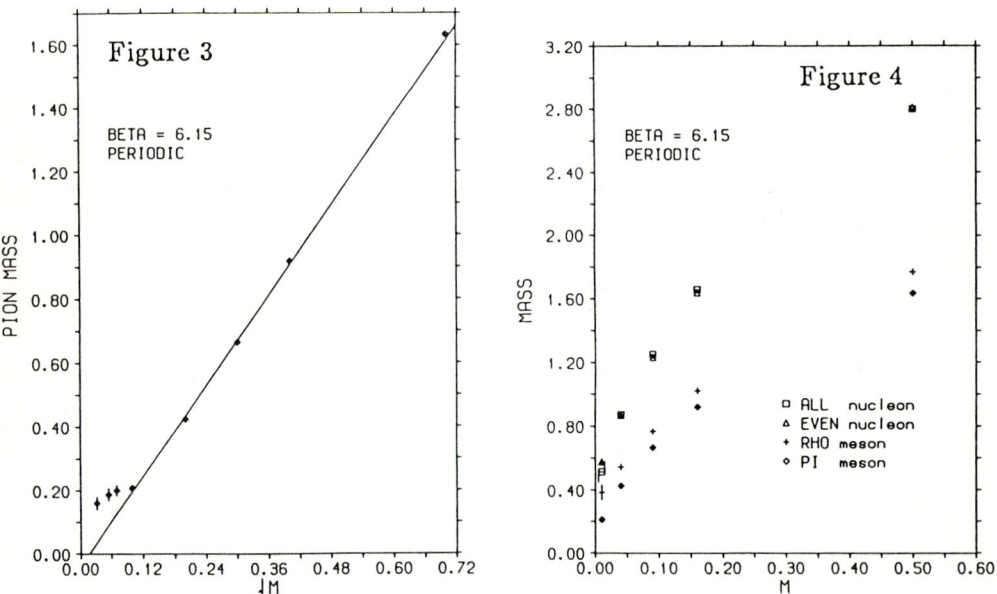

Figure 3: The pion mass in lattice units plotted against the square root of the quark mass. The data is from Bowler et al. [6]

Figure 4: The low lying quenched hadron spectrum as a function of quark mass (Bowler et al. [6])

at which finite size effects appear to be under control is 0.01. An extrapolation is required because the physical u and d quark masses are $\sim$ 7MeV which translates to $\sim$ 0.003 in lattice units. We have used the measured value of the string tension to set the scale. Figure (4) shows the results obtained for a variety of hadrons on the same lattice. The qualitative behaviour is much as one would expect.

Another way of looking at the results is shown in figure (5) where the nucleon mass (normalised to the rho mass) is plotted against the pion mass for a range of quark masses. Assuming that the quenched approximation contains some vestige of truth, then the plotted points should approach the "real world" position as the quark mass is reduced to its physical value. Clearly, an extrapolation with the current data is difficult although the trend does seem to be in the right direction.

In order to try to avoid finite size effects one can reduce beta. This increases the size of the lattice in physical units. However, it turns out that nothing is gained because the results are plagued by lattice artifacts: flavour symmetry is badly broken in the staggered fermion formalism when the lattice spacing is any larger. Current plans of the Edinburgh group are to work on significantly bigger lattices at $\beta = 6.15$ in order to perform simulations at physical quark masses. This requires an enormous increase in computing power. Simulations including the effects of internal quark loops on such large lattices will be very difficult indeed.

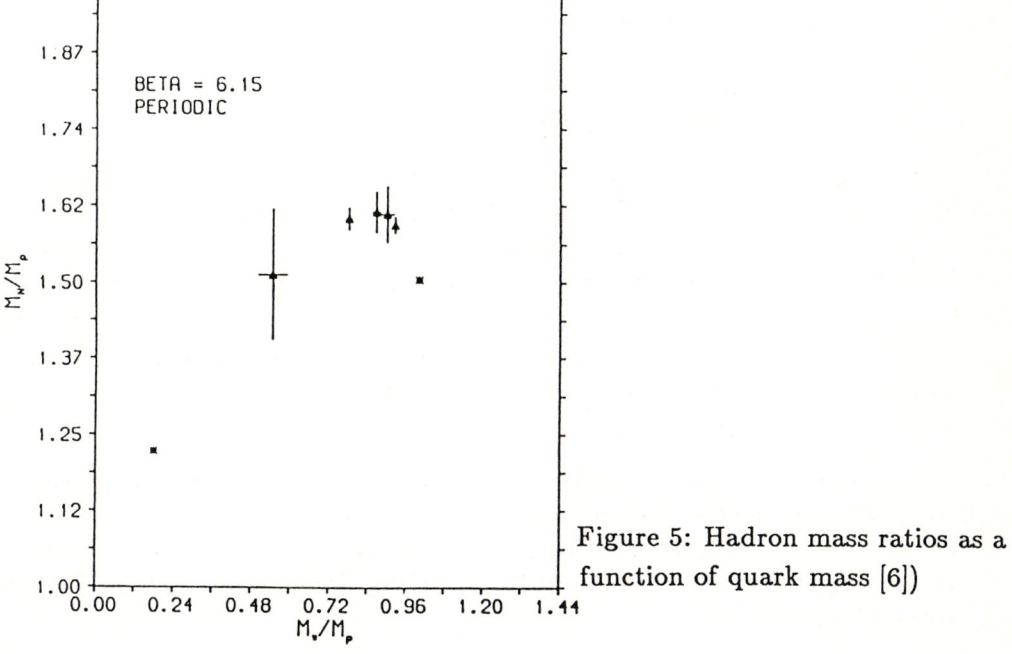

Figure 5: Hadron mass ratios as a function of quark mass [6])

4 Phase Transitions at Finite Temperature and Density

Calculations including dynamical fermions have already begun. One interesting way to study the physics of confinement and chiral symmetry breaking is to study how these properties disappear under extreme conditions. At high temperatures there is expected to be a phase transition from the confining chirally broken phase of QCD to a chirally symmetric quark-gluon plasma.

In the absence of dynamical quarks (corresponding to infinite mass for internal loops) this transition is well understood in terms of an order parameter (the Polyakov line) which is related to the free energy F_q of a static quark source [7]. There is a strong first order phase transition with a large latent heat: F_q is infinite below the transition. Since the transition is first order, it is expected to persist when heavy dynamical quarks are taken into account.

A precise definition of confinement is difficult when light dynamical fermions are included; there is no genuine order parameter which can distinguish between confined and free quarks. A light dynamical quark can bind to a static quark source and the energy of the resulting bound state will always be finite. However, when the quarks are massless there is an exact chiral symmetry for which the condensate $\langle \bar{\psi}\psi \rangle$ is a bona fide order parameter. Theoretical arguments predict a first order transition when the number of flavours is greater than two [8]. Hence any phase transition observed with very light quarks should be related to the restoration of chiral symmetry.

What happens for quarks of intermediate mass is hard to predict; the simplest guess is that nothing will happen. The simulations seem to show that this simple picture is indeed correct. Our task is to discover what happens for the case of two light flavours (the up and down quarks) and one flavour of intermediate mass (the strange quark). Heavier quarks can safely be ignored.

Most recent simulations have used equation of motion methods based on the Langevin Equation [9], Molecular Dynamics [10], or a hybrid of both [11]. The problem of non-locality is avoided by making simultaneous small changes in all link variables at the cost of a slower evolution through phase space. Discretisation errors appear because we can only integrate the equations of motion approximately on a computer. Recently, we have shown that both of these disadvantages are avoidable if one uses a hybrid algorithm to guide a Monte Carlo simulation in phase space [12].

Non-zero temperature is realised on Euclidean lattices by imposing periodicity in (imaginary) time. The temperature T is given by $T = (N_t a)^{-1}$ where N_t is the temporal periodicity. Clearly, for fixed N_t, the temperature can be raised by decreasing the lattice spacing $a(g^2)$. From (7), a reduction in $\beta = \frac{6}{g^2}$ heats up the system.

The most convincing results have been obtained with four flavours of light staggered fermions [13,14,15]. Figure (6) clearly shows the existence of metastable states; the quark gluon plasma coexisting with the low temperature hadronic gas. This is good evidence for a first order chiral symmetry restoration transition.

At the time of writing, the situation for two light flavours is less clear [14,16]. Figure (7) shows a sharp transition between the low and high temperature regimes. However, there is no sign of metastability (Fig.(8)). This indicates either a higher order transition or merely a very rapid crossover.

Much work remains to be done. In the real world there are $2\frac{1}{2}$ flavours of light quarks; the effect of the strange quark may not be negligible. One must also check that the nature of the phase transition survives the continuum limit and measure physical quantities such as the latent heat. This will take an enormous amount of computer time but an impressive start has been made [17].

The best laboratory for studying the quark-gluon plasma is in relativistic collisions of heavy ions. Here, as well as high temperatures, one has to contend with the effects of finite baryon density [18]. These are also expected to drive a phase transition to chirally symmetric quark-gluon matter.

In the continuum, finite baryon density can be handled by adding a term containing a chemical potential κ to the action. On the lattice the time-like term in the fermion action becomes

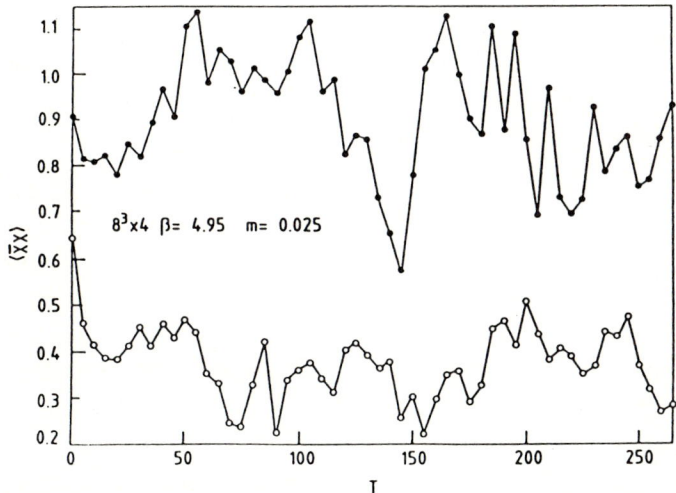

Figure 6: Phase coexistence at the critical temperature for four light flavours. The chiral condensate is shown as a function of simulation time for two different runs. The upper (lower) curve represents the chirally broken (symmetric) phase. The data is from Karsch et al.[15]

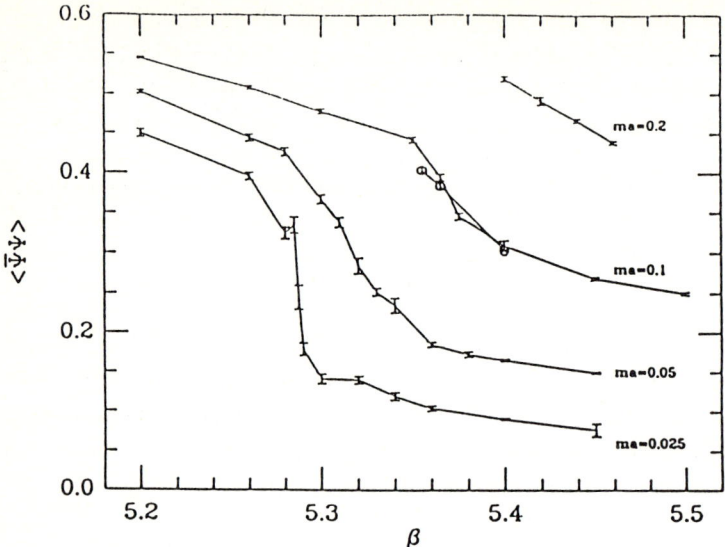

Figure 7: The chiral condensate as a function of β for two flavours of various mass. The data is from Gottlieb et al.[14]

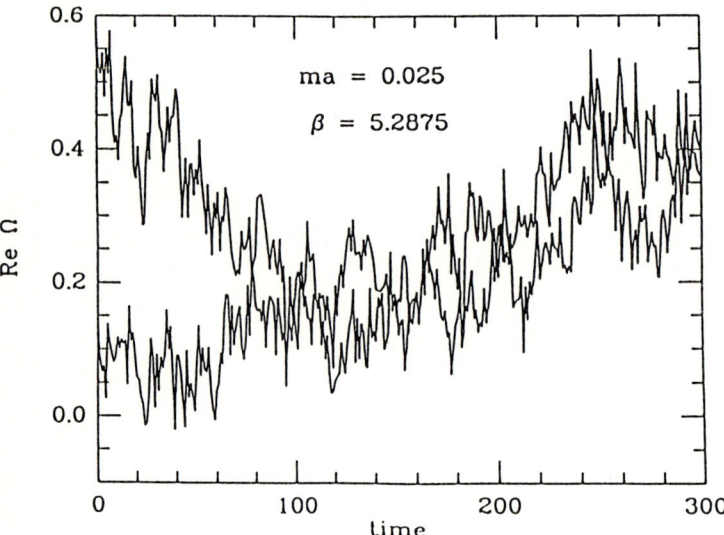

Figure 8: Absence of metastable states for two flavours of quarks. The Polyakov Loop is shown as a function of simulation time for two runs with different starts. The data is from Gottlieb et al.[14]

$$m\delta_{x,y} + \frac{1}{2}\left\{U_t(x)\,e^{\kappa}\,\eta_t(x)\delta_{y,x+\hat{t}} - U_t^\dagger(y)\,e^{-\kappa}\,\eta_t(y)\delta_{y,x-\hat{t}}\right\} \qquad (12)$$

and a new problem arises: the fermion determinant in (9) is no longer real, Monte Carlo update probabilities become complex, and the usual simulation methods break down. The simplest way to get around this problem is to use the quenched approximation. The calculation of fermionic observables with a chemical potential in a quenched background presents no problems of principle. However, it has been shown that the quenched approximation is pathological in the interesting region where the chemical potential exceeds half the pion mass [19]. The fermion determinant must be included and several suggestions have been made to try to accommodate it.

One proposal is to update the system of gauge fields with an effective action which includes only the magnitude of the determinant [20]. The phase must be carried as an extra weight when calculating observables. Since this phase is an extensive quantity, its fluctuations are enormous and the signal to noise ratio becomes very small for observables such as the chiral condensate. Nevertheless useful exploratory results are being obtained on small lattices. [21].

Another approach is to use the Langevin equation which, a priori, doesn't require a real effective action [22]. However, such a system doesn't necessarily satisfy the criteria for the Langevin process to converge. Again, useful results have been obtained for some model systems [23] although the real problem of QCD seems more difficult.

5 Future Prospects

Numerical simulations of lattice QCD have provided us with useful results which could not have been obtained with any other currently known method. Even qualitative results make us confident that our understanding of the long distance behaviour of QCD is correct. Nevertheless, it will be some time before we can confidently quote the QCD prediction for the low lying hadron spectrum to within a few percent. This is something that we know anyway! The most exciting prospect is that we may be able to predict the properties of quark-gluon matter before our experimental colleages can produce it in the laboratory.

6 Acknowledgements

I would like to thank Jean-Marc Richard for the opportunity of presenting this talk and for a very enjoyable stay at Les Houches. Useful discussions with I. Barbour, F. Karsch, L. McLerran, and R. D. Kenway are gratefully acknowledged. Thanks also to K. C. Bowler for reading the manuscript.

References

[1] P. Hasenfratz. Nonperturbative methods in quantum field theory. In *Proceedings of 22nd International Conference on High Energy Physics*, page 169, World Scientific, 1987.

[2] M. Lüscher. *Nucl. Phys.* B232, 445 (1983).

[3] J. M. Drouffe and J. B. Zuber. *Phys. Rep.* 130, 217 (1986).

[4] S. Itoh, Y. Iwasaki, and T. Yoshie. *Phys. Lett.* 185, 390 (1987).

[5] S. Gottlieb et al. *Phys. Rev. Lett.* 55, 1958 (1985).

[6] K. C. Bowler, C. B. Chalmers, R. D. Kenway, D. Roweth, and D. Stephenson. Quenched hadron mass calculations using staggered fermions at $\beta = 6.15$ and 6.3. Edinburgh Preprint 87/403.

[7] B. Svetitsky. *Phys. Rep.* 132, 1 (1986).

[8] R. Pisarski and F. Wilczek. *Phys. Rev.* D29, 338 (1983).

[9] G. G. Batrouni, G. R. Katz, A. S. Kronfeld, G. P. Lepage, B. Svetitsky, and K. G. Wilson. *Phys. Rev.* D32, 2735 (1985).

[10] J. Polonyi and H. W. Wyld. *Phys. Rev. Lett.* 49, 2257 (1983).

[11] S. Duane. *Nucl. Phys.* B257 [FS14], 652 (1985).

[12] S. Duane, A. D. Kennedy, B. J. Pendleton, and D. Roweth. Hybrid Monte Carlo. Edinburgh preprint 87/402, FSU preprint FSU-SCRI-87-27.

[13] R. Gupta et al. *Phys. Rev. Lett.* 57, 2621 (1986).

[14] S. Gottlieb, W. Liu, D. Toussaint, R. L. Renken, and R. L. Sugar. Chiral symmetry breaking in lattice QCD with two and four flavors. Santa Barbara preprint. 1987.

[15] F. Karsch, J. B. Kogut, D. K. Sinclair, and H. W. Wyld. First order chiral phase transition in lattice QCD. ANL-HEP-PR-86-137.

[16] M. Fukugita, S. Ohta, Y. Oyanagi, and A. Ukawa. Numerical evidence for a first order chiral transition in lattice QCD with two light flavours. Kek preprint 86-104. 1987.

[17] E. V. E. Kovacs, D. K. Sinclair, and J. B. Kogut. Return of the finite temperature phase transition in the chiral limit of lattice QCD. ANL-HEP-PR-86-137. 1987.

[18] J. Cleymans, R. Gavai, and E. Suhonen. *Phys. Rep.* 130, 217 (1986).

[19] P.Gibbs. *Phys. Lett.* 182B, 369 (1986).

[20] C. Baillie, K. C. Bowler, P. E. Gibbs, I. M. Barbour, and M. Rafique. The chiral condensate in SU(2) QCD at finite density. Edinburgh Preprint 87/400.

[21] I. Barbour. Private communication.

[22] F. Karsch and H. W. Wyld. *Phys. Rev. Lett.* 55, 2242 (1985).

[23] F. Karsch, J. Kogut, and H. W. Wyld. *Nucl. Phys.* B280, 289 (1987).

From Lattice QCD to Nuclear Physics

H.J. Pirner

CERN, Geneva, Switzerland
Present address: Institut für Theoretische Physik, Philosophenweg 19,
D-6900 Heidelberg, Fed. Rep. of Germany

1. GENERAL INTRODUCTION

Perturbative quantum chromodynamics (QCD) has been extensively studied and agrees well with experiment. As an example, let us look at the angular distribution of two-jet events measured in the UA1 experiment at CERN (see Fig. 1). The experimental data [1] follow the $1/(1-\cos\theta)^2$ behaviour of Rutherford scattering. Even small visible corrections from scaling violations (Q^2-dependence) in the structure functions are calculable. The gluon exchange manifests itself in the same way as photon exchange in electrodynamics. However perturbative QCD only becomes applicable at length scales $1/Q^2 \leqslant (0.1 \text{ fm})^2$. The interesting physics of interacting hadrons and nuclei, is given by a scale of 1 fm to 10 fm. It therefore needs non-perturbative methods to tackle the QCD Lagrangian. Various approaches have been proposed. Since an expansion in the strong coupling constant $\alpha_s(Q^2)$ of QCD is not possible, an expansion in $1/N_c$ has been suggested, where N_c is the number of colours [2,3]. One obtains a theory of weakly interacting mesons with couplings $g_M = O(1/N_c)$ and masses $O(1)$. Baryonic solutions have masses $O(1/g_M) = O(N_c)$, which possibly indicates that they are soliton solutions of the meson-field equations. The existence of a gluon master field proposed by Witten [4] in the $N_c \to \infty$ limit would allow quantitative calculations, but up to now no results in four dimensions are available. There are calculations trying to relate the QCD Lagrangian to a Skyrme-type effective action assuming chiral symmetry breaking [5]. But most of the progress in this field has been made in applications of Skyrme-type models [6].

A different mostly numerical approach to non-perturbative QCD has been followed during the last years, namely lattice QCD [7]. By discretizing continuous space–time on a finite lattice one considers a finite number of gauge-field configurations, which can be simulated numerically. The finite lattice constant regularizes an otherwise diverging field theory. The lattice method has developed into a field where the computational possibilities play an important part in the discussion. However there are also serious, not yet understood, physics problems concerning the treatment of systems with finite fermion density [8].

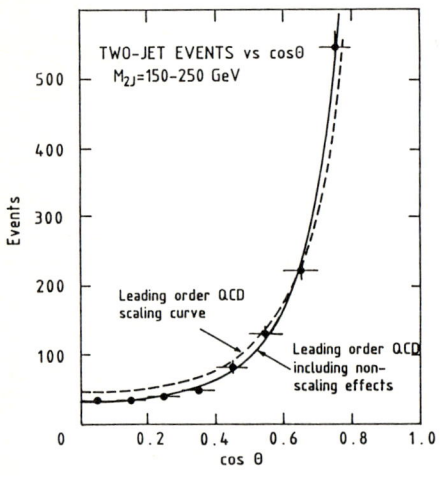

Fig. 1 Two-jet events as a function of $\cos\theta$ from Ref. 1. The $qq \to qq$, $gg \to gg$ and $gq \to gq$ processes give the Rutherford scattering distribution (dashed line) $d\sigma \propto 1/(1-\cos\theta)^2$. The full line includes the scaling violations in the structure functions.

The work I am going to describe here has involved a moderate numerical effort up to now. Its aim is primarily conceptual. Can we define an effective QCD theory, where the short-wavelength gluon degrees of freedom are eliminated in favour of a colour-neutral field characterizing the vacuum and giving confinement. The role of the polarizable medium is played by the high-frequency parts of the gauge fields. Since this effective theory has the same appearance as electrodynamics of dielectric media, it is called colour dielectrics.

From the 'macroscopic' Hamiltonian

$$H = \int \sum_{i=1}^{8} 1/\epsilon [\vec{D}_i^2 + \vec{H}_i^2] d^3x$$

it is easily seen that colour displacement fields $\vec{D}_i$ and magnetic fields $\vec{H}_i$ cannot subsist in a medium where $\epsilon = 0$, because the energy associated with them would be infinite. The vacuum therefore has $\epsilon = 0$. The dynamics of confinement is reduced to find the value of the dielectric field in every point in space around the colour charges. The region with $\epsilon = 0$ forms a bag around the $\epsilon \neq 0$ region inside the hadron, where colour is present and not influenced by vacuum fluctuations.

My paper will consist of three parts. In Section 2 I give an introduction to the main physics of lattice gauge theory. Section 3 gives an outline of the colour dielectric model and first numerical results on the effective action after one block-spinning step. This part of the paper is based on joint work with J. Wroldsen [9]. Section 4 is devoted to a report about phenomenological work, done in collaboration with G. Chanfray and A. Schuh, which uses the colour dielectric model as a starting point for calculations of the free nucleon and nuclear matter or finite nuclei.

2. INTRODUCTION TO CONCEPTS OF LATTICE GAUGE THEORY

For simplicity we are going to discuss non-Abelian SU(2) chromodynamics of gluon fields. That means the gluon fields have a 2 × 2 matrix representation by the Pauli matrices $\vec{\sigma}_i$ (i = 1, 3) as

$$A_\mu(x) = 1/2 \sum_{i=1}^{3} A_\mu^i \sigma^i .$$

Having discretized the three space and one time dimensions we introduce as relevant degrees of freedom the link variables

$$U_\mu(x) = \exp i g \int_x^{x+e_\mu} A_\nu d\xi_\nu = a_0 + i\vec{\sigma}\vec{a} , \quad (1)$$

which symbolize the integral of the gauge field between a site x of the lattice to the neighbouring site $x + e_\mu$, one lattice spacing away from x in the $\mu = (1, 4)$ direction. This link variable has a 2 × 2 matrix representation $U_\mu = a_0 + i\vec{\sigma}\vec{a}$ given by four numbers $(a_0, \vec{a})$. It is unitary, which means that the product of a link in the positive μ direction with the same link traversed in the opposite μ-direction gives the unity operator

$$U_\mu(x) U_\mu^+(x) = U_\mu(x) U_{-\mu}(x+e_\mu) = 1 . \quad (2)$$

Unitarity restricts the four quantities $(a_0, \vec{a})$ to the unit circle $a_0^2 + \vec{a}^2 = 1$. The effect of a gauge transformation G(x) on $U_\mu(x)$ can be seen by expanding U_μ for small gauge fields and using the standard continuum gauge transformation $A_\mu(x) \to A'_\mu = (1/i) [\partial_\mu G(x)]G^{-1}(x) + G(x)A_\mu(x)G^{-1}(x)$. We get

$$\begin{aligned}&1 + i A_\mu dx_\mu \to \\ &1 + \{[G(x+dx) - G(x)]/dx_\mu G^{-1}(x) + i G(x) A_\mu(x) G^{-1}(x)\} dx_\mu \\ &\approx G(x+dx)(1 + i A_\mu dx_\mu) G^{-1}(x) .\end{aligned} \quad (3)$$

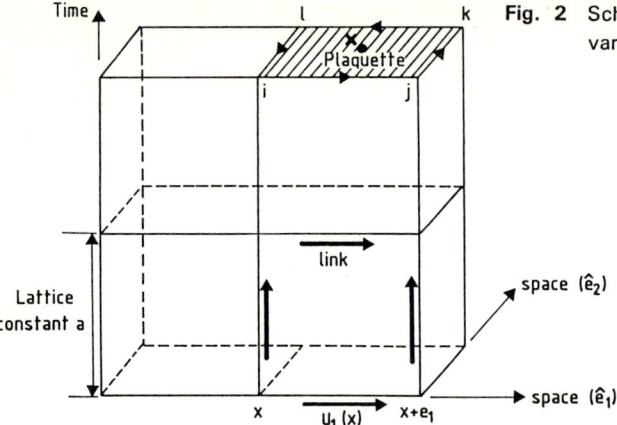

Fig. 2 Schematic picture of a lattice with the link variables and a basic plaquette.

Therefore a link $U_\mu(x)$ transforms by applying the gauge transformations G and G^{-1} on the end points $(x + e_\mu)$ and x of the link. Consequently, the natural choice for a gauge-invariant action is made from closed loops, the smallest of which are plaquettes with four links. Indeed this is the form of the Wegner–Wilson [10,11] action S(U) (see Fig. 2)

$$S(U) = \sum_{\text{all plaquettes}} [1 - 1/2 \, \text{tr} \, (U_{i\ell} U_{\ell k} U_{kj} U_{ji})]$$

$$= \sum S_\Box \quad . \tag{4}$$

The partition function is given by the sum over all possible link configurations weighted with the 'Boltzmann factor'

$$Z = \int \mathcal{D}U \, e^{-\beta S(U)} \, . \tag{5}$$

Here the integral can always be thought of as a multiple integration over the constrained vector variables $a_0, \vec{a}$ at all links, i.e.

$$\int DU = \int \prod_{\text{links}} da_{0,\ell} \, d\vec{a}_\ell \, \delta(a_{0,\ell}^2 + \vec{a}_\ell^2 - 1) \, . \tag{6}$$

The QCD coupling constant g enters into the expression $\beta = 4/g^2$. In the strong coupling limit, i.e. for small β, we see that the exponential allows values of $S_\Box \to 2$ which are related to strong fluctuations of the links, e.g.

$$(U\,U\,U\,U) = \begin{pmatrix} -1 & 0 \\ 0 & -1 \end{pmatrix} \, .$$

For weak coupling or large β the 'Boltzmann' weight prefers $S_\Box \to 0$, i.e. trivial link variables near the identity, namely

$$(U\,U\,U\,U) = \begin{pmatrix} 1 & 0 \\ 0 & 1 \end{pmatrix} \, .$$

Part of the normal numerical problem is associated with the necessity of working in the weak-coupling domain, where the fluctuations of the gauge fields are suppressed and the 'thermalization' of the system takes very long times.

An important point concerns the continuum limit of the action given in Eq. (4), when the lattice spacing a goes to zero. Take a plaquette in the $\hat{e}_1, \hat{e}_2$ plane and expand all link variables around the centre of the plaquette; then one gets (Fig. 2)

$$S_\square = 1 - 1/2\, \text{tr}\, [\exp i g \bar{A}_1 (x - 1/2\, a\hat{e}_2)\, a$$
$$\exp i g \bar{A}_2 (x + 1/2\, a\hat{e}_1)\, a$$
$$\exp -i g \bar{A}_1 (x + 1/2\, a\hat{e}_2)\, a \qquad (7)$$
$$\exp -i g \bar{A}_2 (x - 1/2\, a\hat{e}_1)\, a]\ .$$

Let us use the Baker–Hausdorf formula $e^x e^y = e^{x+y+1/2[x,y]}$ to derive

$$S_\square = 1 - 1/2\, \text{tr}\, \exp \{ia^2 g\, (\partial_1 A_2 - \partial_2 A_1) - a^2 g^2 [A_1, A_2]\}$$
$$= 1 - 1/2\, \text{tr}\, \exp\, ia^2 g\, F_{12}\ , \qquad (8)$$

where the commutator term does not vanish, because of the non-Abelian nature of the gauge fields. For small lattice spacing a, we can expand the exponential. The trace of the first-order term $\text{tr}\, ia^2 g F_{12}$ vanishes when we integrate over all group elements, and the second-order term gives the desired continuum action for $\beta = 4/g^2$:

$$\beta \Sigma S_\square = 1/8\, \text{tr}\, \beta g^2 a^4 \sum_x F_{\mu\nu}(x)\, F_{\mu\nu}(x)$$
$$= g^2 \beta/4 \cdot \int d^4 x\, \text{tr}\, 1/2\, F_{\mu\nu} F_{\mu\nu}\ . \qquad (9)$$

Asymptotic freedom can be expressed in this language as a relation between the QCD coupling constant g and the lattice spacing a, which has the following form for SU(2):

$$g^2/4\pi = 3\pi/[(11/2)\ln(1/\Lambda^2 a^2)]\ . \qquad (10)$$

Here Λ is the dimensional size parameter of SU(2) QCD, i.e. Λ_{QCD}. In order to obtain the continuum limit one has to make the lattice spacing a sufficiently small, which makes g also small. Then one is faced with the problem of slowing down the equilibration of the system together with a real lattice size of small dimensions (1–2 fm) [4]. Recently calculations with $(24-32)^4$ lattices have been started. Turning the above formula (10) around, a numerical simulation of a physical dimensional parameter m^2 is considered as scaling when it behaves like the physical measurable QCD constant, namely as

$$m^2 = \text{const.}\, \Lambda^2 = \text{const.}\, 1/a^2 \exp(-6\pi^2 \cdot \beta/11)\ . \qquad (11)$$

Taking a numerical expectation value at several values of β, a comparison with Eq. (11) establishes whether one is close enough to the continuum limit. Clearly numerical lattice simulations can only give us physical quantities expressed in units of Λ, which has to be taken from experiment.

The numerical technique we use is the heat-bath method [12]. Working our way through the lattice we generate random four vectors $a_0, \vec{a}$ for each link according to the weight $e^{-\beta S(U)}$ which is determined by the neighbouring links. Once the system is in equilibrium we can evaluate any desired correlation function.

3. COLOUR DIELECTRICS

Colour dielectrics is based on regarding the QCD vacuum as a medium characterized by a dielectric field. It is a theoretical approach which promises to construct an effective action for a bag model directly from the QCD Lagrangian. Clearly for many experiments in nuclear physics at higher energies we need a theoretical description which at the same time contains the long-range features of confinement without giving up the elementary quark and gluon degrees of freedom. After all it is the search for these QCD degrees of freedom which motivates the experimental work. Studying the transition from asymptotically free QCD at short distances to confining hadron dynamics at long distances is the exciting task we are faced with. For this problem a transformation is ideal which maps a theory with a lattice spacing a into one with lattice spacing 2a. Since such a mapping or block-spin transformation can be carried out repeatedly we are led to an effective model Lagrangian for long-distance hadron physics. We define a chain of field theories at $a_0, 2a_0, 4a_0, ..., 2^n a_0$, then the continuum limit $a_0 \to 0$ is achieved by letting $n \to \infty$ such that $\lim 2^n a_0 \to$ finite.

In Nielsen and Patkos [13] a perturbative calculation of one block-spinning step towards the colour dielectric Lagrangian is made. Mack [14] introduces models of effective actions and a definition of the colour dielectric field, which is used in our numerical work. For a general discussion of the renormalization group method in Monte Carlo calculations we refer the reader to the article of Wilson [15].

Constructing the block variables for the effective action is a somewhat arbitrary process. We proceed in the following way. First we select every second hypercube on the lattice. Then we make gauge transformations on all sites of a chosen hypercube in order to make its link variables $\tilde{U}$ close to the identity, i.e. $1/2 \, \text{Tr} \, \tilde{U} \approx 1$. We iterate this gauge transformation program approximately 5 to 9 times for each hypercube. Once this is done all the information is in the links U_i which connect the individual hypercubes. In 4 dimensions there are 8 interconnecting links, the average of which defines the new link variables V times the colour dielectric field χ. This scalar field χ is necessary, otherwise the new link variables V would not be unitary.

$$\chi \cdot V = 1/8 \sum_{\substack{i=1 \\ \text{connecting}}}^{8} U_i. \tag{12}$$

It can easily be shown that for unitary matrices $V = b_0 + i\vec{\sigma} \cdot \vec{b}$ and $U_i = a_0^i + i\vec{\sigma} \cdot \vec{b}_i$ we get

$$\det(\chi b_0 + \chi i\vec{\sigma}\vec{b}) = \chi^2 (b_0^2 + \vec{b}^2) = \chi^2$$

and

$$\det(1/8 \, (\Sigma \, a_0^i + i\vec{\sigma} \, \Sigma \, \vec{a}^i))$$
$$= (1/8 \, \Sigma \, a_0^i)^2 + (1/8 \, \Sigma \, \vec{a}^i)^2 \leq \max_{\forall i} (a_0^2 + \vec{a}_i^2) = 1. \tag{13}$$

Therefore, without limiting our generality, χ will take values between 0 and 1.

For the new effective action we assume the following form:

$$e^{-S_{\text{eff}}(\chi \cdot V)} =$$
$$\exp\left[-\beta' \sum_{\square} (1 - 1/2 \, \text{tr} \, (\chi^v)(\chi^v)(\chi^v)(\chi^v)) - \sum_{x} (m^2 \chi^2 + \lambda \chi^4)\right]. \tag{14}$$

The new plaquettes are now of length 2a, and if it is natural to decorate each new link V with the colour dielectric field χ, $0 \leq \chi \leq 1$. That is how we averaged the interconnecting gauge configurations anyway. In the continuum limit the above action will give us the desired result of a dielectric theory,

$$L = 1/2 \, \text{tr} \, B_{\mu\nu} B_{\mu\nu} \chi^4(x) + L(x)$$
$$= 1/2 \sum_{i=1}^{3} (\vec{E}_i^2 - \vec{B}_i^2) \cdot \epsilon + L(x). \tag{15}$$

where

$$\epsilon = \chi^4,$$

i.e. the colour dielectric constant $\epsilon(x) = \chi^4(x)$ modifies the electric and magnetic field strengths $\vec{E}_i^2$ and $\vec{B}_i^2$. In simulations on an 8^4 lattice using ~ 600 sweeps through the lattice, we obtained preliminary results (Fig. 3) for the dielectric constant $\langle \epsilon \rangle = \langle \chi^4 \rangle$. For strong coupling ϵ approaches zero, whereas, at $\beta = 2$, ϵ increases drastically. From simulations of the string tension it is well known that at $\beta = 2$ a cross-over transition takes place, which is reflected in the behaviour of ϵ. Using Schwinger–Dyson [16] equations, we derive the new effective couplings β', m^2, and λ. The self interaction of the colour dielectric field is shown in Fig. 4. It documents the tendency for the effective action to assume its

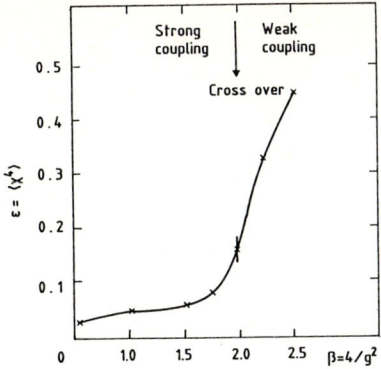

Fig. 3 Preliminary results for the dielectric constant $\langle \epsilon \rangle = \langle \chi^4 \rangle$ in an 8^4 lattice for different coupling constants $\beta = 4/g^2$.

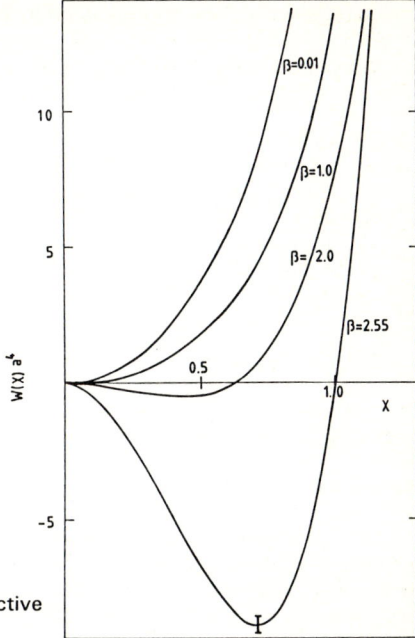

Fig. 4 Preliminary results for the χ-effective action $W(\chi)a^4$ as a function of β.

minimum near $\chi = 0$ when the coupling g is large. This supports the basic hypothesis of the dielectric theory of confinement. Colour sources are the source of electric inductions $\vec{D}_i$, which are related to the colour electric fields $\vec{E}_i$ by $\vec{D}_i = \chi^4 \vec{E}_i$. Therefore, the $\vec{D}_i$ fields can only be non-zero where $\epsilon \neq 0$. If $\epsilon = 0$ is the unique classical vacuum then $\chi \neq 0$ costs energy and this prevents the $\vec{D}$ fields from spreading. As a result, a flux tube will form between quarks of opposite charges in the region where $\epsilon \neq 0$ only. Colour is confined. In the next section we will see some phenomenological applications of this mechanism. The numerical calculations [9] confirm the preliminary observed trends.

4. PHENOMENOLOGY OF COLOUR DIELECTRICS

The original starting point for a colour dielectric quark model came from the need to understand nuclear binding and the EMC effect in the framework of a bag-like model [17]. In order to understand the colour dielectric properties of nuclei, one has to introduce the coupling of the dielectric field to the quarks. In the lattice formulation the quark–gluon term is

$$S = 1/2 \, a^3 \sum_{i,j} \bar{\psi}_i \gamma_\mu e_\mu U_{ij} \psi_j \,, \tag{16}$$

which destroys a quark at the lattice site j and connects it with a link U_{ij} to the site i where it is created. Averaging over different links will replace U by χV or expanding $V = 1 + igaB_\nu e_\nu$, we get, collecting terms $\propto a^4$,

$$\begin{aligned}
S_{\text{eff}} &= a^3 \sum_i \bar{\psi}_i \gamma_\mu e_\mu \chi_i (1 + igaB_\nu e_\nu)(\psi_i + e_K \partial_K \psi_i) \\
&\quad - a^4 \sum_i \bar{\psi}_i \psi_i m_q \\
&\rightarrow \int d^4x \, \{\bar{\psi}(x)\chi(x)(\gamma\partial + ig\gamma B)\psi(x) - m_q \bar{\psi}\psi\} \,.
\end{aligned} \tag{17}$$

Using the covariant derivative

$$D_\mu = \partial_\mu + igB_\mu$$

the total phenomenological Lagrangian has the form

$$\mathcal{L} = \mathcal{L}(\chi) + i\chi\bar{\psi}\gamma_\mu(\partial^\mu + igB^\mu)\psi - \bar{\psi}\psi m_q - 1/4 g^2 \chi^4 B_{\mu\nu} B_{\mu\nu} ,\qquad(18)$$

where

$$B_{\mu\nu} = (\partial_\mu + igB_\mu)B_\nu - (\partial_\nu - igB_\nu)B_\mu$$

and $\mathcal{L}(\chi) = T - W$ is parametrized with

$$W(\chi) = B\chi^2 (A + 6 - 2(A + 4)\chi + (A + 3)\chi^2)\qquad(19)$$

and

$$T(\chi) = 1/2\, \sigma_V^2 (\partial_\mu \chi)^2 .\qquad(20)$$

This phenomenological form of $W(\chi)$ is chosen in accordance with our previous work [18] on the pure gluon theory, where we set $\chi = (\sigma_V - \sigma)/\sigma_V$. Note that the effective pure gluonic Lagrangian of Eq. (18) is similar to the Friedberg–Lee soliton model [19,20] in this case. The dielectric constant is defined in the Lagrangian of Eq. (18) as $\epsilon = \chi^4$.

The fermion part of the colour dielectric model in Eq. (18), however, is totally different from the Friedberg–Lee model. In particular we note that the colour dielectric field leads to absolute quark confinement, since with $\langle\chi\rangle = 0$ in the vacuum the quarks can no longer propagate, i.e. their effective mass becomes infinite. To demonstrate this we introduce a new effective fermion field $\tilde{\psi}$ by rescaling the original field ψ

$$\tilde{\psi} = \sqrt{\chi}\, \psi \qquad(21)$$

which shifts the χ-dependence from the kinetic term in Eq. (18) to the mass term $(m_q/\chi)\bar{\tilde{\psi}}\tilde{\psi}$.

As a first approximation, we consider the B_μ's as perturbative gluon fields, dropping the non-linear parts of the Lagrangian. Phenomenologically we consider the χ-part as sufficient to describe the 0^{++} glueball and associate the χ-propagator with the glueball mass, i.e. $W(\chi)$ at $\chi = 0$ can be expanded as

$$1/2 (\partial^2 W/\partial\chi^2)_{\chi=0} = 1/2\, m_{GB}^2 \sigma_V^2 \chi^2 = B(A + 6)\chi^2 .\qquad(22)$$

In an earlier paper, we have used the parameters $A = -2$, $\sigma_V = 160$ MeV, and $B^{1/4} = 203$ MeV, which reproduce rather well the following physical observables [18]: the transition temperature $T_c = 160$ MeV to the gluon plasma, a glueball mass $m_{GB} = 720$ MeV, and a string tension $\sqrt{E/\ell} = 450$ MeV. In this calculation of the nucleon we have used the same parameters A, $B^{1/4}$, and σ_V. We note in passing that we also tried to calculate the nucleon with these parameters in the Friedberg–Lee soliton bag model [20] and always obtained nucleon masses $m_N \geq 2$ GeV. This is related to a general problem of the original bag model which must use different bag constants B for the heavy and light quark systems, typically $B^{1/4}(c\bar{c}) = 200$ MeV [21] and $B^{1/4}(qqq) \simeq 150$ MeV [22]. We will show how different colour dielectric fields in the interior describe both systems with the same effective Lagrangian.

The only remaining input parameter is the current quark mass on a hadronic size scale. We have argued [17, 23] that the average light quark mass should lie between 20 MeV and 30 MeV, i.e. it is somewhat larger than in the estimates based at $Q^2 = 1$ GeV^2. Different parameters have been chosen by Broniowsky et al. [24] and Thomas et al. [25]. Especially in the last paper the whole region of parameter space has been systematically explored.

The equations of motion for the χ-field and the rescaled ψ-field (now written without $\sim$), are derived from the effective action for N quarks of equal mass m_q

$$(\vec{\alpha}\cdot\vec{p} + m_q/\chi)\psi = E\psi \qquad(23)$$

and

$$-\sigma_V^2 \vec{\nabla}^2 \chi + \partial W/\partial \chi = N \bar{\psi}\psi \, m_q/\chi^2. \tag{24}$$

Numerically, we solve the equations for the potential $\widetilde{W}(\chi) = W(\chi-\delta)$, which has a minimum at $\chi = \delta \simeq 10^{-3}$ such that $m_q/\delta \geq 20$ GeV. Decreasing δ even more we do not find any variation in the results. Qualitatively, one understands the solution of the χ-field in the presence of the quark fields by regarding the effective potential

$$W_{\text{eff}}(\chi) = W(\chi) + N \bar{\psi}\psi \, m_q/\chi \,, \tag{25}$$

which has a minimum away from $\chi = 0$.

Approximating the single-particle densities as

$$\bar{\psi}\psi \approx 1/2 \, \bar{\psi}\gamma_0\psi = 1/2 \, / \, (4\pi/3 \, R^3) \,, \tag{26}$$

we estimate the scalar density $\bar{\psi}\psi = 0.12$ fm^{-3} for a hadron size of $R = 1$ fm. Then we get for $\bar{\chi}$ as the minimum of $W_{\text{eff}}(\chi)$ ($m_q = 20$ MeV, $m_{GB}^2 \sigma_V^2 = 0.013$ GeV^4):

$$\chi(0) = \bar{\chi} = 0.16 \,. \tag{27}$$

In Fig. 5 we show the exact solutions of the coupled equations for $\chi(r)$. Indeed the value of χ inside the hadron is not very different from zero [$\chi(0) = 0.17$]. Comparing this value with $\chi(0) = 0.6$ of the $3\bar{3}$ flux tube [18] solution, we see even less change of the vacuum value of χ. The corresponding quark wave functions (Fig. 6) have a reduced small component compared to the MIT bag model, since the effective quark mass inside the bag amounts to $m_q/\bar{\chi} \approx 120$ MeV. In effect the colour dielectric model interpolates between the constituent quark model $m_{\text{eff}} \simeq 300$ MeV and the original MIT bag model $m_q = 0$.

The r.m.s. radius of the quark distribution is 1.2 fm, which is larger than the experimental r.m.s. radius of 0.83 fm. The magnetic moment of 2.66 μ_N ($\mu_p = 2.79$ μ_N) and the ratio of $g_A/g_V = 1.32$ are excellent results. Since $\bar{\chi}$ inside the bag is rather small it is clear that the shape of the effective potential $W(\chi)$ near $\chi = 1$ does not influence the nucleon mass. Calculations with different sets of parameters B and A do not change the result as long as $m_{GB}^2 \sigma_V^2$ is kept constant. The shape of the effective potential W around $\chi = 0$ determines the dynamics. This is however different for the two phase solutions of Ref. 25, where in the interior of the bag $\chi \to 1$. It is noteworthy that the energy of the bag behaves as $E_{\text{bag}} \approx 8.5 * (m_q \, \sigma_v \cdot m_{GB})^{1/3}$. The quark mass is given as $m_q = m_0 + \text{const.} \, a^2 \langle \bar{\psi}\psi \rangle$, if the resolution $1/a < 1$ GeV. In the limit of $m_0 \to 0$ the bag mass behaves as $E_{\text{bag}} \propto \langle \bar{\psi}\psi \rangle^{1/3}$ similar to the result of QCD-sum rule calculations for the nucleon.

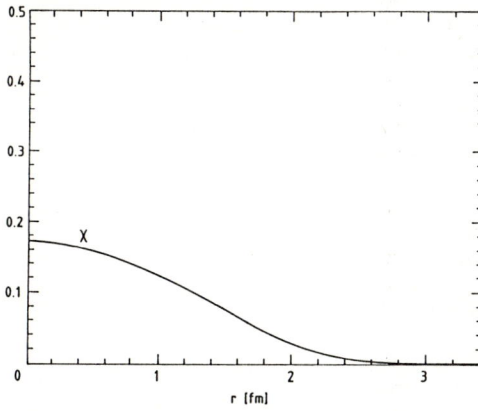

Fig. 5 The colour dielectric field $\chi(r)$ for the nucleon.

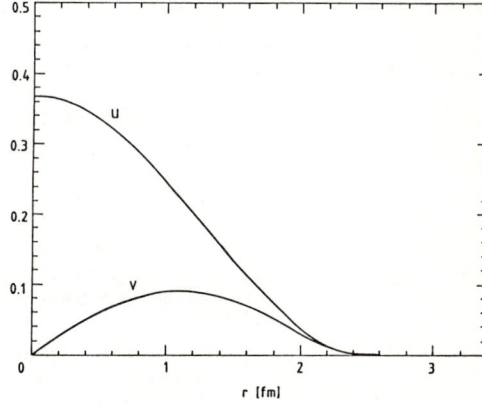

Fig. 6 The big u(r) and small v(r) components of the quark wave functions of the nucleon.

In the same model as used in Ref. 23 we have calculated [26] soliton matter; i.e. we have looked for quark and χ-field solutions which are periodic at $r = r_0$, the size of a Wigner–Seitz cell. Let us recall that the nuclear matter density ϱ and r_0 are related as

$$1/(4\pi/3\, r_0^3) = \varrho\,,$$

i.e. r_0 gives the radius of the free spherical cell available to a nucleon in nuclear matter. We take $(\partial\chi/\partial r)/r_0 = 0$ and $\partial(\psi^\dagger\psi)/\partial r = 0$ as new boundary conditions at the surface of the spherical cell. The fermion boundary condition allows two solutions, which correspond in the non-relativistic limit to the symmetric and antisymmetric solutions. They give the energies at the lower- and the upper-band edge. The shape of the quark densities is rather different for both cases. The lowest energy quarks are *strongly delocalized*; they would belong to deeply bound 'nucleons'. At the top of the band the quarks are *well localized*, even compressed in comparison to their free valence quark wave function. This model is certainly very crude, but it could tell us that the quark distributions in nuclei are much less obvious than a naïve interpretation of standard nuclear physics results can teach us. Interestingly enough most existing data cover only valence nucleons such as the famous mapping out of the $3s_{1/2}$ nucleons in Tl and Pb with electron scattering. In a schematic model [27] similar to that described in our earlier paper [17], we have also calculated the gluon cloud corrections to the nucleon properties in nuclear matter. The non-vanishing colour dielectric field allows the quarks to leak out of their bags reducing the kinetic energy of quarks, whereby nucleons get bound, and in addition it allows a gluon cloud outside the bag. We have calculated in this more extensive model the nuclear Nolen–Schiffer anomaly. This anomaly is traced back to the deconfinement of up and down quarks in the nuclear environment which is different owing to their different masses. A crystal of nucleons (soliton matter) does not, of course, correspond to our understanding of nuclei as a liquid or a gas of nucleons. Therefore we have investigated [28] what kind of structures develop when we allow a random distribution of the colour dielectric field with two values χ_{in} and χ_{out} characterizing the values of χ inside and outside the nucleons. Such a problem is analogous to percolation studies of electrical conductivity in amorphous materials. We have used such a model before to study the interaction radius in ^{3}He [29]. It has now been applied by many others to nuclei and deep inelastic scattering. Summarizing the results of this work, we show in Fig. 7 how the effective confinement zone $\langle r_{eff}^2\rangle$ increases with the hard-core nucleon radius R_c, inside which $\chi = \chi_{in}$. For the deuteron we see the physics of a dilute bound state, whereas for the heavier nuclei around $R_c = 1$ fm a transition takes place from the nuclei as colour insulators to colour conductors.

To conclude, I would like to summarize the three main theoretical points of this paper:
i) Integration over high-frequency gluon configurations generates an effective colour dielectric model Lagrangian, characterized by a few coupling constants, which can be calculated numerically.
ii) This Lagrangian generates bag-like solutions for the nucleon.
iii) This Lagrangian allows a first attempt at determining quark wave functions in nuclei.
I think that progress along these lines in non-perturbative QCD will make experiments accessible to a

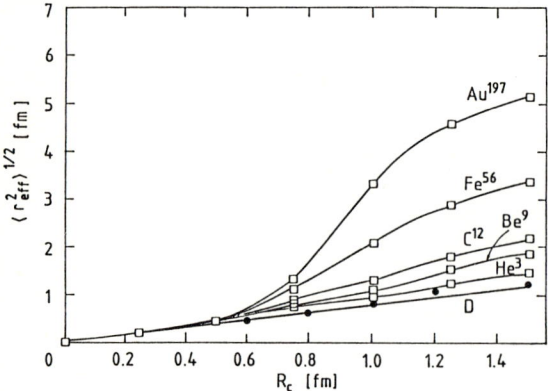

Fig. 7 The effective confinement radius $\langle r_{eff}^2\rangle^{1/2}$ in nuclei as a function of the hard core size of the nucleon R_c.

theoretical interpretation. Here I am mostly thinking of deep inelastic scattering experiments done at the upper edge of the available energy in the scaling region with large x. Existing inclusive lepton–nucleus experiments in the intermediate x-region have given us a glimpse of quark degrees of freedom in nuclei. Nuclear binding and a softening of the valence quark distribution are connected. New experiments at small and large x will illuminate the role of sea quarks, soft gluons, and the motion of quarks in nuclei. Recently [30], we have calculated the cross section $\pi^- A \to \mu^+ \mu^- + X$ taking into account the EMC effect and parton multiple scattering in the nucleus. In the $p_\perp$-distributions of the lepton pair there may be an experimental possibility to measure the color dielectric constant of nuclei.

I should like to thank the organizers of this Workshop for their kind invitation. Our work on colour dielectric lattice theory [9] has greatly benefited from discussions with Guido Martinelli, who also supplied us with an SU(3) version of the block transformation program.

REFERENCES

[1] G. Arnison et al., Phys. Lett. **158B**, 494 (1985).
[2] G. 't Hooft, Nucl. Phys. **B72**, 461 (1974).
[3] E. Witten, Nucl. Phys. **B160**, 57 (1979).
[4] E. Witten, *in* Recent Developments of Gauge Theories (eds. G. 't Hooft et al.) (Plenum, New York, 1980).
[5] I.J.R. Aitchison and C.M. Fraser, Phys. Lett. **146B**, 63 (1984).
P. Simic, The Low Energy Meson Action from QCD, Rockefeller Univ. preprint RU/85/124 (1985).
[6] G.E. Brown and M. Rho, Comments Nucl. Part. Phys. **15**, 245 (1986).
[7] C. Rebbi (ed.), Lattice Gauge Theories and Monte Carlo simulations (World Scientific, Singapore, 1983).
[8] I. Barbour, N.E. Behill, E. Dagotto, F. Karsch, A. Moreo, M. Stone and H.W. Wyld, University of Illinois preprint ILL–(TH)86/23 (1986).
[9] H.J. Pirner, J. Wroldsen and M. Ilgenfritz, Construction of an Effective Color Dielectric Lattice Action from QCD by Blocking, CERN TH-preprint.
[10] F. Wegner, J. Math. Phys. **12**, 2259 (1971).
[11] K.G. Wilson, Phys. Rev. **D10**, 2445 (1974).
[12] M. Creutz, Phys. Rev. **D21**, 2308 (1980).
[13] H.B. Nielsen and A. Patkos, Nucl. Phys. **B195**, 137 (1982).
[14] G. Mack, Nucl. Phys. **B235**, 197 (1984).
[15] K.G. Wilson, *in* Recent Development of Gauge Theories (eds. G. 't Hooft et al.) (Plenum Press, New York, 1980).
[16] M. Falcioni, G. Martinelli, M.L. Paciello, G. Parisi and B. Taflienti, Nucl. Phys. B, **265** [FS 15] 187 (1986).
[17] G. Chanfray, O. Nachtmann and H.J. Pirner, Phys. Lett. **147B**, 249 (1984).
[18] M. Rosina, A. Schuh and H.J. Pirner, Nucl. Phys. **A448**, 557 (1986).
[19] R. Friedberg and T.D. Lee, Phys. Rev. **D15**, 1694 (1977).
[20] R. Goldflam and L. Wilets, Phys. Rev. **D25**, 1951 (1982).
[21] J. Baacke, Y. Igarashi and G. Kasperidus, Z. Phys. **C113**, 119 (1982).
[22] T. DeGrand, R.L. Jaffe, K. Johnson and J. Kishis, Phys. Rev. **D12**, 2060 (1975).
[23] A. Schuh and H.J. Pirner, Phys. Lett. **173B**, 19 (1986).
[24] W. Broniowski, M.K. Banerjee and T.D. Cohen, University of Maryland preprint PP No. 86–081, ORO–5126–273 (1986).
[25] A.G. Williams, L.R. Dodd and A.W. Thomas, Phys. Lett. **B176**, 158 (1986) and A Numerical Study of a Confining Colour Dielectric Soliton Model (preprint).
[26] A. Schuh and H.J. Pirner, in preparation.
[27] G. Chanfray and H.J. Pirner, Quark Confinement and Nuclear Binding, Phys. Rev. **C35**, 760 (1987).
[28] F. Güttner and H.J. Pirner, Nucl. Phys. A **457**, 555 (1986).
[29] H.J. Pirner and J.P. Vary, Phys. Rev. Lett. **46**, 1376 (1981).
[30] P. Chiappetta and H.J. Pirner, Nuclear Effects in the production of high mass Drell-Yan pairs by $\pi^- W$ and $\pi^- D$ interactions, CERN-TH 4646/87 to be published in Nucl. Phys. B.

Frontiers of the Quark Model[*]

H.J. Lipkin

Department of Nuclear Physics, Weizmann Institute of Science,
Rehovot 76100, Israel

1. INTRODUCTION

I am very pleased to be back here at Les Houches for the third time. My first visit was in 1958 when I presented the first example of what is now called a spectrum generating algebra. My lectures in the proceedings were the first publication showing how a non-compact algebra generates the spectrum of a model Hamiltonian [1]. My second visit in 1968 was just after I discovered what is now called the "Large N limit" and the proceedings give the first publication showing that a system of what are now called colored quarks and antiquarks interacting with a one-gluon-exchange potential becomes a system of noninteracting bound quark-antiquark pairs in the limit where the number of colors becomes infinite [2]. I am afraid that I will not meet these standards in this talk.

There are many exciting frontiers in the quark model today, and I shall not be able to cover all of them. Some of them are:

Why does the simple naive quark model work as well as it does?

1. Where are the exotic states which are not $q\bar{q}$ or qqq ?
2. How can the quark model be used to explain and predict experimental results in high-energy spin physics?
3. What is the meaning of empirical corrections for high-energy scattering beyond the additive quark model?
4. Can the quark model give the nucleon-nucleon force and other properties of nuclei?
5. What exactly is the OZI rule and how is it obtained from QCD?

2. THE SAKHAROV-ZELDOVICH MODEL

2.1 The Simple Naive Model

As a dramatic example of the frontier of finding an explanation from QCD of why the quark model works so well, consider the success of the naive model first proposed by Sakharov and Zeldovich [3] and later refined by DeRujula, Georgi and Glashow [4]. and by Cohen and Lipkin [5]. This model successfully describes hadron masses with a very simple mass formula.

$$M = \sum_i m_i + \sum_{i>j} \frac{\vec{\sigma}_i \cdot \vec{\sigma}_j}{m_i m_j} v_{ij}, \qquad (1)$$

---

[*] Supported in part by the Minerva Foundation, Munich, Germany

where m_i is an effective quark mass, $\vec{\sigma}_i$ is a quark spin operator and v_{ij} is a hyperfine interaction.

2.2 The Nucleon Magnetic Moment

The remarkable success of this formula can be seen from the following derivation of the sum of the nucleon magnetic moments. The magnetic moment of any composite system with constituents of charge q_i and mass m_i is

$$\mu = \left\langle \sum_i \left(\frac{q_i \hbar}{m_i c}\right) g_i J_{z_i} \right\rangle, \tag{2a}$$

where g_i is the g factor for the constituent. Since hadron magnetic moments are expressed in terms of the nuclear magneton, $(e\hbar/M_p c)$ we can write

$$\mu = \frac{M_p}{m_i} \left\langle \sum_i \left(\frac{b_i}{2} + t_{zi}\right) g_i J_{zi} \right\rangle \left(\frac{e\hbar}{M_p c}\right), \tag{2b}$$

where we have introduced the Gell-Mann–Nishijima formula for the charge and b_i and t_{z_i} are the baryon number and z-component of isospin.

If we average over isospin, assume $g_i = 2$, $b_i = \frac{1}{n}$, where n is the number of constituents in the proton ($n = 3$ for the quark model) and that m_i is the same for all constituents,

$$\mu_p + \mu_n = \frac{M_p}{n m_i} \left\langle 2 J_{Z_i} \right\rangle \left(\frac{e\hbar}{M_p c}\right) = \frac{M_p}{n m_i} \left(\frac{e\hbar}{M_p c}\right). \tag{3}$$

The value of m_i is now obtained in terms of hadron masses directly from the mass formula (1) by using the conventional quark model wave functions for the nucleon and Δ to cancel out the contribution of the unknown hyperfine interaction,

$$n m_i = \frac{1}{2}(M_N + M_\Delta). \tag{4a}$$

Thus

$$0.88 = \mu_p + \mu_n = \frac{2 M_p}{M_N + M_\Delta} = 0.865. \tag{4b}$$

2.3 Hadron Mass Relations and the Effective Quark Mass

The mass formula (1) can be interpreted in a language familiar to nuclear physicists as the sum of single-particle energies (called effective masses in this case) and two-body interactions. However, one can say that Sakharov and Zeldovich *anticipated* QCD by using a flavor-dependent hyperfine interaction for the two-body interaction and assuming that all other contributions are included in the effective mass terms.

With these assumptions they obtained a relation between meson and baryon masses in surprising agreement with experiment. By taking linear combinations of masses in which the hyperfine interaction cancels out, they were able to calculate the difference between the effective masses of the strange and nonstrange quarks from both baryons and mesons,

$$m_s - m_u = M_\Lambda - M_N = 177 \,\text{MeV} = \frac{3}{4}\left(M_{K^*} - M_\rho\right) + \frac{1}{4}\left(M_K - M_\pi\right) = 180 \,\text{MeV}. \quad (5)$$

This striking evidence that mesons and baryons are made of the same quarks was overlooked for amusing reasons [6] and rediscovered in 1978 [7].

Another relation between meson, baryon and quark masses is obtained by taking linear combinations of masses in which the quark mass terms cancel and only the hyperfine terms contribute:

$$\frac{M_\Delta - M_N}{M_{\Sigma^*} - M_\Sigma} = \frac{M_\rho - M_\pi}{M_{K^*} - M_K} = \frac{m_s}{m_u}. \quad (6)$$

The ratio of baryon mass differences is 1.53; the ratio of meson mass differences is 1.61. The small discrepancy was later explained by a refinement of the model involving the different sizes of baryon and meson wave functions. functions [5]. An equivalent formula for the quark mass ratio was noted by Sakharov [6] along with the comment that the masses are of course effective masses.

2.4 Further Refinements of the Model and the Λ Moment

DeRujula et al. added input from QCD that the hyperfine interaction is produced by one gluon exchange and inversely proportional to quark masses. This immediately led to the prediction of the correct sign of the $N - \Delta$ and $\rho - \pi$ mass splittings. Their assumption that the same quark mass parameter appears in the color magnetic moments and the electromagnetic moments of the quarks led to a very successful prediction of the Λ magnetic moment from eq.(6) [4]

$$\mu_\Lambda = -\frac{\mu_p}{3}\frac{m_u}{m_s} = -\frac{\mu_p}{3}\frac{M_{\Sigma^*} - M_\Sigma}{M_\Delta - M_N} = -0.61. \quad (7)$$

The subsequently measured experimental value -0.61 n.m. is in exact agreement with this prediction.

Cohen and Lipkin then added the assumption that the effective mass parameter m_i appearing in both terms in eq. (1) was the same and justified this with a simple model calculation [5].

This immediately leads to another independent prediction for the magnetic moment of the Λ [8]

$$\mu_\Lambda = -\frac{M_p}{3m_s} = \frac{-2M_p}{M_N + M_\Delta + 6\left(M_\Lambda - M_p\right)} = -0.58. \quad (8)$$

This is in good agreement with the experimental value -0.61 n.m. Exact agreement is obtained both with experiment and with the DGG value (7) if the experimental value of μ_p is used as input.

The basic physics expressed by these successful relations is in the three different roles of the quark mass parameter m_i. The same value is used for both terms in eq. (1) and for the Dirac moment of the constituent quark. Why this should be valid remains to be explained by QCD, although some indications of the underlying physics has been given in one simple model [6].

3. EXOTIC HADRONS

The possible existence of exotic hadrons remains a principal frontier in hadron spectroscopy and the application of QCD to hadron physics [9]. It is also at the frontier between particle and nuclear physics. Particle physics considers systems of quark-antiquark pairs and three quarks. Nuclear physics considers systems of three-quark clusters which interact and bind. So far there is little connection between the two; the properties of the nucleon are sufficient to describe nuclear data without mentioning the internal quark structure.

One possible connection between particle and nuclear physics may be revealed in the properties of systems containing four, five or six quarks, if bound states or resonances exist. Until recently there has been no convincing evidence for such exotic states, but there are now some new possibilities.

3.1 New $\phi\pi$ and $\eta\pi$ Resonances

Recent experiments suggest the presence of four-quark states in the 1.5 GeV region. The observation at Serpukhov [10] of a $\phi - \pi$ resonance at 1480 keV seems to confirm the Close-Lipkin prediction [11] of an isovector vector meson whose charged state would have the quark constituents $(u\bar{d}s\bar{s})$. However, this state would not be produced by the pion exchange reaction observed at Serpukhov. One might expect considerable flavor mixing in such four quark states, since the $s\bar{s}$ pair in the four-quark state can be in a color octet part of the time and be converted into a $d\bar{d}$ pair via a single gluon intermediate state [12].

We now note that the $(u\bar{d}d\bar{d})$ state goes into $(u\bar{d}u\bar{u})$ under the G-parity transformation, and that the sums and differences of these two states give eigenstates of G with opposite eigenvalues and the same isospin, $I = 1$. Such G-parity doublets occur in the 10-10* representations of flavor SU(3) which seems to be the natural classification for the Serpukhov state.

The exotic $\eta - \pi$ resonance reported by GAMS at 1400 MeV has exactly the properties required [13] for the G-doublet companion of the Serpukhov state. It would therefore be extremely interesting to obtain further experimental verification for these two states as well as for the other exotic states expected in the same 10-10* multiplet.

3.2 The U strange meson

The recently observed strange baryon-antibaryon resonances (U) [14] at 3.1 GeV with exotic isospin 3/2 can be described in a diquark-antidiquark model with a relative orbital angular momentum between them providing a centrifugal barrier to prevent decay by rearrangement into two mesons [7]. For sufficiently large orbital angular momentum the diquark and antidiquark are spatially separated and the color field between them resembles that in a meson. Color sextet diquarks which have led to "baryonium" predictions in disagreement with experiment are discussed elsewhere [7] and not considered here. The high-spin diquark-antidiquark system can decay into a baryon-antibaryon pair via a diagram exactly analogous to ordinary $q\bar{q}$ meson decays; the diquark emits a gluon which creates a quark-antiquark pair. Predictions for masses and widths of these states can be obtained by extrapolating known information about mesons and baryons. The lowest configuration with the required quantum numbers is a spin-1 flavor-symmetric nonstrange diquark and a spin-0 flavor-antisymmetric strange antiquark or vice versa. These exotic states must also be in the 10 and 10* representations of flavor SU(3), since these are the only exotic representations appearing in the products $(6) \otimes (3)$ and $(6^*) \otimes (3^*)$ describing this diquark-antidiquark coupling.

This model predicts a rich spectrum of these diquark-antidiquark states with high spins and flavors corresponding to a cryptoexotic nonet for "rotating ground state configurations" and additional exotic 10-10*-plets for "rotating configurations" with one excited diquark or antidiquark. A state of a doubly strange diquark and a spin-0 nonstrange antidiquark would have the SU(3) quantum numbers of the Ω^- baryon and have the analogous property of being the only state in an SU(3) decuplet which cannot cascade down to a lower state by pion emission.

3.3 The H Dibaryon

One of the most interesting candidates for a bound exotic is the H dibaryon, a bound state of two Λ's. Although Jaffe's original calculation [15] and subsequent work [17] indicate a gain in hyperfine interaction energy by recoupling color and spins in the six quark system over the two-Λ system, a lattice gauge calculation [17] indicates that the H is unbound and well above the $\Lambda\Lambda$ threshold. Furthermore, although hyperfine binding calculations [16] indicate sensitivity of the hyperfine energy to flavor-$SU(3)$ symmetry breaking, the lattice results are insensitive to the strange quark mass and $SU(3)$ breaking [18].

This difference in the effects of $SU(3)$ breaking suggests that different physics dominates lattice gauge and bag or potential model calculations. Which calculations include the important physics is not obvious. However, the lattice calculation shows a repulsive Λ-Λ interaction generated by quark exchange [18] which is not included in bag model calculations and could well prevent the six quarks from coming close enough together to feel the additional binding of the short range hyperfine interaction [19]

3.4 The Anti-Charmed Strange Baryon (Pentaquark)

It is therefore of interest to look for other cases of hyperfine binding where such a repulsive exchange force may not be present. One case is an anticharmed strange baryon ($\bar{c}uuds$), which has a hyperfine binding roughly equal to that of the H [20,21,22], but which has no possibility of a quark exchange force in the lowest decay channel FN [19,22] We denote the anticharmed strange baryon as $P_{\bar{c}s}$ for pentaquark. If this five quark system breaks up into an F and a nucleon, there is no possible quark exchange between the two hadrons without flavor exchange, and therefore no diagonal matrix element of the one-gluon-exchange interaction that could give rise to a short range repulsion.

Anticharmed baryons were suggested as good candidates for possible bound exotics at the 1980 baryon conference [23]. However, the nonstrange anticharmed baryon was not bound by the hyperfine interaction and the strange anticharmed baryon was very remote from experiment and not pursued seriously. Now, however, that charmed strange baryons have been produced in hadronic experiments which also produce charmed antiquarks, such experiments may also produce the anticharmed strange baryon if it is bound.

4. DETAILED ANALYSIS OF THE H AND THE PENTAQUARK

4.1 Hyperfine Binding – The H and the $P_{\bar{c}s}$

Examination of the color-spin hyperfine interaction in the H and $P_{\bar{c}s}$ using Jaffe's color-spin algebra [15] shows that both states have very similar properties and indicate that the $P_{\bar{c}s}$ may well be bound. For any multiquark system the hyperfine interaction is maximized by choosing the best posssible color-spin coupling for the system, subject to two constraints:

1. All quarks of a given flavor must satisfy the Pauli principle.
2. The multiquark state must be a color singlet.

The second constraint is relevant for the Λ but not for any of the other stable mesons and baryons. All other states in the baryon octet which have two quarks symmetric in flavor are required by the Pauli principle to be either in the 70 or 20 dimensional representations of the color-spin SU(6). The 56-dimensional representation which is totally symmetric is forbidden by the Pauli principle. The 70 and 20 contain a spin-1/2 color singlet and a spin 3/2 color singlet respectively and give a unique classification for the baryon octet and decuplet, except for the Λ with no further restrictions from the second constraint.

The Λ, which has three quarks of different flavors can be placed without restrictions from the Pauli principle in the symmetric 56 representation of SU(6). This would optimize the hyperfine interaction. However, the 56 does not contain a color singlet, and therefore the Λ must be classified the 70 with the other octet baryons.

The stability against breakup of an exotic multiquark system can be examined by checking whether hyperfine energy can be gained by recoupling the color and spins of the lowest lying two-hadron threshold [24]. The above argument shows that the most favorable cases are those in which a Λ is present in the breakup threshold. The requirement that the three quarks in the Λ must be in a color singlet is immediately relaxed when additional quarks are present. Thus the multiquark system has additional freedom not present in other cases to gain hyperfine energy by recoupling color and spins.

4.2 Explicit Calculation of the Hyperfine Binding

We now summarize the results of explicit calculations of the hyperfine interaction for the $P_{\bar{c}s}$ [20,21,22]. We first consider the limit where the charmed quark has infinite mass and its hyperfine interaction energy is neglected, and SU(3) flavor symmetry is assumed for the light quarks. We compare this with the similar calculation for the H dibaryon. This calculation applies also for analogous states with heavier antiquarks which are equally attractive candidates for bound exotics. We use a variational approach with a wave function in which the two-body density matrix is the same for all pairs as in a baryon, and can then use the experimental $N - \Delta$ mass splitting to determine the strength of the hyperfine interaction energy [22].

A simplified form of the color-spin hyperfine interaction [15] can be used for systems containing only quarks and no active antiquarks:

$$V = -(v/2)[C_6 - C_3 - (8/3)S(S+1) - 16N], \qquad (9)$$

where v is a parameter defining the strength of the interaction, C_6 and C_3 denote the eigenvalues of the Casimir operators of the $SU(6)$ color-spin and $SU(3)$ color groups respectively, S is the total spin of the system and N is the number of quarks in the system.

The hyperfine interaction (9) is easily evaluated for the states of interest by substituting the eigenvalues of the Casimir operators [25]

$$M(\Delta) - M(N) = V(\Delta) - V(N) = 16v, \qquad (10a)$$
$$B(H) = V(H) - 2V(\Lambda) = -8v = -(1/2)[M(\Delta) - M(N)], \qquad (10b)$$
$$B(P_{\bar{c}s}) = V(P_{\bar{c}s}) - V(\Lambda) = -8v = -(1/2)[M(\Delta) - M(N)], \qquad (10c)$$

where B(X) denotes the difference in hyperfine energy between the state X and the relevant threshold. We have chosen to classify the four quarks in the $P_{\bar{c}s}$ in the 210 of $SU(6)$ to optimize the hyperfine interaction, and disregard the charmed antiquark which has no hyperfine interaction. The gain in hyperfine interaction for the $P_{\bar{c}s}$ over the NF or ΛD threshold (degenerate in this symmetry limit) is equal to the gain for the H over the relevant $\Lambda\Lambda$ threshold and is just half the $\Delta - N$ mass splitting.

At this level it appears that the $P_{\bar{c}s}$ is an equally attractive candidate for hyperfine binding as the H dibaryon.

4.3 The Effect of SU(3) Symmetry Breaking

SU(3) symmetry breaking reduces the stability of the H [16] H and the $P_{\bar{c}s}$. The hyperfine binding energies of both are reduced by reducing the color-magnetic interaction of the strange quark. But the breakup thresholds are unaffected because the strange quark plays no role in the color-magnetic interactions of the $\Lambda\Lambda$, ΛD and NF final states.

The broken-$SU(3)$ hyperfine interaction can be written

$$V_{br} = V(1-\delta) + \delta V_n, \tag{11}$$

where δ is a parameter expressing the suppression of the strange quark hyperfine interaction and V_n is the hyperfine interaction (9) acting only in the space of the nonstrange quarks.

The eigenfunctions of the interaction V_n are states in which the colors of the three nonstrange quarks are coupled to either a singlet or an octet. These states, denoted respectively by $|n_1;s\rangle$ and $|n_8;s\rangle$, can be expressed as linear combinations of two $SU(6)$ eigenstates $P_{\bar{c}s}$ and a state denoted by $P'_{\bar{c}s}$ in the 105' of $SU(6)$ [22]:

$$|n_1;s\rangle = (1/\sqrt{2})(P_{\bar{c}s} + P'_{\bar{c}s}), \tag{12a}$$
$$|n_8;s\rangle = (1/\sqrt{2})(P_{\bar{c}s} - P'_{\bar{c}s}). \tag{12b}$$

The interactions V and V_n are each diagonal in one of the two bases related by the transformation (12), with eigenvalues given by eq. (9). Their 2×2 matrices in this subspace are easily constructed in either basis and substituted into eq. (11) to give the matrix V_{br} [22]:

$$\langle P_{\bar{c}s} | V_{br} | \alpha P_{\bar{c}s} \rangle = -(16 - 11\delta)v, \tag{13a}$$
$$\langle P'_{\bar{c}s} | V_{br} | \alpha P'_{\bar{c}s} \rangle = 5\delta v, \tag{13b}$$
$$\langle P_{\bar{c}s} | V_{br} | \alpha P'_{\bar{c}s} \rangle = 3\delta v. \tag{13c}$$

The same matrix was obtained in an equivalent treatment [20,21] using a diquark-diquark decomposition of the four quark state and flavor $SU(3)$ instead of our 3-1 decomposition and color-spin $SU(6)$ [22].

The results of diagonalizing this matrix which gives the hyperfine energy of the broken $SU(3)$ eigenstates are discussed in detail by Gignoux et al. [20,21]. The lowest eigenvalue is shown to vary monotonically from the value $-16v$ in the symmetry limit where $\delta = 0$ to $-8v$ which is just the threshold energy in the case $\delta = 1$ where the strange quark hyperfine interaction is zero. The reduction of binding by symmetry breaking is less serious for the $P_{\bar{c}s}$ than for the H. A similar analysis to include the effect of the hyperfine interaction of a charmed antiquark with finite mass shows a very small change in the overall hyperfine energy relative to the NF threshold.

4.4 A Possible Molecular State

The exact mass of the $P_{\bar{c}s}$ is not easily estimated because there are too many uncertain factors. The gain in potential energy from the hyperfine interaction may not be enough to overcome the kinetic energy needed to keep the constituents confined in our trial wave function. Whether a better wave function and possibly a molecular-type two-cluster wave function [26] would be barely bound may be calculable with sufficiently large lattices. The uncertainties in the color electric interaction and possible repulsive interactions must also be considered.

A rough estimate of the binding of a molecular type wave function extending over a distance large compared with the range of the hyperfine interaction is obtained from a model describing such states by a two-body Schroedinger equation with a short range potential proportional to the parameter $B(X)$, the gain in hyperfine energy by recoupling color and spin [27] This equation has a bound state if B(X) is greater than a critical value depending upon hadron masses and the strength of the interaction. The ratios of the predicted critical values for the H and $P_{\bar{c}s}$ to similar predictions for $K\bar{K}$ molecular states [26,27] show that the H and $P_{\bar{c}s}$ [22] satisfy the condition for being more strongly bound than $K\bar{K}$ molecules:

$$(7/8) = B(H)/B(K\bar{K}) > M(K)/M(\Lambda) = 0.44 \qquad (14a)$$
$$(7/8) = B(P_{\bar{c}s})/B(K\bar{K}) > M(K)[M(N) + M(F)]/2M(N)M(F) = 0.39, \qquad (14b)$$

where $B(K\bar{K})$ is defined by analogy with eqs. (10), the right-hand sides are the ratios of the relevant reduced masses and the values of (7/8) are obtained from previous calculations [24] in the SU(3) symmetry limit and where the meson and baryon masses are related by assuming the same two-body radial density matrices for meson and baryon wave functions. The conditions (14a) and (14b) will still be satisfied if corrections from SU(3) breaking and other effects are less than a factor of two. Thus the H and $P_{\bar{c}s}$ are excellent candidates for weakly bound molecular states, particularly if the δ and S^* mesons are indeed $K\bar{K}$ molecules [26,27].

4.5 Guides to further Investigations of H and $P_{\bar{c}s}$

There is therefore interest both in experimental searches for the $P_{\bar{c}s}$ and in lattice gauge calculations. The simplest lattice calculation with an infinitely heavy charmed antiquark and four light quarks $uuds$, can easily be done in parallel with the more complicated H calculation both in the symmetry limit where all light quarks have the same mass and with $SU(3)$ symmetry breaking. Comparing the results for these cases may provide considerable insight into our understanding of the physics of QCD in multiquark systems even if the $P_{\bar{c}s}$ is not found as a physical bound state in experiment. There is however a difficulty in treating loosely bound molecular states on the lattice, since these are sensitive both to the details of the short range hyperfine interaction and the long range part of the wave function. A proper treatment of both these effects may require a rather large lattice.

The experimental search for the $P_{\bar{c}s}$ has a completely different character from the search for the H, which has no easily detected signature, requires a sophisticated single-purpose experiment for its production and identifies it only by missing strangeness and missing mass. The $P_{\bar{c}s}$ has several distinctive signatures in decay modes detectable against a multiparticle background. Any multipurpose or exploratory experiment which produces charmed pairs in the presence of strange baryons can be used also to look for the $P_{\bar{c}s}$

In particular, an experiment which produces charmed strange baryons will also produce charmed antiquarks, and there may be a possibility that a strange baryon will pick up a charmed antiquark as well as a charmed quark in a hadronization process.

Some examples of decay modes which provide a good signature are

$$P_{\bar{c}s} \to p\phi\pi^- \tag{15a}$$

$$P_{\bar{c}s} \to p\eta\pi^- \tag{15b}$$

$$P_{\bar{c}s} \to \Lambda K^+\pi^-\pi^- \tag{15c}$$

$$P_{\bar{c}s} \to pF^- \tag{15d}$$

$$P_{\bar{c}s} \to \Lambda D^- \tag{15e}$$

where the two decay modes (15d) and (15e) can occur if the $P_{\bar{c}s}$ is a resonance above the relevant threshold, or if it is a weakly bound "molecule". In the latter case the charmed meson would be "off-shell" and have an apparently lower mass.

Note that there must always be associated production of a charmed particle together with the $P_{\bar{c}s}$. This can also be used to optimize signal to noise.

5. ESTIMATES OF THE U MASS IN A DIQUARK-ANTIDIQUARK MODEL

If the U particle is a state of a spin-1 nonstrange diquark(antidiquark) and a spin-0 strange antidiquark(diquark), in a state of high angular momentum, its mass can be estimated [7] from known meson masses. We use a variational approach [28] with a four-body trial wave function which separates into the internal motions of the quark and diquark and the relative motion of the two. The internal motions are taken to be the same as that of the ground state mesons, denoted by M, and the relative motion the same as that of the appropriate excited strange mesons, except for radial scaling factors chosen to minimize the energy:

$$\psi_{4q,L}(r_D, r_{\bar{D}}, r_{rel}) = \psi_{M,0}(r_D/\sqrt{2})\psi_{M,0}(r_{\bar{D}}/\sqrt{2})\psi_{M,L}(\sqrt{2}r_{rel}) \tag{16}$$

where $\psi_{4q,L}(r_D, r_{\bar{D}}, r_{rel})$ denotes the trial four-body wave function with orbital angular momentum L and $\psi_{M,L}(r)$ denotes the spatial wave function of a meson with orbital angular momentum L. If we use the masses of the nonstrange L=0 mesons and the strange orbitally excited K^* mesons to get a state with one strange and three nonstrange quarks, we obtain

$$M_{4q,L} = (3/4)M_\rho + (1/4)M_\pi + M_{K_L^*} . \tag{17}$$

Substituting the latest K^* data [29] we obtain $M_{4q,L=4} = 2991 MeV$. This is close enough to the U mass in this crude calculation to suggest that the U might indeed be an orbitally excited state of the diquark-antidiquark configuration.

6. ACKNOWLEDGEMENT

Discussions with J.D. Bjorken, N. Isgur, J.M. Richard and H. Thacker are gratefully acknowledged. After this work was completed and presented briefly at Moriond [19], the author learned of similar independent work of Gignoux et al. [20,21]. The notation of the present paper has been modified to conform to their usage, and the detailed calculation of the energy as a function of the symmetry-breaking parameter δ which duplicates their results was omitted. It is a pleasure to thank J.M. Richard for a prepublication copy of their provisional manuscript.

References

1. H.J. Lipkin, in: "The Many Body Problem", Les Houches 1958, ed. by C. de Witt and P. Nozieres, Dunod, Paris (1959); S. Goshen and H.J. Lipkin, Ann. Phys. (N.Y.) $\underline{6}$ (1959) 301
2. H.J. Lipkin, in: "Physique Nucleaire, Les Houches 1968," ed. by C. de Witt and V. Gillet, Gordon and Breach, New York (1969) p. 585
3. Ya.B. Zeldovich and A.D. Sakharov, Yad. Fiz $\underline{4}$ (1966) 395; Sov. J. Nucl. Phys. $\underline{4}$ (1967) 283
4. A. DeRujula, H. Georgi and S.L. Glashow, Phys. Rev. $\underline{D12}$ (1975) 147
5. I. Cohen and H.J. Lipkin, Phys. Lett. $\underline{93B}$ (1980) 56
6. A.D. Sakharov, private communication; Harry J. Lipkin, Annals of the New York Academy of Sciences 452 (1985) 79, and London Times Higher Education Supplement, January 20 (1984) p.17
7. H.J. Lipkin, Phys. Lett. $\underline{74B}$ (1978) 399
8. H.J. Lipkin, Phys. Rev. Lett. $\underline{41}$ (1978) 1629
9. H.J. Lipkin, in: Intersections Between Particle and Nuclear Physics, Proc. Conf. on The Intersections Between Particle and Nuclear Physics, Lake Louise, Canada, 1986, Edited by Donald F. Geesaman, AIP Conference Proceedings No. 150, p. 657
10. S.I. Bityukov et al., Serpukhov Preprint IHEP 86-110
11. F.E. Close and H.J. Lipkin, Phys. Rev. Lett. $\underline{41}$ (1978) 1263
12. F.E. Close and H.J. Lipkin Weizmann Institute Preprint WIS/87/8/Feb-PH Submitted to Hadron '87, Int. Conf. on Hadron Spectroscopy.
13. M. Boutemeur, to be published in Hadrons, Quarks and Gluons, Proc. of the XXII Rencontre de Moriond
14. M. Bourquin et al., Phys. Lett. $\underline{B172}$ (1986) 1133
15. R.L. Jaffe, Phys. Rev. Lett. $\underline{38}$ (1977) 195
16. J.L. Rosner, Phys. Rev. $\underline{D33}$ (1986) 2043
17. P. MacKenzie and H. Thacker, Phys. Rev. Lett. $\underline{65}$ (1985) 2539
18. H. Thacker, private communication
19. Harry J. Lipkin, Argonne preprint ANL-HEP-CP-87-51, to be published in Hadrons, Quarks and Gluons, Proc. of the XXII Rencontre de Moriond
20. J.M. Richard, these proceedings
21. C. Gignoux, B. Silvestre-Brac and J.M. Richard, Paris preprint, PAR-LPTHE 87-17, submitted to Phys. Lett.
22. H.J. Lipkin, Weizmann preprint, WIS/87/32/May-PH, submitted to Phys. Lett.
23. Harry J. Lipkin, In Baryon 1980 (Proc. IVth Int. Conf. on Baryon Resonances, Toronto, Canada, 14-16 July 1980), ed. by Nathan Isgur (University of Toronto, 1981) p. 461

24. N. Isgur and H.J. Lipkin, Phys. Lett. $\underline{99B}$ (1981) 151
25. H. Högaasen and P. Sorba, Nucl. Phys. $\underline{B145}$ (1978) 119
26. J. Weinstein and N. Isgur, Phys. Rev. Lett. $\underline{48}$ (1982) 659; Phys. Rev. $\underline{D27}$ (1983) 588.
27. H.J. Lipkin, Phys. Lett. $\underline{124B}$ (1983) 509
28. H.J. Lipkin, Phys. Lett. $\underline{B172}$ (1986) 242
29. S. Suzuki, to be published in Proc. of Hadron '87, Second Int. Conf. on Hadron Spectroscopy, KEK, Tsukuba, Japan (1987)

Heavy Multiquark States

L. Heller

Theoretical Division, Los Alamos National Laboratory,
Los Alamos, NM 87545, USA

I. Introduction

In a number of papers [1-5] it has been suggested that the dimeson ($Q^2\bar{q}^2$) is stable against breakup into the two ($Q\bar{q}$) mesons provided the mass m of Q is large. Since some of these papers make purely phenomenological assumptions about the nature of the interaction in the four quark system, the physical basis for the result is not obvious. We have recently argued [6] that for sufficiently large m the dimeson *must* be bound, and in Section II we show how this result follows from minimal assumptions that are consistent with Quantum Chromodynamics.

To actually decide if a particular dimeson is bound, it is necessary to make an assumption about the form of the confining interaction. Our method for deriving the static part of the potential energy is discussed in Section III; it is the fact that the two-body and four-body systems are treated on the same footing that enables us to calculate and compare the energies of single mesons and dimesons. In Section IV the method for solving the four-body problem, and the results, are presented.

II. Qualitative discussion of dimesons

We shall consider the four-body system ($Q^2\bar{q}^2$) composed of two quarks and two antiquarks for two special cases of the quark masses.

A. $m \to \infty$ with $\bar{m}$ fixed

A very simple physical picture emerges in this limit because all the relative momenta in the problem remain finite *except* for that between the two heavy quarks. Their relative motion is dominated by their Coulomb attraction in the color $\bar{3}$ state, and their relative wave function becomes hydrogenic with a reduced mass $\mu = m/2$ and an effective coupling constant $\alpha_{EFF} = 2\alpha_s/3$. The Bohr radius of this pair is

$$a = \frac{3}{m\alpha_s} \tag{2.1}$$

and the energy associated with their relative motion is

$$E(Q^2) = -\frac{1}{9} m\alpha_s^2 . \tag{2.2}$$

We know that none of the other energies (apart from rest mass) in the four-body system is proportional to m for the very same reason that the energy of a single meson (Q$\bar{q}$) is not proportional to m, namely, the kinetic energy is governed by m in the large m limit. The only assumption about the confining potential that is needed to complete this argument is the trivial one that it remains finite when two particles come close together. Finally,

$$2M(Q\bar{q}) - M(Q^2\bar{q}^2) = \frac{1}{9} m\alpha_s^2 + O(m^\circ) \tag{2.3}$$

which proves that for sufficiently large m there must be a bound exotic dimeson.

B. $m = \bar{m}$

All the relative momenta are comparable in this case, and since the $Q\bar{q}$ Coulomb attraction in the color singlet state is stronger (2x) than the QQ attraction in the $\bar{3}$ state, the $Q\bar{q}$ pairing is the preferred one. While this does not prove that there is no bound dimeson for this system, it is consistent with the fact that we have not found any [6].

III. The Potential Energy

To derive the static part of the potential energy for a system of quarks and/or antiquarks we use a Born-Oppenheimer approximation to the MIT bag model. The quarks are treated as static, localized sources of the glue field, and the latter is required to satisfy the bag model boundary condition $n^\mu F_{\mu\nu} = 0$ on the surface. After solving for the glue field and the correct bag surface, the energy is regarded as the potential energy in which the quarks move. In Fig. 1 this approximation is contrasted with another extreme approximation to the bag model, the 'cavity' approximation, which has been used to describe systems containing only light quarks. In that approach it is the boundary condition on the quark field that determines their allowed modes; and the total energy is regarded as the actual mass of the hadron.

If one tries to treat heavy quarks in the cavity approximation it is necessary to mix high mode numbers into the wave function because the small increase in kinetic energy can be more than compensated for by the gain in color Coulomb attraction. On the other hand, if light quarks are treated in the Born-Oppenheimer approximation, there may be important non-static terms in the potential energy. We assume that this is not the case and will calculate systems like (b^2u^2) just using the static potential.

We start at the classical level with a set of quarks at position $\vec{r}_i$ with color charges F_i^a enclosed in a bag described by surface parameters S_α. Since color magnetic moments are neglected at this stage, the only boundary condition is

CAVITY APPROXIMATION		BORN-OPPENHEIMER APPROXIMATION
u,d	s	c,b,

FIXED spherical cavity.   FIXED quark positions.

Boundary condition on QUARK field.   Boundary condition on GLUE field.

Quarks in lowest mode.   Solve for glue field and bag boundary.

E = Hadron mass energy.   E = static potential

Fig. 1. Comparison of two extreme approximations to the MIT bag model.

$\hat{n} \cdot \vec{E}^a = 0$, and this Neumann problem can be solved for an arbitrary surface. Adding together the color electrostatic energy and the bag volume energy gives

$$W(\vec{r}_i, F_i^a, S_\alpha) = \int dV \left[\frac{1}{2} \sum_a \vec{E}^{a\,2} + B \right]. \tag{3.1}$$

Since the surface parameters are not dynamical variables they must be eliminated, leading to

$$\frac{\partial W}{\partial S_\alpha} = 0, \tag{3.2}$$

which is called the equation of constraint or the pressure balance condition. When the solution of this equation is inserted into (3.1) the result is the potential energy

$$V = W\left(r_{ij}, F_i \cdot F_j, S_\alpha(r_{ij}, F_i \cdot F_j)\right). \tag{3.3}$$

Since the bag forms around the quarks, the potential energy is translation invariant.

After the kinetic energy of the quarks is added to the potential energy (3.3) quantization is carried out, and this leads to V becoming a matrix in color space. For the system of two quarks and two antiquarks it is a 2×2 matrix since there are two independent color singlet states.

To actually implement the program described in (3.1) - (3.3) requires solutions of the bag model with deformed surfaces. If we were dealing with molecular-type states in which one (Q$\bar{q}$) pair were rather well separated from the other, then bag deformation would be an important consideration [see Fig. 2]. As the states discussed in this paper are not of this type, we expect a spherical approximation to the bag to be good for the parts of the wave function having large probability, and since an analytic Green's function is known for a sphere it is possible to write down an analytic expression [7] for W in (3.1). If the interparticle separations are small, $r_{ij} \lesssim 1$ Fm, then it is sufficient to keep just the dipole term from the homogeneous part of the Green's function, and this leads to [7]

$$V = \sum_{i>j}^{N} \alpha_s \frac{F_i \cdot F_j}{r_{ij}} + \frac{k}{\sqrt{2}} \left[(\vec{D}^a)^2 \right]^{1/2} \tag{3.4}$$

where

$$\vec{D}^a = \sum_i^N F_i^a \vec{r}_i \tag{3.5}$$

is the color dipole moment operator. The string tension is given by

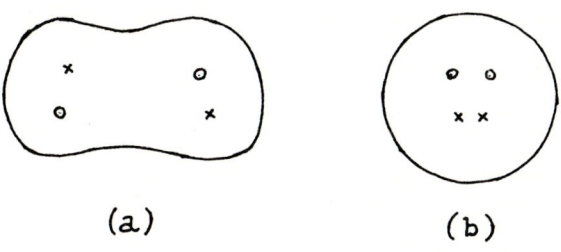

(a)   (b)

Fig. 2. Bag shapes for the two-quark two-antiquark system. (a) A deformed bag that would be important for a molecular type state. (b) If all interparticle spacings are comparable a spherical bag is expected to provide a good approximation.

$$k = \left(\frac{32\pi}{3} B\alpha_s\right)^{1/2} \tag{3.6}$$

and the bag radius, which is also an operator in color space, has the expression

$$R = \left[\frac{3\alpha_s}{4\pi B}(\vec{D}^a)^2\right]^{1/6} . \tag{3.7}$$

There are two noteworthy features of this potential energy. First of all, (3.4) has the same structure for any number N of quarks and/or antiquarks; we can use it, therefore, to calculated both mesons and dimesons to see if the latter are stable against breakup into the former. The second point to note is that the second term in (3.4), which is the confining potential, is a many-body operator. We have argued on theoretical grounds that the confining interaction must be a many-body potential [8]. In addition, phenomenology does not tolerate a sum of two-body potentials because it gives rise to unphysical van der Waals forces between hadrons.

When (3.4) is specialized to the $(Q\bar{q})$ system it becomes

$$V^{(2)} = -\frac{4}{3}\frac{\alpha_s}{r} + \left(\frac{2}{3}\right)^{1/2} kr , \tag{3.8}$$

which has the same Coulomb plus linear structure as the potential derived from lattice gauge theory and also the phenomenological potentials that have been used to fit
ψ ($c\bar{c}$) and Υ ($b\bar{b}$) spectra. Note that there is a confining term even at small distances, but the slope is only ~0.8 of its value at large distances, $r > 1$ Fm, where the bag develops into a tube of flux [7,9].

For the $(Q^2\bar{q}^2)$ system the 2×2 matrix $V^{(4)}$ determined from (3.4) is written out in [10] in the singlet-singlet, octet-octet representation defined by the couplings

$$\psi_1 = |[(12)^1(34)^1]^1\rangle$$
$$\psi_8 = |[(12)^8(34)^8]^1\rangle , \tag{3.9}$$

where particle 1 and 3 are quarks, and 2 and 4 are antiquarks.

Since some of the four-body wave function extends into the region where the spherical approximation to the bag breaks down, we use some physical arguments to write down the potential there. The most important region is the one in which one $(Q\bar{q})$ pair starts to separate from the other. If the separation R between the (12) pair and the (34) pair becomes large, in the representation (3.9) we require that

$$V^{(4)} \to \begin{pmatrix} V^{(2)}(r_{12}) + V^{(2)}(r_{34}) & 0 \\ 0 & C_8 R \end{pmatrix} . \tag{3.10}$$

The significance of the various matrix elements is as follows. The diagonal element in the singlet-singlet state is the sum of the two potential energies within the individual $(Q\bar{q})$ pairs with no interaction between them. The diagonal element in the octet-octet state is the confining part of the bag model potential energy between two octets. Concerning the off-diagonal element O, it must fall off sufficiently rapidly with distance so that there are no van der Waals forces, and this means at least exponentially. We actually use a Gaussian form, $\exp(-R^2/d^2)$, to make a smooth transition from (3.4) to large distances, so O is itself a Gaussian and we expect the parameter d to be approximately 1 fm. The details are given in [11].

IV. Meson and Dimeson Energies

Given the potential energy described in Section III, the next step is to solve the Schrödinger equation for the meson and dimeson problems and compare their energies. We did this initially using the nonrelativistic expression for the kinetic energy of the quarks, but since the light quarks (m = 0.35 GeV) are quite relativistic all the results reported here were obtained with $\sum_i (\vec{p}_i^2 + m_i^2)^{1/2}$ as the kinetic energy operator. These calculations that use the static potential energy and the relativistic expression for the kinetic energy are referred to as "semirelativistic".

To actually solve for the energy of the ground state of each system we have performed variational calculations and also Green's function calculations. The latter also start with a trial wave function, ψ_v, but then let it evolve in (imaginary) time, and the ground state energy is projected out via

$$E_o = \lim_{\tau \to \infty} \frac{\langle \psi_v | H e^{-\tau H} | \psi_v \rangle}{\langle \psi_v | e^{-\tau H} | \psi_v \rangle} . \tag{4.1}$$

The Monte Carlo technique for doing this is described in [11].

The variational wave function for a single meson is chosen to be

$$\psi_v = \exp(-\alpha r) , \tag{4.2}$$

and for the dimeson

$$\psi_v = \{\exp[-\alpha_c(r_{12} + r_{34}) - \alpha_{13} r_{13}] + [1 \leftrightarrow 3]\} \psi_3$$

$$+ c_m \{\exp[-\alpha_c(r_{12} + r_{34}) - \alpha_{13} r_{13}] - [1 \leftrightarrow 3]\} \psi_6 , \tag{4.3}$$

where α, α_c, α_{13}, and c_m are the variational parameters. The color states ψ_3 and ψ_6 are defined by the couplings

$$\psi_3 = |[(13)^{\bar{3}} (24)^3]^1 \rangle$$
$$\psi_6 = |[(13)^6 (24)^{\bar{6}}]^1 \rangle \tag{4.4}$$

and are antisymmetric and symmetric, respectively, under the interchange of either the two quarks or the two antiquarks. According to the discussion in Section II we expect the term in (4.3) containing ψ_3 to be dominant for large quark mass, and that is why it was assigned a symmetric spatial wave function.

The correlations that are built into (4.2) and (4.3) contain sufficient flexibility to describe, on the one hand, the expected large m limit in which [see (2.1)] $\alpha_{13} = \bar{m} \alpha_s/3$; and also the limit of two separated mesons, which corresponds to $\alpha_{13} = 0$ and $\alpha_c = \alpha$, with $c_m = -\sqrt{2}$ making each meson a color singlet. An additional correlation between the two light quarks would complicate the variational calculation, but is not essential for the Green's function calculation.

For the bag model parameters we choose $\alpha_c = 0.370$ and $B^{1/4} = 0.245$ GeV, which leads to the string tension having the value $\bar{k} = 1.07$ GeV/Fm. In conjunction with the quark masses $m_c = 1.364$ GeV and $m_b = 4.781$ GeV, this results in a good fit to the $c\bar{c}$ and $b\bar{b}$ spectra.[6] In Table I we show the masses of four mesons obtained by solving the semi-relativistic Schrödinger equation with the potential (3.8). As the hyperfine interaction is not included at this stage the comparison with experiment is made in terms of the appropriately weighted masses of the vector and pseudoscalar particles. Since the mass of the η_b has not yet been measured, an estimate of the hyperfine splitting has been made for that system.

It is seen in Table I that the calculated ground state masses of the $(c\bar{u})$ and $(b\bar{u})$ systems are too large. If the mass of the light quark is reduced from the value 0.350 GeV that was used there, then the discrepancy with experiment is reduced, but even with $m_u = 0$ the mass of $(b\bar{u})$ is still too large by about 0.13 GeV. While this indicates some deficiency in the present approach to mixed light-heavy systems, nevertheless, we proceed to calculate the dimeson systems and expect that there is some cancellation of the error in the energy difference.

Table II shows the results of our calculations for the energies of three dimesons. These were obtained from (4.1) again using the semirelativistic Hamiltonian described in Section III. The dimeson binding energy is with respect to the two separated mesons

$$\text{Binding Energy} = M(Q\bar{q}) + M(Q\bar{q}') - M(QQ\bar{q}\bar{q}') . \qquad (4.5)$$

Table I. The masses of some mesons. α is the variational parameter in (4.2), and E is the eigenvalue of the semirelativistic Schrödinger equation with the potential (3.8). An estimate has been made of the hyperfine splitting in the Υ-η_b system. All energies are in GeV. The parameters are $\alpha_s = 0.370$, $B^{1/4} = 0.245$ GeV, $m_c = 1.364$, $m_b = 4.781$, and $m_u = m_d = 0.350$.

Quark Content	$\alpha(\text{fm}^{-1})$	E	M	M_{XPT}
$c\bar{c}$	4.1	0.32	3.05	$\frac{3}{4}(\psi) + \frac{1}{4}(\eta_c) = 3.07$
$b\bar{b}$	8.2	-0.13	9.43	$\frac{3}{4}(\Upsilon) + \frac{1}{4}(\eta_b) \approx 9.44$
$c\bar{u}$	3.3	0.54	2.25	$\frac{3}{4}(D*) + \frac{1}{4}(D) = 1.97$
$b\bar{u}$	4.0	0.43	5.56	$\frac{3}{4}(B*) + \frac{1}{4}(B) = 5.31$

Table II. Dimeson energies. E is the eigenvalue of the semirelativistic Schrödinger equation; the statistical uncertainty in the Green's function Monte Carlo calculation is ±10 MeV. The binding energy as defined in (4.5) is shown without and with the inclusion of the hyperfine interaction. The final column gives the spin quantum numbers of the lowest energy dimeson state.

Quark Content	E	Binding Energy [GeV] No Hyperfine	With Hyperfine	Spin State (S_{QQ} $S_{\bar{q}\bar{q}'}$, S)
$bb\bar{u}\bar{d}$	0.77	0.09	0.07	1 0 1
$bb\bar{u}\bar{u}$			0.03	1 1 0
$bc\bar{u}\bar{d}$	0.87	0.09	0.00	1 1 0
$bc\bar{u}\bar{u}$				
$cc\bar{u}\bar{d}$	0.99	0.08	(-0.07)	1 1 0
$cc\bar{u}\bar{u}$				

The column labelled "No Hyperfine" omits this interaction in both the mesons and the dimeson, while the column "With Hyperfine" includes it in *both*. Note that before the hyperfine interaction is turned on there is very little difference in the binding energies of the various dimesons, there being a spread of only 10 MeV between the values in the three rows. This already shows that the b quark is not heavy enough for the term in (2.3) that is linear in m (the Coulomb term) to completely dominate. This is consistent with the known fact that the confining potential is quite important in the $(b\bar{b})$ system.

After the hyperfine interaction is turned on every one of the dimesons in Table II becomes less bound; this is because the hyperfine attraction in the mesons is larger than in the dimeson[12]. In fact, those containing two c quarks become unbound and those with one b and one c are marginal.

The final column of Table II contains the spin quantum numbers of the two heavy quarks, S_{QQ}, the two light antiquarks, $S_{\bar{q}\bar{q}}$, and the total spin S of the lowest energy state. A dimeson containing two *distinct* light quarks ($\bar{u}$ and $\bar{d}$) does not have a Pauli principle restriction on their spin state and may, therefore, have a lower energy than the corresponding dimeson with identical light quarks. This is indeed the case for the dimesons containing two b quarks, for which $(bb\bar{u}\bar{d})$ is bound by approximately 40 MeV more than $(bb\bar{u}\bar{u})$. The details of the hyperfine calculation are given in [11].

V. Conclusions

We have shown that for sufficiently large quark mass m and fixed antiquark mass $\bar{m}$ the dimeson $(Q^2\bar{q}^2)$ *must* be stable against all strong decays, due to the color Coulomb attraction of the two quarks in the color $\bar{3}$ state. Eq. (2.3) is the mathematical statement of this result.

Using the confining potential derived from a Born-Oppenheimer approximation to the MIT bag model, we have obtained the ground state energy of a number of dimesons. Those containing two b quarks and two light antiquarks are indeed energetically bound against decay into two mesons, but the binding energy is not great. In the most favorable case, $(bb\bar{u}\bar{d})$ is bound by 70 MeV with respect to the two mesons $(b\bar{u}) + (b\bar{d})$. The corresponding system containing two c quarks is not bound, and the mixed system with one b and one c is borderline. The numerical results show that the mass of the b quark is not large enough for the bb Coulomb attraction to completely dominate the energy. For the t quark this would indeed be the case.

This work was performed in collaboration with J. Carlson and J. A. Tjon, and a more complete account will be found in [11]. This research was supported by the U. S. Department of Energy.

References

1. J. P. Ader, J. M. Richard, P. Taxil: Phys. Rev. D <u>25</u>, 2370 (1982).
2. J. L. Ballot, J. M. Richard: Phys. Letts. <u>123</u>B, 449 (1983).
3. L. Heller: in <u>Workshop on Nuclear Chromodynamics, Quarks, and Gluons in Particles and Nuclei</u>, Edited by S. Brodsky and E. Moniz, World Scientific (1986), p. 306.
4. C. Zouzou *et al.*: Z. Phys. C <u>30</u>, 457 (1986).
5. H. J. Lipkin: Phys. Letts. <u>172</u>B, 242 (1986).
6. L. Heller, J. A. Tjon: Phys. Rev. D <u>35</u>, 969 (1987).
7. A. T. Aerts, L. Heller: Phys. Rev. D<u>23</u>, 185 (1981); and Phys. Rev. D<u>25</u>, 1365 (1982).
8. L. Heller: in <u>Quarks and Nuclear Forces</u>, Springer Tracts in Modern Physics, <u>100</u>, ed. by D. C. Fries and B. Zeitnitz, Springer-Verlag (1982), p. 145.
9. K. Johnson: in <u>Current Trends in the Theory of Fields</u>, AIP Conference Proceedings No. 48, edited by J. E. Lannutti and P. K. Williams (1978), p. 112.
10. L. Heller, J. A. Tjon: Phys. Rev. D<u>32</u>, 755 (1985).
11. J. Carlson, L. Heller, J. A. Tjon: to be published.
12. H. Lipkin pointed out that the mass of the 1^+ dimeson $(bb\bar{u}\bar{d})$ should be compared with $M(B) + M(B^*)$, not $2M(B)$. If this is done the binding energy of 0.07 GeV in Table II changes to 0.13 GeV.

Stable Multiquark States

C. Gignoux, B. Silvestre-Brac, and J.M. Richard
Institut des Sciences Nucléaires, 53, av. des Martyrs,
F-38026 Grenoble Cedex, France

A review is presented of stable multiquark states bound either by flavour independent confining forces or by the chromomagnetic interaction. In the first category, one finds states of the type $(QQ\bar{q}\bar{q})$ with a large M/m mass ratio. In the second one, the H = (ssuudd) is controversial, whereas the P = $(\bar{Q}$uuds), its isospin partner $(\bar{Q}$ddus) and the P' = $(\bar{Q}$ssud) are more likely to be stable against dissociation into a meson and a baryon.

I Introduction

There are several possible multiquark states. In this contribution, however, we will not discuss resonances such as dibaryons or baryonia, which could show up as bumps in experimental observables or govern the microscopic dynamics of hadron-hadron interactions. We instead concentrate on multiquarks which are stable under strong interactions and, in particular, cannot split into two ordinary hadrons. Such states, if they exist, would be as important for the quark dynamics as the hydrogen molecule for atomic physics or the alpha particle for nuclear physics.

We shall review here some recent studies done in potential models. It will be shown that both the central and the spin-spin components of the force offer the possibility of binding multiquark systems below their dissociation threshold. It remains, however, that stable multiquark states are extremely rare, since their binding requires very particular flavour and spin configurations.

II Central Potential

The most striking success of potential models is their ability to describe simultaneously different families such as $c\bar{c}$, $b\bar{b}$, $c\bar{s}$, etc... Flavour independence, indeed, is expected in QCD, since the gluons couple to the colour charge of the quark, which is the same for all flavours.

In a flavour-independent potential, heavy quarks experience more binding. For instance, the quarkonium ground state energy fulfills

$$E(t\bar{t}) < E(b\bar{b}) < E(c\bar{c}) . \qquad (1)$$

Meanwhile, when going from Q=c to Q=t, the reduced mass and, hence, the binding energy in $Q\bar{q}$ or $\bar{Q}q$ does not vary much. The inequality (1) explains why, in the $^3S_1$

channel, there are two narrow charmonium states, and three narrow $b\bar{b}$ states, whereas at least seven are expected for $t\bar{t}$, depending on the t quark mass.

In the meson sector, the reduced hamiltonian depends linearly on the inverse reduced mass μ^{-1}. The lowest binding energy is then a concave function of μ^{-1} i.e.: (there is a generalization involving the sum of the first n levels)[1]

$$\overline{QQ} + q\bar{q} < Q\bar{q} + Q\bar{q}. \qquad (2)$$

The generalization to baryons, which is rather delicate, has been discussed recently[2]. In short, for any plausible potential, one gets

$$QQq + qqq < Qqq + Qqq. \qquad (3)$$

Similarly, one obtains the following inequality[3]:

$$\overline{QQQ} + qqq < 3\,\overline{(Qq)} \qquad (4)$$

provided the mass ratio M/m is large enough (this means, for instance, that a $\bar{t}\bar{t}\bar{t}$ antibaryon does not annihilate on ordinary matter). For equal quark masses, however, the above inequality is reversed, namely[4]

$$2\,(qqq) > 3\,(q\bar{q}). \qquad (5)$$

The inequality (5) can be proved easily for any additive potential fulfilling the popular "1/2" rule $V_{QQ} = 1/2\ V_{Q\bar{Q}}$, or in the case of linear confinement with "Y-shape", i.e, $V_{Q\bar{Q}} = \lambda\,r$ and $V_{QQQ} = \lambda\,(d_1 + d_2 + d_3)$, where d_i means the distance of the ith quark to a junction whose location minimizes the potential.

The message coming from these inequalities is rather clear The general situation, corresponding to (5), is that quarks are heavier in baryons than in mesons. This rule holds also for multiquark states, which are usually very unstable since they break spontaneously into lighter hadrons. For instance $qq\bar{q}\bar{q} \rightarrow (q\bar{q}) + (q\bar{q})$. There is, however, a net tendency for heavy quarks to cluster together, as shown in the inequalities (3-5).

This is illustrated in the case of a system made of two quarks and two antiquarks interacting through flavour-independent forces, without hyperfine corrections. For equal masses, the ground state is made of two separated mesons, $(q\bar{q})+(q\bar{q})$. For the cryptoexotic $(\overline{QQ}q\bar{q})$, the threshold $(\overline{QQ})+(q\bar{q})$ benefits of the deeper binding of the simple systems and of the heavy quark clustering, and thus no collective binding below this threshold is conceivable. For the genuine exotic $(QQ\bar{q}\bar{q})$ with flavour number 2, the situation is, however, different: the threshold 2 $(Q\bar{q})$ is obviously favoured by the kinetic energy, but the collective state $(QQ\bar{q}\bar{q})$ benefits of the deep binding of the heavy quarks. This is why, if the mass ratio M/m is large enough, it becomes stable, i.e.

$$QQ\bar{q}\bar{q} < Q\bar{q} + Q\bar{q} . \tag{6}$$

This analysis was carried out in ref.[5] with an explicit study of the four-body problem, using as a prototype an additive potential

$$V = -3/16 \sum \lambda_i \lambda_j \, v(r_{ij}) . \tag{7}$$

The same qualitative conclusion was also reached by LIPKIN [6] and by HELLER and TJON [7], who used a slightly different approach.

III Chromomagnetic Interaction

While the additive ansatz (7) for the central potential is far from being fully satisfactory, the analog form for the spin-spin potential

$$V_{SS} = -\sum K_{ij} \lambda_i \lambda_j \, \boldsymbol{\sigma}_i \boldsymbol{\sigma}_j \tag{8}$$

is very commonly accepted. In (8), K_{ij} is a short-range operator, which appears as a recoil correction to the static potential. A Breit-Fermi form $K_{ij} \propto \delta(r_{ij}) / (m_i m_j)$ has been often used in the literature. On this basis, JAFFE [8] argued that the state H = (uuddss) could well be bound with respect to its lowest threshold $\Lambda \Lambda$. This is, however, controversial, as explained by CARBONELL [9] in this volume. ROSNER [10], indeed, introduced explicitly the SU(3)$_F$ breaking in the form

$$V_{SS} = -a \sum \lambda_i \lambda_j \, \boldsymbol{\sigma}_i \boldsymbol{\sigma}_j / m_i m_j \tag{9}$$

and showed that the chromomagnetic binding of the H is very sensitive to the mass ratio m_s/m_u.

Recently, new stable configurations were proposed by GIGNOUX et al [11], and independently, by LIPKIN [12]. These are the heavy baryon P = ($\bar{Q}$uuds), its isospin partner ($\bar{Q}$ddus) and P' = ($\bar{Q}$ssud), where Q denotes an heavy quark. For Q=∞, and the SU(3)$_F$ limit $m_u = m_s$, the P's have the same chromomagnetic binding $-8a/m^2$ with respect to their threshold ($\bar{Q}$q) + (qqq) than the H with respect to $\Lambda \Lambda$. The P states, however, survive much better than the H the breaking of SU(3)$_F$, so they have better chances to be stable.

IV Conclusion

The quark model works rather well for describing mesons and baryons. When it is reasonably extrapolated to more complicated systems, there is no proliferation of stable multiquark states in the (theoretical) spectrum. For most configurations, the ground state is made of a juxtaposition of ordinary mesons and baryons. In a few cases, however, corresponding to very particular spin and flavour configurations, stable multiquarks are expected.

Present calculations are still rather crude. Sometimes, they simply consist of a diagonalization of the colour spin operator (8-9). At best, they are based on variational calculations in line with simple potential models. Hopefully, the multiquark potential energy will be calculated in the future by lattice QCD methods and will be used in improved few-body calculations. The physics of multiquark states is, of course, intimately correlated to the microscopic calculation of interhadronic forces in the quark model.

Acknowledgments

We thank J. Carbonell, S. Fleck and H.J. Lipkin for enjoyable and useful discussions.

References

1   R. Bertlmann, A. Martin, Nucl. Phys. B168, 111 (1980)
2   E. Lieb, Phys. Rev. Lett. 54, 1987 (1985)
    A. Martin, J.M. Richard, P. Taxil, Phys. Lett. 176, 224 (1986)
3   C. Gignoux et al, in Proc. Symp. on Production and Decay of Heavy Flavours, Heidelberg, Mai 86, ed. K.R. Schubert and R. Waldi (Univ. of Heidelberg)
4   J.P. Ader, J.M. Richard, P. Taxil, Phys. Rev. D25, 2370 (1982)
    S. Nussinov, Phys. Rev. Lett. 51, 2081 (1983)
    A. Martin, J.M. Richard, Phys. Lett. B185, 426 (1987)
5   S. Zouzou et al, Z. Phys. C30, 457 (1986) and refs. therein
6   H.J. Lipkin, Phys. Lett. B172, 242 (1986)
7   L. Heller, J.A. Tjon, Phys. Rev. D35, 969 (1987)
    see, also, the contribution by L. Heller at this workshop.
8   R.L. Jaffe, Phys. Rev. Lett. 38, 195 (1977)
9   J. Carbonell et al, contribution at this workshop
10 J.L. Rosner, Phys. Rev. D33, 2043 (1986)
11 C. Gignoux et al, "Possibility of stable multiquark baryons", Phys. Lett. B, (in press)
12 H.J. Lipkin, contribution at this workshop and Proc. 1987 Rencontre de Moriond, ed. J. Tran Thanh Van (editions Frontieres, 1987)

Strong Decay of Baryons

Fl. Stancu and P. Stassart

Institut de Physique B5, Université de Liège,
Sart Tilman, B-4000 Liège, Belgium

We calculate nonstrange baryon resonances pion decay widths. Resonance wavefunctions are described by a semi-relativistic QCD-inspired flux-tube model while for the decay process we assume a quark-antiquark pair creation model. Results are compared with experimental data and previous theoretical results.

1. INTRODUCTION

Some time ago, a so-called "naive" quark pair creation model (QPC) had been introduced to describe strong decay of hadrons [1]. In that work, mesons and baryons were described by gaussian wavefunctions. Attempts to provide a better picture of baryon wavefunctions have led to flux tube variational wavefunctions [2]. Carlson, Kogut and Pandharipande (hereafter abbreviated as CKP) use a semi-relativistic hamiltonian including 2- and 3-body contributions to a confining term. Parameters of the model are found by minimizing the energy. One gluon exchange terms have been added in Ref. [3]. This latter model has been used, together with a pseudoscalar emission one, to describe the decay of baryon resonances.

In the present work, quark pair creation and flux-tube wavefunctions are brought together to calculate resonances decay widths. The quark model is summarized in next section, and so is QPC model in section 3. In that section we also derive the expressions of the transition amplitude. Results and discussions are presented in section 4.

2. FLUX TUBE BARYON WAVEFUNCTIONS

We use the semi-relativistic Hamiltonian of Ref. [2]

$$H = \sum_i (p_i^2 + m_i^2)^{\frac{1}{2}} + V(\vec{r}_1, \vec{r}_2, \vec{r}_3) + E_o \quad . \tag{2.1}$$

The short range behavior of the potential V is that of the two-body color Coulomb interaction. The long range one is linearly confining

$$V_{LR} = \sum_i \sqrt{\sigma} \, r_{i4} \quad , \tag{2.2}$$

where r_{i4} is determined [2] by the condition that it minimizes the static energy. Using parametrized wavefunctions, one finds their parameters by minimizing the expectation value of (2.1). The baryon wavefunctions have the form

$$\psi_n(\vec{r}_1, \vec{r}_2, \vec{r}_3) = F_{123} \prod_{i<j} f(r_{ij}) \, \Phi_n(\vec{r}_1, \vec{r}_2, \vec{r}_3) \quad , \tag{2.3}$$

where f is parametrized as

$$f(r_{ij}) = \exp\left[-W(r_{ij}) \lambda_1 r_{ij} - (1 - W(r_{ij})) \lambda_{1.5} r_{ij}^{1.5}\right]$$

$$W(r_{ij}) = \frac{1 + \exp(-r_o/a)}{1 + \exp((r_{ij} - r_o)/a)} \tag{2.4}$$

and F_{123} contains the 3-body contribution

$$F_{123} = 1 - \beta\sqrt{\sigma}\, (\sum_i r_{i4} - \tfrac{1}{2} \sum_{i<j} r_{ij}) \quad . \tag{2.5}$$

$\sqrt{\sigma}$ has the conventional value of the string tension constant

$$\sqrt{\sigma} = 1 \text{ GeV/fm} \quad .$$

Results of the variational calculations of the parameters λ_1, $\lambda_{1.5}$, r_0, a, β can be found in ref. [2]. The emitted meson wavefunctions are taken as

$$\psi(r_{ij}) = f(r_{ij})\, r_{ij}^\delta \tag{2.6}$$

to deal with the singularity of the $L = 0$ $q\bar{q}$ system at the origin. Using Jacobi coordinates

$$\vec{\rho} = \frac{1}{\sqrt{2}}(\vec{r}_1 - \vec{r}_2) \quad ; \quad \vec{\lambda} = \frac{1}{\sqrt{6}}(\vec{r}_1 + \vec{r}_2 - 2\vec{r}_3) \tag{2.7}$$

one can write [4,5] the $\Phi_n = \Phi_{LM}^{sym}(\vec{\rho},\vec{\lambda})$, to be inserted in (2.3).

The various baryon states are described in a SU(6) spin-flavor basis. To split them from one another with respect to their quantum numbers, one includes the hyperfine interaction consisting of a spin-spin and a tensor component [6]. Singularities are removed by a form factor $\exp(-\tfrac{1}{2}\Lambda^2 q^2)$ where Λ can be viewed as the "size of the quark", in order to cope with the deviation from one gluon exchange due to the cloud of virtual particles surrounding the quark [7].

The states considered in this work are N and Δ's, with up to 2 units of angular momentum or 1 unit of radial excitation (i.e. those pertaining to the $56(0^+,2^+)$, $56'(0^+,2^+)$, $70(0^+,1^-,2^+)$, $20(1^+)$ SU(6) supermultiplets). Mass spectra and mixing angles between states of common J^π can be found in refs. [3,8] for some values of (m,Λ).

3. QUARK PAIR CREATION AND TRANSITION AMPLITUDE

The modelization of the decay will be based on the creation of a quark-antiquark pair somewhere within the hadronic matter. This process will be described by a purely phenomenological constant γ : dynamics of quarks and gluons inside the hadron is not taken into account here. Other quarks are assumed to be spectators and the quantum numbers of the pair are those of the vacuum, i.e. it is in a $^3P_0$ state because of parity and charge conjugation conservation [1].

To calculate the transition amplitude, let us define the $\hat{T}$ matrix :

$$S = 1 - 2\pi i \delta(E_f - E_i)\, \hat{T} \quad ; \quad \hat{T} = I_1 \times I_2 \times I_3 \times \hat{T}_{vac} \tag{3.1}$$

$$\langle q_4 \bar{q}_5 | \hat{T}_{vac} | 0 \rangle = \delta(\vec{k}_4 + \vec{k}_5)\gamma \sum_m \langle 11m-m|0\,0\rangle\, Y_1^m(\vec{k}_4 - \vec{k}_5)\chi_1^{-m}\Phi_0, \tag{3.2}$$

where χ_1^{-m} is the spin triplet wavefunction and Φ_0 the flavor singlet one. The transition amplitude is then equal to

$$\langle NM|\hat{T}|R\rangle = \gamma \sum_m \langle 11m-m|0\,0\rangle\langle \Phi_N \Phi_M | \Phi_R \Phi_{vac}^{-m}\rangle\, I_m(R;N,M) \quad , \tag{3.3}$$

where Φ's are spin-flavor wavefunctions and

$$I_m(R;N,M) = \frac{1}{3\sqrt{3}} \delta(\vec{k}_N + \vec{k}_M) \int d^3k_\rho\, d^3k_\lambda\, Y_1^m(-2(\vec{k}_M + \sqrt{2/3}\,\vec{k}_\lambda))$$

$$\times \psi_R(\vec{k}_\rho, \vec{k}_\lambda) \psi_N^*(\vec{k}_\rho, \vec{k}_\lambda + \sqrt{2/3}\,\vec{k}_M) \psi_M^*(\vec{k}_M + 2\sqrt{2/3}\,\vec{k}_\lambda) \quad , \tag{3.4}$$

as can be found from momentum conservation, $\vec{k}_\rho$, $\vec{k}_\lambda$ being momentum Jacobi coordinates.

A Fourier transform brings us back to configuration space coordinates. This will allow us to use the flux tube derived baryon wavefunctions of section 2 :

$$I_m(R;N,M) = -\sqrt{3/4\pi} \cdot \frac{1}{(2\pi)^{3/2}} \gamma \, \delta(\vec{k}_N + \vec{k}_M) \int d^3\rho \, d^3\lambda \, d^3x$$
$$\times \psi_R(\vec{\rho}, \vec{\lambda} + 2\sqrt{2/3}\,\vec{x}) \psi_N^*(\vec{\rho},\vec{\lambda}) e^{i\vec{k}_M(\sqrt{2/3}\,\vec{\lambda} + \vec{x})} \vec{\varepsilon}_m \cdot (\vec{k}_M + i\vec{\nabla}_x) \psi_M^*(2\vec{x}), \quad (3.5)$$

where $\vec{x}$ is the Jacobi coordinate related to the emitted meson.

4. RESULTS AND DISCUSSION

We calculate the 9 dimensional integral (3.5) using a Monte Carlo method. The decay width Γ is obtained from the following expression [3]:

$$\Gamma = \frac{1}{\pi} \frac{|\langle NM|T|R\rangle|^2}{2J_R + 1} \frac{k_M E_N}{m_R} (\langle I_N \, I_M \, I_{3N} \, I_{3M} | I_R \, I_{3R} \rangle)^{-2},$$

where J is the total angular momentum, I the isospin and I_3 its third component.

In Table 1, we display the square root of nonstrange baryon widths obtained (i) in ref. [3], using the mass spectrum and mixing angles labelled set II, (ii) in this work, using that same set, without and with configuration mixing, (iii) from experiment [9]. The resonances included in this Table are those classified as two star or more by the Particle Data Group [9].

These results have been obtained using a Monte Carlo integration program. Since computation time grows by a power of 5 while precision increases by a power of 2, our numerical results have been limited here to a 20% precision range for the least favourable cases. More accurate results are expected soon.

Table 2 shows a comparison between these various results using a χ^2 formula :

$$\chi^2 = \Sigma \, (\frac{\Gamma^{1/2}_{theory} - \Gamma^{1/2}_{exp}}{\Delta \, \Gamma^{1/2}_{exp}})^2 \quad .$$

In ref. [3], the two parameters of the decay process model have been fitted by minimizing that χ^2. In the present work, the unique decay parameter γ has been fixed by requiring that resonance P33(1232) (lowest Δ resonance) has its decay width reproduced. Since this resonance is the one known with least uncertainty, our results and theirs can be compared easily. Notice that agreement with experiment would require that $\chi^2 \leq N - 1 = 17$.

These comparisons indicate that the pair creation model describes the decay better than the pseudoscalar emission one, having used identical flux-tube wavefunctions in both cases. The number of parameters has been reduced from 2 to 1 while the results went closer to experiment. (One could also compare them with a prior 4 parameter model [10]).

Still, the agreement is just fair. One shouldn't expect to get perfect agreement with a model about which several questions can arise : is the pion correctly described in such a model where its special character due to chiral symmetry breaking has not been taken into account ? The validity of the model might be tested further by calculating the N + ρ channel decay widths. One should also notice the dependence on hyperfine splitting modelization. Though no fit has been made therewith, this has some impact on our results : one can question how closer agreement between that model and experimental mass spectra and mixing angles could change our decay re-

Table 1. Square root of baryon decay width, $\Gamma_{N\pi}^{1/2}$ in MeV$^{1/2}$

Column 2 : ref. [3], mixing, set II
Column 3 (4) : this work, without (with) configuration mixing
Column 5 : experiment [9]

Resonance	Ref. [3], mix	No mix	Mix	Experiment [9]
($7/2^+$ N) F17(1990)	0.5	1.4	1.0	4.2 ± 2.0
($7/2^+$ Δ) F37(1950)	3.6	8.9	6.9	9.8 ± 2.0
($5/2^+$ N) F15(1680)	3.6	9.3	6.8	8.7 ± 0.85
($5/2^+$ Δ) F35(1905)	1.1	4.5	2.2	5.8 ± 1.6
($3/2^+$ N) P13(1720)	8.6	2.4	7.0	5.4 ± 1.8
($3/2^+$ Δ) P33(1232)	10.9	10.7	10.7	10.7 ± 0.2
($3/2^+$ Δ) P33(1600)	10.6	4.9	7.9	7.0 ± 2.1
($3/2^+$ Δ) P33(1920)	4.1	8.0	7.7	6.5 ± 1.25
($1/2^+$ N) P11(1440)	1.0	9.1	6.8	10.9 ± 4.0
($1/2^+$ N) P11(1710)	6.0	6.8	4.8	4.0 ± 1.05
($1/2^+$ Δ) P31(1910)	6.4	2.4	5.6	7.0 ± 1.9
($5/2^-$ N) D15(1675)	3.4	5.7	4.9	7.4 ± 1.25
($3/2^-$ N) D13(1520)	5.0	10.0	8.5	8.3 ± 1.05
($3/2^-$ N) D13(1700)	2.3	2.6	3.4	3.2 ± 0.7
($3/2^-$ Δ) D33(1700)	2.9	4.9	4.2	6.1 ± 1.65
($1/2^-$ N) S11(1535)	9.4	9.4	5.9	8.3 ± 2.65
($1/2^-$ N) S11(1650)	8.9	6.4	10.1	9.5 ± 2.0
($1/2^-$ Δ) S31(1620)	3.7	4.9	4.2	6.5 ± 1.0

Table 2. χ^2 over Table 1 resonances

Ref. [3]	No mix	Mix
92	38	31

sults. To learn more about all this, strange and charmed sectors should be investigated as has been done already for pion decay [11, 12].

Finally, one should not forget that the coupling constant γ is taken as a fitted parameter here. Within this framework, we can let it depend on where the pair creation takes place within the hadron [11,12] but ultimately, the answer to what γ is lies in the dynamics of the process, not in any kinematical picture.

The above numerical results should be regarded as preliminary.

ACKNOWLEDGEMENT

We would like to thank Dr. R. Sartor for fruitful discussions.

REFERENCES

1. A. Le Yaouanc, L. Oliver, O. Pène, J.C. Raynal : Phys. Rev. D8, 2223 (1973)
2. J. Carlson, J. Kogut, V.R. Pandharipande : Phys. Rev. D27, 233 (1983)
3. R. Sartor, Fl. Stancu : Phys. Rev. D34, 3405 (1986)
4. N. Isgur, G. Karl : Phys. Rev. D19, 2653 (1979)
5. N. Isgur, G. Karl : Phys. Rev. D20, 1191 (1979)
6. A. De Rújula, H. Georgi, S.L. Glashow : Phys. Rev. D12, 147 (1975)
7. J. Carlson, J. Kogut, V.R. Pandharipande : Phys. Rev. D28, 2807 (1983)
8. R. Sartor, Fl. Stancu : Phys. Rev. D31, 128 (1985)
9. Particle Data Group : Phys. Lett. B170 (1986)
10. R. Koniuk, N. Isgur : Phys. Rev. D21, 1868 (1980)
11. R. Kokoski : Ph.D. Thesis, University of Toronto (1984)
12. R. Kokoski, N. Isgur : Phys. Rev. D35, 907 (1987).

Resonating Group Method Applied to Hadrons

P. Bicudo

Centro de Física da Matéria Condensada, Av. Prof. Gama Pinto 2,
P-1699 Lisboa, Portugal

The R.G.M. (Resonating Group Method) constitutes an important tool for studying hadronic physics. We outline its basic ideas and techniques and test current non-relativistic quark-quark potentials in the scattering K^+-P.

1. INTRODUCTION

The R.G.M. was first applied to atomic physics /1/, then to nuclear physics, and since 1980 it has been applied to hadronic physics /2/. The method is devoted to studying hadronic phenomena (scattering, decays, multiquarks, ...) in the framework of hadrons' supposed constituents, the quarks.

For instance, The N-N potential exhibits a short-range repulsive core which can be explained by the interplay between quark interchange and the interquark potential (Fig.1). Other explanations for the incompressibility of nuclear matter also exist in the literature, notably the Skyrmeon explanation.

The basic idea is the following. If strongly bound clusters (**groups**) of quarks (hadrons) are allowed to scatter (or form **resonating** states), then the Pauli principle, together with the cluster nature of the asymptotic particles, induces non-local potentials which seem, so far, to describe fairly well some low-energy hadronic features.

In Sect.2 we set out the framework of the method. In Sect.3 we briefly describe the techniques that allow us to determine the effective non-local potential between the hadrons. In Sect.4 we test the quark-quark potential in the K^+-P scattering. Sect.5 is devoted to the discussion of the consistency of the R.G.M.'s results.

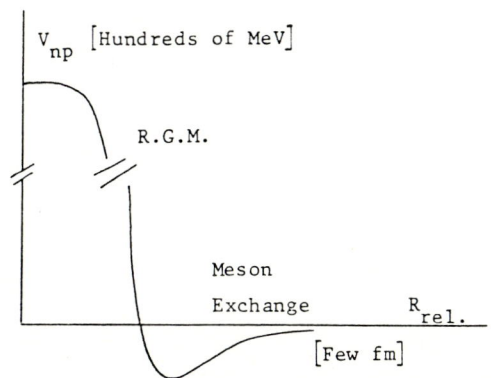

Fig.1. The N-N potential exhibits a short-range repulsive core, which can be explained by the interplay between quark interchange and the interquark potential.

2. QUARKS and HADRONS in R.G.M.

Quark (a^+, a) and antiquark (b^+, b) operators appear directly in the hamiltonian operator, when written in the second quantization formalism:

$$H = H(a^+_i, a_j, b^+_k, b_l) \; ,$$

where H includes the kinetic energy, the potential and the $q,\bar{q}$ creation and annihilation terms. If we project in the Schroedinger formalism, we obtain

$$(H - E) \; A|\Psi> = 0 \; ,$$

where the antisymmetrizer is made of exchange operators. We note that gluons are not accounted for explicitly, but they are hidden in the microscopic potential. In this way the $q_i.q_j$ potentials are chosen to be color dependent, usually in the form $\lambda_i.\lambda_j$.

Hadrons are the simplest bound state eigensolutions (mesons or baryons) of the microscopic hamiltonian. When studying scattering problems, we freeze the degrees of freedom of the incoming and outgoing clusters (hadrons) - as previously stated the hadron wave functions should be eigensolutions of the microscopic hamiltonian - and allow the relative wave function between these clusters to vary (R.G.M. ansatz).

It is convenient to use Jacobi coordinates, which are chosen to be intracluster, relative and center of mass coordinates. See Fig.2. In this way we obtain the R.G.M. equations in the relative coordinate only:

$$H_R |\Phi_R> = E |\Phi_R> \; .$$

At this point we have to use a non-relativistic model (as we don't know in relativity how to derive a kinetic energy which is decomposable in cluster terms and in relative momentum terms). Although there is no strong evidence against non-relativistic quark models, the development of a relativistic approximation would be quite relevant to enlarge the applications of R.G.M.

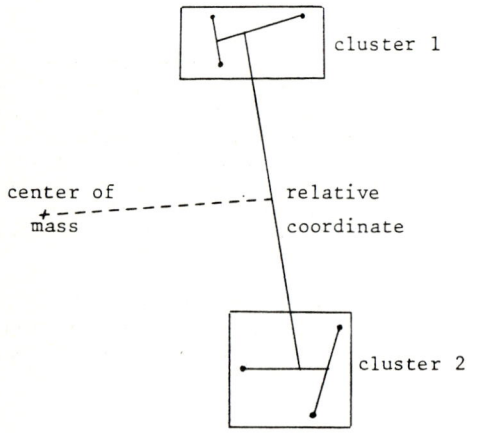

Fig.2. Jacobi coordinates.

3. SEPARABLE EFFECTIVE POTENTIALS and OVERLAP KERNELS

The effective R.G.M hamiltonian H_r is obtained after folding $H-E$ (integrating out coordinates) with the clusters' wave functions $|\phi\rangle$:

$$H_R - E_R = \langle \Phi_{c1} | \langle \Phi_{c2} | (H-E) A | \Phi_{c1} \rangle | \Phi_{c2} \rangle ,$$

where, using the previously defined Jacobi coordinates, it is straightforward to isolate the relative kinetic energy and the clusters' masses:

$$H_R - R_P = T_R + V_R - (E - m_{c1} - m_{c2}) ,$$

The remaining operator is then the effective R.G.M. potential V_R. It is clear that the reduced mass appearing in T_R is due to the constituent quarks' masses. If we work with the π, a very light meson, this poses a problem.

If we have more than one channel, we have to fold with different clusters' wave functions, according to the channel, and we obtain a system of coupled equations. For instance, with two channels the R.G.M. equations read

$$\begin{bmatrix} V_{AA} + T_A - E_A & V_{AB} \\ V_{BA} & V_{BB} + T_B - E_B \end{bmatrix} \begin{bmatrix} |\chi_A\rangle \\ |\chi_B\rangle \end{bmatrix} = 0 .$$

As hadrons are colour singlets, there are no direct potential contributions for the hadron-hadron potential. In fact, V_{eff} is the sum of all terms with quark exchange (overlap kernels):

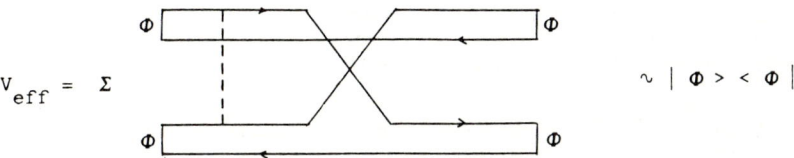

$$V_{eff} = \Sigma \qquad \sim |\phi\rangle\langle\phi|$$

These kernels can, for any accuracy, be written as separable potentials of finite rank, as if the clusters' wave functions $|\phi\rangle$ (which are superimposed) propagate, via the exchange operator, from the intracluster coordinates to the relative coordinate. The kernels are easy to calculate with the help of Ribeiro's graphical rules /3/ or Moshinsky's rotations /4/, and converge very fast in the number of radial excitations as long as we use the harmonic oscillator basis to describe $|\phi\rangle$, which is quite all right as $|\phi\rangle$ is a product of bound states.

As V is a rapidly convergent separable potential in the harmonic oscillator basis, the T matrix, and hence the observables, are easy to calculate even if we work with many coupled channels.

4. K⁺-P TESTS of the R.G.M.

In order to trust R.G.M. predictions for multiquarks, annihilation, quark-gluon plasma ... one should first test it carefully. An excellent test of the potential used in the non-relativistic quark model is provided by the K⁺-P scattering:
- u $\bar{s}$ u u d is an exotic flavour, thus no first-order annihilation reactions may occur.
- π does not couple to K (pseudoscalars), thus O.P.E.P. does not contribute.
- the first coupled channel (K-Δ) is 300 MeV above threshold.
- the reduced mass given by quarks is not far from the real one.

Hence low-energy scattering should be described with one channel and we need not get concerned about annihilation.

In a good first approximation, R.G.M. produces the following potential:

$$V_R(E_R) = -\frac{1}{3}(T_{00} + \frac{7}{3}V_{00} - E_R) \times |\Phi_{000}\rangle\langle\Phi_{000}| \quad,$$

where colour, flavour, spin and angular momentum have already contributed and T_{00}, V_{00} are the two-body radial matrix elements of the kinetic operator and of the hyperfine $S_1 \cdot S_3$ potential.

Experimentally the scattering length is $a(K^+P) = 3$fm which is much smaller than T_{00} or V_{00}. In fact, see Fig.3, the experimental /5/ S-wave phase shifts (K⁺-P) are described by the potential $V(E) = (E/3)|\Phi_{000}\rangle\langle\Phi_{000}|$. This suggests that

$$T_{00} + \frac{7}{3}V_{00} \sim 0 \quad.$$

If we use the one-gluon exchange hyperfine potential obtained from naive Q.C.D. and the simple confining harmonic oscillator potential /5/ we have from spectroscopy the relation

$$(1.5 \pm 0.5) T_{00} + V_{00} \sim 0,$$

which is inconsistent with the scattering result.

However, if we use a phenomenologically broadened O.G.E. potential together with a linear confining potential /6/, then the spectroscopic results agree with scattering /7/.

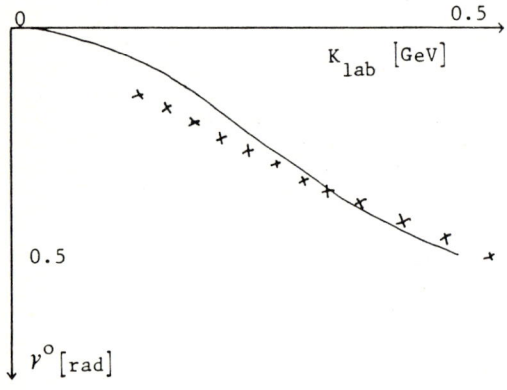

Fig.3. K⁺-P S-wave experimental phase shifts (x x x) against phase shifts produced (-----) by the potential $V(E) = (E/3)|\Phi_{000}\rangle\langle\Phi_{000}|$.

5. CONCLUSION: PHENOMENOLOGICAL VERSUS CONSISTENT POTENTIAL

According to the preceding test, the naive potentials seem to fail, while a potential which phenomenologically fits spectroscopy yields good results, at least in K^+P scattering. This is reassuring, although there is still a lot of work to be done in this direction, as scattering is a more sensitive place to test hadronic physics than spectroscopy. Spin-orbit and tensor quark-quark potentials as well as quark-antiquark annihilation processes have still to be tested in a consistent way.

We are also researching in another direction, trying to obtain the good quark hamiltonian, not yet from Q.C.D., but already from a theory with a non-trivial vacuum. In fact, if we suppose that a confining quark-antiquark potential exists which leads to $3P_0$ bound states, q and $\bar{q}$ get mixed and the vaccum condenses /8/. Then many interesting new features arise:
- q gains a momentum-dependent mass.
- the small π mass is understood as a quasi Goldstone boson mass.
- a quark-antiquark creation term appears /9/. See Fig.4.

Whether the new vacuum is diacoloured (confining), still remains to be shown.

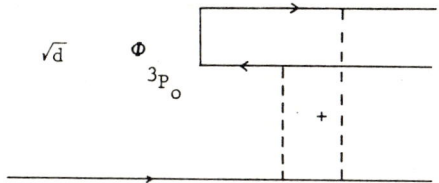

Fig.4. $q\bar{q}$ creation amplitude, where d is the density of the condensate and ϕ_{3P_0} is the wave function of the condensated bosons.

ACKNOWLEDGEMENT

I am very grateful to O. Pene, H. Rubinstein, H. Lipkin, B. Sylvestre-Brac, J. Carbonell, E. Werner and H. Nielsen for several discussions during this workshop.
I also wish to express my gratitude to Emílio Ribeiro for his helpful suggestions.

REFERENCES

/1/ J. A. Weeler, Phys. Rev. 52, 1083 (1937).
/2/ J. E. F. T. Ribeiro, Z. Phys. C5, 27 (1980).
/3/ J. E. F. T. Ribeiro, Phys. Rev. D25, 2046 (1982).
    E. van Beveren, Z. Phys. c17, 135 (1983).
    P. Bicudo, M.S. Thesis, Lisboa (1986).
/4/ M. Moshinsky, Nuc. Phys. 13, 104 (1959); Cargèse lect. 3 (1969).
/5/ I.Bender, H.Dosh, H.Pirner, H.Kruse, N. Phys. A414, 359 (1984).
/6/ R. Badhuri, L. Cohler, T. Nogami, N. Cim. 65A, 376 (1981).
/7/ P. Bicudo, J. Ribeiro, to be published in Z. Phys. C.
/8/ le Yaouanc, L.Oliver, O.Pene, J.Raynal, Phys.Rev.D33,3898(1986).
    P. Ferstl, M. Schaden, E. Werner, N. Phys. A452, 680 (1986).
/9/ L. Keldysh, A. Koslov, Sov. Phys. JETP 27, 521 (1968).

Quark Confinement and Nuclear Dynamics*

E.J. Moniz

Center for Theoretical Physics, Department of Physics and
Laboratory for Nuclear Science, Massachusetts Institute of Technology,
Cambridge, MA 02139, USA

The quantitative success of conventional descriptions of nuclear structure and dynamics in terms of hadronic degrees of freedom and effective interactions is rather surprising in light of the modern view of hadrons as extended composite systems. We are still quite far from a QCD-based understanding of the "duality" between hadronic and quark descriptions and thus of the limits of an effective hadronic representation of strong-interaction physics. Here, we will pursue the more modest goal of demonstrating how, within a quark model, the general features of quark confinement and exchange symmetry can in fact lead to many of the qualitative phenomena observed in low-energy hadronic physics. The work is based on the model introduced and studied in collaboration with Lenz, Londergan, Rosenfelder, Stingl and Yazaki [1]. Recent extensions will also be discussed.

A second major goal of this model study is that of eliciting characteristic signatures of the underlying quark substructure and of exploring the consequences of "hidden color dynamics" for nuclear observables. Such dynamics (the term will be defined precisely below in the context of our specific model) cannot be studied for an isolated hadron.

I. The Model

We formulate our quantum mechanical model for the $q^2\bar{q}^2$ system [1]. This two-meson system is the simplest to deal with and encompasses the essential physics. Extension to the other systems is straightforward.

Formulating the model for one isolated hadron is, of course, quite simple. The color-singlet requirement means that the color structure of the quark interaction potential can be absorbed into the strength parameter:

---

*Supported in part through funds provided by the U.S. Department of Energy under contract DE-AC02-76ER)3069.

$$H = k_q + k_{\bar{q}} + v_{q\bar{q}} ,\qquad(1)$$

where v is a confining potential. We take this opportunity to note that we shall work below with a harmonic potential $v=(1/2)\mu\omega^2 r^2$ with dimensionless parameters $2m=1$, $b_0^2=(\mu\omega)^{-1}=1$, $\omega=4$.

The long-range nature of the force is crucial for the multi-hadron problem. We insist on saturation of the color forces; i.e., outside the interaction region (characterized by hadron extension), the confining forces operate only between neighboring quarks in color singlets. This rules out the standard construction in terms of two-body potentials, since the approach necessarily leads to strong interaction van der Waals forces [1,2]. To enforce the color singlet condition in a manner consistent with quark exchange symmetry, we must simultaneously consider color and configuration space. First we consider a basis in color space. Two complete bases in the space of overall color singlets are

$$|0>_c \equiv |(1\bar{1})_0 \otimes (2\bar{2})_0>$$
$$|1>_c \equiv |[(1\bar{1})_{(N_c^2-1)} \otimes (2\bar{2})_{(N_c^2-1)}]_0>\qquad(2)$$

and

$$|\tilde{0}>_c \equiv |(1\bar{2})_0 \otimes (2\bar{1})_0>$$
$$|\tilde{1}>_c \equiv |[(1\bar{2})_{(N_c^2-1)} \otimes (2\bar{1})_{(N_c^2-1)}]_0> .\qquad(3)$$

The set in Equation (2) is based upon the pairing of $(1\bar{1})$ and $(2\bar{2})$; the set in Equation (3) is based on the $(1\bar{2})$ and $(2\bar{1})$ pairing. These bases are related by a unitary transformation

$$\begin{pmatrix}0\\1\end{pmatrix} = \begin{pmatrix}\alpha_N & -\sqrt{1-\alpha_N^2}\\-\sqrt{1-\alpha_N^2} & -\alpha_N\end{pmatrix}\begin{pmatrix}\tilde{0}\\\tilde{1}\end{pmatrix}\qquad(4)$$

$$\alpha_N = 1/N_c$$

where N_c is the number of colors. Clearly, each of the bases is "natural" for one of the two ways in which mesons can be separated. This suggests that the best framework for setting up the $q^2\bar{q}^2$ potential will be provided by joint color-configuration space projectors. To this end, we introduce a convenient set of coordinates

$$\vec{R} = \frac{1}{4}(\vec{r}_1 + \vec{r}_2 + \vec{r}_{\bar{1}} + \vec{r}_{\bar{2}})$$

$$\vec{x} = \frac{1}{2}(\vec{r}_1 + \vec{r}_2) - \frac{1}{2}(\vec{r}_{\bar{1}} + \vec{r}_{\bar{2}})$$

$$\vec{y} = \frac{1}{2}(\vec{r}_1 + \vec{r}_{\bar{1}}) - \frac{1}{2}(\vec{r}_2 + \vec{r}_{\bar{2}}) \quad (5)$$

$$\vec{z} = \frac{1}{2}(\vec{r}_1 + \vec{r}_{\bar{2}}) - \frac{1}{2}(\vec{r}_2 + \vec{r}_{\bar{1}}).$$

The c.m. coordinate $\vec{R}$ is basically irrelevant. The coordinate $\vec{x}$ corresponds to the separation of quarks from antiquarks. Confinement implies that the wavefunction is strongly damped as $|\vec{x}|$ becomes large compared to the confinement scale. The coordinates $\vec{y}$ and $\vec{z}$ serve as channel coordinates for mesons grouped according to the color basis states in Equations (2) and (3), respectively. This leads naturally to a partitioning of the color-configuration space through the orthonormal projectors

$$P_\alpha = \theta(y^2 - z^2)|\alpha\rangle\langle\alpha|$$

$$\tilde{P}_\alpha = \theta(z^2 - y^2)|\tilde{\alpha}\rangle\langle\tilde{\alpha}| \quad (6)$$

$$\sum_{\alpha=0,1}(P_\alpha + \tilde{P}_\alpha) = 1$$

$$P_\alpha P_\beta = \delta_{\alpha\beta} P_\alpha, \quad P_\alpha \tilde{P}_\beta = 0.$$

Basically, we have partitioned space along the rearrangement surface y=z and use the color basis appropriate to the asymptotic state in each partition.

Saturation of the color forces implies that

$$P_0 V P_0 = [v(1\bar{1}) + v(2\bar{2})]$$
$$= 4(x^2 + z^2)$$

$$\tilde{P}_0 V \tilde{P}_0 = [v(1\bar{2}) + v(2\bar{1})] \quad (7)$$
$$= 4(x^2 + y^2)$$

$$P_1 V P_1 \to \infty$$
$$\tilde{P}_1 V \tilde{P}_1 \to \infty \tag{8}$$

outside the interaction volume. Obviously, this is not enough to fully specify the potential. We construct a "minimal" model by taking Equation (7) to hold true throughout space and by assuming harmonic forces for the "color-hidden" potentials:

$$\tilde{P}_1 V \tilde{P}_1 = P_1 V P_1 \equiv 4[x^2 + \nu_c^2 (y^2 + x^2)]. \tag{9}$$

One parameter, ν_c, is introduced in Equation (9) to represent the relative strength of confining forces in the color-hidden and singlet configurations. Apart from this one parameter, the dynamics are specified entirely by the confining forces operative in isolated hadrons and by requiring continuity of Ψ, Ψ' at the rearrangement surface. This, in turn, can be cast into a boundary condition on the problem treated in one partition of configuration space. The number of colors, N_c, enters directly into the continuity condition:

$$\left[\frac{\Psi_0}{\Psi_0}(n) = -\lambda \frac{\Psi_1}{\Psi_1}(n)\right]_{y=z}$$

$$\Psi_i \equiv P_i \Psi \tag{10}$$

$$\Psi_i^{(n)} \equiv (\partial/\partial y - \partial/\partial z) \Psi_i$$

$$\lambda \equiv \frac{N_c s + 1}{N_c s - 1}, \tag{11}$$

where $s = +1(-1)$ for symmetry (antisymmetry) under exchange of spatial coordinates. In the limit $N_c = 1$, $\lambda = \infty$ for $s = +1$ and $\lambda = 0$ for $s = -1$; in the limit $N_c = \infty$, $\lambda = 1$. Thus, λ can be viewed as a parameter which takes its extreme values for $N_c = 1$ and interpolates between these extremes as N_c varies. We shall take the unconventional (that is, with respect to various studies of QCD) approach of starting with $N_c = 1$ and indeed of devoting most of our discussion to that limit. This small-N_c limit is in many ways the more interesting limit for studying strong interaction dynamics.

II. U(1)

The Hamiltonian for $N_c=1$ takes on a particularly simple form. From Equations (7) and (8), we see that the $\vec{x}$ coordinate decouples and that a straightforward partial wave reduction can be accomplished [1]. The L=0 wavefunction can be written as

$$\Psi_{L=0}(\vec{x},\vec{y},\vec{z}) = \phi_0(x) \frac{\psi(y,z)}{yz}, \tag{12}$$

where $\phi_0(x)$ is the ground state oscillator wavefunction and ψ satisfies the Schrödinger equation

$$[-\partial^2/\partial y^2 - \partial^2/\partial z^2 + 4\min(y^2,z^2)]\psi = E\psi, \tag{13}$$

$$E = 6 + k_0^2,$$

where k_0 is the channel momentum. The boundary conditions become

$$\psi(y,z)\rfloor_{y=z} = 0, \text{ antisymmetric}$$

$$\psi^{(n)}(y,z)\rfloor_{y=z} = 0, \text{ symmetric} \tag{14}$$

at the rearrangement surface and

$$\psi \underset{y\to\infty}{\to} \phi_0(z)e^{-ik_0 y} - \sum_n \phi_n(z)S_n e^{ik_n y} \tag{15}$$

$$k_n^2 = k_0^2 + (6-\varepsilon_n),$$

where k_n is the channel momentum with internal excitation of the mesons. The methods used to solve this problem (and that for $N_c>1$) are instructive and are discussed in depth in Reference 1.

A. Bound State

The U(1) symmetric problem has a bound state with binding energy $B=0.04\omega$, which is about 1% of the meson-meson zero-point energy $E_0=3\omega$. This is a quantum-mechanical effect arising from the exchange process summarized in the rearrangement surface boundary condition. The average separation of mesons in the ground state is $d=4.7\bar{R}$, where $R=\sqrt{3/8}$ is the meson root-mean-square radius. Clearly, this is a loosely bound, deuteron-like state. A quantitative measure is provided by the probability distribution $P(\vec{R},\vec{\rho})$ for finding a quark at the distance $\vec{\rho}$ from the meson-meson center-of-mass for a fixed separation $\vec{R}\equiv\vec{y}$. This is shown in Figure 1 for $\vec{\rho}\|\vec{R}$ and R=d. This profile is indistinguishable from that obtained by placing two free-space mesons at the separation d. This suggests that it

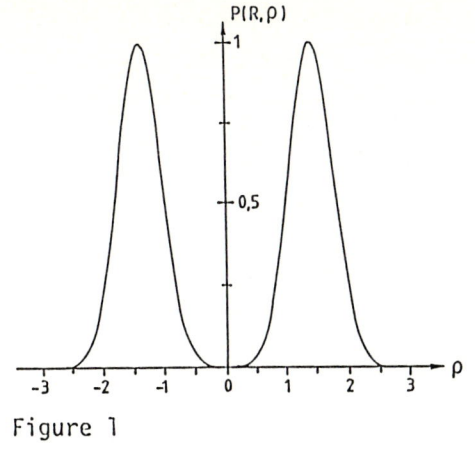
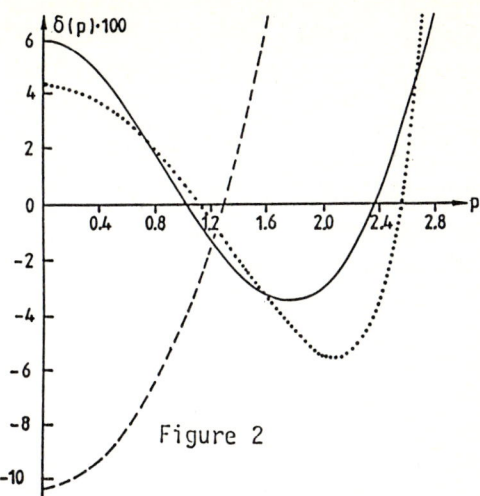

Figure 1. Bound state probability distribution for finding a quark a distance ρ from the overall CM for fixed separation R=d of the two mesons, evaluated for $\vec{\rho} \parallel \vec{R}$.

Figure 2. Difference between quark momentum distribution in the bound state, $n(p)$, and in a free hadron, $n_0(p)$, where $\delta(p)\equiv(n(p)-n_0(p))/n_0(p)$. Dashed curve is the result for the effective hadron theory; dotted curve is the effective hadron result with the hadron size parameter increased by 5.5%.

should be possible to describe the system rather well in terms of hadronic rather than quark degrees of freedom, even though the interactions are driven entirely by quark exchange. We return below to a systematic discussion of this point and will confirm this suggestion.

The quark momentum distribution in the bound state proves to be more interesting. This would be measured in deep inelastic electron scattering. If the mesons in the bound state are thought of as inert "packages" of quarks, then a simple convolution argument implies that

$$\langle p^2 \rangle_B > \langle p^2 \rangle_{hadron}. \tag{16}$$

The increase in $\langle p^2 \rangle$ from the free-hadron value comes from the relative motion of the two mesons. This inequality is <u>not</u> satisfied in the quark-exchange model. The virial theorem with the somewhat unusual Hamiltonian of Equation (13) yields the standard result

$$\langle T \rangle = \langle V \rangle = 1/2 \, E_B \tag{17}$$

requiring that

$$\langle p^2 \rangle_B < \langle p^2 \rangle_{hadron}. \tag{18}$$

The full momentum distribution difference is shown in Figure 2. The inequality Equation (18), together with the integral constraint and the

61

large-p increase caused by "Fermi motion", leads to the extra node at small momentum. This is a signature of the quark-exchange dynamics. To reproduce this effect with the package model, one could increase the meson size so as to reduce the zero point motion. Such "inflated hadron" approaches have received considerable phenomenological attention. However, our model suggests that this idea, while reproducing n(p) by parameter-fitting, is not useful in correlating different observables [1].

B. Effective Hadron Theories

We discuss here the methodology for "hadronizing" the quark model and apply the results to the bound state structure. We have already suggested that an effective hadron theory should work reasonably well for the large-distance or low-energy properties; nevertheless, the description must become more complicated at very short distances. Further, the momentum distribution makes it clear that inert, ground-state mesons cannot reproduce all qualitative features.

The construction of an equivalent hadronic model necessarily involves identifying and projecting out the hadron internal degrees of freedom. The procedure is far from unique. This conclusion follows from the fact that no unique description of a hadron exists when hadrons "overlap"; i.e., one cannot unambiguously identify which hadron contains which quark. We describe the approach of Gardner [3].

An equivalent hadron theory clearly entails specification of the Hamiltonian as a coupled-channel problem, with the channels corresponding to the various hadronic states allowed asymptotically. However, it is important to realize that all operators in the theory also must be expressed in the hadronic basis. We will focus on the charge operator as an example, by examining the bound state form factor:

$$F(q) = \langle \psi_B | \hat{\rho}(q) | \psi_B \rangle , \qquad (19)$$

where $\hat{\rho}$ is the quark model charge operator. Introducing the projectors P_n over internal coordinates, we have

$$F(q) = \Sigma_{nm} \langle X_n | (P_n \hat{\rho}(q) P_m) | X_m \rangle . \qquad (20)$$
$$| X_m \rangle = P_m | \psi_B \rangle ,$$

where $| X_m \rangle$ is the effective hadronic coupled-channel wavefunction. A sum over all terms in the hadronic expansion will yield the exact quark model result independent of the specific choice of projectors. The interest lies in determining whether a truncation at a very few internal degrees of freedom leads to a reasonable and qualitatively accurate model (as appears to be the case in the real world).

One can straightforwardly "hadronize" the model by picking either y or z as the confined coordinate:

$$P_n(y,z;y'z') = \delta(y-y')\phi_n(z)\phi_n(z'), \quad (21)$$

where the ϕ_n are the free space oscillator wavefunctions. The effective charge operator is then just

$$[\rho_q(y)]_{nm} = f_{00}(q)f_{nm}(q)j_0(qy/2)$$
$$f_{nm}(q) = \int_0^\infty dz \phi_n(z) j_0(qz/2) \phi_m(z). \quad (22)$$

This is just the standard one-body charge operator including the free-space hadron form factor. Similarly, the coupled channel Hamiltonian has a very simple local form with this projection scheme; for example, the ground state channel Hamiltonian reads [3]

$$h_{00}(y) = -\partial^2/\partial y^2 + V_{eff}(y)$$
$$V_{eff}(y) = -4\int_y^\infty dz(z^2-y^2)\phi_0(z)^2. \quad (23)$$

The potential is everywhere attractive and falls off with a Gaussian dependence for $y\to\infty$. However, this approach, which implicitly assigns specific quarks into hadrons, gives extremely poor convergence to the quark model results. For example, the lowest truncation does not lead to a bound state.

Apparently, it is crucial that the underlying quark exchange dynamics be somehow reflected in the choices of projectors. This can be done by working only in the region y>z and imposing a boundary condition at the rearrangement surface. In doing so, the definition of a hadron is, in effect, modified in the interaction region. As a first example, we introduce a set of functions $\phi_n(z;y)$ for y>z, such that

$$(-\partial^2/\partial z^2 + 4z^2)\phi_n(z;y) = \varepsilon_n(y)\phi_n(z;y)$$
$$[\partial/\partial z\ \phi_n(z;y)]_{z=y} = 0 \quad (24)$$
$$\int_0^y dz \phi_n(z;y)\phi_m(z;y) = \delta_{nm}$$

and define projectors

$$P_n(yz;y'z') = [\phi_n(z;y)\phi_n(z';y')\delta(y-y')\theta(y-z)\theta(y'-z')$$
$$+ \phi_n(y;z)\phi_n(y';z')\delta(z-z')\theta(z-y)\theta(z'-y')]. \quad (25)$$

The ground state channel potential then reads [3]

$$V_{00} = \varepsilon_0(y) - \varepsilon_0(\infty) + \phi_0(y;y)\partial_y\phi_0(y;y) + \int_0^y dz[\partial_y\phi_0(z;y)]^2. \qquad (26)$$

This potential is repulsive at small distances and diverges $\sim y^{-2}$; at large separation, it is attractive and falls with a Gaussian dependence. The potential is shown in Figure 3(a). The channel potential corresponding to the lowest instrinsic excitation is also shown. It is qualitatively similar except that the repulsive core sticks out farther. The channel coupling potential is non-local

$$V_{01}(y) = V_{01}^{(1)}(y) + V_{02}^{(2)}(y)\vec{\partial}_y \qquad (27)$$

and is also shown in Figure 3(a). A crucial point is that the lowest truncation in this scheme already gives a bound state, with binding energy $B^{(0)}=0.022\omega$. Recall that the zero point energy is 6, so this result is remarkably close to the total energy; the inclusion of two more channels increases the binding energy to 0.032ω.

To calculate the form factor, we write the charge operator in hadronic variables as

$$P_n\hat{\rho}_q P_m \rightarrow [\rho_q(y)]_{nm} \equiv [\rho_q^{1-body}(y) + \rho_q^{exch}(y)]_{nm}, \qquad (28)$$

where the 1-body piece is that written out in Equation (27). There is in addition a term which corresponds to interaction with a quark exchanged

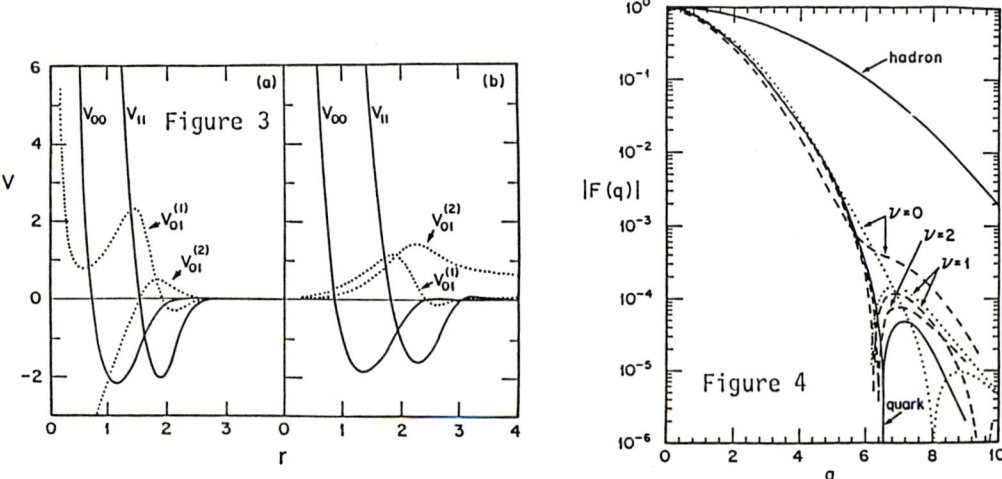

Figure 3. Effective hadronic potentials in two different schemes.

Figure 4. Effective form factor at different levels of truncation. Dashes and dots correspond to schemes (a) and (b), respectively, in Figure 3. The form factor for an isolated hadron is also shown.

between the two hadrons. With the projectors of Equation (25), we have [3]

$$[\rho_q^{exch}]_{nm} = f_{00}(q)j_0(qy/2)[\int_0^y dz \phi_n(z;y) j_0(qz/2) \phi_m(z;y) - f_{nm}(q)]. \quad (29)$$

This operator explicitly goes to zero for q=0 and, for finite q, is non-vanishing only in the interaction region. Figure 4 shows the bound state form factor for low-order truncation of the effective hadron theory in comparison with the exact quark model result. It is impressive that with only two states (i.e., the $\nu=1$ truncation) the diffraction minimum is reproduced almost perfectly and the secondary maximum is off by only a factor of two when the form factor itself is less than 10^{-4}. Clearly, this scheme has "hadronized" the quark model in a manner which has potentials qualitatively similar to those obtained phenomenologically in nuclear physics and which reproduces observables rather well with few effective degrees of freedom. It is also interesting that the agreement with the form factor requires a delicate cancellation between the one-body and exchange charges. This is shown in Figure 5.

As stressed above, there are many other ways to incorporate a rearrangement surface boundary condition in the projectors. See Reference 6 for a full discussion. The results of a second, rather different scheme are shown in Figures 3-5. Note that the channel interactions are quantitatively different, although the key feature of short-range repulsion and

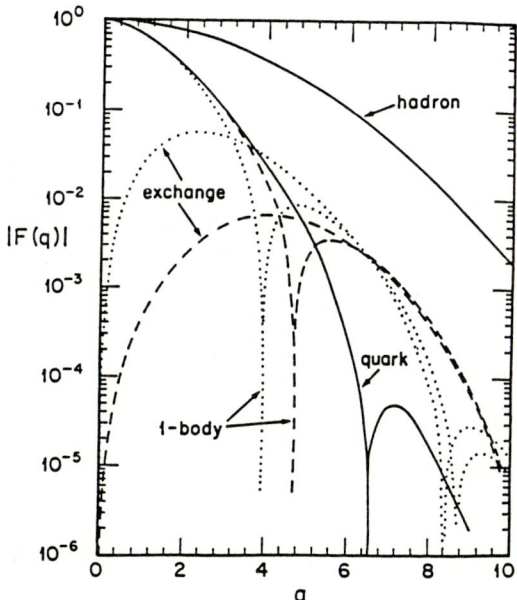

Figure 5. One-body and exchange contribution to the effective form factor.

medium-range attraction is preserved. Further, the balance between one-
body and exchange contributions is very different. Nevertheless, convergence to the observable is rather similar.

C. Y-scaling

The nuclear dynamic response function provides the most direct insight
into single particle propagation in the many-body system. At very large
momentum and energy transfer, the distribution functions of the elementary
constituents are studied. This is the regime of the EMC effect. Within
our simple model context, our discussion of the quark momentum distribution
is germane to this situation. Another interesting regime is that of large
momentum but small energy transfer (i.e., Bjorken -x much greater than
one). In this regime, y-scaling [4] has been interpreted as showing that
the nuclear response is dominated by quasifree nucleon knockout [5].
Further, the scaling variable y is the longitudinal momentum of the struck
nucleon (assuming quasifree kinematics). The scaling then implies that the
nucleon momentum distribution is being measured in a regime dominated by
the short-range repulsive NN interaction. Clearly, the issue of whether or
not a nucleons-only interpretation is valid at large momentum transfer is
an important one. Kumano [6] has investigated this within the framework of
the quark-exchange model.

The model considered differs from that described above in two ways.
First, the antisymmetric case is considered, so that the rearrangement surface boundary condition becomes

$$\phi_n(z;y)_{z=y} = 0. \qquad (30)$$

With this boundary condition, there is no bound state; indeed, the effective potential generated by the projection operator is purely repulsive and
diverges like $1/r^2$. The second difference is that the two hadrons are
placed in an overall "nuclear" harmonic oscillator potential, with oscillator frequency $\Omega \equiv 5\omega$. Thus, we have a simple pair model. The response function is then calculated both in the underlying quark model and in the
effective hadron model truncated so as to include only ground state
hadrons. The quantity

$$\tilde{R}(q,\omega) \equiv \frac{qR(q,\omega)}{[f(q)]^2} \qquad (31)$$

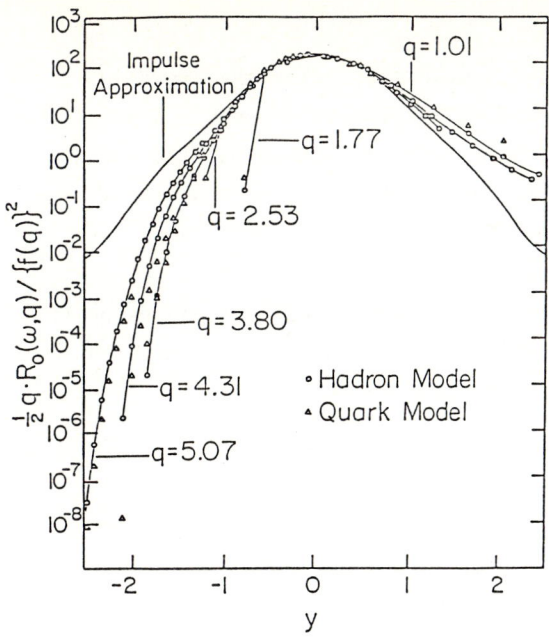

Figure 6. y-scaling in the quark model and in the effective hadron model.

is shown in Figure 6 as a function of

$$y = \vec{p} \cdot \hat{q} = \frac{\omega - q^2/2}{q}. \qquad (32)$$

Also shown is the scaling curve

$$P(y) = \lim_{q \to \infty} \tilde{R}(q,\omega) = \int \frac{d^2 p_\perp}{(2\pi)^2} n(\sqrt{p_\perp^2 + y^2}) \qquad (33)$$

calculated directly from the hadron momentum distribution in the effective hadron model. For $|y| < 1$, the scaling curve is reached for rather low momentum transfer. This is not the case for large $|y|$, as might be expected [7] because of the strongly repulsive hadron-hadron interaction. However, the more important observation is that the effective hadron model results are reasonably close to those of the quark model. Consequently, explicit effects of quark substructure are not evident. The traditional framework based upon hadron degrees of freedom and effective interactions is rather accurate, lending credence to the idea of extracting nucleon momentum distributions through y-scaling analyses.

D. Scattering

The elastic channel S-matrices for S-wave scattering are shown in Figure 7 for both spatially symmetric and antisymmetric cases. The inelastic thresholds are denoted by arrows. The antisymmetric case shows little coupling to inelastic channels. The phase shift is consistent with strong repulsion, consistent with our discussion in the last subsection. By contrast, the symmetric S-matrix displays strongly inelastic resonances near the inelastic thresholds. There are closed-channel resonances built upon the hadronic internal excitations. The "binding" mechanism is basically the same as that leading to the "nuclear" bound state.

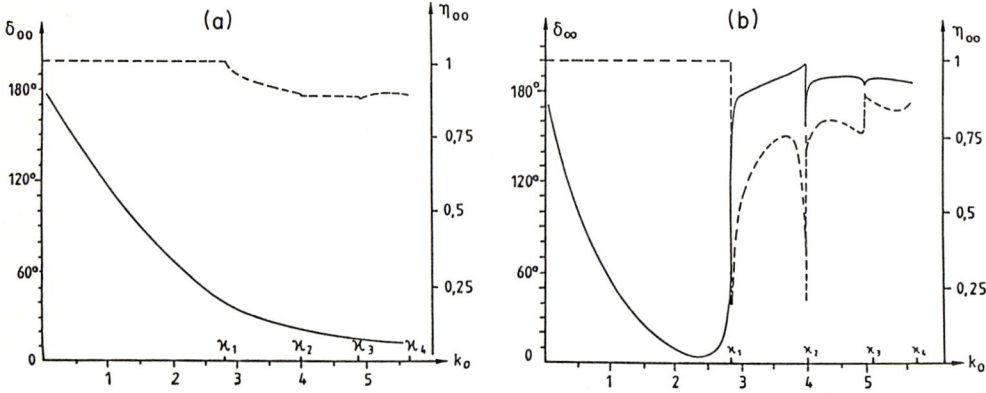

Figure 7. Elastic S-matrix $S_{00} = \eta_0 \exp(2i\delta_{00})$ for $U(1)$ scattering. (a) Antisymmetric. (b) Symmetric. The nth inelastic channel threshold is denoted by K_n.

III. $N_c > 1$

With the explicit inclusion of color, we have two parameters at our disposal λ and ν_c (see Equations (9) and (11)). The parameter λ is determined by N_c and controls the relative importance of propagation in the color singlet and "hidden color" modes. The parameter ν_c determines the properties of the hidden color mode. It will be useful to divide our discussion into three categories depending on the strength of ν_c. Recall that our rationale for introducing this parameter is the fact that we know little about the operation of confining forces in the color nonsinglet configurations and thus hope to use observables to constrain the phenomenology of such forces. We shall refer to the hidden color spectrum as the discrete spectrum which results from the interaction of Equation (9) without coupling to the color singlet configurations.

For $\nu_c < 1/2$, the hidden color ground state is below elastic scattering threshold and the density of hidden color states is high even in the low-energy regime. We then expect a very significant effect on the elastic scattering amplitude. This is seen in Figure 8 for the case $\nu_c=0.4$, $\lambda=1.2$. The phase shift has risen by 5π below the second inelastic threshold. Further, one finds bound states with binding energies comparable to hadronic excitation energies. In such a case, the presence of the hidden degree of freedom is rather clear. Of course, this has no resemblance to the real world and such soft color modes can be ruled out.

For $\nu_c > 7/6$, the hidden color ground state is above the inelastic threshold. This leads to a strong effective repulsion, as seen in Figure 9. In this regime, the internal hadron structure basically leaves no signature in strong interaction phenomena. This could provide a justification of the traditional picture of strong interactions if supplemented with attractive forces from the exchange of mesons. This is not terribly appealing, since features such as the very small deuteron binding energy do not then have a very natural explanation. Nevertheless, it cannot be ruled out.

A more satisfactory situation arises for intermediate ν_c. One of the few theoretical arguments for a specific range of ν_c is provided by the flux tube notion that the restoring force on a quark pulled away from a $q\bar{q}^2$ cluster should be the same as that operative in a single meson. This gives

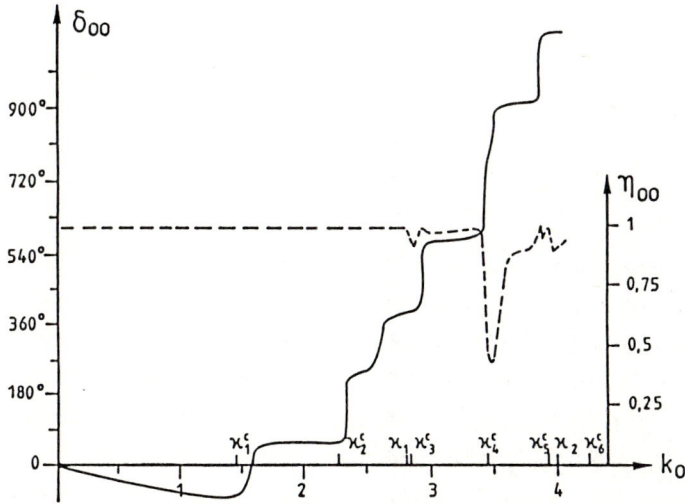

Figure 8. Elastic antisymmetric S-matrix for $N_c=3$ and $\nu_c=0.4$. The nth hidden color threshold (n=0,1,...) is denoted by K_n^c.

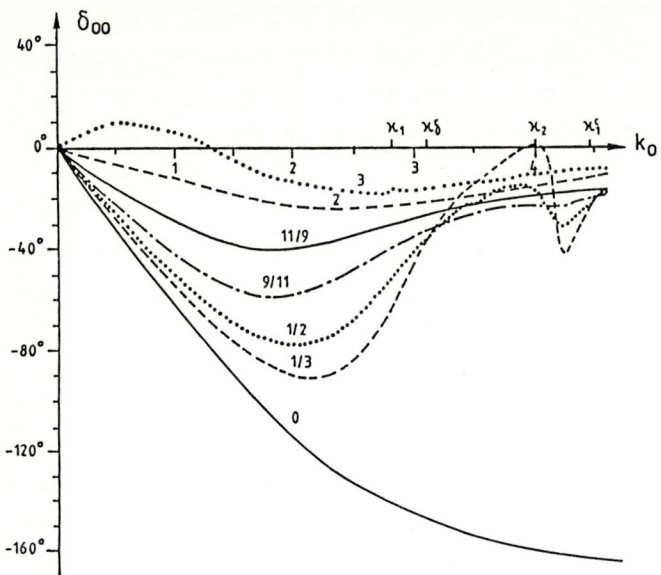

Figure 9. Elastic S-matrix for different values of λ and ν_c=1.3.

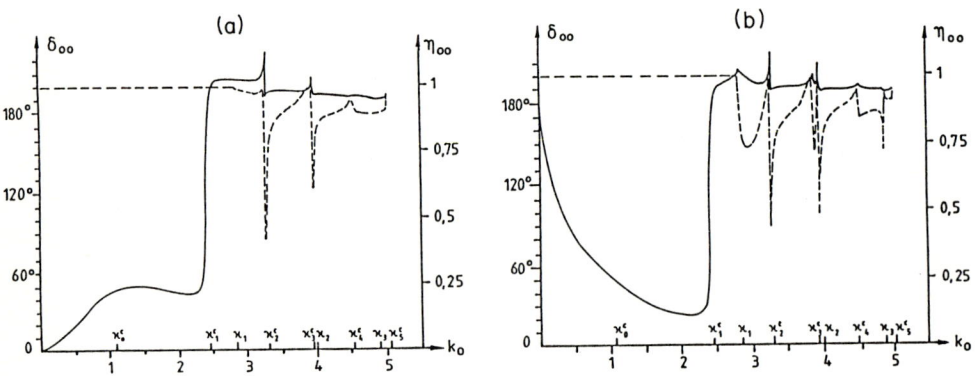

Figure 10. Elastic S-matrix for N_c=3 and ν_c=0.6. (a) Antisymmetric. (b) Symmetric.

$\nu_c=1/\sqrt{2}$ in our model. In this case the color hidden ground state lies between elastic and inelastic threshold, with the color hidden excited states above inelastic threshold. It is found [1] that the important qualitative results found for the U(1) case are preserved in this case. Some additional phenomena appear. The elastic channel S-matrix is shown in Figure 10 for ν_c=0.6 and N_c=3. A resonance associated with the color hidden ground state appears in the antisymmetric case. However, the width is comparable to the hadron internal excitation scale, and such a broad structure would be difficult to isolate phenomenologically. At higher energies, the excited hidden color configurations may lead to narrow structures. If energetically possible, the resonances decay predominantly into excited mesons with low relative energy. However, the location and properties of these "di-hadrons" depend intricately on the details of the model (e.g., relative position of the color singlet thresholds and hidden color states). Consequently, our conclusion is that a "reasonable" value for ν_c together with the "minimal" confining dynamics leads to typical low-energy strong-interaction phenomena and holds out the possibility that structures directly associated with the color degree of freedom may be found at relatively high energies. Unfortunately, a more faithful representation of QCD is needed for reliably guiding us to them.

References

1. Lenz, F., Londergan, J. T., Moniz, E. J., Rosenfelder, R., Stingl, M., and Yazaki, K.: Ann. Phys. 170 (1986) 65.
2. Greenberg, O. W. and Lipkin, H. J.: Nucl. Phys. A370 (1981) 349.
3. Gardner, S. and Moniz, E. J.: MIT preprint CTP #1483.
4. West, G. B.: Phys. Repts. 18C (1975) 264.
5. Sick, I., Day, D. and McCarthy, J. S.: Phys. Rev. Lett. 45 (1980) 871.
6. Kumano, S.: MIT Ph.D. thesis (1985); Kumano, S. and Moniz, E. J.: to be published.
7. Weinstein, J. J. and Negele, J. W.: Phys. Rev. Lett. 49 (1982) 1016.

The Cheshire Cat Principle Applied to Hybrid Bag Models

H.B. Nielsen[1] and A. Wirzba[2]

[1]The Niels Bohr Institute, University of Copenhagen,
Blegdamsvey 17, DK-2100 Copenhagen Ø, Denmark
[2]NORDITA, Blegdamsvej 17, DK-2100 Copenhagen Ø, Denmark

Here is argued for the Cheshire Cat point of view according to which the bag (itself) has only notational, and no physical significance. In a 1+1 dimensional exact Cheshire Cat model it is explained how a fermion can escape from the bag by means of an anomaly. We also suggest that suitably constructed hybrid bag models may be used to fix such parameters of effective Lagrangians that can otherwise be obtained only from experiments. This idea is illustrated in a calculation of the mass of the pseudoscalar η' meson in 1+1 dimensions. Thus there is hope of finding a constructional principle for a phenomenologically sensible model.

1. Introduction

Essentially, this is a review of work done by one of us (H. B. N.) in collaboration with S. Nadkarni and I. Zahed [1-3] and with some ideas from K. Johnson. It has some overlap with reference [4].

The hybrid bag model [5] (sometimes also called the chiral or topological or "Brown" bag model) has in the interior of the bag quarks and gluons as fields in the same way as the MIT bag model [6]. However, the "outside" region is not without degrees of freedom, but there are in addition fields like e.g. in the nonlinear σ-model [7] or more sophisticated Skyrme models [8,9]. Thus in the "outside" region of the hybrid model, there are e.g. a σ-field $\sigma(x,t)$ and an isospin-triplet of pion fields $\vec{\pi}(x,t)$ obeying the nonlinear constraint

$$\sigma(x,t)^2 + \vec{\pi}(x,t)^2 = f_\pi^2 \,, \tag{1.1}$$

where f_π is the pion decay constant ($f_\pi = 93$ MeV experimentally). The model is called the "hybrid" bag model because it is a hybrid or combination of two different models: The MIT bag model [6] on the one hand, and a σ-model [7] or Skyrme model [8,9] on the other hand. Of course these hybrid models are not limited just to the form mentioned above. For example, the nonlinear σ-model outside the bag can very well be replaced by a more sophisticated version involving vector meson fields such as ρ and ω [10].

The purpose of the present work is twofold: First, we would like to present and argue for the point of view that the bag (wall) of hybrid bag models has no physical significance. It merely separates regions in spacetime in which different field descriptions of the same phy-

Fig. 1. The (existing) smile of a (nonexisting) Cheshire Cat

sics are used. Secondly, we would like to suggest that the Cheshire Cat point of view can serve as the guiding principle which allows one to link models which are otherwise only experimentally determinable effective models to the underlying theory, i.e. QCD. In this way one might rescue too ill-determined models for the phenomenology. In other words, by insisting on the Cheshire Cat principle, one has so many relations between the parameters of the hybrid model at hand that the whole model eventually becomes fully determinable.

The expression "Cheshire Cat" has its origin in the fable of Lewis Carroll about "Alice in Wonderland" and "Through the Looking Glass" [11]: Quotation: "Well, I've often seen a cat without a grin", thought Alice, "but a grin without a cat! It is the most curious thing I ever saw in my life!" - Lewis Carroll (1865) - (see Fig.1). The Cheshire Cat point of view is to postulate that the bag itself of the hybrid bag model does not exist (is not observable), in analogy to the non-existence of the Cheshire Cat in the Lewis Carroll fable. Only the grin of the Cheshire Cat and the formalism of the bag exist.

In Sect. 2, we shall present the two alternative points of view on the (hybrid) chiral bag model: The traditional one that the separation between the inside and outside regions is physically measurable, and the Cheshire Cat one that the bag has no physical significance. Then in Sect. 3, we shall review the remarkable fact that in 1+1 dimensions there exists an exact Cheshire Cat bag formalism. In Sect. 4, we will elaborate on a most intriguing point: how can a quark "escape" from its confining bag "jail"? We will report in a kind of "crime story" that quantum mechanics, and in this special case the anomaly, "spirits" these fermions away from the bag. However, the quark itself is "tricked" by the anomaly since the latter has merely "drowned" the quark in the Dirac sea when this "prepares" to pass through the bag wall. After that, in Sect. 5, we will suggest possible applications of the Cheshire Cat formalism for practical calculations. In Sect. 6 we present the program that this principle is useful for linking otherwise only experimentally constrainable effective Lagrangians to underlying theories and of fitting the parameters of the former. This is illustrated by the 1+1 dimensional example in Sect. 7 where we compute the mass of an η'-field by insisting on the Cheshire Cat principle when a little "test"-bag has been inserted into a static η'-field configuration. In Sect. 8, we will resume the "escape story" of the quark from the bag and discuss the way confinement appears in the language of a Cheshire Cat bag model. Section 9 contains the conclusion.

2. Two Ways of Looking at a Hybrid Bag Model

There are two philosophies of the hybrid type of bag models: (A) The bag does really exist physically, i.e. it is possible - in principle at least - to make some measurement which settles whether a given point (event) or small region is inside or outside the bag (or just on the surface). This is illustrated by Fig. 2, where a little experimentalist has an apparatus which can measure a quantity telling whether the apparatus is inside or outside the bag. (B) There is in reality no such thing as a bag: Whatever field configuration one considers, say somewhere outside the "bag", it can be simulated by some configuration of the inside field types in such a way that all measurements would give the same result. In this interpretation, the bag model of the hybrid type is just a way of dividing spacetime into two regions - the "inside" and the "outside"- and then describing the same physics (at the end really QCD) in different approximations and different language in these two regions.

It is this latter point of view (B) which we call "the Cheshire Cat point of view", because the bag has disappeared physically leaving only a formal track (analogous to the Cheshire Cat smile) behind.

Of course, these two points of view differ in their physics: in principle the question, whether it is possible to distinguish an inside and an outside region without imposing any restriction on the fields, could be settled by measurement. However, there is one caveat: It will always be possible to assign a physical difference between inside and outside, if one restricts the allowed field configurations in these regions in a special way. Such a possibility [12] of making a distinction by imposing extra requirements on the field should not be considered as a violation of the Cheshire Cat principle: If one for instance postulates that the $\vec{\pi}$-field outside the bag should be zero (so that there is only the σ-field, $\sigma = f_\pi$ say), then one has excluded the possibility of describing regions in which there are - in the physical situation considered -

Fig. 2. The traditional point of view that the bag really exists. A little experimentalist can measure whether he is inside or outside the bag

nonzero pion fields "outside". So, under such a restriction - as $\vec{\pi} = 0$ outside - one cannot freely choose where the bag wall shall be. We may say that by imposing field restrictions - at best both inside and outside - one can reduce a Cheshire Cat model into an ordinary bag model. By the pion-field-equal-to-zero restriction one gets the MIT bag model [6]. The latter one is in all likelihood not a Cheshire Cat model. The "outside" region of the MIT bag model can only be the (pure) vacuum state, which has too few degrees of freedom for even an approximate simulation of the "QCD" theory in the "inside". Under such restrictions of the fields it may - even in a Cheshire Cat model - be meaningful to consider vibrations [12,13] in the bag model. In a Cheshire Cat type model without field restrictions, however, these bag surface vibrations would be physically meaningless.

The interpretation of hybrid bag models in the Cheshire Cat way (B) is consistent with the fact that one gets good descriptions of e.g. the baryon properties with a lot of different values of the bag radius (see Fig. 3). Indeed, in the limit of infinite bag radius, all of space is treated as "inside", in other words as QCD. Since the latter one is considered as the correct theory of hadron physics in nature, the limit of infinite bag radius should obviously agree with nature. On the other hand, in the limit of zero bag radius, all of space is treated as "outside" and the hybrid bag models coincide with models of the σ- or Skyrmeon-type. These are considered to describe nature (a baryon say) almost correctly [8,9,14]. Thus in some way, there should be a linkage to the radius $R = \infty$ description. Furthermore, the hybrid bag models themselves get a good description of a baryon in nature by some finite nonzero radius R_0 [5]. So we see that several different radius values ∞, R_0 and 0 are expected to give valid approximate descriptions of the same baryon. The bag model pictures illustrating these three ways of describing a baryon are shown in Fig. 3 using crosshatching to denote the "outside region".

Actually, there are even calculations [15,16] showing that some simple properties (e.g. lowest moments of baryon number, energy or isospin distributions) are almost independent of the bag radius chosen (see Fig. 4). However, there seem to be problems with higher moments [16].

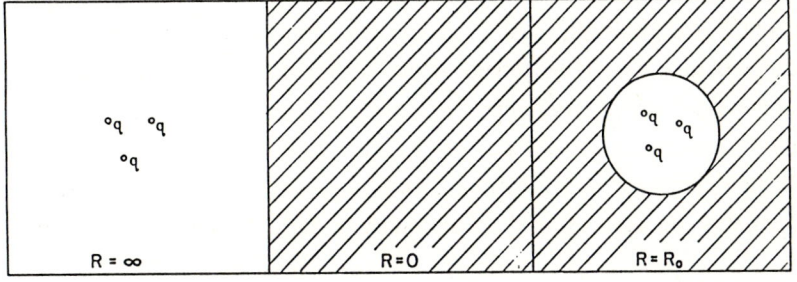

Fig. 3. A baryon described in three different ways: an infinitely big bag (bag radius $R = \infty$), no bag at all, i.e. a Skyrmeon type description ($R = 0$), and a hybrid bag model with finite nonzero bag radius ($R = R_0$)

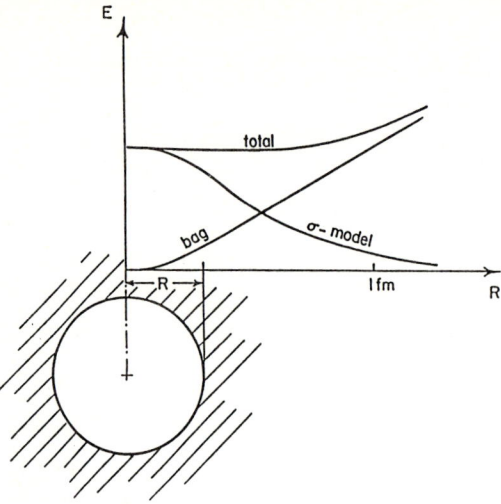

Fig. 4. The energy of a baryon calculated in a hybrid bag model as a function of the bag radius R (see [15]). The energy of the degrees of freedom of the "inside" (bag) and of the "outside" (σ-model) are shown as function of the bag radius R in addition to the energy of the total system

Ignoring higher moments, we may say there are good reasons to believe that the hybrid bag model describes a baryon approximately correctly for all values of the bag radius, provided one performs the difficult QCD calculation in the case of large radii - say by lattice Monte Carlo methods. This argumentation obviously supports the Cheshire Cat principle in the sense that the value of the bag radius has no physical significance and can be changed at will provided one arranges the fields appropriately in the region over which the bag wall is moved. With less support from these arguments we would like to generalize the suggestion that the bag radius has no physical significance to the assumption that the bag wall position has no physical significance at all. This is the Cheshire Cat principle [1-4].

However, it should be stressed that we cannot expect the Cheshire Cat principle to be more than approximately valid in the usual 3+1 dimensional hybrid bag model. If it were exactly true, it would in fact mean that σ- or Skyrme-type models should provide exact models for QCD. This is, however, not the case [14]. The deviations of the hybrid bag model from the Cheshire Cat principle should be linked to the deviations of the σ- or Skyrmeon model from QCD and thereby from nature. Hence we should consider such deviations a measure of the unreliability of the model to describe nature. In other words, finding, e.g., that some quantity calculated in the hybrid bag model varies a bit with the radius of the bag chosen, we should consider this variation an estimate of the uncertainty in the prediction of the quantity in question.

Very strong support of the validity of the Cheshire Cat principle, in addition to the above-mentioned more phenomenological arguments, is the existence of exact Cheshire Cat bag models in 1+1 dimensions [1-3].

3. Exact Cheshire Cat Bag Model in 1+1 Dimensions

We shall now show how it is possible to construct a bag model in a 1+1 dimensional world which obeys the Cheshire Cat principle exactly. This relies upon the fact that in 1+1 dimensions one is able to describe the same physics in two different complementary languages, using fermion or boson fields respectively, i.e. that one has complete bosonization and fermionization [17,18].

Let us start by considering a massless free boson field $\phi(x,t)$ described by the action

$$S = \int dx dt \, \tfrac{1}{2} \partial_\mu \phi \, \partial^\mu \phi \,. \tag{3.1}$$

This theory can be "fermionized" into a free massless Dirac fermion theory, see [17]. In the following we will briefly sketch the main idea of how one gets anticommuting objects (fermion fields) from a pure boson language. In fact, we may construct a Dirac field (in the Weyl representation)

$$\psi = \begin{pmatrix} \psi_L \\ \psi_R \end{pmatrix} \tag{3.2}$$

by

$$\begin{pmatrix} \psi_L(x,t) \\ \psi_R(x,t) \end{pmatrix} \propto \exp\left[-i\sqrt{\pi}\int_{-\infty}^{x} d\xi \left[\Pi(\xi,t)\pm\phi'(\xi,t)\right]\right] = \exp\left[-i\sqrt{\pi}\int_{-\infty}^{x} d\xi\,\Pi(\xi,t) \mp i\sqrt{\pi}\,\phi(x,t)\right], \tag{3.3}$$

where $\Pi(x,t)$ is the canonical conjugate field of ϕ, i.e. $[\phi(x,t),\Pi(y,t)] = i\delta(x-y)$. Now, one can use that

$$\exp\left[-i\sqrt{\pi}\int_{-\infty}^{x} d\xi\,\Pi(\xi,t)\right] \tag{3.4}$$

is a translation operator in the space of ϕ values, i.e.

$$\exp\left[-i\sqrt{\pi}\int_{-\infty}^{x} d\xi\,\Pi(\xi,t)\right] i\sqrt{\pi}\,\phi(y,t) \exp\left[+i\sqrt{\pi}\int_{-\infty}^{x} d\xi\,\Pi(\xi,t)\right]$$

$$= i\sqrt{\pi}\,\phi(y,t) - i\pi\Theta(x-y) \,. \tag{3.5}$$

Thus it can easily be seen that

$$\psi_{L/R}(x,t)\,\psi_{L/R}(y,t) = -\psi_{L/R}(y,t)\,\psi_{L/R}(x,t) \quad \text{for } x \neq y \tag{3.6}$$

(remember that $\exp\left[\pm i\pi\{\Theta(x-y)+\Theta(y-x)\}\right] = -1$ for $x \neq y$). From (3.3) one can deduce furthermore that - expressed in the boson language - $\psi(x_0,t)$ (annihilates or) creates a soliton at the position x_0, see Fig.5.

It is nice to postulate that ϕ is really an angle variable and that $U(x,t) \equiv \exp\left[i\sqrt{4\pi}\,\phi\right]$ has physical significance rather than $\phi(x,t)$ itself [2]. Then the action (3.1) becomes

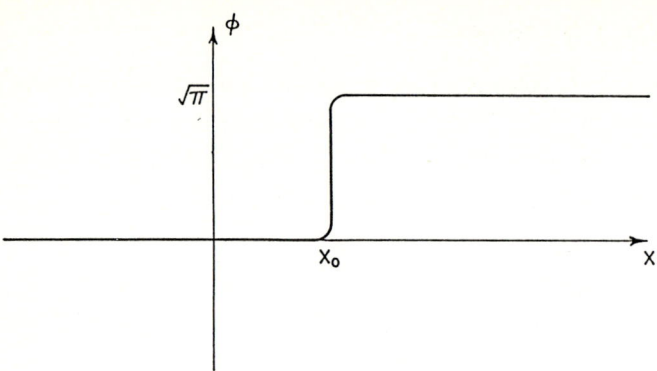

Fig.5. A kink or soliton configuration in the ϕ field at the position x_0 of the height $\sqrt{\pi}$ as created by the $\psi_\pm$ field

$$S = \int dx dt \, \frac{f_\pi^2}{2} \partial_\mu U \, \partial^\mu U^\dagger \tag{3.1a}$$

where

$$f_\pi \equiv \frac{1}{2\sqrt{\pi}} \tag{3.7}$$

is the 1+1 dimensional analogue of the pion decay constant of the nonlinear σ-model in 3+1 dimensions [19,1]. In this nonlinear (U) representation, it is obvious that there is a topological winding number (and therefore solitons in 1+1 dimensions) and that this winding number shifts by one unit if $\phi \to \phi \pm \sqrt{\pi}$. Now, the fermion operators simply become

$$\psi_{L/R}(x,t) = W^{\frac{1}{2}}(x,t) \, U^{\mp\frac{1}{2}}(x,t) \tag{3.8}$$

where

$$W(x,t) = \exp\left[-\int_{-\infty}^{x} d\xi \, U^{-1}(\xi,t) \, \partial_0 U(\xi,t)\right]. \tag{3.9}$$

One may note that actually

$$\left[W^{\frac{1}{2}}(x,t), U(y,t)\right] = 0 \quad \text{for } x \neq y \tag{3.10}$$

so that $W^{\frac{1}{2}}(x,t)$ is a local operator when conceived as a bosonic one, i.e. it has no effect on the bosonic configurations except in the vicinity of x. In the language of the U's, the fermion operators look formally completely local from the bosonic point of view and only a little non-locality "sneaks in" because the square root $U^{\mp\frac{1}{2}}(x,t)$ of $U^{\mp 1}(x,t)$ has a sign ambiguity which is determined by the continuity of $U^{\mp\frac{1}{2}}(x,t)$ as function of x. Thereby the sign of ψ is determined in a nonlocal way and thus it becomes possible that $W^{\frac{1}{2}}(x,t) U^{\mp\frac{1}{2}}(x,t)$ (see (3.8)) can anticommute in spite of being built from local boson fields. The topological dependence of the sign allows it to switch the sign at one position x by a modification of the boson field at another point $y \neq x$.

Detailed calculations [2,17] on the fermion fields constructed in (3.3) show that these behave precisely as free massless Dirac fields and yield the following equations between the boson and the fermion fields:

$$\partial_\mu \phi = \sqrt{\pi}\, \overline{\psi} \gamma_\mu \gamma_5 \psi \,. \tag{3.11}$$

Note that we use the following conventions: $(g_{\mu\nu}) \equiv diag\,(1,-1)$ as metric where $\mu,\nu = 0,1$; $\gamma_0 = \gamma^0 \equiv \sigma_1$, $\gamma_1 = -\gamma^1 \equiv -i\sigma_2$, $\gamma_5 = \gamma^5 \equiv \sigma_3$ for the γ-matrices (Weyl representation) where σ_i are the conventional Pauli $SU(2)$ matrices. The left/right Dirac projection matrices are $P_{L/R} = \frac{1}{2}(1 \pm \gamma_5)$.

Now it is possible to develop a formalism in which one uses the fermion fields $\psi(x,t)$ in part of space time called the "inside" of the bag region, while one uses the boson field $\phi(x,t)$ in the rest of space time (the "outside" of the bag region) [1-4]. We can in fact construct an action

$$S = S_V + S_{\overline{V}} + S_{\partial V} \tag{3.12}$$

consisting of following contributions:

$$S_V = \int_V dx dt \; i\frac{1}{2}\, \overline{\psi}\, \overleftrightarrow{\partial}_\mu \gamma^\mu \psi \,, \tag{3.13a}$$

which describes massless Dirac fermion fields inside the bag space time volume V. Secondly there is

$$S_{\overline{V}} = \int_{\overline{V}} dx dt \; \frac{1}{2}\, \partial_\mu \phi\, \partial^\mu \phi \,, \tag{3.13b}$$

which represents the boson field ϕ in the "outside" region $\overline{V}$, the complement of V. Finally we have the boundary term

$$S_{\partial V} = \frac{1}{2} \int_{\partial V} d\Sigma_\mu \; n^\mu\, \overline{\psi}\, e^{i\sqrt{4\pi}\phi \gamma_5}\, \psi \,, \tag{3.13c}$$

where ∂V denotes the boundary. The integral $\int d\Sigma_\mu$ runs along the bag wall, $d\Sigma_\mu$ denotes the twovector normal, and $n^\mu = g^{\mu\nu} n_\nu$ the normal pointing inward (into the bag). The total action (3.12) has been arranged so cleverly that the Euler-Lagrange equations (or "equations of motion") on the boundary turn out to be part of the bosonization relations. In fact, the action (3.12) has not only the usual Euler-Lagrange equations

$$\partial\!\!\!/\, \psi = 0 \tag{3.14a}$$

and

$$\partial_\mu \partial^\mu \phi = 0 \tag{3.14b}$$

in the regions V and $\overline{V}$ respectively, but also leads to the following stationarity equations at the boundary:

$$\left[i\not\partial\psi - n^2 e^{i\sqrt{4\pi}\phi\gamma^5}\psi\right]|_{boundary} = 0 \qquad (3.14c)$$

and

$$\left[n^\mu\partial_\mu\phi - \sqrt{\pi}\,\overline{\psi}\not n\gamma_5\psi\right]|_{boundary} = 0 \ . \qquad (3.14d)$$

The second of these two boundary equations is derived from the ϕ-variation and is merely one of the components (i.e. the one normal to the bag wall) of the bosonization relation (3.11). The component along the bag wall time track is

$$t^\mu\partial_\mu\phi = \sqrt{\pi}\,\overline{\psi}t^\mu\gamma_\mu\gamma_5\psi = -\sqrt{\pi}\,\overline{\psi}\not n\psi \qquad (3.15)$$

where t_μ is the tangent vector to this time track. This equation (3.15) is fulfilled only due to the (quantum mechanical) anomaly. We shall consider this in detail in the next section. Classically, namely, one finds just

$$\overline{\psi}\not n\psi|_{boundary} = 0 , \qquad (3.16)$$

instead of (3.15). The boundary condition (3.14c) resulting from the fermion field ψ variation can be easily identified as part of the "fermionization" relation (3.3).

It was shown in reference [2] that the boundary part $S_{\partial V}$ (3.13c) of the action is uniquely determined by a series of requirements such as symmetries: chiral symmetry, C, P and T. Furthermore the "Cheshire Cat criterion" (CCC) itself is of significance to this uniqueness question. The CCC states that the boundary equations must not become so restrictive that they do not allow the fields and their derivatives to take values as freely as if there were no boundary. This means that e.g. the matrix (note that $n^2 = -1$)

$$in_\mu\gamma^\mu + e^{i\sqrt{4\pi}\gamma^5\phi} \equiv in_\mu\gamma^\mu + U^{\gamma_5} \qquad (3.17)$$

in the boundary condition (3.14c) has to be of rank 1 only, rather than 2, as is generic for a 2×2 matrix. Otherwise we would namely get in general $\psi = 0$ at the boundary, and that would certainly be too strong a condition.

Once we have found the bosonization (and fermionization) relations and once we know that we have a case of bosonization, it must follow that the bag model given by (3.12) does indeed completely describe the bosonic as well as the equivalent fermionic theory. This means in particular that the bag wall has no physical significance, so that there is a case of exact "Cheshire Catness". In reference [2] this fact was made absolutely unambiguous by the construction of the Hamiltonian and the momentum generator and by the evaluation of their mutual commutation. However, it should logically be sufficient just to have derived the bosonization at the boundary and to know that the bosonization equivalence between the "inside" and the "outside" theories exists.

4. The Escape of the Quark from the Bag Jail - A 'Crime Story'

There is one very curious phenomenon in connection with the exact Cheshire Cat bag in 1+1 dimensions that even applies to the case of only an approximate Cheshire Cat bag in 3+1 dimensions: To a large extent the original MIT bag model [6] was constructed in order to represent quark confinement in an understandable manner. So, we naively expect any bag model to incorporate confinement. However, in the preceding section, we were able to construct a bag model which exactly (!) describes a free (massless) fermion theory. Now, in conventional bag models, the gluon fields, especially the color Coulomb fields, play a major role by preventing the bag from breaking into pieces which contain quarks of noncompensating color-charges. However, just the quark boundary condition

$$-i\not{n}\psi|_{boundary} = U^{\gamma_5}\psi|_{boundary} \tag{4.1}$$

(see Sect. 3 for definitions) can prevent the quark from leaving the bag and therefore from appearing in the "outside" region. In spite of this, we know that the theory, which we have considered in the previous section, is exactly the one of free massless fermions in 1+1 dimensions. So, there is an apparent contradiction: The "quark" seems to be safely confined to the bag "jail" in spite of being exactly free! The "guard" of the "jail" (=the bag) can use the bag boundary condition (4.1) and its hermitian conjugate

$$\overline{\psi}\not{n}i|_{boundary} = \overline{\psi}U^{\gamma_5}|_{boundary} \tag{4.2}$$

to deduce after multiplying (4.1) with $\overline{\psi}$ from the left and (4.2) with ψ from the right the following equations - valid on the boundary:

$$-\overline{\psi}i\not{n}\psi|_{boundary} = \overline{\psi}U^{\gamma_5}\psi|_{boundary} \tag{4.3a}$$

$$\overline{\psi}i\not{n}\psi|_{boundary} = \overline{\psi}U^{\gamma_5}\psi|_{boundary}. \tag{4.3b}$$

After a subtraction, the "guard" would establish that there is no flow of (isoscalar vector) quark current through the boundary, i.e.

$$n_\mu \overline{\psi}\gamma^\mu \psi|_{boundary} = 0. \tag{4.4}$$

So, the "guard" finds the quark safely confined in the (really nonexisting) bag "jail". Note that the gluon fields do not play a role for the question whether a quark is confined to the bag or not. They might be helpful in preventing the bag from splitting into pieces (but actually as we will see in Sect. 8 in the case of a more complete Cheshire Cat bag, which even allows for gluon-like fields, the bag can still be split). Of course, the gluon fields are essential for the quark-confinement itself, but not for the confinement of quarks to the bag. In the following, we will therefore limit our discussion to the case of a truly free quark, i.e. no gluon fields are present.

The "escape agent", who offers to help the quark out of the bag "jail" is the anomaly (see Fig.6). An anomaly is a quantum effect that violates some symmetry or equivalently a conservation law that was present in the classical analogue of the system under consideration. Here, the anomaly (which helps in the escape of the quark from the bag "jail") is the one that violates the conservation of fermion number or baryon number [20,4]. It is operating at the bag surface only. As we shall explain, this anomaly may be understood as a pumping up or down of the Dirac sea of the quarks [21]. It will turn out that there is a production of (extra) fermion number at the (right) bag wall, say, with a rate $\dot{\phi}/\sqrt{\pi}$ where $\dot{\phi}$ is the time derivative of the "η' - field" ϕ at the bag wall. In order to understand and calculate this anomaly effect we shall first see what happens naively when a quark hits the bag wall - say the right one - attempting to escape.

A (right-moving) wave packet representing the quark comes along, see Fig.7a. Since the quarks are free and massless, the wave packet simply moves undisturbed in shape with the speed of light. When it hits the bag wall (Fig.7b), it is reflected as a left-moving wave packet (Fig.7c).

In fact, it can easily be seen that the boundary condition (4.1) relates the left and right moving γ^5-projections to each other because the matrix $n_\mu \gamma^\mu$ is off-diagonal in the Weyl representation. The reflected wave packet obtains the same temporal shape as the incoming wave

Fig. 6. The quark captured in the bag "jail". The anomaly, the "escape agent", is already visible trying to "dig" it out

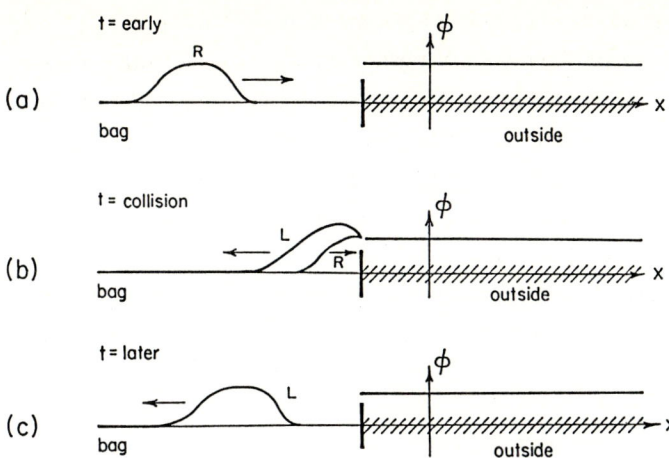

Fig. 7. Time sequence of the reflection of a quark wave packet at the right boundary of a 1+1 dimensional bag model in the approximation of a constant φ field at the boundary and in the outside. The incoming right-moving right handed quark wave packet given at early times (a) is reflected at the bag wall at collision time (b) into a left-moving, left handed quark which travels at later times (c) with the same shape and speed as the incoming

$$\psi_L(t+x) = \psi_R(t+x-2x_{boundary}) \quad (4.5)$$

provided the interaction matrix U^{γ_5} in (4.1) can be considered constant in time. If U^{γ_5} varies with time, the reflected wave packet will get modified by an extra time and space dependent phase factor relative to the incoming one. If the wave packet can be considered short in comparison to the time scale of the variation of φ at the boundary, we may use a Taylor expansion for φ as a function of time for the period of reflection of the wave packet, i.e.

$$\phi(t, x_{boundary}) \approx \phi(t_0, x_{boundary}) + \dot{\phi}(t_0, x_{boundary})(t-t_0). \quad (4.6)$$

Thus a factor $\exp(-i\sqrt{4\pi}\,\dot{\phi}(t+x))$ appears in the reflected wave function. Thus the particle represented by this wave has an extra momentum

$$\Delta p = -\sqrt{4\pi}\,\dot{\phi} \quad (4.7a)$$

and an extra energy

$$\Delta E = \sqrt{4\pi}\,\dot{\phi}. \quad (4.7b)$$

This means that during the reflection the particle (quark) has obtained an (extra) "kick" transferring energy and momentum by an amount $2\sqrt{\pi}\,\dot{\phi}$ to it (or removing from it). Here $\dot{\phi}$ is the rate of variation at the boundary at the "instant" of reflection of the wave (≈ particle). This kick is in addition to the main effect of the boundary: the reflection, i.e. the turning of the sign of the momentum.

The anomaly of the fermion number at the bag wall can be understood as the result of this kick which causes quarks to emerge from or to be pushed into the Dirac sea depending on the sign of $\dot{\phi}$. With inclusion of the proper signs, the number of quarks pumped out of the Dirac sea near the bag wall is

$$\dot{Q}_{anomaly} = \frac{\sqrt{4\pi}\,\dot{\phi}}{h} = \frac{2\sqrt{\pi}\,\dot{\phi}}{2\pi} = \frac{\dot{\phi}}{\sqrt{\pi}}. \tag{4.8}$$

Here, we made use of the fact that the number of particles reflected per unit time at the bag wall corresponds to the number present in a region of length unity. This follows because the particles move with the velocity of light, which is used as velocity unit here. Thus, the number of single fermion states (filled or not) which are pushed in unit time from negative to positive energy due to the kick (resulting from the ϕ-time variation) is given by the number of states in unit length of space and in a region of momentum space $\sqrt{4\pi}\,\dot{\phi}$. The derivation of (4.8) used that the number of states per unit phase space is $1/h = 1/2\pi$ where $h = 2\pi$ is Planck's constant (note that $\hbar \equiv 1$). In summary, there will appear $\dot{\phi}/\sqrt{\pi}$ more particles per unit time above the Dirac sea or correspondingly the same number of holes (= antiparticles) fewer in the Dirac sea. It is this appearance of new (extra) particles or the disappearance of holes - computed with proper signs - which makes up the (quantum) anomaly. We see now that the anomaly may have the possibility of hiding a quark by "drowning" it in the Dirac sea. Furthermore, the anomaly (4.8) can be expressed in terms of a (formal) inflow current from the bag wall, i.e. instead of the classical result (4.4) (which fooled the "guard of the bag jail"), we have

$$-d\Sigma\,\bar{\psi}\slashed{n}\psi|_{boundary} = \dot{Q}_{anomaly}\,dt \neq 0. \tag{4.9}$$

This result has the following covariant form:

$$-\bar{\psi}\slashed{n}\psi|_{boundary} = \frac{1}{\sqrt{\pi}}\,t^{\mu}\partial_{\mu}\phi|_{boundary}, \tag{4.10}$$

where t^{μ} denotes the "tangential", i.e. (in 1+1 dimensions) timelike direction at the bag surface. Equation (4.10) is precisely the missing component of the bosonization relations (3.15) that could not be obtained just from the stationarity boundary equations (3.14d) of the action S (3.12). Thus the "standard" hybrid bag model with the usual boundary condition (4.1) has built-in the complete bosonization relations (3.11) - provided quantum effects are properly taken into account.

Having discovered the anomaly, this quantum effect, is, however, only part of the "story". An answer is still missing to the following problem: Not only the "guard of the bag jail" was fooled, but also the quark, because the anomaly (our "escape agent") does not really liberate the quark, but merely drowns it in the Dirac sea. Thus there must be some way to carry the information into the "outside" that in reality (i.e. in the free fermion theory) a quark there is a kink - a bosonic field configuration - moving in the "outside" region of the bag pretending to be the quark, but is it? The original quark that has just become one of the Dirac sea quarks has to "dive" back to the left with negative energy (see Fig.8c).

Let us give an resumé of the "escape story" of the quark as seen from the bag "jail": A right-moving quark runs towards the right end of a bag (Fig.8a). It hits the wall and transfers its $U(1)_R$ charge (its Q_R value) - a conserved charge - to the η'-field ϕ outside the bag (Fig.8b). In this ϕ-field a kink-shaped signal runs out towards the right (Fig.8c). This kink has to carry the unit of $U(1)_R$ charge transferred. In order to support $Q_R = 1$ and to be purely right moving, the ϕ-field kink must drag along a region of space where ϕ is shifted by the amount $\sqrt{\pi}$. This shift in ϕ at the bag surface causes the anomalous disappearance of one quark (namely $\sqrt{\pi}/\sqrt{\pi} = 1$), i.e. the anomaly pushes one quark into the Dirac sea, presumably the very quark that hit the bag boundary to start with. So, the quark is not truly reflected by the boundary. Instead it is drowned in the Dirac sea and has now to "dive" back to the left while a kinklike signal is traveling to the right. In a fermionization language, this ϕ-field kink represents a fermion. Thus - after a "re-fermionization" of the bosonic "outside" region - a possible interpretation of the net result would be that quark was modified ("disguised") into a kinklike signal in the ϕ-field running to the right outside the bag. In our two-phase language, however, the quark was "tricked" by the anomaly. It was drowned in the Dirac sea, thus it never escaped. Rather a soliton, a "strange" motion in the η'-field ϕ was sent out "pretending" (to some extent) to be the quark.

5. Applications of the Cheshire Cat Principle

Let us presume that it is correct to interpret a hybrid bag model in the Cheshire Cat sense. Are there now practical applications of this principle, i.e. in terms of technical calculational projects?

As an elementary consequence, we have a way of checking practical calculations by redoing them with bags of different sizes and/or shapes. Regardless of the bags chosen, the result of any calculation of any hadronic quantity has to be the same in principle if the Cheshire Cat interpretation is assumed to be true. In practice, the deviations from this agreement are now linked to the approximations used and serve as a measure of the validity of the hybrid bag model itself.

Furthermore, we may learn from, say, the exact 1+1 dimensional Cheshire Cat bag models how to do calculations which are of fundamental difficulty: e.g., in any bag model there is the inherent problem of what happens to the bag if it has to disappear in physical reactions with only leptons left in the final state, as e.g. in $\rho^0 \rightarrow e^+e^-$. A ρ^0 - decay to a positron-electron pair [22] is naturally conceived of (in the bag model) as due to the annihilation of a quark q and antiquark $\bar{q}$ of the ρ^0 into a virtual photon which successively becomes the e^+e^--pair. But, then there is an empty bag left over and the whole decay can only take place if the overlap of the empty bag with the vacuum is different from zero. In other words, the decay amplitude $\rho^0 \to e^+ e^-$ acquires a factor $<0|empty\ bag>$. Now, if one takes the point of view that the bag itself has a physical significance, i.e. is in principle determinable by experiment, this overlap amplitude $<0|empty\ bag>$ will converge to zero when the discrimination between "inside" and "outside" becomes a locally measurable quantity. In the Cheshire Cat picture on the other hand where every bag state can be represented in terms of the "outside" fields, the overlap $<0|empty\ bag>$ should have a meaningful - presumably nonzero - value. In our simplest 1+1 dimensional Cheshire Cat model the fermion vacuum happens (?) to coincide with the outside or bosonic vacuum. This is of course not really an accident at all since it simply reflects the fact that the vacuum is the lowest energy state of the field theory, whether one considers it a bosonic or a fermionic field theory.

Finally, we might use the Cheshire Cat principle to fix parameters, thus allowing a much more complicated "outside" theory.

6. The Development of a Refined Hybrid Bag Model with Many Fields in the Outside Region

It is the last application mentioned in the previous section that we would like to advocate in this and the following section. In our opinion it is really a program in its own right and worthwhile to pursue and develop further than we will have space and time to do here.

The main content of this program of fixing parameters of the theory is to replace as much phenomenological input as possible by use of the Cheshire Cat principle. In this way one might hope to achieve the following: one might reduce the number of free parameters that must be fitted to experimental data. One might even find a connection between a complicated effective model (which describes hadron physics at the phenomenological level) and the underlying microscopic theory, QCD. The essential idea is first to calculate some quantity in the language of the effective theory in an "outside" region of space. Secondly, to redo this calculation when a "test"-bag (of any given size and/or shape) has been inserted into this region. Thirdly, in insisting on the Cheshire Cat principle, to vary the parameters in the effective theory until both results coincide. The equations obtained by identifying the two results might be used for fixing a parameter of the effective model.

In replacing the simple nonlinear σ-model by a more complicated model in the "outside", one would presumably encounter more free parameters. If one introduces e.g. bosons with higher spin there will be more coupling constants and masses that have to be determined either from experimental input or from some principle. One of the most promising features of this program is that it might be possible to produce so many relations between the parameters (masses and coupling constants) - by inserting or not inserting bags and equating the corresponding results - that one can end up with a phenomenologically sensible model. One may even think about the application of a relatively complicated Cheshire Cat hybrid bag model as calculational machinery for QCD calculations: the Cheshire Cat principle might provide enough relations between the parameters such that the parameters in the "outside" hadronic field theory can be determined in terms of those of the "inside" QCD theory.

In practice the Cheshire Cat principle is not fulfilled exactly in 3+1 dimensional models. Therefore the relations obtained between parameters by imposing the principle can only be satisfied approximately. Thus in using all these relations, one would most likely encounter contradictions. In other words, the parameters would be overdetermined. On the other hand, this would give rise to the hope that by a clever selection one might even have a chance of determining all the parameters. This means that the use of an approximate Cheshire Cat model as a calculational technique for QCD might be a bit of an art. One must select the sensible relations out of an infinite set of contradictory restrictions. Nevertheless, used with some judgement, such a scheme of calculation might be promising.

7. Illustration of the Parameter Fitting Idea

In the following we would like to illustrate how the Cheshire Cat principle can be applied to fitting parameters of the theory. A rather simple example is the determination of the coefficient $f_\pi^2/2$ of the kinetic term (3.1a) of the boson field (see Krakow lecture [4]). The main idea is that a little bag is put in a region of space where the η'-field ϕ has slowly been varying. If the Cheshire Cat principle is valid, the profile of the η' field in the region outside the bag is not allowed to change whether a "test"-bag has been put in or not. Now, the coefficient $f_\pi^2/2$ is determined by identifying the energy of this "test"-bag with the energy of the former η' field in the same "volume" (length).

Here, we would like to present a more involved example to illustrate the idea of parameter fitting under the Cheshire Cat principle: the determination of the η' mass in 1+1 dimensional massless electrodynamics (i.e. Schwinger model [23]). If we add to a massless free Dirac field theory an electromagnetic interaction, which in 1+1 dimensions is just the Coulomb interaction

$$-\frac{1}{4}\int dxdy \, e\bar{\psi}(x,t)\gamma^0\psi(x,t)|x-y|e\bar{\psi}(y,t)\gamma^0\psi(y,t) , \qquad (7.1)$$

we obtain the Schwinger model which can be bosonized into a free, but now massive (pseudo) scalar boson theory [17]. In fact, it can easily be shown that the interaction Hamiltonian of the Schwinger model is equivalent to the mass term of the boson field after insertion of the bosonization relation (use (3.11) and $\gamma_1\gamma_5 = \gamma^0$ valid in 1+1 dimensions)

$$\sqrt{\pi}\bar{\psi}\gamma^0\psi = \partial_1\phi \qquad (7.2)$$

and some partial integrations:

$$-\frac{1}{4}\int dxdy \; e\bar{\psi}(x,t)\gamma^0\psi(x,t)|x-y|e\bar{\psi}(y,t)\gamma^0\psi(y,t)$$

$$=-\frac{1}{4}\frac{e^2}{\pi}\int dxdy \, \partial_1\phi(x,t)|x-y|\partial_1\phi(y,t) = +\frac{e^2}{4\pi}\int dxdy \, \phi(x,t)\varepsilon(x-y)\partial_1\phi(y,t)$$

$$=\frac{e^2}{2\pi}\int dxdy \, \phi(x,t)\delta(x-y)\phi(y,t) \qquad (7.3)$$

$$=\frac{e^2}{2\pi}\int dx \, \phi(x,t)^2 = \frac{1}{2}m^2\int dx \, \phi(x,t)^2 \; .$$

The last equation in (7.3) is valid provided we take

$$m^2 = \frac{e^2}{\pi} \; . \qquad (7.4)$$

Note that in 1+1 dimensions the charge e has the dimension of mass. In order that the boson mass (η' mass) acquires the value listed above, it is necessary that the fermion theory can be bosonized exactly. Our central idea is now to compute this relation in a different way that does not directly rely on bosonization but rather uses the requirement of the existence of a Cheshire Cat bag model based on this bosonization. Thus one could perform this type of calculation even in the case when no exact bosonization is valid anymore. That means one might generalize the calculation of the boson mass (here the η' mass) even to the 3+1 dimensional case of the hybrid bag model where the bosonization is at most approximate [14].

The principle behind our η'-mass calculation is first to consider a static situation of the η' field $\phi(x,t)$. Secondly, one inserts a small "test"-bag in such a situation. Thirdly, one has to insist that the inserted bag is static, too. Especially, the bag should not get filled up by an increasing amount of positive or negative chiral charge. As a prerequisite for this calculation, we first note that the existence of an anomaly was of great importance for the escape of the quark from the bag. This implies that there has to be a boundary interaction term between the η'-field $\phi(x,t)$ and the electromagnetic potential A_μ. Note that in our 1+1 dimensional example the latter represents the gluon (subject here to an $U(1)$ rather than to a nonabelian gauge group). In fact, when the anomaly at the boundary forces the "quark" to disappear (by drowning it in the Dirac sea), the electric charge of the "quark" vanishes too. Then however, the electric charge would become nonconserved and hence gauge invariance would be broken unless one introduces a compensating charge. In this case there are two alternatives: On the one hand the electric field could couple to the soliton (which represents the "quark" outside the bag), on the other hand an extra surface current could be introduced in the model that compensates the anomaly contribution. If we wish to identify the Schwinger model "photon" with a gluon, we should not allow it to exist outside the bag. Therefore, we can only choose the second alternative: we must try to construct a compensating charge on the bag surface! The anomalous charge produced at the bag wall is proportional to the time derivative $\dot\phi$ of the η'-field (see Sect. 4). Thus, the total charge at the resting bag wall, which has been summed until a given moment of time t

$$e \int^t dt' \frac{\dot\phi}{\sqrt{\pi}} \Big|_{r.b.w.} \tag{7.5}$$

has to be proportional to the η'-field ϕ itself. So, we should have an extra charge on the right bag wall (abbreviated $r.b.w.$)

$$Q_{extra} = -e \frac{\phi(t,x)}{\sqrt{\pi}} \Big|_{r.b.w.} \tag{7.6}$$

to compensate (7.5). This means there should be an extra term in the electromagnetic action

$$\int dt \, eA_0 \frac{\phi}{\sqrt{\pi}} \Big|_{r.b.w.} \, . \tag{7.7}$$

Covariantly this becomes

$$\int_{\partial V} d\Sigma \, \frac{e}{\sqrt{\pi}} \varepsilon^{\mu\nu} n_\nu A_\mu \phi \tag{7.8}$$

where $\varepsilon^{01} = 1$, $\varepsilon^{\mu\nu} = -\varepsilon^{\nu\mu}$, and $n^\mu = g^{\mu\nu} n_\nu$ is the inward pointing normal.

In summary, the bag model has the extended action (as compared to (3.12)):

$$S = S_V + S_{\bar V} + S_{\partial V} \tag{7.9}$$

where

$$S_V = \int_V dxdt \left[\bar\psi(x,t) \left[i \frac{1}{2} \overleftrightarrow{\partial}_\mu - eA_\mu \right] \gamma^\mu \psi(x,t) + \frac{1}{2} E^2 \right] \tag{7.10a}$$

$$S_{\bar V} = \int_{\bar V} dxdt \left[\frac{1}{2} \partial_\mu \phi(x,t) \partial^\mu \phi(x,t) - \frac{1}{2} m^2 \phi(x,t)^2 \right] \tag{7.10b}$$

$$S_{\partial V} = \int_{\partial V} d\Sigma \left[\frac{1}{2} n^2 \bar\psi e^{i\sqrt{4\pi}\gamma^5 \phi} \psi - \frac{e}{\sqrt{\pi}} \varepsilon^{\mu\nu} n_\mu A_\nu \phi \right] . \tag{7.10c}$$

Here V denotes the spacetime region inside the bag, $\bar V$ that outside, and ∂V the boundary. E is the electric field strength defined below in (7.12).

Now consider the following situation (see Fig.9): A little test-bag is being inserted into a static field configuration of the η'-field ϕ obeying

$$\partial_1^2 \phi - m^2 \phi = 0 \, . \tag{7.11}$$

Under the assumption that the bag is so small that the field value is approximately the same at both walls one finds opposite and numerically equal extra charges on the two walls: $\pm e\phi/\sqrt{\pi}$.

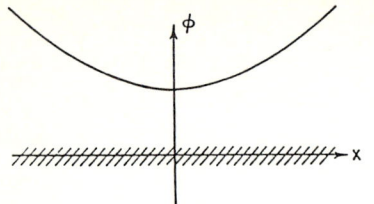 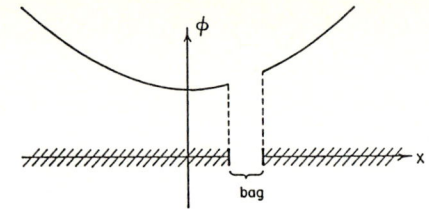

Fig. 9. A little "test"-bag inserted into a background of a static weakly space dependent η'-field ϕ. Note that the ϕ field in the "outside" region is totally undisturbed by the inserted bag

These charges give rise to an electric field (this electric field plays the role of the color Yang Mills field in our model) inside the bag of strength

$$E \equiv F_{01} \equiv \partial_0 A_1 - \partial_1 A_0 = e \frac{\phi}{\sqrt{\pi}} \ . \tag{7.12}$$

Now there exists another anomaly, namely the Adler-Bell-Jackiw anomaly [24] which in 3+1 dimensions tells that the axial charge

$$Q_5 = \int d^3x \, j^{50}(x) \tag{7.13}$$

is nonconserved by

$$\partial_\mu j^{5\mu}(x) = -\frac{\alpha}{4\pi} \varepsilon_{\mu\nu\rho\sigma} F^{\mu\nu} F^{\rho\sigma} \tag{7.14}$$

where $\alpha = e^2/4\pi$. In 1+1 dimensions the corresponding Adler-Bell-Jackiw anomaly has following form:

$$\partial_\mu j^{5\mu}(x) = -\frac{2}{2\pi} eE = \frac{e}{\pi} F_{10} \tag{7.15}$$

in the case of a single Dirac "quark" species, i.e. two Weyl ones (see Fig.10). Note that we use the convention $j^{5\mu} \equiv j_L{}^\mu - j_R{}^\mu$ for the 1+1 dimensional expressions.

This anomaly causes a destruction or creation of chiral charge (depending on the sign of ϕ) in the test-bag with a rate of

$$\dot{Q}_5 = -\frac{e}{\pi} \frac{e}{\sqrt{\pi}} \phi L = -\frac{e^2}{\sqrt{\pi^3}} \phi L \tag{7.16}$$

units of chiral charge per unit time. Here, L denotes the length of the bag. In order that even after the insertion of the test-bag the situation is still static, the chiral charge variation must be compensated by an inflow of chiral charge from the bag walls. The inflow of chiral charge into the bag is

$$\frac{\partial_1 \phi}{\sqrt{\pi}}\bigg|_{r.b.w.} - \frac{\partial_1 \phi}{\sqrt{\pi}}\bigg|_{l.b.w.} \ . \tag{7.17}$$

This can easily be derived from the boundary condition at the bag wall (see 3.14d):

$$\sqrt{\pi} \bar{\psi} n_\mu \gamma^\mu \gamma^5 \psi = n_\mu \partial^\mu \phi \tag{7.18}$$

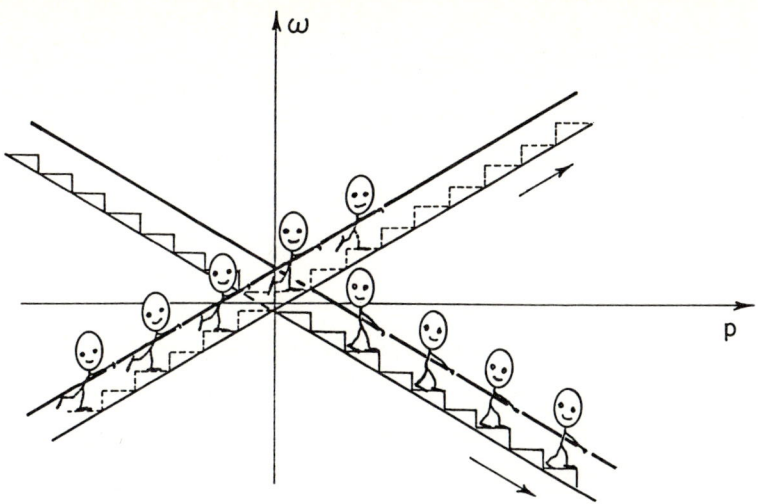

Fig. 10. The Adler-Bell-Jackiw anomaly [23] (in 1+1 dimensions) explained by a pumping up or down of right and left handed (= right and left moving) fermions respectively [21]. The electric field E, pointing in the positive x direction, pulls both right and left movers towards the right. Thereby the energy and the momenta of the fermions are changed as symbolized by the moving staircases in the figure. The axial charge is given by the difference in the number of right and left movers. Actually, the number of occupied and empty steps should be infinite (representing the infinite number of positive and negative fermion states). Essentially, this is the reason why quantum-mechanically the axial charge is not conserved in an electric field (in 1+1 dimensions) whereas it is on the single particle level. Note that the (normal) charge (i.e. the sum of occupied right and left moving fermion states) remains at a constant value while the axial charge (i.e. the difference) increases more and more in magnitude.

which holds even when there is an electromagnetic field included. (This is the case because the fermion dependent boundary conditions result from the kinetic terms of the action (7.9), which are independent of A_μ). Equating the inflow (7.17) to the production $-\dot{Q}_5$ yields

$$\frac{\partial_1 \phi}{\sqrt{\pi}}\Big|_{r.b.w.} - \frac{\partial_1 \phi}{\sqrt{\pi}}\Big|_{l.b.w.} = \frac{e^2}{\sqrt{\pi}^3} \phi L \ . \tag{7.19}$$

Now by a Taylor expansion in the short (=small) bag approximation

$$\frac{\partial_1 \phi}{\sqrt{\pi}}\Big|_{r.b.w.} - \frac{\partial_1 \phi}{\sqrt{\pi}}\Big|_{l.b.w.} = \frac{L \partial_1^2 \phi}{\sqrt{\pi}} \tag{7.20}$$

we obtain the following stability criterion (that the bag is e.g. not emptied of chiral charge):

$$\frac{L \partial_1^2 \phi}{\sqrt{\pi}} = \frac{e^2}{\sqrt{\pi}^3} L \phi \tag{7.21}$$

or

$$\partial_1^2 \phi - \frac{e^2}{\pi} \phi = 0.$$ (7.22)

This agrees with the static part of the Klein-Gordon equation

$$\partial_1^2 \phi - m^2 \phi = 0,$$ (7.23)

provided that

$$m = \frac{e}{\pi}.$$ (7.24)

Thus, we have reobtained the result for the mass parameter m of the "outside-the-bag" theory already known from the bosonization, however, without directly using the latter in the derivation.

We have presented the above in order to stress that one can make such a fit of a parameter (in this example m) by comparing two situations of different bag configurations (in this example, a small bag with no bag at all, see Fig.9). Note that we could perform such a calculation even in a case where the Cheshire Cat principle was not exact. Then, however, there would be no guarantee that the resulting value of the parameter would be independent of, say, the size of the test-bag used in the calculation. Nevertheless, if there is an approximate Cheshire Cat principle at work, the parameter fitted (e.g. m) has to have a rather flat dependence on the bag size (e.g. L). In the exact Cheshire Cat case - as we have just seen - the size L drops out of the expressions in the calculation and therefore the parameter m is predicted to be independent of L.

8. Confinement - An Epilog to the 'Crime Story'

From the "crime story" of Sect. 4 we learned that the bag does not confine quarks, because the anomaly can cause them to disappear (from the bag). At the same time, kinks have to appear in the "outside" region. Since these can be interpreted as the original quarks if one "re-fermionizes" the bosonic "outside" region of the bag model, the net effect is that the quarks can pass through the bag wall, completely untouched.

In the case of a truly free fermion theory, the fermion is of course not confined. By definition there is no confinement at all. However, if we consider the Schwinger model [23], we will have confinement: In 1+1 dimensions, the Coulomb field is a linear potential. Thus even pure electrodynamics is confining. How does this fact correspond to the "escape story" that the bag cannot really confine the quarks? At this juncture we should issue a warning: The Cheshire Cat principle does not mean that the quark would not be finally confined by other means (here the electric field of the 1+1 dimensional Schwinger model). With respect to confinement, it only means that the bag does not do the job. Now, the point is that the emitted kinks will eventually be caught because of the mass term in the ϕ-field theory [17]. In fact, this mass term is the bosonized formulation of the Coulomb interaction between the quarks (see (7.3)). Since it is this Coulomb interaction which causes the confinement of quarks, it is not surprising that it is its bosonic analog, the mass term, that finally confines the kinks (the bosonic equivalents of quarks). Now, the kink corresponds to a shift of the field ϕ by the amount $\sqrt{\pi}$ as it passes by. Therefore, it pulls along a one-dimensional region in which the field is either lifted by $\sqrt{\pi}$ or lowered by the same amount. This, however, costs energy because of the mass term. This energy is

$$E = L_{shifted} \frac{1}{2} m^2 \left[\sqrt{\pi}\right]^2, \tag{8.1}$$

where $L_{shifted}$ is the length of the region of space over which the field ϕ is shifted by $\pm\sqrt{\pi}$, i.e. the length spanned by the kink and antikink which form the ends of the "hadron". Such a linear energy dependence of the distance between kink and antikink suggests confinement of the kinks.

In reality, when a kink and an antikink move away from each other, the field spanned between the two does not stay at a constant nonzero value. Again the mass term $\frac{1}{2} m^2 \phi^2$ drives the ϕ-field - located in between - towards zero. In the fermionic language, this effect represents the splitting of the string (where a quark-antiquark pair is created) well known to be of importance for realistic four-dimensional models [25]. In the boson language, this splitting of the string is given by the collapse of the ϕ field towards zero.

The splitting of the bag, however, is possible irrespective of the splitting of hadrons, since the bag has no physical significance. In fact, we can have a bag containing a single quark, because its charge can be compensated by the surface charge (7.6). A bag at the site of a meson can therefore be cut between the quark and the antiquark. The price will be though that the surface charges (7.6) imply the existence of a non-trivial η'-field in the gap region between the two bag "bits".

We should repeat again: the Cheshire Cat principle does not at all exclude confinement, however, it excludes that the bag itself plays a role in this question. Actually - as mentioned in Sects. 2 and 6 - even in an approximate Cheshire Cat model, one can recover the complete QCD in the limit of an infinite bag radius. It is obvious that in this case there is confinement and that the bag has nothing to do with this fact.

Since there is confinement in QCD and since therefore there should exist a confinement scale, one might think that in practical calculations a good choice for the bag radius of a hybrid bag model would be just the value of this confinement scale. However, in such a case any perturbative treatment in the interior region of the bag would be most certainly doomed from the very start. Rather, for that purpose, one should choose values of the bag radius which allow a limited perturbative treatment and which lie in a region where the results of the lowest moments (energy density, etc.) show approximate radius independence (see Sects. 2 and 6).

9. Conclusions

The main points of the present work have been the presentation of "the Cheshire Cat point of view" of the hybrid bag model and the suggestion of a new scenario for applying hybrid-bag-like models as a method of attempting to do QCD calculations in the low energy or strong coupling regime. The idea is to apply the bag model in a way completely different from the way used in the early times of the bag model. That means our Cheshire Cat philosophy is quite distinct from what one may call "the-bag-really-exists" point of view. We hold that one may very healthily decide to work with a bag model of the Cheshire Cat type even if - in a physical sense - no bag exists at all: Suppose we have some true description of some natural phenomenon by a field theory model, e.g. we believe hadron physics is described by the field theory QCD. Suppose we have technical problems in calculating with this theory, but that we may - under some conditions - approximate or perhaps even totally replace the latter by a seemingly totally different field theory. Then it might be a good idea to try to find a model that approximates or replaces the original field theory in some regions of spacetime but not in others. That means we want a model in which spacetime is divided into two regions, say, which we may - for historical reasons - call "inside" and "outside" and then we take e.g. in the "inside" the full theory QCD and in the "outside" the approximation or replacement, a σ-type model for instance. The idea is now to construct (formally) such a two-phase model so that it looks like a bag model. The boundary terms in the action should be constructed in such a way that the Euler-Lagrange equations at the boundary develop the relations between the fields of the "inside" region and the "outside" needed for the implementation of the approximation or replacement scheme.

Such a model has an action of the general form $S = S_V + S_{\bar{V}} + S_{\partial V}$ consisting of three contributions: one from each of the regions, S_V from the "inside" and $S_{\bar{V}}$ from the "outside", and then a contribution $S_{\partial V}$ from the border surface between the two regions V and $\bar{V}$. Formally, this action looks precisely like the one of a hybrid bag model. If the "approximation" used in the "outside" region is a good one to the full theory wanted (i.e. to QCD), the results of a calculation with this baglike model should be independent of where one chooses to place the boundary surface ∂V (i.e. of counting a region as the "inside" or "outside").

We proposed to use this requirement as a new way of fixing the parameters of the approximate field theory in the "outside" region in terms of the parameters of the exact field theory (e.g. QCD) defined in the "inside". In the present work, we illustrated this idea by adjusting what we called the "η'-mass". But actually, we only considered the case of 1+1 dimensional massless electrodynamics (Schwinger model). In fact, we compared the possible static η'-field configuration in the following two situations: 1) no bag at all, i.e. all of space-time belongs to the "outside", and 2) with a small (test-)bag inserted (i.e. a region of rather small extent in space, but of infinite extent in the time, was declared as "inside"). It turned out that we could calculate (or estimate) the second-order derivative of the η'-field $\partial_x^2\phi$ divided by ϕ itself needed in the small bag case by use of the Adler-Bell-Jackiw anomaly. A key point in this calculation was to realize that - in order to construct a good Cheshire Cat bag model out of a Schwinger model - one has to add an extra coupling term in $S_{\partial V}$

$$-\int_{\partial V} d\Sigma \, \frac{e}{\sqrt{\pi}} \, \varepsilon^{\mu\nu} n_\mu A_\nu \phi$$

that compensates for an anomaly in the electric charge conservation at the bag surface. This term in turn gave rise to an electric field inside the inserted small test-bag that was proportional to the ϕ-field at the site of this bag. Due to this field there was a chiral charge loss resulting from the Adler-Bell-Jackiw anomaly that had to be compensated by an inflow of chiral charge into the bag. This led to the $\partial_x^2\phi$-estimate such that the η'-mass could be determined.

It is our suggestion that other parameters might be determined along similar lines. In our example of the 1+1 dimensional model, the "approximation" used in the "outside" happened to be exact, but in 3+1 dimensions this can presumably hardly be hoped for. In this case, one has to be satisfied with an approximate Cheshire Cat model, in the sense that the fitted parameters can only be approximately independent of the bag radius for instance. Having determined the parameters, we would be in possession of a model which would allow us to choose freely the regions of spacetime where one would use the approximation and the ones where one would find it profitable to apply QCD using a more direct (approximation) technique. In such a way, the model might be used for calculations of physically interesting quantities. In this sense, we suggest that the construction of a Cheshire Cat bag model can be considered as a calculational technique for QCD with a huge amount of flexibility. One might even apply these kinds of ideas to lattice Monte Carlo calculations for smaller spacetime regions - say the interior of a hadron.

Formally, the technique proposed closely resembles a bag model. However, let us stress again the philosophy behind it: There is no requirement that the bag should be represented by anything physical! In other words, the Cheshire Cat philosophy is: The bag does not exist physically, it is only a technical device for calculations! So one could sensibly apply the idea of a Cheshire Cat bag model without having any indication that there are (truly) existing bags. However, it is still too early to predict whether the proposed technique will turn out to be successful in practice.

A major point in the present work was the apparent discrepancy between the following statements: On the one hand, the bag model was originally constructed in order to implement confinement, and on the other hand, it is possible to construct a Cheshire Cat bag model as a calculational device to study a model that does not confine at all. For instance, we considered an exact Cheshire Cat bag model based on free massless Dirac fermions. This paradox turned out almost as a "crime story" about how the quark - which in principle was known to be just a free massless fermion - could escape from the bag ("jail"). The "escape agent" was the anomaly at the surface ∂V of the bag. Now anomalies may - at least sometimes - be interpreted as a means for pumping fermions up from or down into the Dirac sea. Therefore, what the anomaly - here the baryon number anomaly - actually did to get the quark out the bag was to push it down into the Dirac sea. At the same time a kink emerged from the bag surface and moved on, representing the quark "in reality" in the "outside" region.

In the model based on the totally free Dirac particle, the kink moves on unconfined. In the case of a Cheshire Cat bag model based on the Schwinger model, the quark still escapes from the bag disguised as a kink in the η'-field. However, it eventually gets confined as kink. So in our 1+1 dimensional Cheshire Cat bag model based on the Schwinger model, confinement is finally there, but it has nothing to do with the bag. Of course this is as it should be: Since the bag in the Cheshire Cat philosophy has no physical significance, it should not confine the quark even when the quark happens to be confined for other reasons, say in the Schwinger model by the Coulomb field.

So indeed, our Cheshire Cat type of bags are in philosophy and physics rather different from the old M.I.T. quark confining bag. Even though this is the case, this does not mean that the Cheshire Cat point of view makes older bag model calculations irrelevant or obsolete - at least not those based on hybrid bag models. One may rather use and interpret them as calculations in the Cheshire Cat bag technique. In fact, the Cheshire Cat bag was historically developed from these hybrid bag models and their phenomenological suggestion that the baryon properties are rather insensitive to a bag radius variation.

We hope for a promising future for the Cheshire Cat bag model as a tool for QCD calculations consisting in successively more and more developed and complicated QCD-approximating models for the "outside" region!

Acknowledgement

It is a pleasure to thank Niels Brene for comments to and criticism of the manuscript. One of us (H.B.N.) would like to acknowledge discussions at the Krakow conference and the present one, especially with Marek Jezabek, Maciek A. Novak and Michael Praszalowicz.

References

1   S. Nadkarni, H.B. Nielsen, and I. Zahed: Nucl. Phys. B253, 308 (1985).
2   S. Nadkarni and H.B. Nielsen: Nucl. Phys. B263, 1 (1986).
3   S. Nadkarni and I. Zahed: Nucl. Phys. B263, 23 (1986).
4   H.B. Nielsen: "The Cheshire Cat Principle for Hybrid Bag Models", The Workshop on Skyrmions and Anomalies, (Krakow, Febr. 22-24, 1987) and Niels-Bohr-Institute Preprint, NBI-HE-87-26 (1987).

5   A. Chodos and C.B. Thorn: Phys. Rev. D12, 1833 (1975);
    T. Inoue and T. Maskawa: Prog. Theor. Phys. 54, 1833 (1975);
    C. Callan, R. Dashen, and D. Gross: Phys. Lett. 78B, 307 (1978);
    G.E. Brown and M. Rho: Phys. Lett. 82B, 177 (1979);
    N.I. Kochelev: Yad. Fiz. 39, 729 (1984);
    The model type we have in mind does *not* include the cloudy bag, *see ref. 26.*

6   A. Chodos, R.L. Jaffe, K. Johnson, C.B. Thorn, and V.F. Weisskopf: Phys. Rev. D9, 3471 (1974);
    A. Chodos, R.L. Jaffe, K. Johnson, and C.B. Thorn: Phys. Rev. D10, 2599 (1974);
    T. de Grand, R.L. Jaffe, K. Johnson, and J. Kiskis: Phys. Rev. D12, 2060 (1975);
    K. Johnson: Acta Phys. Pol. B6, 865 (1975).

7   M. Gell-Mann and M. Lévy: Nuovo Cim. 16, 705 (1960);
    for a review see: B.W. Lee: Chiral Dynamics (Gordon and Breach Science Pub., New York, 1960).

8   T.H.R. Skyrme: Proc. Roy. Soc. (London) A260, 127 (1961); Nucl. Phys. 31, 556 (1962).

9   A.D. Jackson and M. Rho: Phys. Rev. Lett. 51, 751 (1983);
    G. Adkins, C. Nappi, and E. Witten: Nucl. Phys. B228, 552 (1983).

10 Ö. Kaymakcalan, S. Rajeev, and J. Schechter: Phys. Rev. D30, 594 (1984);
    Ö. Kaymakcalan and J. Schechter: Phys. Rev. D31, 1109 (1985);
    U.-G. Meissner, N. Kaiser, A. Wirzba, and W. Weise: Phys. Rev. Lett. 57, 1676 (1986);
    Y. Brihaye, N.K. Pak and P. Rossi: Nucl. Phys. B254, 1109 (1985);
    T. Fujiwara, Y. Igarashi, A. Kobayashi, M. Otsu, T. Sato, and S. Sawada: Prog. Theor. Phys. 74, 128 (1985).

11 Lewis Carroll: Alice's Adventures in Wonderland (Macmillan Publ., London, 1865).

12 L.V. Laperashvili and H.B. Nielsen: Nucl. Phys. B276, 93 (1986).

13 G.E. Brown, J.W. Durso, and M.B. Johnson: Nucl. Phys. A397, 447 (1983).

14 E. Witten: Nucl. Phys. B160, 57 (1979).

15 G.E. Brown, A.D. Jackson, M. Rho, and V. Vento: Phys. Lett. 140B, 285 (1984);
    M. Rho: "Cheshire Cat Phenomena and Quarks in Nuclei", Saclay Preprint, SPhT/86-159 (1986);
    M. Jezabek: Phys. Lett. 174B, 429 (1986).

16 M. Jezabek: "Anomalous Charges of Fermion Vacuum in Chiral Bags", Max-Planck-Institute Preprint, MPI-PAE/PTh 13/87 (1987).

17 S. Coleman: Phys. Rev. D11, 2088 (1975);
    S. Mandelstam: Phys. Rev. D11, 3026 (1975);
    A. Luther: Phys. Rev. B19, 320 (1979), Phys. Rep. C49, 261 (1979).

18 *For an earlier attempt see:* T.H.R. Skyrme: Proc. Roy. Soc. A262, 237 (1961).

19. I. Zahed and D. Klabucar: Phys. Rev. $\underline{D30}$, 2647 (1984).
20. J. Goldstone and R.L. Jaffe: Phys. Rev. Lett. $\underline{51}$, 1518 (1983).
21. H.B. Nielsen and M. Ninomiya: Phys. Lett. $\underline{130B}$, 389 (1983).
22. G.A. Kozlov et al.: In <u>Special Topics in Gauge Field Theories</u>, Proc. of the XVIII International Symposium, Ahrenshoop, GDR (1984).
23. J. Schwinger: Phys. Rev. $\underline{128}$, 2425 (1962).
24. S.L. Adler: Phys. Rev. $\underline{177}$, 2426 (1969);
    J.S. Bell and R. Jackiw: Nuov. Cim. $\underline{60A}$, 107 (1969).
25. E.G. Gurvich: Phys. Lett. $\underline{67B}$, 358 (1979);
    A. Casher, E. Neuberger, and S. Nussinov: Phys. Rev. $\underline{D20}$, 179 (1979); $\underline{D21}$, 1966 (1980);
    H. Bohr and H.B. Nielsen: Niels-Bohr-Institute Preprint, NBI-HE-78-3 (1978).
26. S. Théberge, A.W. Thomas, and G.A. Miller: Phys. Rev. $\underline{D24}$, 2838 (1983).

Pion and Nucleon Structure: Low Energy Aspects*

W. Weise

Institute of Theoretical Physics, University of Regensburg,
D-8400 Regensburg, Fed. Rep. of Germany

1. INTRODUCTION AND MOTIVATION

What are the relevant degrees of freedom which govern the strong interactions of hadrons at low and intermediate energies (i.e. at energies and momentum transfers up to about 1 GeV) ? This question is evidently fundamental to nuclear physics: how do we understand the structure and dynamics of nuclear constituents ? Furthermore, it is a major conceptual and theoretical challenge: we are confronted with the full complexity imposed by the non-perturbative nature of Quantum Chromodynamics (QCD) at low energy.

QCD has inspired some nuclear physicists to think of nuclei and the nucleon-nucleon interaction explicitly in terms of quarks and gluons, not only in high energy processes such as deep inelastic lepton scattering where perturbative approaches are justified, but also at low energies where perturbative methods are bound to fail. One should keep in mind that modern meson exchange potentials /1/ have provided us with a remarkably successful phenomenology of nuclear forces without any explicit reference to quark degrees of freedom. The theme we would like to develop here is that the basic features of meson theory are in fact consistent with the principles of low-energy (non-perturbative) QCD, at least down to length scales of about half a Fermi. We shall be guided by the conjecture that, in the limit of large number of colors N_c, QCD reduces to a non-linear meson theory /15,16/. In the long-wavelength limit, the corresponding effective Lagrangian based on Chiral Symmetry is expressed entirely in terms of Goldstone pions as the relevant collective degrees of freedom. Vector mesons are shown to appear quite naturally in this scheme, and baryons emerge as solitons. We shall discuss in detail the electromagnetic form factors of the nucleon. The outcome is a hierarchy of nucleon radii which differ according to the probing field by which they are measured: electromagnetic radii are larger than the radius associated with the nucleon axial current, which is in turn larger than the radius of the baryon number distribution inside the nucleon. For the latter we find characteristic values of about 0.5 fm. Such small sizes can be accommodated quite well in the established wisdom of nuclear physics.

2. FACTS AND PHENOMENOLOGY

Throughout these notes, we shall concentrate on the physics with u- and d-quarks, i.e. the flavor SU(2) (isospin) sector of QCD. This is the one relevant for the physics of non-strange hadrons and nuclei. We consider the region of energies and momentum transfers up to about 1 GeV (i.e. four-momentum transfers $q^2 = \omega^2 - \vec{q}^2$ ranging between timelike values $q^2 < 1$ GeV² and spacelike values $q^2 > -1$ GeV²). In this domain we expect that the physics is governed by the lowest states of the meson and nucleon spectrum: the pion and the ω(770) and ρ(783) vector mesons; the nucleon and the Δ(1232) isobar.

---

*) Work supported in part by Deutsche Forschungsgemeinschaft, grant We 655/9-1, and by Bundesministerium für Forschung und Technologie, grant MEP 0234 REA.

2.1 Facts from Nuclear Physics

It is well established that a very good description of a wide variety of nuclear phenomena is achieved in terms of nucleons and mesons, rather than quarks and gluons. The meson exchange phenomenology of the nucleon-nucleon interaction /1,2/ is highly successful down to distance scales of about 1/2 fm. One-pion exchange is established quantitatively at large distances. At intermediate distances dipersion theoretic methods have been applied with considerable success to describe two-pion exchange processes /2/ and ρ meson exchange. A good part of the short range repulsion can be understood in terms of ω meson exchange.

Another well-established fact is the existence of meson exchange currents in few-body systems. A particularly interesting example is the deuteron electrodisintegration d(e,e´)pn close to threshold but at large momentum transfers up to $|\vec{q}| \approx$ 1 GeV/c /4/. This reaction probes the spatial distribution of the isovector meson exchange current down to distances $r \lesssim 1$ fm between nucleons /5,6/. At such distances one would naively expect that a picture based on the exchange of pointlike pions between almost pointlike nucleons breaks down and dissolves into a complicated quark-gluon problem. This is not the case. As shown in Figure 1, traditional meson theory works down to distances which are significantly smaller than the pion Compton wave length. Even more important is the observation that Chiral Symmetry governs the dynamics of such pionic phenomena. The curve labelled "soft pions" in Fig. 1 is the result of a calculation /5/ based entirely on this symmetry. A "full" calculation (see Fig. 1a) taking into account additional short-distance effects such as vector meson exchange together with meson-nucleon form factors with a typical range of about 0.5 fm gives an almost identical result /6/. Similar conclusions are also reached for the ^{3}He magnetic form factor (see Fig. 1b). Again, a calculation based on Chiral Symmetry and the resulting "soft pion" theorems accounts for the data up to remarkably large momentum transfers /7/.

The observation that meson theory works down to length scales considerably smaller than the QCD scale parameter $1/\Lambda(QCD) \sim 1$ fm is highly non-trivial in view of the relatively large electromagnetic hadron radii: the measured proton and pion r.m.s. charge radii are about 0.85 and 0.7 fm, respectively. These apparent radii as seen

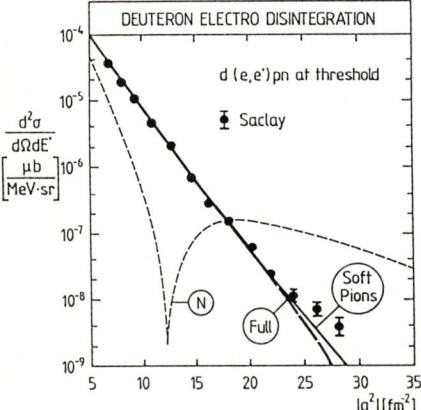
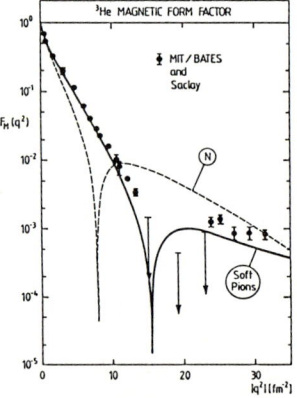

Figure 1a: Differential cross section for deuteron disintegration at threshold. The data are taken from /4/. Short-dashed curve (N): one-nucleon currents only; Solid curve (soft pions): based on Chiral Symmetry with point nucleons and pions /5,6/. "Full" curve includes nucleons and Δ´s, pion and ρ meson exchange

Figure 1b: Meson exchange effects in the ^{3}He magnetic form factor. Data from /8/. The curves are results of a realistic three-body calculation without (N) and with (soft pions) exchange currents based on Chiral Symmetry with pointlike pion-nucleon coupling

by electromagnetic probes have frequently (and erroneously) been identified with typical radii for quark confinement in non-strange hadrons. Such an interpretation is clearly incompatible with basic facts from nuclear physics. The average distance between nucleons inside nuclei is d ≃ 1.8 fm. Confinement radii of about 1 fm would imply that the quark cores of nucleons frequently overlap in nuclei. The idea of nucleons as basic nuclear constituents would then be questionable even at large distances.

2.2 The Pion Form Factor and Vector Meson Dominance

Let us briefly discuss the pion electromagnetic form factor $F_\pi(q^2)$, for which the empirical data /9/ are shown in Fig. 2. The following observations are relevant:

(a) The pion r.m.s. charge radius is

$$\sqrt{\langle r^2 \rangle} = \sqrt{6} \left[\frac{dF_\pi(q^2)}{dq^2} \right]^{1/2}_{q^2=0} = (0.66 \pm 0.01) \text{ fm}. \tag{1}$$

(b) The form factor in the timelike region is dominated by the ρ meson which appears as a pronounced resonance at $q^2 = m_\rho^2 \simeq (770 \text{ MeV})^2$ with the ρ → ππ decay width $\Gamma_\rho \simeq 150$ MeV at resonance.

Both features find a natural interpretation in the Vector Meson Dominance (VMD) model of photon-hadron interactions /10/. It assumes that the electromagnetic current is dominated by vector mesons (ρ, ω, ...). Hence the photon sees the pion only via an intermediate ρ^0 meson. Even with an intrinsically pointlike pion, this picture leads to the following characteristic q^2 dependence of the pion form factor:

$$F_\pi(q^2) = \frac{m_\rho^2}{m_\rho^2 - q^2 - im_\rho \Gamma_\rho(q^2)}, \tag{2}$$

with the r.m.s. radius

$$\sqrt{\langle r^2 \rangle_\pi} = \sqrt{6}/m_\rho \simeq 0.63 \text{ fm}, \tag{3}$$

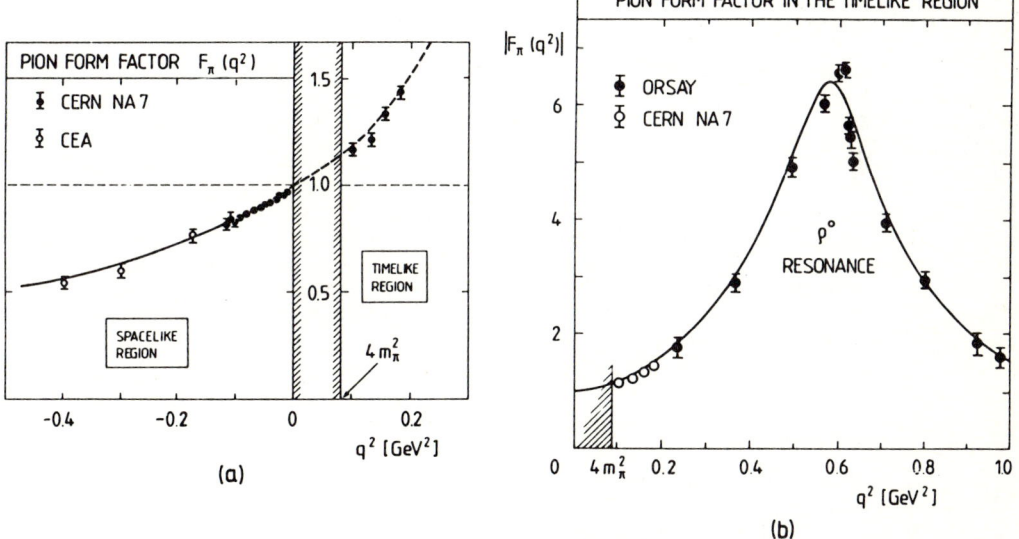

Figure 2: Pion electromagnetic form factor in the spacelike (a) and timelike (b) regions of q^2. The curve is based on the q meson dominance model modified to include a small intrinsic pion size of about 1/3 fm /11/

which is remarkably close to the measured value (1). One concludes that the apparent radius as seen by the photon is determined almost completely by the intermediate ρ meson: the "intrinsic" radius of the pion may be considerably smaller than the measured charge radius. In fact, the phenomenological fit in Fig. 2 is obtained with an intrinsic pion radius of about 1/3 fm. For most nuclear physics purposes, the pion can therefore be considered as intrinsically pointlike.

It is instructive to see the Vector Meson Dominance idea in analogy with dipole polarization phenomena in many-body systems. The vector mesons have the "dipole" spin and parity $J = 1^-$; the ρ and ω mesons are just the lowest dipole excitations of the QCD vacuum.

3. ELEMENTS OF LOW ENERGY QCD

At length scales of about 1/10 of a Fermi, QCD is a perturbative theory of quarks and gluons. At length scales of order 1 fm, QCD manifests itself in terms of composite hadrons. A systematic understanding of the connection between these two domains has not been achieved so far, despite the progress of lattice calculations, and there is little hope for a breakthrough in the nearest future. Much emphasis has therefore been focused on the development of an appropriate effective Lagrangian for low energy hadron physics which is based on QCD and its symmetries, but which uses the "proper" degrees of freedom: mesons rather than quarks and gluons. We will now review briefly some of the steps and observations which support this procedure.

3.1 QCD with massless Quarks: Chiral Symmetry

It is generally accepted that the masses of current quarks in the isospin sector with $N_f = 2$ flavors are small on hadronic scales: the up and down quark masses m_u and m_d at $|q^2| \sim 1$ GeV are found to be of order 10 MeV, i.e. almost negligible as compared to the typical QCD scale $\Lambda(QCD) \sim 200$ MeV. In the limit $m_{u,d} \to 0$, the QCD Lagrangian in the SU(2) flavor sector is invariant under global chiral rotations of the quark fields:

$$q(x) \to \exp[i\gamma_5 \vec{\tau}\cdot\vec{\theta}/2] \, q(x) \qquad (4)$$

(Chiral $SU(2)_L \times SU(2)_R$ symmetry).

3.2 States of the Vacuum

One can think heuristically of the weak and strong coupling domains of QCD as two phases: the high energy phase of quarks and gluons, and the low energy phase of composite hadrons. These phases correspond to different states of the vacuum in QCD: the perturbative vacuum in which quarks abd gluons propagate almost freely, and the physical, non-perturbative vacuum which excludes free quark and gluon fields. This latter state of the vacuum is the one relevant for low energy hadron and nuclear physics. In the physical vacuum the mass of an isolated quark is infinite (provided there is absolute confinement), whereas in the perturbative vacuum the quarks have their small current masses. The physical, non-perturbative vacuum is characterized by the existence of condensates, i.e. non-zero vacuum expectation values $<\bar{q}q>$ and $<G_{\mu\nu}G^{\mu\nu}>$ of the quark and gluon fields. In our context the quark pair condensate with its empirical value /12,13/

$$<\bar{u}u> = <\bar{d}d> = -(245 \pm 25 \text{ MeV})^3 \qquad (5)$$

is of particular importance. The non-vanishing $<\bar{q}q>$ implies that Chiral Symmetry is spontaneously broken.

3.3 The Pion

The very special features of the pion are supposed to be directly related to Chiral Symmetry and its spontaneous breaking. The pion is identified with the associated Goldstone Boson which is massless in the exact chiral limit. The actual pion mass $m_\pi \simeq 140$ MeV (a small number on hadronic scales of 1 GeV) is understood in terms of the non-zero but small current quark masses. Current algebra gives the relation

$$2 m_\pi^2 f_\pi^2 = -(m_u + m_d) \langle \bar{q}q \rangle , \qquad (6)$$

which involves the pion decay constant f_π= 93.3 MeV and reflects the fact that $m_\pi \to 0$ as $m_{u,d} \to 0$. The pion should not be viewed as bound pair of a single valence quark and antiquark, but as a collective mode built on the non-perturbative QCD vacuum. The concepts developed in this context have a farreaching analogy /14-16/ with the description of low energy collective excitations in strongly interacting many-body systems.

3.4 Summary: the Low Energy Program

The previous discussion has outlined the most important non-strange low energy modes of QCD with masses below 1 GeV:
(a) The pion as the isospin triplet of $J^P = 0^-$ Goldstone Bosons of spontaneously broken SU(2) X SU(2) Chiral Symmetry. Due to the small pion mass which is related to the u- and d-quark masses of order 10 MeV, the pion is expected to be the most important collective degree of freedom in the low energy, long wavelength limit;
(b) the $J^P = 1^-$ isovector and isoscalar vector mesons which represent the lowest dipole excitations of the QGD ground state.

It is then natural to aim at a description of low energy hadron physics in terms of the pion together with the ρ and ω meson as the relevant degrees of freedom. These mesons can be thought of as the leading terms of a multipole expansion in the meson angular momentum J.

4. CHIRAL EFFECTIVE LAGRANGIAN

4.1 Non-linear σ-Model and Wess-Zumino term

Consider QCD with N_f massless quark flavors and N_c colors. The color gauge group is $SU(N_c)$, and we have a global $U(N_f)_L \times U(N_f)_R$ chiral symmetry group. The aim is to derive a chiral effective Lagrangian under the following conditions:

(a) Let $N_c \to \infty$; i.e. keep only terms of leading order in N_c;
(b) assume that Chiral Symmetry is spontaneously broken, with the pseudoscalar mesons as Goldstone Bosons;
(c) take the low-energy and long-wavelength limit in which the massless Goldstone Bosons dominate the physics.

Under these assumptions it is shown in ref./17/ how to derive a non-linear chiral Lagrangian, including an anomalous (Wess-Zumino) term /18/ and a term responsible for the splitting of the π° and η' (Adler-Bell-Jackiw term) related to the Abelian anomaly /19/. (Later on, we shall reduce the scheme to the chiral $SU(N_f) \times SU(N_f)$ group with $N_f = 2$ and the pions as Goldstone Bosons, but for the moment we keep $N_f = 3$ in mind).

The technical steps which lead to this effective Lagrangian are the following: the meson fields π^a correspond to the composite quark operators $\bar{q}\gamma_5 t^a q$, where t^a are the generators of the group $U(N_f)$. One introduces the π^a in the functional integral by an appropriate change of variables. The Jacobian of this substitution generates the anomalous Wess-Zumino and Adler-Bell-Jackiw terms. One then integrates over all color fields to obtain the effective action in terms of color singlet fields. In the low energy limit this action is expressed in terms of the π^a only. In the reduction to $SU(N_f = 2)$, the relevant field variable is

$$U(x) = \exp[i\vec{\tau}\cdot\vec{\pi}(x)/f_\pi] , \qquad (7)$$

where the pion decay constant f_π = 93.3 MeV appears as the only scale. The original

SU($N_F=3$) scheme replaces $\vec{\tau}\cdot\vec{\pi}$ by $\lambda^a \pi^a$ where π^a is now the octet of pseudoscalar mesons ($\pi^\pm, \pi^0$, $K^\pm$, K^0, $\bar{K}^0$, η), and the λ^a are standard Gell-Mann matrices.

Omitting the Abelian anomaly term for simplicity, the resulting low energy chiral effective Lagrangian is obtained as /16,17/

$$\mathcal{L}_{eff} = \mathcal{L}_o + \mathcal{L}_{WZ} + \ldots . \tag{8}$$

The first term is the well-known non-linear sigma model:

$$\mathcal{L}_o(x) = \frac{f_\pi^2}{4} \text{tr}[\partial_\mu U(x) \, \partial^\mu U^+(x)] . \tag{9}$$

It represents the prototype of any meson theory which embodies Chiral Symmetry. The second term in eq.(8) is the anomalous Wess-Zumino term /18/. This is the part of the effective Lagrangian which still "remembers" the quarks, even when they have been eliminated as explicit degrees of freedom: it represents the chiral anomalies of the Fermions in QCD at the level of the effective non-linear meson theory. (See ref./2o/ for reviews).

4.2 The anomalous (baryon) current

The Wess-Zumino term generates an associated current

$$B^\mu = \frac{\varepsilon^{\mu\nu\lambda\sigma}}{24\pi^2} \text{tr}(U^+ \partial_\nu U \, U^+ \partial_\lambda U \, U^+ \partial_\sigma U) \tag{10}$$

with very special properties: by virtue of the $\varepsilon^{\mu\nu\lambda\sigma}$ tensor, it is conserved,

$$\partial_\mu B^\mu(x) = 0 ,$$

irrespective of the detailed dynamics, i.e. independent of the detailed form of the Lagrangian. The charge

$$B = \int d^3x \, B_0(x) \tag{11}$$

associated with this current is a topological quantity (winding number). It was conjectured long ago by SKYRME /21/ and later confirmed /21/ the B_μ should be identified with the baryon current, i.e. B is the baryon number. The sector with B = 0 (which hosts e.g. mesons, baryon-antibaryon pairs etc. apart from the vacuum) is topologically disconnected from the one with B = 1 (which hosts one baryon together with any number of mesons), and so forth.

4.3 Vector Mesons

The phenomenological success of the Vector Dominance Model has already pointed to the importance of vector mesons in a description of hadron physics down to length scales of about 1/2 fm. Before going into details, let us recall that, apart from the conserved axial current, the theory based on the Lagrangian (8) has two types of conserved vector currents: an isovector current $\vec{J}^\mu(x)$ which is the symmetry current of isospin transformations, and the anomalous isoscalar current $B^\mu(x)$. To lowest order in the pion field (from here on we stay in the $N_F = 2$ sector with underlying SU(2) × SU(2) chiral symmetry), $\vec{J}^\mu(x) = \vec{\pi}(x) \times \partial^\mu \vec{\pi}(x) + \ldots$ involves at least a pair of pions. Similarly, the isoscalar current (10) has at least three pions when expanding $U(x) = 1 + i\vec{\tau}\cdot\vec{\pi}(x) + \ldots$. It is then plausible that there should be two composite $J^P = 1^-$ modes: the isovector $\vec{\rho}^\mu(x)$ and the isoscalar $\omega^\mu(x)$ fields, which couple to these currents. In fact, the ρ meson does decay strongly into two pions, while the ω meson decay goes into three pions.

Our starting point is the observation /22/ that the non-linear sigma model (9) has a hidden local SU(2) × U(1) symmetry in addition to the global SU(2) × SU(2) chiral symmetry already discussed. Writing $U = \xi \cdot \xi$ (in the unitary gauge), with

$$\xi(x) = \exp[i\vec{\tau}\cdot\vec{\pi}(x)/2f_\pi] , \tag{12}$$

one finds that $\mathcal{L}_0$ is invariant under local transformations

$$\xi(x) \rightarrow h(x)\,\xi(x) \quad \text{with} \quad h \in SU(2) \times U(1) \ . \tag{13}$$

This introduces the gauge-covariant derivative

$$\partial_\mu \xi(x) \rightarrow D_\mu \xi(x), \quad D_\mu = \partial_\mu - (i/2)g(\vec{\tau}\cdot\vec{\rho}_\mu + \omega_\mu) \tag{14}$$

with the SU(2) and U(1) gauge fields $\vec{\rho}_\mu$ and ω_μ to be identified with the ρ and ω mesons. The essential physical assumption is that the underlying QCD or meson dynamics generates the ρ and ω mesons as physical objects, so that kinetic terms are added to the Lagrangian.

4.4 The Effective Lagrangian of the coupled $\pi\rho\omega$ - system

We summarize now the resulting non-linear effective Lagrangian of the interacting system of pions, ρ and ω mesons in compact form with $U = \xi^2 = \exp(i\vec{\tau}\cdot\vec{\pi}/f_\pi)$:

$$\mathcal{L}_{eff} = (f_\pi^2/4)\,\text{tr}(\partial_\mu U \partial^\mu U^\dagger) - (1/2g^2)\,\text{tr}(F_{\mu\nu}F^{\mu\nu})$$
$$- \alpha(f_\pi^2/4)\,\text{tr}[D_\mu\xi\cdot\xi^\dagger + D_\mu\xi^\dagger\cdot\xi]^2 + (N_C/2)g\omega_\mu B^\mu + \mathcal{L}_{\pi\rho\omega} \ , \tag{15}$$

where
$$D_\mu = \partial_\mu - (i/2)g(\vec{\tau}\cdot\vec{\rho}_\mu + \omega_\mu) = \partial_\mu - iV_\mu \ ,$$
$$F_{\mu\nu} = \partial_\mu V_\nu - \partial_\nu V_\mu - i[V_\mu, V_\nu] \ . \tag{16}$$

The canonical choice $\alpha = 2$ identifies the gauge coupling g with $g_{\rho\pi\pi} \simeq 5.9$ as determined from the $\rho^0 \rightarrow \pi^+\pi^-$ decay width, and yields the KSFR relation /23/ for the vector meson masses:

$$m_\rho = m_\omega = \sqrt{2}\,g_{\rho\pi\pi}\,f_\pi \ . \tag{17}$$

The last two terms in eq.(15) represent the leading couplings of the ω meson which are determined by the gauged form of the Wess-Zumino term. The $\pi\rho\omega$ coupling Lagrangian is

$$\mathcal{L}_{\pi\rho\omega} = (N_C/64\pi^2)\,g^2\,\varepsilon^{\mu\nu\lambda\sigma}\,\omega_{\mu\nu}\,\text{tr}\{\vec{\tau}\cdot\vec{\rho}_\lambda[i(U^\dagger\partial_\sigma U + \partial_\sigma U U^\dagger) + (g/2)U^\dagger\vec{\tau}\cdot\vec{\rho}_\sigma U]\} \ . \tag{18}$$

This term generates the dominant decay mechanism of the ω meson via $\omega \rightarrow \rho\pi \rightarrow 3\pi$.

To complete the scheme, we add a small pion mass term

$$\mathcal{L}_{SB} = (1/2)\,f_\pi^2 m_\pi^2\,\text{tr}(U - 1), \tag{19}$$

which breaks Chiral Symmetry explicitly.

Apart from the pion decay constant $f_\pi = 93.3$ MeV and the pion mass $m_\pi = 139.6$ MeV, the gauge coupling $g = g_{\rho\pi\pi} \simeq 5.9$ is the only additional parameter. Its universality in both the SU(2) and U(1) sector of the hidden local symmetry implies equal ρ and ω meson masses as in eq.(17). The coupling strength related to the $\omega_\mu B^\mu$ interaction is

$$g_{\omega NN} = (3/2)g \simeq 8.9 \ . \tag{20}$$

Since B_μ represents the baryon current, this $g_{\omega NN}$ is to be identified with the meson-nucleon coupling constant. We have $g_{\omega NN}^2/4\pi \simeq 6.2$, which is about half of what is required in phenomenological one-boson exchange potentials of the nucleon-nucleon interaction.

4.5 Electromagnetic interactions: Vector Meson Dominance

Finally, let us incorporate electromagnetic interactions by gauging the subgroup generated by the quark charge operator $Q = (1/3 + \tau_3)/2$. This leads to an extended covariant derivative in $\mathcal{L}_{eff}$ of eq.(15):

$$D_\mu \xi = (\partial_\mu - iV_\mu)\xi + ieA_\mu \xi Q \quad , \tag{21}$$

where A_μ is the photon field. The physical implications become transparent in the weak field limit by expanding $\xi(x) = 1 + (i/2f_\pi)\vec{\tau}\cdot\vec{\pi}(x) + \ldots$; one finds the following photon-meson couplings:

$$\begin{aligned}\mathcal{L}_{\gamma\pi\pi}(x) &= e(1 - \alpha/2)(\vec{\pi} \times \partial^\mu\vec{\pi})_3 A_\mu(x) \quad, \\ \mathcal{L}_{\gamma\rho}(x) &= -e\alpha\, g\, f_\pi^2\, \rho_3^\mu(x)\, A_\mu(x) \quad, \\ \mathcal{L}_{\gamma\omega}(x) &= -(e/3)\,\alpha\, g\, f_\pi^2\, \omega^\mu(x)\, A_\mu(x) \quad.\end{aligned} \tag{22}$$

One observes that for $\alpha = 2$, the Vector Meson Dominance principle (VMD) is realized, with

$$\begin{aligned}\mathcal{L}_{\gamma\rho} &= -(em_\rho^2/g)\,\rho_3^\mu A_\mu \quad, \\ \mathcal{L}_{\gamma\omega} &= -(em_\omega^2/3g)\,\omega^\mu A_\mu \quad,\end{aligned} \tag{23}$$

and no direct photon-pion coupling. What has been reached at this point is a unification of the current algebra and PCAC ideas on one hand with the VMD phenomenology on the other hand.

The result is conveniently expressed in terms of the current-field identities

$$\begin{aligned}J_{I=0}^\mu(x) &= -(m_\omega^2/3g)\omega^\mu(x), \\ J_{I=1}^\mu(x) &= -(m_\rho^2/g)\rho_3^\mu(x) + \ldots \quad,\end{aligned} \tag{24}$$

for the isoscalar and isovector electromagnetic currents, respectively. The dots in the isovector current indicate higher order terms which arise at short distance scales due to the non-Abelian structure of the ρ meson field (not present for the ω meson). As an immediate consequence of the underlying VMD, the pion form factor is obtained in the form (2) with a charge radius $\sqrt{\langle r_\pi^2\rangle} = \sqrt{6}/m_\rho = \sqrt{3}/gf_\pi = 0.62$ fm.

5. NUCLEONS AS SOLITONS

5.1 Soliton solutions

Let us now investigate the soliton solutions of the non-linear effective meson Lagrangian (15). For simplicity, we omit in most of this discussion the $\pi\rho\omega$-coupling term (18) and refer to this as the "minimal" model. We will also give results for the "complete" model (with inclusion of the $\pi\rho\omega$-term), but refer to /24,25/ for further details.

Solitons are classical, localized and stable field configurations. We search for them with the following ansatz of maximal symmetry ("hedgehog" ansatz):

$$\begin{aligned}U(\vec{r}) &= \exp[i\,\vec{\tau}\cdot\hat{r}\, F(r)] \quad; \\ \omega_0(\vec{r}) &\equiv \omega(r)\,, \quad \omega_i(\vec{r}) = 0 \quad; \qquad (i=1,2,3). \\ \rho_i(\vec{r}) &= (\hat{r}\times\vec{\tau})_i\, (G(r)/gr)\,, \quad \rho_0(\vec{r}) = 0 \quad;\end{aligned} \tag{25}$$

The spherically symmetric profiles $F(r)$, $G(r)$ and $\omega(r)$ represent the pionic, ρ and ω mesonic content of the soliton. They are obtained as solutions of the coupled field equations derived by minimizing the static energy. The input parameters f_π, g and m_π remain fixed at their empirical values throughout the procedure. The boundary conditions are $F(0) = \pi$, $F(\infty) = 0$ (as implied by the baryon number $B = 1$), $G(0) = -2$, $G(\infty) = 0$ and $\omega'(0) = \omega(\infty) = 0$. The meson profiles of the $B = 1$ soliton are shown in Fig. 3. Note that these profiles are concentrated on a scale substantially smaller than 1 fm. The soliton mass is $M_0 \simeq 1.5$ GeV in this model. Following eq.(10), the baryon number distribution of the soliton is obtained as

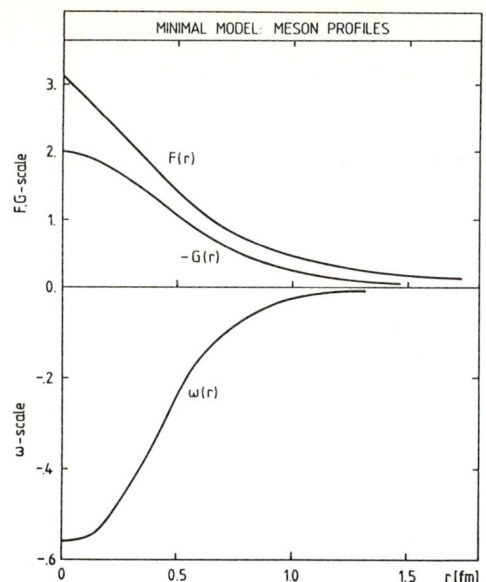

Figure 3: Radial profiles of the $B = 1$ chiral soliton based on the effective Lagrangian (15) (Minimal model: no $\pi\rho\omega$ coupling)

$$B_0(r) = -(F'(r)/2\pi^2 r^2)\sin^2 F(r) . \tag{26}$$

Evidently, the baryon number is

$$B = 4\pi \int_0^\infty dr\, r^2\, B_0(r) = (2/\pi) \int_0^\pi dF \sin^2 F = 1 . \tag{27}$$

The mean square radius of the baryon number distribution becomes

$$\langle r_B^2 \rangle = 4\pi \int_0^\infty dr\, r^4\, B_0(r) \simeq 0.25 \text{ fm}^2 . \tag{28}$$

The size scale $\langle r_B^2 \rangle \simeq 1/2$ fm is typical for this model. We will see later how the explicit presence of vector mesons connects the small baryon size with the much larger electromagnetic radii.

5.2 Collective Quantization

The soliton described so far has neither good spin $\vec{J}$ nor good isospin $\vec{I}$: only the sum of both is a conserved quantity with $[\vec{I}+\vec{J},U] = 0$. In particular, the hedgehog configuration represents a superposition of baryon states with $I = J$. The degeneracy of the states with $I = J = 1/2; 3/2; \ldots$ (the nucleon, the Δ isobar,...) can be removed by an adiabatic isospin rotation

$$A(t) = \exp[i\vec{\tau}\cdot\vec{\Omega}t] \tag{29}$$

with a (small) rotational frequency $|\vec{\Omega}|$. The adiabatic rotation of the meson fields introduces the time component of the ρ meson and the space components of the ω meson, which are not present at the classical level. The time-dependent Lagrange function becomes

$$L(t) = -M_0 + \theta\, \text{tr}(\dot{A}\dot{A}^+), \tag{30}$$

where M_0 is the classical soliton mass and θ is the moment of inertia which is explicitly given in ref./25/ as a function of the classical profiles F, G, ω and the excitations introduced by the adiabatic rotation. The new mass spectrum is

$$M = M_0 + \frac{I(I+1)}{2\theta} \tag{31}$$

with J = I. The Δ-N mass splitting is $M_\Delta - M_N = 3/2$ and turns out to be about 350 MeV in this model. The soliton mass ($M_0 \simeq 1.5$ GeV) is on the large side, but one expects that quantum fluctuations (soft-pion corrections) are likely to bring this mass down by several hundred MeV /26/.

6. ELECTROMAGNETIC AND AXIAL PROPERTIES OF THE NUCLEON

6.1 Electromagnetic Currents and Form Factors

Let us now examine the isoscalar and isovector currents of the nucleon in more detail, starting from the elctromagnetic couplings derived in section 4.5. The isoscalar current is just the $U(1)_V$ symmetry current, and its previously established VMD structure holds also for the nucleon current. The isovector current becomes

$$\vec{J}^\mu_{I=1} = i(f^2_\pi/4) \, tr[\vec{\tau}(U^+\partial^\mu U - \partial^\mu U U^+) - 4\vec{\tau}(\xi^+\partial^\mu\xi - \partial^\mu\xi\xi^+) \\ + 2ig\vec{\tau}(\xi^+\vec{\tau}\cdot\vec{\rho}^\mu\xi + \xi\vec{\tau}\cdot\vec{\rho}^\mu\xi^+)] \tag{32}$$

plus an additional term which couples the ω field to $\vec{\tau}$ times a combination of $U^+\partial^\mu U$ in second order /25/. Given these currents, one can explicitly calculate the nucleon electric and magnetic form factors given in the Breit frame (in which the photon has no energy transfer) as

$$<N_f(\vec{q}/2)|J_0(0)|N_i(-\vec{q}/2)> = G_E(\vec{q}^2) \, x^+_f x_i \, , \\ <N_f(\vec{q}/2)|\vec{J}(0)|N_i(-\vec{q}/2)> = (G_M(\vec{q}^2)/2M_N) \, x^+_f i\vec{\sigma}\times\vec{q}x_i \, , \tag{33}$$

where $|N(\vec{p})>$ denotes nucleon states with momentum $\vec{p}$; here $x_{i,f}$ are two-component Pauli spinors, $\vec{q}$ is the momentum transfer. The isoscalar and isovector form factors are related to the ones for proton and neutron by

$$G^{p,n}_{E,M} = G^S_{E,M} \pm G^V_{E,M} \, ,$$

with the normalizations $G^p_E(0) = 1$, $G^n_E = 0$, $G^p_M(0) = \mu_p = 2.79$, $G^n_M(0) = \mu_n = -1.91$.

6.2 Results: Nucleon Electromagnetic Sizes

We have already emphasized the smallness of the size related to the baryon number distribution inside the nucleon, $<r^2_B> \simeq 0.5$ fm. Vector Meson Dominance relates this size to the radius of the isoscalar charge distribution probed by the photon. In fact VMD implies that the isoscalar photon sees the ω meson content of the soliton, rather than the baryon density $B_0(r)$. The static ω meson field equation is

$$(\vec{\nabla}^2 - m^2_\omega)\omega(r) = (3g/2)B_0(r) = -(3g/4\pi^2 r^2)F'(r) \sin^2 F(r) \tag{34}$$

(omitting the πρω-coupling term which is irrelevant for this discussion). Following eq.(24), the isoscalar charge form factor becomes

$$G^S_E(q^2) = -(m^2_\omega/3g)\int d^3r \, e^{i\vec{q}\cdot\vec{r}} \, \omega(r) = \frac{1}{2} \frac{m^2_\omega}{m^2_\omega+\vec{q}^2} \, 4\pi \int_0^\infty dr \, r^2 j_0(qr) \, B_0(r). \tag{35}$$

It is given here in two equivalent ways: either directly in terms of $\omega(r)$, the meson profile of the nucleon, or in terms of the baryon density distribution $B_0(r)$, but then multiplied by the ω meson propagator $(\vec{q}^2 + m^2_\omega)^{-1}$. These two forms are related by the field equation (34). It is now evident that the isoscalar charge radius $\sqrt{<r^2_E>}$ is significantly larger than the r.m.s. radius of the baryon density. One finds

$$<r^2_E> = -12(dG^S_E(\vec{q}^2)/d\vec{q}^2)_{\vec{q}^2=0} = (6/m^2_\omega) + <r^2_B> \simeq (0.8 \text{ fm})^2 \, . \tag{36}$$

It is instructive to visualize this significant difference between charge and baryon radius by the corresponding r-space distributions, see Fig. 4.

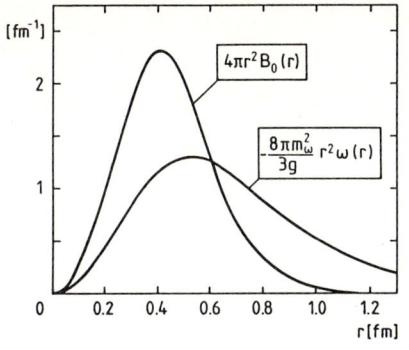

Figure 4: Isoscalar charge and baryon number distributions of the nucleon multiplied by $4\pi r^2$ (taken from ref./25/)

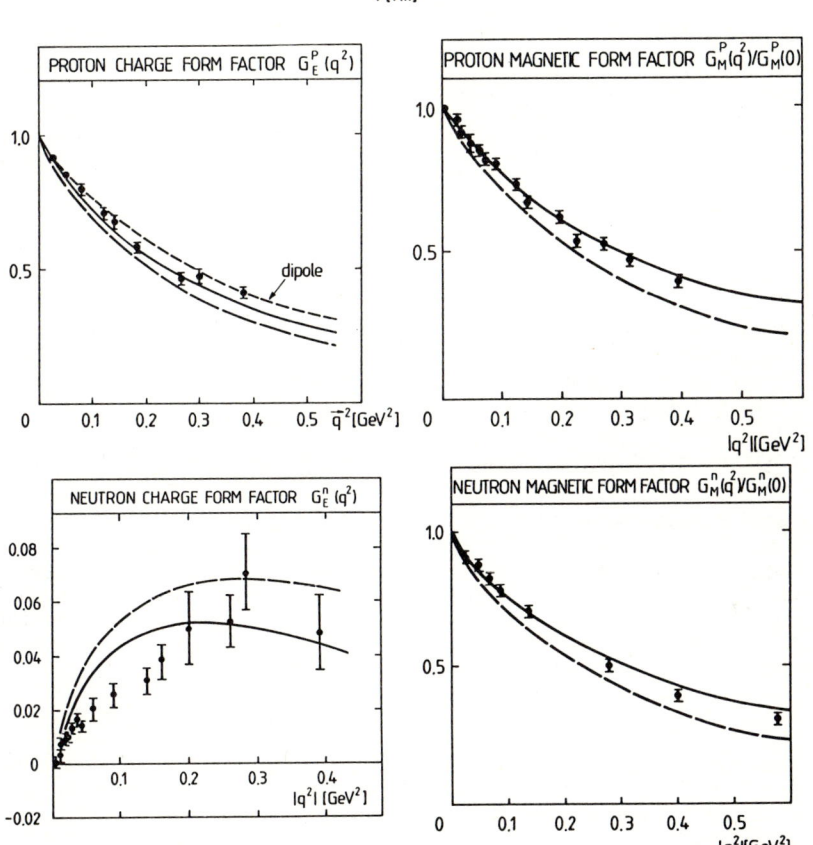

Figure 5: Proton and neutron charge and magnetic form factors. The solid line is the result of the "minimal" model; the dashed line shows the effect of additional $\pi\rho\omega$ couplings. For the proton charge form factors, the canonical dipole fit $(1 + \vec{q}^2/0.71 \text{ GeV}^2)^{-2}$ is shown for comparison.

The isovector electromagnetic form factors have a similar leading $(\vec{q}^2 + m_\rho^2)^{-1}$ dependence: the isovector photon sees primarily the ρ meson component of the $B = 1$ soliton. However, there are deviations from exact VMD in this model at distances smaller than about $\sqrt{6}/m_\rho$. They are related to the non-abelian structure of the ρ meson couplings, as a consequence of which the isovector source function in the ρ meson field equations depends on the ρ field itself. We summarize in Fig.5 the results/25/ for the nucleon charge and magnetic form factors. The overall agreement with the low $\vec{q}^2$ behaviour of the empirical form factors is indeed quite satisfactory. Magnetic moments and electromagnetic radii for the "minimal" and "complete" model are compared in table 1. Note that the $\pi\rho\omega$ coupling plays an important role in improving the proton and neutron magnetic moments.

Table 1: Static electromagnetic properties of the nucleon (from ref./25/)

Quantity	Minimal Model	Complete Model	Experiment
$\langle r_E^2 \rangle_p^{1/2}$ (fm)	0.93	0.97	0.86 ± 0.01
$\langle r_E^2 \rangle_n$ (fm^2)	-0.22	-0.25	-0.119 ± 0.004
$\langle r_M^2 \rangle_p^{1/2}$ (fm)	0.84	0.94	0.86 ± 0.06
$\langle r_M^2 \rangle_n^{1/2}$ (fm)	0.85	0.94	0.88 ± 0.07
μ_p	3.36	2.77	2.79
μ_n	-2.57	-1.84	-1.91
$\lvert \mu_p/\mu_n \rvert$	1.31	1.51	1.46

6.3 The Axial Form Factor

The axial current $A_\mu^a(x)$ is of special interest and importance in models based on the underlying Chiral Symmetry of QCD: it is a conserved current in the chiral limit $m_\pi \to 0$. The Breit frame matrix elements of $\vec{A}^a(x)$ define the axial and induced pseudoscalar form factors $G_A(\vec{q}^2)$ and $G_P(\vec{q}^2)$ as follows:

$$\langle N_f(\vec{q}/2) | \vec{A}^a(0) | N_i(-\vec{q}/2) \rangle = \chi_f^+ \left[\frac{E}{M_N} G_A(\vec{q}^2) \vec{\sigma}_T + (G_A(\vec{q}^2) - \frac{q^2}{4M_N^2} G_P(\vec{q}^2)) \vec{\sigma}_L \right] \frac{\tau^a}{2} \chi_i , \quad (37)$$

where $\vec{\sigma}_T = \vec{\sigma} - \vec{\sigma}_L$ and $\vec{\sigma}_L = \hat{q}(\vec{\sigma}\cdot\hat{q})$ are the transverse and longitudinal spin operators, M_N is the nucleon mass, $\vec{q}$ is the momentum transfer and $E = \sqrt{M_N^2 + \vec{q}^2/4}$. There are no second-class currents in this scheme, so that matrix elements of the time component of the axial current vanish in the Breit frame. The detailed derivation of the axial current can be found in ref./25/. PCAC relates the induced pseudoscalar form factor $G_P(\vec{q}^2)$ to the pion-nucleon form factor. The result for $G_A(\vec{q}^2)$ is shown in Fig. 6 together with the available data. The empirical mean square axial radius

$$\langle r_A^2 \rangle = -(6/g_A) (dG_A(\vec{q}^2)/d\vec{q}^2)_{\vec{q}^2=0} \quad (38)$$

is quite well reproduced, as demonstrated in table 2. The axial vector coupling constant $g_A = G_A(0)$ reaches a value close to unity in the complete model. Here the inclusion of the $\pi\rho\omega$ coupling term derived from the Wess - Zumino action gives an important improvement over previously calculated values of g_A in the simple Skyrme model /27/, which come out systematically too low (see table 2).

Table 2: Axial properties of the nucleon in the minimal and complete model (without and with $\pi\rho\omega$ coupling term; see ref./25/). The results obtained in the conventional Skyrme model /27/ are also shown.

Quantity	Minimal Model	Complete Model	Conventional Skyrme Model	Exp.
g_A	0.87	0.99	0.65	1.25 ± 0.01
$\langle r_A^2 \rangle^{1/2}$ (fm)	0.64	0.62	0.35	0.68 ± 0.02

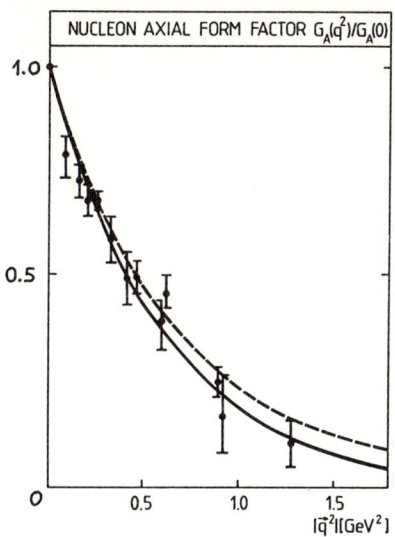

Figure 6: Nucleon axial form factor calculated in the "minimal" (solid line) and "complete" model (dashed line); taken from ref./25/

7. CONCLUDING REMARKS

A significant observation is the hierarchy of nucleon sizes which differ according to the field by which they are probed: the radius $\sqrt{\langle r_B^2 \rangle} \simeq 0.5$ fm of the baryon number density inside the nucleon is smaller than the axial radius $\sqrt{\langle r_A^2 \rangle} = (0.6 - 0.7)$ fm, which is in turn smaller than the typical electromagnetic radii $\sqrt{\langle r_{E,M}^2 \rangle} \simeq 0.85$ fm. Hence a small baryon radius of about 1/2 fm is consistent with considerably larger proton charge and magnetic radii. What has been achieved at this point is a unification of Chiral Symmetry and Current Algebra with the Vector Meson Dominance idea and with the Soliton concept of Skyrme and Witten.

From a nuclear physics point of view, the picture that has emerged is quite satisfactory: the small baryon size is consistent with the observation that nuclei at low energy are composites of nucleons interacting through their meson fields, rather than a soup of quarks. Even at the classical level, non-linear meson theory already permits to establish interesting connections with the one-boson exchange phenomenology of the nucleon-nucleon force. In particular, this scheme provides a consistent way to calculate meson-nucleon form factors /28/. The corresponding cutoff masses in equivalent monopole form factors are typically of order 1 GeV /28/, quite close to the one usually used as free parameters in OBE potentials. Some features of these potentials are described quite well, in particular those related to pion and rho meson exchange at distances around 1 fm and larger, such as the isovector tensor NN interaction. On the other hand, the classical omega meson field accounts for only about half of the repulsion in the central nucleon-nucleon potential. Furthermore, the intermediate range attraction related to scalar-isoscalar two-pion exchange is absent at the classical level. Efforts have to be made to incorporate this important degree of freedom into the scheme.

The author is grateful to Norbert Kaiser and Ulf-G. Meißner who have done most of the work related to this paper. Stimulating discussions with G.E. Brown, A.D. Jackson, M. Rho and A. Wirzba are also gratefully acknowledged.

References

1. R. Machleidt, Ch. Elster and K. Holinde, Phys. Reports 149, 1 (1987)
2. G.E. Brown and A.D. Jackson, The Nucleon-Nucleon Interaction, North-Holland, Amsterdam (1976);
   R. Vinh Mau, in: Mesons in Nuclei, M. Rho and D.H. Wilkinson, eds. , Vol. I, p. 151, North-Holland, Amsterdam (1979)
3. T.E.O. Ericson and W. Weise, Pions and Nuclei, Clarendon Press, Oxford (1987), in print
4. B. Frois, in: AIP Conf. Proc. 142, 416 (1986)
5. D.O. Riska, in: Mesons in Nuclei, M. Rho and D.H. Wilkinson, eds., Vol. II, p.755, and refs. therein, North-Holland, Amsterdam (1979)
6. J.F. Mathiot, Nucl. Phys. A 412, 201 (1984)
7. P.U. Sauer, private communication
8. J.M. Cavedon et al., Phys. Rev. Lett. 49, 986 (1982)
   P. Dunn et al., Phys. Rev. C 27, 71 (1983)
9. S.R. Amendolia et al., Phys. Lett. 138 B, 545 (1984), Nucl. Phys. B 277, 168 (1986)
10. J.J. Sakurai, Currents and Mesons, Univ. of Chicago Press (1969)
11. G.E. Brown, M. Rho and W. Weise, Nucl. Phys. A 454, 669 (1986)
12. J. Gasser and H. Leutwyler, Phys. Reports 87, 74 (1982)
    L.J. Reinders, H.R. Rubinstein and S. Yazaki, Phys. Reports 127, 1 (1985)
13. M.A. Shifman, A.I. Vainstein and V.I. Zakharov, Nucl. Phys. B 147, 385, 448 (1979)
14. Y. Nambu and G. Jona-Lasinio, Phys. Rev. 122, 345 (1961), 124, 246 (1961)
15. S.L. Adler and A.C. Davis, Nucl. Phys. B 244, 469 (1984)
16. V. Bernard et al., Nucl. Phys. A 412,349 (1984), A 440, 605 (1985);
    W. Weise, Nucl. Phys. A 434, 685 (1985); Int. Rev. Nucl. Phys. 1, 57 (1984), World Scientific, Singapore (1985)
17. N.I. Karchev and A.A. Slavnov, Teor. Mat. Fiz.(USSR) 65, 1099 (1986)
18. J. Wess and B. Zumino, Phys. Lett. 37 B, 95 (1971)
19. S. Adler, Phys. Rev. 117, 2426 (1969); J. Bell and R. Jackiw, Nuovo Cim. 60 A, 47 (1969)
20. I. Zahed and G.E. Brown, Phys. Reports 142, 1 (1986); U.-G. Meißner and I. Zahed, Adv. Nucl. Phys. 17, 143 (1986)
21. T.H. Skyrme, Proc. Roy. Soc. (London) 260, 127 (1961);
    A. Balachandran et al., Phys. Rev. Lett. 49, 1124 (1982);
    E. Witten, Nucl. Phys. B 223, 422 (1983)
22. M. Bando, T. Kugo, S. Uehara, K. Yamawaki and T. Yanagida, Phys. Rev. Lett. 54, 1215 (1985); M. Bando, T. Kugo and K. Yamawaki, Nucl. Phys. B 259, 493 (1985)
23. K. Kawarabayashi and M. Suzuki, Phys. Rev. Lett. 16, 255 (1966);
    Fayyazuddin and Riazuddin, Phys. Rev. 147, 1071 (1966)
24. U.-G. Meißner, N. Kaiser, A. Wirzba and W. Weise, Phys. Rev. Lett. 57, 1676 (1986)
25. U.-G. Meißner, N. Kaiser and W. Weise, Nucl. Phys. A
26. I. Zahed, A. Wirzba and U.-G. Meißner, Phys. Rev.D 33, 830 (1986);
    J.P. Blaizot, M. Chemtob and V. Pasquier, Nucl. Phys. A 451, 605 (1986)
27. G.S. Adkins and C.R. Nappi, Phys. Lett. 137 B, 251 (1984)
28. N. Kaiser, U.-G. Meißner and W. Weise, preprint TPR-87-14, submitted to Phys. Lett. B

Chiral Symmetry and Light Mesons

A. Le Yaouanc, L. Oliver, O. Pène, and J.-C. Raynal

Laboratoire de Physique Théorique et Hautes Energies,
Université de Paris, Bâtiment 211,
F-91405 Orsay Cedex, France

1. INTRODUCTION

The purpose of this lecture is to consider the relations between two concepts which play an essential role in the study of hadrons and their characteristics : the quark model and chiral symmetry. We will mainly consider the mesons formed of u and d quarks.

1.1 The Quark Model

The quark model describes hadrons as quark bound states (see Fig. 1 for the example of isovector mesons). The calculations describe rather satisfactorily the masses of hadrons with different interquark potentials. But over all, it explains all the known hadrons except a few controversial states (glueball or hybrid candidates).

In the quark model, a mass of about 300 MeV is usually associated to the quarks u and d. The π meson differs from the ρ meson only in the fact that the spins of the quarks are parallel for the ρ and anti-parallel for the π. In a first approximation, the masses of these two mesons are equal. Their mass difference is attributed to the spin-spin strength.

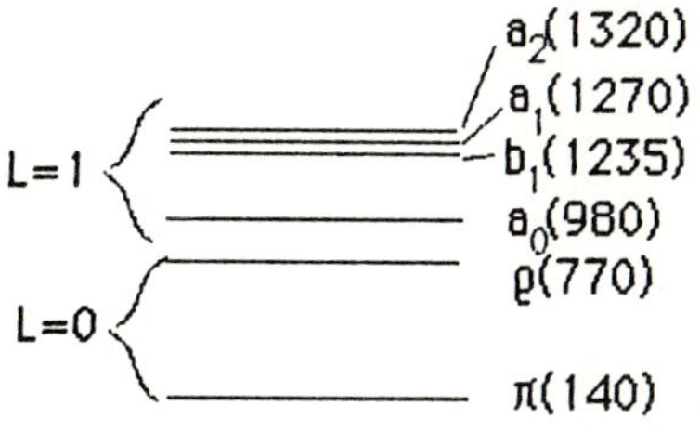

$$L=1 \left\{ \begin{array}{l} a_2(1320) \\ a_1(1270) \\ b_1(1235) \\ a_0(980) \end{array} \right.$$

$$L=0 \left\{ \begin{array}{l} \rho(770) \\ \pi(140) \end{array} \right.$$

FIG. 1

1.2 The Theory of Partial Conservation of Axial Current (PCAC)

The theory of partial conservation of the axial current (PCAC) describes the pion as a Goldstone boson, which clearly distinguishes it from the ρ. In a first approximation (exact chiral symmetry) the pion has a zero-mass, in distinction to the ρ. The "current" masses of the quarks are considered in this approach to be $m_u = 4$ MeV and $m_d = 7$ MeV. Many soft pion theorems are derived within the framework of PCAC, for example the Goldberger-Treiman theorem, the Adler condition, the Adler-Weissberg relation and in general the relations between the processes differing from each other by a pion in the initial or final state. In general, these relations are satisfied within 10 %.

What is the connection between these two approaches ? We will present a treatment of the quark model which agrees with chiral symmetry, in the spirit of the pioneering work of Nambu and Jona-Lasinio [1].

2. CHIRAL SYMMETRY

Axial (or chiral) transformations are defined by

$$q(x) \to \exp\{i(a^i \lambda^i_F + a^0)\gamma_5 / 2\} q(x) , \qquad (1)$$

where

$$q(x) = \begin{pmatrix} u(x) \\ d(x) \\ s(x) \end{pmatrix} \qquad (2)$$

a^i's, $i = 0...8$ are real parameters and λ^i_F's are the Gell-Mann matrices of the flavour space.

The key point is that the Lagrangian of QCD is approximately invariant for these transformations. This is obvious for the purely gluonic part and this is also true for

$$\bar{q}(x) (i\partial + g A^a \lambda^a_c / 2) q(x) , \qquad (3)$$

where the λ_c's are now in colour space.

Only the mass terms

$$m \bar{q}(x) q(x) \qquad (4)$$

are not invariant. If the masses are very small (4 and 7 MeV for u and d) the chiral symmetry will then be satisfied to a very good approximation.

The generators of axial transformation are the axial charges

$$Q_5^i = \int dx\, q^+(x) \gamma_5 \lambda^i / 2\, q(x) ,$$
$$Q_5 = \int dx\, q^+(x) \gamma_5\, q(x) . \qquad (5)$$

3. GOLDSTONE THEOREM

Two cases are possible:

a) The Wigner mode in which the vacuum is chirally symmetric:

$$Q_5^i |\Omega\rangle = 0 , \qquad (6)$$

where $|\Omega\rangle$ is the vacuum state.

b) The Goldstone mode, or the mode of spontaneous breaking of chiral symmetry

$$Q_5^i |\Omega\rangle \neq 0 . \qquad (7)$$

In the case a) one can predict parity doublets, i.e. the hadrons are gathered into states of opposite parity with equal masses. For instance, the pion and the $a_0(980)$ meson masses, or the ρ mass and the $a_1(1270)$, or else the nucleon mass and the N* (1535) mass would be equal. One can easily see that this is not the case, even in an approximate way.

In the case b) one predicts Goldstone bosons, that is $m_\pi = 0$. This is close to experiment since the pion is very light compared to other hadrons. We will then study the Goldstone mode. In addition to (7), it is characterized by a non-zero vacuum expectation value of $\bar{q}q$:

$$\langle\Omega| \bar{q}(0) q(0) |\Omega\rangle \neq 0 \ . \tag{8}$$

The equations (7) and (8) are related since $\bar{q}q$ is not chiral invariant:

$$[Q_5^i, \bar{q}q] = 2\bar{q}\, \gamma_5\, \lambda^i/2\, q \neq 0 \ . \tag{9}$$

One can easily visualize the Goldstone theorem from Fig. 2. At the origin of the coordinates there is a state which would be chiral symmetric. But it is not the minimum of energy. The minimum of energy is not unique because of chiral symmetry: all the chiral transformations of a minimum have the same energy. These chiral transformations form an orbit drawn in dotted lines in Fig. 2. The generators Q_5^i move the system on this orbit. It must then create zero energy excitations. We will sketch a more rigorous proof of the Goldstone theorem.

(Simplified) Proof of the Goldstone theorem

Hypothesis: Equation (8) and the relations

$$[Q_5^i, \bar{q}\lambda^i/2\, \gamma_5 q] = 1/3\, \bar{q}q + \sum_i 1/2\, d_{iij}\, \bar{q}\lambda^i/2\, q \tag{10}$$

derived from (1) hold, and

$$[Q_5^i, H] = 0 \ , \tag{11}$$

where H is the Hamiltonian. The latter equation exhibits the chiral symmetry in the dynamics of the system.

Conclusion

Let us consider the matrix element of (10) in the vacuum:

$$\langle\Omega| [Q_5^i, \bar{q}\lambda^i/2\, \gamma_5 q] |\Omega\rangle = 1/3 \langle\Omega|\bar{q}q|\Omega\rangle$$

$$= 1/3 \sum \langle\Omega|Q_5^i|n\rangle \langle n|\bar{q}\lambda^i/2\, \gamma_5 q|\Omega\rangle - \langle\Omega|\bar{q}\lambda^i/2\, \gamma_5 q|n\rangle \langle n|Q_5^i|\Omega\rangle \ , \tag{12}$$

where we have summed over a complete basis of intermediary states $|n\rangle$. The vacuum state is withdrawn from the sum since its contribution is obviously zero. Let us consider the matrix element of (11):

$$\langle\Omega| [Q_5^i, H] |n\rangle = (E_n - E_0) \langle\Omega|Q_5^i|n\rangle = 0 \ . \tag{13}$$

From (13) we can see that if a state $|n\rangle$ has a non-zero matrix element of Q_5^i with the vacuum, then it has necessarily the same energy as the $|\Omega\rangle$ vacuum.

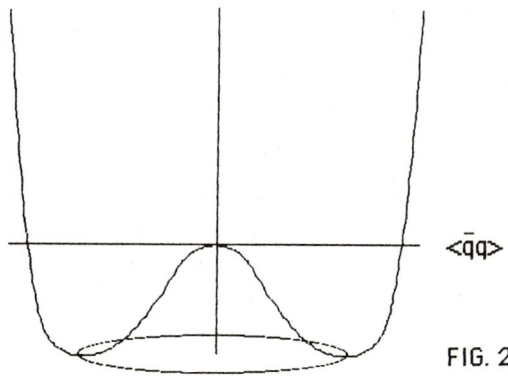

FIG. 2

From (12) it follows that there must exist at least one $|n\rangle$ such that $\langle\Omega|Q_5^i|n\rangle \neq 0$ and which therefore has the same energy as the vacuum.

These states are the Goldstone bosons at rest. From (12) it follows that these states $|n\rangle$ must also satisfy

$$\langle n| \bar{q} \lambda^i/2 \gamma_5 q |\Omega\rangle \neq 0 . \tag{14}$$

Therefore they will be pseudo-scalar mesons such as pions and kaons. As there are 8 matrices λ^i, one expects 8 light pseudoscalar mesons, and actually pions, kaons and anti-kaons and the η constitute these 8 mesons. They all have a mass lower than 500 MeV.

Remark

One may wonder why there is not a ninth Goldstone boson corresponding to the ninth charge of (5), which is a flavour singlet. It turns out that the meson η' is heavy: 958 MeV. The solution of this problem, called the U(1) problem, has been given by t'Hooft and is related to axial anomaly. We will not pursue it here.

4. SPONTANEOUS BREAKING OF CHIRAL SYMMETRY THROUGH QUARKS

From now on, we will not write the flavour indices in order not to make the notation too lengthy. We will also suppose that the quark mass is zero. A small correction due to the current masses can easily be done.

We start from the chiral invariant state $|0\rangle$ corresponding to the vacuum of a free massless quark system. It corresponds to the point at the origin in Fig. 2. The non-zero expectation value of $\bar{q}q$ shows the existence of a $\bar{q}q$ scalar condensate (the equivalent of the Cooper pairs of superconductivity) which make the "vacuum" $|0\rangle$ unstable. One can write in a simplified manner

$$\bar{q}q \sim d_L^+ b_L^+ + d_R^+ b_R^+ . \tag{15}$$

The operator (15) is invariant for rotation, P, C, and so on, but it is not chiral invariant since $b_L d_L$ decreases the chirality by 2 while $b_R^+ d_R^+$ increases it by 2. On the other hand, as

$$[Q_5, d_L^+ b_L^+ + d_R^+ b_R^+] = -2(d_L^+ b_L^+ - d_R^+ b_R^+) \sim \bar{q} \gamma_5 q \tag{16}$$

one can see that the generators Q_5^i create pseudoscalar states, pions and so on.

The dynamics of this spontaneous breaking is the following: quarks and antiquarks attract each other. If this attraction is sufficient to create a bound state of negative energy (the quarks have a zero-mass) the vacuum will become unstable. The bound states (Cooper pairs) will condense as long as the Fermi statistics does not stop the process. We will now give a more quantitative description of the phenomenon.

5. BOGOLIUBOV TRANSFORMATIONS

Our aim is to look for Cooper pair condensation, on the assumption that the state corresponding to the chiral non-invariant vacuum is one of the Bogoliubov transformations of the state $|0\rangle$, i.e. one of the states

$$|\Omega\rangle = \prod_{p,s} \frac{1 + \chi(\vec{p}) b^+(\vec{p},s) d^+(-\vec{p},s)}{\sqrt{1 + |\chi(\vec{p})|^2}} , \tag{17}$$

where $\chi(\vec{p})$ is a function of $\vec{p}$ which we will determine later on. From (17) it follows that

$$(b(\vec{p}, s) - \chi(\vec{p}) \, d^+(-\vec{p}, s)) |\Omega\rangle = 0 \, ,$$
$$(d(\vec{p}, s) - \chi(\vec{p}) \, b^+(-\vec{p}, s)) |\Omega\rangle = 0 \, . \tag{18}$$

The Bogoliubov transformations are, by definition, annihilated by a linear combination of creators and annihilators corresponding to the state $|0\rangle$.

Let us call the solutions of the massless free Dirac equation corresponding to a positive (negative) energy u_L^0 and u_R^0 (v_L^0 and v_R^0). If we express the quark fields in terms of annihilators (18) and their corresponding creators, we find the positive (negative) energy solutions corresponding to the vacuum $|\Omega\rangle$

$u_R(p) + \chi(p) \, v_R(p)$ $E > 0$

$u_R(p) - \chi(p) \, v_L(p)$

$v_L(p) - \chi(p) \, u_R(p)$ $E < 0$

$$v_R(p) + \chi(p) \, u_R(p) \, . \tag{19}$$

We can thus characterize a Bogoliubov transformation by specifying the Dirac spinors (19). It is in fact sufficient to indicate the subspaces of positive and negative energy, i.e. the projectors on these subspaces. We call the projector on the space of positive (negative) energy states $\Lambda^+(p)$ ($\Lambda^-(p)$). We have

$$\Lambda^+(p) + \Lambda^-(p) = 1 \tag{20}$$

$$(\Lambda^+(p))^2 = \Lambda^+(p) \quad \text{and} \quad (\Lambda^-(p))^2 = \Lambda^-(p) \, . \tag{21}$$

Equation (20) states that the sum of the positive and negative energy subspaces span the entire Dirac space, and (21) simply states that the $\Lambda^\pm$'s are projectors.

The projectors corresponding to the vacuum $|0\rangle$ are

$$\Lambda^{\pm 0}(p) = [1 \pm \alpha.p]/2 \tag{22}$$

and for the states $|\Omega\rangle$ defined in (17)

$$\Lambda^\pm(\vec{p}) = [1 \pm (\cos \varphi(p) \, \vec{\alpha}.\vec{p} + \sin \varphi(p) \, \beta)]/2 \, , \tag{23}$$

where $\varphi(p) = 2 \, \text{Arctan}(\chi(p))$. For comparison, let us note that the Dirac solutions for a free massive particle are

$$\Lambda^\pm(\vec{p}) = \{1 \pm (\vec{\alpha}.\vec{p} + \beta m)/(p^2+m^2)^{1/2}\}/2 \, . \tag{24}$$

One can see that the expression $\sin \varphi$ in (23) is related to the mass term of (24) and hence to the spontaneous breaking of chiral symmetry. This is confirmed by

$$[\Lambda^\pm, \gamma_5] = \sin(\varphi) \, \beta \, \gamma_5 \, . \tag{25}$$

If $\sin \varphi = 0$ the solution is chiral invariant and is reduced to the case (22) that is, to the state $|0\rangle$. In the opposite case, $\sin \varphi \neq 0$, we have a spontaneously created mass: although the initial Hamiltonian has no mass, one can see that the particles corresponding to the vacuum $|\Omega\rangle$ have an effective mass. This mass is a gap between the states of negative and positive energy, which is similar to the gap between the quasi-particles in superconductivity.

The problem is now the following : which state, among all those considered above, will have the lowest energy and is therefore the best candidate for the physical vacuum ? The solution suggested by experiment is a chiral symmetry breaking solution. This would explain at the same time the low pion mass and the constituent mass of 300 MeV which the quark model attributes to the quarks u and d, whereas it is granted that in the QCD Lagrangian the quark masses are low. The quarks of the quark model would be the quasi-particles obtained by a Bogoliubov transformation from the Lagrangian quarks. We can now consider the dynamics.

6. A DYNAMICAL MODEL

We start from the Hamiltonian [2]

$$H = \sum_{\vec{x}} q^+(\vec{x})(-i\vec{\alpha}.\vec{\nabla}) q(\vec{x})$$

$$+ 1/2 \sum_{\substack{\vec{x},\vec{y} \\ a=1,8}} v(\vec{x}-\vec{y})(q^+(\vec{x})\lambda^a/2\, q(\vec{x}))(q^+(\vec{y})\lambda^a/2\, q(\vec{y})) \quad . \tag{26}$$

One can easily check that this Hamiltonian is chiral invariant. This is an effective Hamiltonian between quarks. The potential is presumed to reproduce the effect of the gluons. One will take a confining potential

$$v(x-y) = c|x-y|^\alpha \quad , \qquad \alpha > 0 \quad . \tag{27}$$

The potential (26) is coupled to the colour charge density which results in confining colour singlet mesons and baryons and only them. The problem is : Is there a spontaneous breaking of chiral symmetry ? In other words, which is, among the Bogoliubov transformations (17) the state of minimum energy ? We must point out that this criterion is limited since there are a large number of field configurations which are far more complex than the Bogoliubov transformations and which may have a lower energy. Our hope is that this real vacuum is not very different from a Bogoliubov state.

Let us compute the energy density in a state $|\Omega\rangle$ given by

$$E = (1/V)\langle\Omega|H|\Omega\rangle = -6\sum_{\vec{k}} |\vec{k}| \cos\varphi(k) + (2/V)\sum_{\vec{k},\vec{k}'} v(\vec{k}-\vec{k}') \times$$

$$\times (1 - \hat{k}.\hat{k}' \cos\varphi(k')\cos\varphi(k') - \sin\varphi(k)\sin\varphi(k')) \quad , \tag{28}$$

where V represents the space volume.

7. GAP EQUATION

We search for the extrema of energy density, i.e. the states $|\Omega\rangle$ (parametrized by the function $\varphi(k)$) such that

$$\delta E/\delta\varphi(k) = 0 \quad . \tag{29}$$

This equation is called the "gap equation" by analogy with superconductivity. Of course the solution of this equation is not necessarily the minimum. It can be either a relative maximum or a relative minimum. Next we will have to find the minimum among the solutions of (29). From (23) and (26) the gap equation is written

$$k\sin(\varphi(k)) = (2/3V)\sum_{\vec{k}'} v(\vec{k}-\vec{k}')\{\sin(\varphi(k'))\cos(\varphi(k)) - (\hat{k}.\hat{k}')\cos(\varphi(k'))\cos(\varphi(k))\} \quad .$$

$$\tag{30}$$

We can also understand (30) as an Hartree-Fock equation : one computes the effect on the quarks of the Cooper pair condensates (this effect is expressed in relation to the projectors Λ^+ (Λ^-)), one then deduces the Dirac equation to which the quarks are subject and one imposes that the solutions of this Dirac equation precisely determine the subspaces of positive (negative) energy corresponding to Λ^+(Λ^-). One thus exactly recovers equation (30).

It can easily be seen that the function $\varphi = 0$ is always a solution of (30), which means that the vacuum $|0\rangle$ is always an extremum of the energy density. Indeed, in the case where $|0\rangle$ is not the minimum, it is a local maximum (cf. Fig. 2). We can now try to determine whether $|0\rangle$ is a minimum or whether there exists a direction in the functional space near $|0\rangle$ in which the energy decreases. In other words, is the state $|0\rangle$ stable ?

8. THE EQUATION OF INSTABILITY

The energy density is calculated as a functional of $\varphi(k)$ near $\varphi(k) = 0$ to the second order in $\varphi(k)$. The result is

$$\Delta E = 12 \sum_k \varphi^+(k) [k + \frac{2}{3V} \sum V(\vec{k}-\vec{k}')\hat{k}.\hat{k}'] \varphi(k)$$

$$- \frac{8}{V} \sum_{k,k'} \varphi^+(k) V(\vec{k}-\vec{k}') \varphi(k') \quad . \tag{31}$$

The analysis of (31) shows that the term proportional to k corresponds to the kinetic energy of the quarks, the term $\sim V(\vec{k}-\vec{k}')\hat{k}.\hat{k}'$ corresponds to their self-energy and the last term $\sim V(\vec{k}-\vec{k}')$ corresponds to the potential energy.

<u>Theorem</u> : For any potential r^α $\alpha > 0$ the vacuum $|0\rangle$ is unstable.

The main line of the proof [2] is the following. We notice that the kinetic energy is positive. But one can make its contribution negligible by expanding a trial function since the kinetic energy decreases when the radius of a wave function increases. The potential energy term is positive for a potential in r^α. The self-energy term is negative since it is given by the potential term of the Hamiltonian (26) which depends on colour : this term gives a positive potential between a quark and an antiquark of opposite colour and it gives a negative potential between a quark and itself, since they both have the same colour. The proof ends by demonstrating that the negative contribution of the self-energy wins upon the positive contribution of the potential energy.

9 GAP EQUATION FOR $V(r) \propto r^2$

The Fourier transform is simple :

$$V(k) \sim \Delta_K \delta(k) \quad . \tag{32}$$

The gap equation (30) becomes

$$\frac{d}{dK} (k^2 \frac{d\varphi(k)}{dK}) = 2 k^3 \sin \varphi(k) - \sin (2 \varphi(k)) \quad . \tag{33}$$

We have chosen the energy units in terms of the coefficient multiplying r^2 in the potential in order to simplify the expression (33).

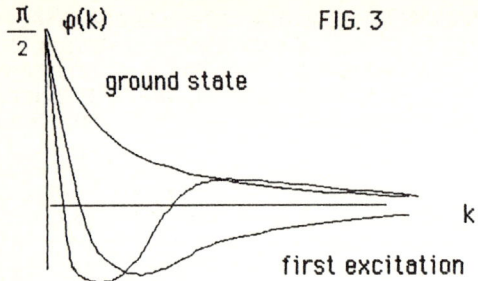

FIG. 3

The solutions of this equation are shown in Fig. 3. There exist an infinite number of solutions corresponding to local minima (these solutions are similar to the spectrum of a hydrogen atom). The solution corresponding to the minimum vacuum energy is the one having no nodes. Note that $\varphi(0) = \pi/2$ for all the solutions. This is a general result for any solution breaking chiral symmetry : it means that

$$\Lambda^+(0) = 1/2\,[1 + \gamma_0] \;. \tag{34}$$

Once the solution is found, we can calculate the expectation value of $\bar{q}q$

$$<\bar{q}(0)\,q(0)> = -\frac{3}{\pi^2} \int k^2\,dk\,\sin\varphi(k) \;. \tag{35}$$

10. LIGHT MESONS SPECTRUM

We solve the Salpeter equation [3]. The wave function $\chi_p(\vec{k})$ is a matrix in the Dirac space and p represents the total momentum of the meson. The wave function of zero momentum satisfies the equation

$$H(\vec{k})\chi(\vec{k}) - \chi(\vec{k})H(\vec{k}) - \frac{4}{3V}\sum_{\vec{k}'} V(\vec{k}-\vec{k}')\,\{\Lambda^+(\vec{k})\chi(\vec{k}')\,\Lambda^-(\vec{k})$$

$$- \Lambda^-(\vec{k})\chi(\vec{k}')\,\Lambda^+(\vec{k})\} = \omega\chi(\vec{k}) \;,$$

$$H(\vec{k}) = \vec{\alpha}\cdot\vec{k} + \frac{4}{3V}\frac{1}{2}\sum_{\vec{k}'} V(\vec{k}-\vec{k}')(\Lambda^+(\vec{k}') - \Lambda^-(\vec{k}')) \;, \tag{36}$$

where the first term represents the quark energy, the second term the antiquark energy, the term containing $V(\vec{k}-\vec{k}')$ represents the binding energy, and ω is the total energy of the meson.

There always exists a solution of zero energy corresponding to the Goldstone boson at rest :

$$\chi_0(k) = i[\Lambda^-(k),\,\gamma_5] = i\,\sin\varphi(k)\,\gamma_5\gamma_0 \;. \tag{37}$$

The fact that $\chi_0(k)$ is solution of (36) is derived from the fact that $\sin\varphi(k)$ is a solution of the gap equation. We use the fact that the energy density of the state $|\Omega>$ is invariant for the chiral transformed states of $|\Omega>$. The chiral transformation acts on the projectors $\Lambda^+(k)$ by the transformation

$$e^{i\theta\gamma_5}\,\Lambda^+(k)\,e^{-i\theta\gamma_5} \;. \tag{38}$$

By differentiating with respect to θ and using the invariance of the energy density one gets

$$[\gamma_5, H(k)] - \frac{4}{3V} \sum_{k'} V(k-k') [\gamma_5, \Lambda^-(k')] = 0 \qquad (39)$$

and therefore

$$[[\gamma_5, H(k)], \Lambda^-(k)] - \frac{4}{3V} \sum_{k'} V(k-k') [[\gamma_5, \Lambda^-(k)], \Lambda^-(k)] \, , \qquad (40)$$

which leads to (36) by using the Jacobi identity with $\omega = 0$ for the wave function (37). The pion thus described is not a simple $q\bar{q}$ state : it is a collective state [4].

The other mesons can be deduced only by a numerical calculation. The results are shown in Fig. 4. As can be seen, the agreement is semi-quantitative. The sequence of the states is maintained (the small mass of the pion would be easy to find out by attributing a small current mass to the quarks). The spacing of the levels a_0, a_1, b_1, a_2 is excessive compared to experiment. This is undoubtedly related to the fact that a vector potential (26) was chosen and the phenomenological studies proved that a vector potential gives a much too large LS spacing. The scalar potential is known to give a better result. But the scalar potential is not chiral invariant and cannot be used in the problem under study.

Once we have determined the energy scale (i.e. the constant multiplying the r^2 potential) by the mesonic spectrum, we can calculate other constants. We find

$f_\pi = 20$ MeV   (experiment : 95 MeV)

$<\bar{q}(0)\,q(0)> = -(120 \text{ MeV})^3$ (experiment : $-(250 \text{ MeV})^3$)

$m_q = 121$ MeV   (experiment : 300 MeV) .   (41)

The predicted difference of energy density with the vacuum $|0>$ is -15 MeV/fm

the velocity of the pion is

$$c_\pi = 3.1 \qquad (42)$$

in units of the velocity of light, whereas we expect 1 for a massless particle : this gives an estimate of the non-covariance of our model. These results are clearly not in agreement with experiment. One can think that this disagreement is due to the harmonic potential used in (32). Nevertheless, the same calculation by Adler and Davis [5] with a linear potential gives values for f_π, m_q and $\bar{q}q$ even less accurate (these authors have not calculated the meson spectrum).

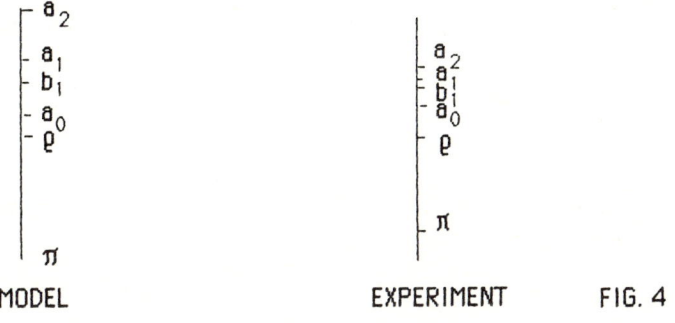

FIG. 4

On the other hand, the PCAC theorems are automatically verified by this model. For instance

$$-2m_q \langle \bar{q}(0) q(0) \rangle = m_\pi^2 f_\pi^2 \ . \tag{43}$$

The latter result is achieved by attributing a small current mass to the quarks.

11 CONCLUSION

We have proposed an approach which reconciles the quark model and chiral symmetry (PCAC).

A large class of confining potentials implies spontaneous breaking of chiral symmetry.

Pions are Goldstone bosons.

The states of the quark model are recovered, but the mass predictions are modified as wanted by PCAC.

The constituent quark mass is created by the spontaneous breaking of chiral symmetry.

For the harmonic potential, the agreement with the meson mass spectrum is semi-quantitative.

There are discrepancies with experiment of a factor of 3 for f_π, $\langle \bar{q}q \rangle$, m_q.

The model is non-covariant.

The PCAC theorems are verified.

The same techniques can be applied with similar results to QCD inspired models on a lattice [6].

REFERENCES

1. Y. Nambu and G. Jona-Lasinio, Phys. Rev. <u>122</u>, 345 (1961) ; <u>124</u>, 246 (1961).
2. A. Le Yaouanc, L. Oliver, O. Pène and J.-C. Raynal, Phys. Rev. <u>D29</u>, 1233 (1984).
3. A. Le Yaouanc, L. Oliver, S. Ono, O. Pène and J.-C. Raynal, Phys. Rev. <u>D31</u>, 137 (1985).
4. V. Bernard, R. Brockmann, W. Weise, Nucl. Phys. <u>A440</u>, 605 (1985).
5. S.L. Adler and A.C. Davis, Nucl. Phys. <u>B224</u>, 469 (1984).
6. A. Le Yaouanc, L. Oliver, O. Pène and J.-C. Raynal, Phys. Rev. <u>D33</u>, 3098 (1986) ; Phys. Letters <u>182</u>, 365 (1986).

Chiral Field Theories as Models for Hadron Substructure

S.H. Kahana

Physics Department, Brookhaven National Laboratory,
Upton, NY 11973, USA

A model for the nucleon as soliton of quarks interacting with classical meson fields is described. The theory, based on the linear sigma model, is renormalizable and capable of including sea quarks straightforwardly. Application to nuclear matter is made in a Wigner-Seitz approximation.

1. THE BASIC MODEL: VALENCE QUARK SOLITONS

Several field theoretic models of the substructure of hadrons now exist, some more phenomenological than others, but most based on some effective Lagrangian. The MIT bag model [1] and the soliton model of FRIEDBERG-LEE [2,3] are the antecedents, but most of the more modern theories [4-8] incorporate the chiral symmetry which is so dominantly a part of low energy particle and nuclear physics. I wish to discuss one version of these effective field theories pursued with George RIPKA [6,9], which allows naturally for the possibility of sea quarks in the nucleon substructure. We begin with the Lagrangian for the linear sigma model [10]

$$\mathcal{L} = \bar{\psi}\,[i\gamma\partial - g(\sigma + i\gamma_5\,\vec{\tau}\cdot\vec{\pi})]\psi + \frac{1}{2}\partial_\mu\sigma\,\partial^\mu\sigma + \frac{1}{2}\partial_\mu\vec{\pi}\cdot\partial^\mu\vec{\pi} - F(\sigma,\vec{\pi})$$

$$F(\sigma,\vec{\pi}) = \frac{\alpha}{4!}(\sigma^2 + \pi^2 - f_\pi^2)^2 \,, \tag{1}$$

where $\psi, \bar{\psi}, \sigma, \vec{\pi}$ are quark scalar and pseudoscalar meson fields.

Baryons appear in this model as static solutions of the hamiltonian

$$H = \int d^3r\,\{\psi^+[\vec{\alpha}\cdot\vec{p} + g\beta(\sigma + i\gamma_5\,\vec{\pi}\cdot\vec{\tau})]\psi + \frac{1}{2}[(\nabla\sigma)^2 + (\nabla\vec{\pi})^2] + F(\sigma,\vec{\pi})\} \tag{2}$$

If $\sigma, \vec{\pi}$ assume classical values through coherent states,

$$\phi^a(x)|\phi_{c1}^a\rangle = \phi_{c1}^a(x)|\phi^a\rangle \,, \quad \phi^a \equiv [\sigma, \pi^i] \,,$$

then the hamiltonian can be diagonalized by seeking solutions of

$$h|\lambda\rangle = e_\lambda|\lambda\rangle$$
$$h = \vec{\alpha}\cdot\vec{p} + g\beta(\sigma + i\gamma_5\,\vec{\pi}\cdot\vec{\tau}) \,. \tag{3}$$

The Dirac sea is constructed by filling all of the negative energy states

$$|D\rangle = \prod_{e_\lambda < 0} a_\lambda^+ |0\rangle \,, \tag{4}$$

and the total energy, with some valence states occupied, can be split into a Dirac and a meson part

$$E = E_D + E_M$$

$$E_D = \Sigma_\lambda e_\lambda$$

$$E_M = \int d^3r \, \frac{1}{2} \left[(\nabla\sigma)^2 + (\nabla\vec{\pi})^2 + F(\sigma,\vec{\pi}) \right] . \tag{5}$$

Self-consistent solutions can be obtained by maximizing E with respect to the scalar fields $\sigma, \vec{\pi}$. The potential function $F(\sigma,\vec{\pi})$ is minimized by remaining on the "chiral circle"

$$\phi^a \phi_a = \sigma^2 + \pi^2 = f_\pi^2 ,$$

and such a choice is tantamount to the limit of large α. We work near such a limit in which case the chiral fields are collectively described by the unitary operator

$$U = \frac{1}{f_\pi} (\sigma + i\gamma_5 \vec{\tau}\cdot\vec{\pi}) = e^{i\gamma_5 \vec{\tau}\cdot\vec{\theta}(\vec{r})} . \tag{6}$$

However, a lowest energy is obtained for the hedgehog configuration [4,11]

$$\vec{\theta}(r) = \frac{\vec{r}}{r} \theta(r) . \tag{7}$$

To properly solve the classical equations of motion and achieve a finite total energy, one must impose

$$\theta \approx \frac{1}{r^2} \to 0 \quad \text{as } r \to \infty \tag{8a}$$

$$\theta \approx n\pi , \quad r = 0 . \tag{8b}$$

The first condition must be altered appropriately for a finite π-meson mass and implies the symmetry-breaking, but translationally invariant vacuum described by $\sigma = f_\pi$, $\vec{\pi}=0$ (that is for $\theta(r) \equiv 0$) is achieved at large distance. In this latter vacuum the quark energies are given by

$$\varepsilon(k) = \sqrt{k^2 + m^2} , \tag{9}$$

and the states are solutions of

$$h_0 = \vec{\alpha}\cdot\vec{p} + \beta m .$$

Chiral symmetry is broken at the minimum in $F(\phi)$, and the fermions acquire the mass $m = gf_\pi$ there. States $\{\lambda\}$ are labelled by the overall spin obtained by combining the broken spin and isospin operators

$$\vec{G} = \vec{J} + \vec{T} . \tag{10}$$

The single fermion spectra obtained for a chiral angle

$$\theta = n\pi e^{-r/R} \tag{11}$$

are displayed in Fig.1 for the "winding number", $n = 1,2$. A localized solution of the system of $\{(2),(3),(7),(11)\}$ exists as soon as the energy ε_0 in Fig.1 drops below gf_π. Baryon numbers 1,2 are associated with $n = 1,2$ through

$$N_B = n \, N_c \left(\frac{1}{N_c}\right) = n \, 3\left(\frac{1}{3}\right) , \quad N_c \text{ the number of colours,}$$

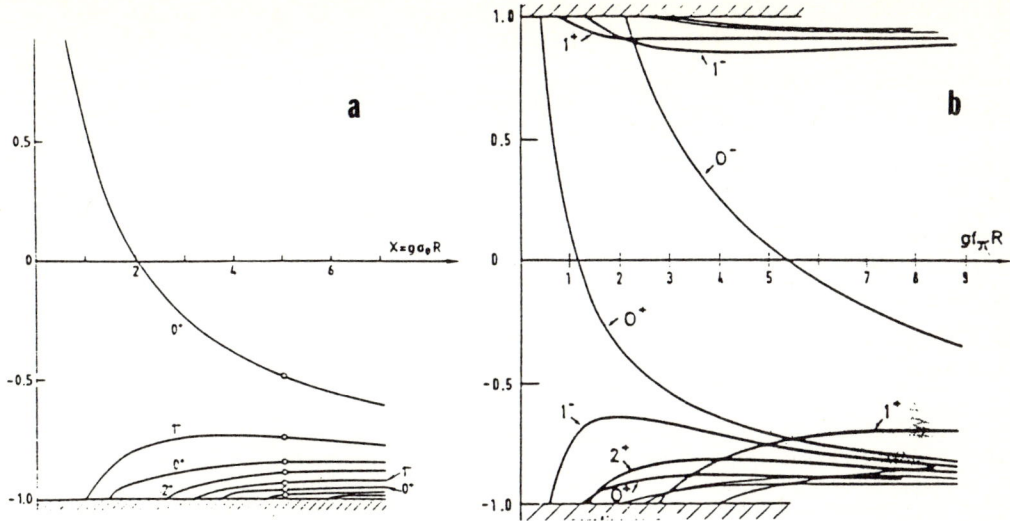

Figure 1

The single fermion spectrum for two cases in which (a) a single $G=0^+$ level with energy ε_0 drops to negative energies or (b) both a $G=0^+$ and a 0^- level drop into the Dirac sea. These correspond to winding number n=1,2 in Eq.(11). For small enough X these valence levels remain at positive energy.

for filling of the n lowest lying valence levels. For sufficiently large value of the dimensionless coupling or size parameter $X = gf_\pi R$, first a $G=0^+$ and then a $G=0^-$ level actually pass through zero energy joining the Dirac sea.

A soliton, constructed from valence quarks only, has an energy approximately given by $E = E_{valence} + E_{meson}$,

$$\frac{E}{gf_\pi} = N_c \left(\frac{a}{X} - b\right) + \frac{4\pi^2 X}{6g^2}, \qquad (12)$$

with, for example, a = 3.12, b = 0.93 for $\theta=\pi(1 - r/R)$ [6] and slightly different values for θ given by (11). Minimizing (12) to obtain the stable soliton size R yields $E_{soliton} = N_c f_\pi [10.6 - 0.94g]$. The stability is destroyed for sufficiently large coupling g = 10.6/0.94, a situation which persists when non-valence quarks are included. Of course, the soliton constructed in this way has broken isospin and is properly an intrinsic state for both the nuclear and delta. Fixing f_π = 93 MeV to the π-meson decay constant and the total soliton mass to an average between m_N and m_Δ yields a radius ≈ 0.6 fm, some 30% too small.

2. SEA QUARKS AND THE CASIMIR ENERGY

The fermion part of the vacuum energy of our soliton, i.e. the Casimir energy, is more properly evaluated relative to the translationally invariant vacuum [12]

$$E_D = \sum_{\lambda<0} \varepsilon_\lambda - \sum_{k<0} \varepsilon(k)$$
$$= \int_{-i\infty}^{i\infty} \frac{d\omega}{2\pi i} \, \text{tr} \, \ln\left[1 + \frac{1}{-\gamma p + m} V_F\right], \quad V_F = \left[g(\sigma + i\gamma_5 \vec{\tau}\cdot\vec{\pi}) - gf_\pi\right]. \qquad (13)$$

Equation (13), when expanded in the fermion perturbation, $V_F = h-h_0$, constitutes the so-called quark one-loop approximation. To this latter energy we could also add the valence energies and obtain the total baryon energy, except in the case for $n = 1$ with $\varepsilon(0^+)<0$, when (13) already includes the valence energy.

As it stands (13) is divergent and must be renormalized. An alternative form for (13) can be written in terms of the one-fermion hamiltonian [13,16]

$$E_D = -\frac{1}{2} \text{tr}\left(\sqrt{h^2} - h\right) - \frac{1}{2} \text{tr}\left(\sqrt{h_0^2}\right) = -\frac{1}{2} \int_{-i\infty}^{i\infty} \frac{d\omega}{2\pi i} \text{tr} \ln(1 + GV), \tag{14}$$

where the squared hamiltonians are given by

$$h^2 = -\nabla^2 + g^2(\sigma^2 + \vec{\pi}^2) + ig\vec{\gamma} \cdot [\nabla\sigma + i\gamma_5 \vec{\tau} \cdot \nabla\vec{\pi}]$$

$$= h_0^2 + V = (-\nabla^2 + g^2 f_\pi^2) + V, \quad G = (h_0^2 + \omega^2)^{-1}$$

$$V = g^2(\sigma^2 + \vec{\pi}^2 - f_\pi^2) + ig f_\pi \vec{\gamma} \cdot \nabla U. \tag{15}$$

On the chiral circle the perturbation potential in (14) and (15) is a function only of field derivatives, leading to the derivative expansion of the energy and other observables [7,8,13,15,16]. The quark current operator is the simplest such operator expressible in these terms, and is given in the lowest order in these derivatives by

$$J^\mu(x) = \langle\bar{\psi} \gamma^\mu \psi\rangle \approx \frac{1}{2\pi^2} \varepsilon^{\mu\alpha\beta\gamma} \varepsilon_{dabc} \phi^d \partial_\alpha \phi^a \partial_\beta \phi^b \partial_\gamma \phi^c. \tag{16}$$

This yields the following approximate relation for the baryon density [15]

$$J^0(x) = -\frac{1}{2\pi r^2} \sin^2\theta \frac{d\theta}{dr}, \tag{17}$$

and the baryon number

$$N_B = \int 4\pi r^2 J_0(x) dr = \frac{1}{\pi}\left[\theta - \frac{1}{2}\sin 2\theta\right]_0^\infty = n, \tag{18}$$

relating our model density to the anomalous density of SKYRME [11].

These approximate relationships become exact in the limit $X = gf_\pi R$. Figure 2 shows the actual densities arising from both volume and sea quarks and a comparison of the summed densities to the right-hand side of (17), i.e. to the baryon density in a Skyrmion [11,14,15]. This comparison of the derivative expanded chiral theory [7,8,16] to the Skyrme theories is extended further in the following. The contribution of sea quarks to baryon structure can clearly be appreciable. The perturbation expansion of (13) can be used to define the necessary counterterms to render E_D finite. The first two terms are in fact quadratically and logarithmically divergent in terms of a cutoff momentum and yield as counterterms

$$\Delta E_D^{(1)} = -\frac{1}{2} \int \frac{d\omega}{2\pi i} \text{tr } GV = -\frac{g^2}{4} \frac{2\nu}{\Omega} \langle\sigma^2+\pi^2-f_\pi^2\rangle \sum_k \frac{1}{\varepsilon(k)}, \tag{19a}$$

$$\Delta E_D^{(2)} = \frac{1}{16} \langle g^4(\sigma^2+\vec{\pi}^2-f_\pi^2)^2 + g^2[(\nabla\sigma)^2+(\nabla\vec{\pi})^2]\rangle \frac{2\nu}{\Omega} \sum_k \frac{1}{\varepsilon(k)^3}, \tag{19b}$$

where ν is the total Coulomb spin-isospin degeneracy of each quark level, and $\frac{1}{\Omega}\langle\rangle \equiv \frac{1}{\Omega} \int d^3\underline{r}$. These counterterms are clearly proportional to field quantities already defined in the Lagrangian (1). Thus the renormalized Dirac energy is expressible as

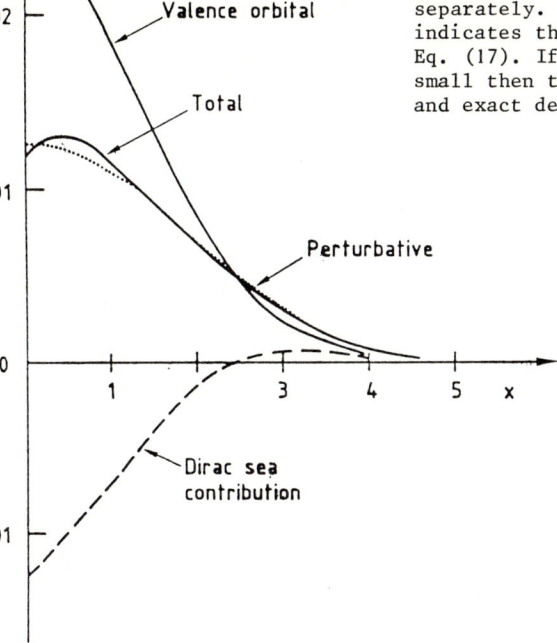

Figure 2

The quark density distribution in the baryon treated as soliton. The valence orbital, Dirac sea, and total densities are shown separately. For comparison, the dotted line indicates the anomalous density given in Eq. (17). If the soliton were sufficiently small then the deviation between perturbative and exact densities would be larger.

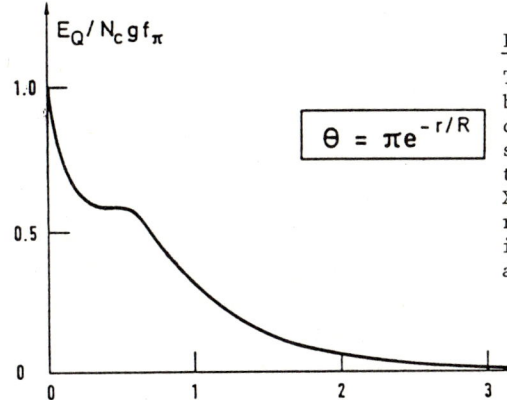

Figure 3

The renormalized Dirac energy including both valence and sea contributions as calculated "exactly" from (20) by actually summing over eigenvalues of h. The calculation cannot be carried to arbitrarily large $X=gf_\pi R$. However, the derivative expansion result of (21) then applies, and the curve in this figure will drop slightly below the axis.

$$E_D^R = -\frac{1}{2} \text{tr} \left(\sqrt{h_0^2+V} - \sqrt{h_0^2}\right) + \frac{1}{4} \text{tr } V \frac{1}{|h_0|} - \frac{1}{16} \text{tr } V^2 \frac{1}{|h_0^3|} \; . \tag{20}$$

Fig. 3 indicates the results of a detailed non-perturbative calculation of the spectrum [13,16].

3. STRONG COUPLING LIMIT: DERIVATIVE EXPANSION

The use of a derivative expansion for "smooth" slowly varying chiral fields $\phi^a(x)$ and also of the limit $g \to \infty$ (or $gf_\pi R \to \infty$) have been referred to above. These limits are in effect equivalent and lead to several interesting results [13]. First, the gap between occupied negative energy levels and unoccupied valence orbits in Fig. 1 will in this limit become extremely large. In such a limit then, the present theory acquires a form of confinement. Clearly any nondegenerate level at negative energy not completely occupied by N_c quarks shall possess infinite energy in the limit $g \to \infty$.

The strong coupling limit can then be shown, by integrating out all fermion variables, to have a combined meson plus Dirac energy

$$E_{TOT} = \frac{1}{2} f_\pi^2 \langle \phi_i^2 \rangle + \frac{\nu N_c}{384 \pi^2} \langle 3 \phi_i^2 \phi_j^2 - 2(\phi_i \cdot \phi_j)^2 - 2 \phi_{ij}^2 \rangle , \qquad (21)$$

with $\phi_i \cdot \phi_j = (\partial_i \phi^a)(\partial_j \phi^a)$ etc.

Two comments can be made about this limit: first, the quartic terms in (21) differ from those in the Skyrme theory; second, the theory is unstable. This can be demonstrated by explicit evaluation of the total meson (first term) plus Dirac (second term) energy in (21) using (11) for $\Theta(r)$:

$$E = 30.82 \, f_\pi^2 R - \frac{0.017}{R} . \qquad (22)$$

The total Dirac mass acquired in symmetry breaking is more than cancelled by the binding to mesons yielding the above negative second term, which in turn yields $E \to -\infty$ as $R \to 0$.

A stable theory can be recovered by including in this strong coupling limit a vector meson with energy

$$E_{vector} = \frac{1}{2} \frac{g_\omega^2}{2\pi} \int d^3 r_1 d^3 r_2 \, n_B(\vec{r}_1) \frac{e^{-m_\omega r_{12}}}{r_{12}} n_B(\vec{r}_2) \approx \frac{g_\omega^2}{m_\omega^2 R^3} . \qquad (23)$$

4. APPLICATIONS

4.1 Phenomenology

The baryon phenomenology of the model presented above can be obtained only after introducing some form of symmetry restoration. D. E. KAHANA, G. RIPKA and A. JACKSON [18] have put forward an isospin cranking picture which allows the proper definition of physical variables. In this approach, for example, a quark moment of inertia,

$$I_q = 2/3 \sum_{p,h} \frac{\langle h|\vec{t}|p \rangle \cdot \langle p|\vec{t}|h \rangle}{\epsilon_p - \epsilon_n} , \qquad (24a)$$

is introduced and added to the meson moment of inertia to yield a rotational spectrum for the nucleon and delta,

$$E_T = E_0 + \frac{T(T+1)}{2(I_{meson} + I_q)} . \qquad (24b)$$

4.2 Nuclear Matter

A second application can be made to nuclear matter [19], treating the latter as a solid or gas of solitons. A very simple approach making use of the derivative expansion, and performing the many-body calculation in a Wigner-Seitz approximation has been done. Such a limiting calculation possesses the virtue of an underlying confinement for the quarks in individual baryons.

In place of the chiral angle in (11), one uses

$$\theta(r) = \pi\, e^{-r/R_0} - \pi\, \frac{r}{R}\, e^{-R/R_0} \; , \tag{25}$$

where the soliton size parameter is R_0, and the average Wigner-Seitz cell radius R is related to the nuclear density

$$\rho = \frac{4}{4\pi/3\; R^3} \; .$$

At the edge of the cell $\theta(r) = 0$ in (25), guaranteeing that a single baryon is contained in each cell. By using (21) and (22) one arrives at a soliton energy, in matter, of the form

$$M_{soliton} = A\, R_0 - \frac{B}{R_0(R)} + \frac{C}{R_0^3(R)} \; ,$$

and on reminimizing obtains a relation between soliton sizes at zero and finite density

$$R_0(R) = R_0(\infty)\left(1 + \eta(R)\right) \; , \quad R_0(\infty) = R_0 \; .$$

For low density $R = 3R_0$, $\eta = -0.03$, and for near nuclear density $R \approx 2R_0$, $\eta = -0.05$. This apparent decrease in soliton size on being immersed in nuclear matter is obtained leaving out the usual decrease in f_π found in other models, but is a warning that the usual "swelling" need not obtain. The EMC effect can then be explained, for example, by allowing for an effective rest mass perhaps 5% reduced from the normal value [19].

More recent work [20] has suggested that the $N_B = 0$ sector of the above theories is unstable in the limit in which boson (quantum) fluctuations are omitted. Such a situation follows from the nature of the counterterms in (11), and may force one to use models such as defined here as the only effective low energy theories. Efforts to understand this interesting point are under way.

ACKNOWLEDGEMENT

This manuscript has been authored under Contract No. DE-AC02-76CH00016 with the U. S. Department of Energy.

REFERENCES

1. A. Chodos, R.L. Jaffe, K. Johnson, D.B. Thorn and V.F. Weisskopf: Phys.Rev. D9 (1974) 3471
2. R. Friedberg and T.D. Lee: Phys. Rev. D15, 1694,(1977); D16, 1096 (1977)
3. R. Goldflam and L. Wilets: Phys. Rev. D25, 1951 (1982)
4. G.E. Brown, A.D. Jackson, M. Rho, and V. Vento: Phys. Lett. B140, 285 (1984)
5. H.C. Birse and M.K. Banerjee: Phys. Lett B136, 284 (1984)
6. S. Kahana, G. Ripka, and V. Soni: Nucl. Phys. A415, 351 (1984)
7. I.J.R. Aitchison and C. M. Fraser: Phys. Lett. B146, 63 (1984)

8. R. MacKenzie: Santa Barbara preprint NSF-ITP-84-135 (1984)
9. S. Kahana and G. Ripka: Nucl. Phys. A429, 462 (1984)
10. M. Gell-Mann and M. Levy: Nuovo Cim. 16, 705 (1960)
11. T.H.R. Skyrme: Nucl. Phys. 31, 556 (1962); Journal Math. Phys. 12, 1735 (1971)
12. S. Kahana and G. Ripka: In AIP Conf. Proc. No. 110 (1984)
13. S. Kahana, R. Perry, and G. Ripka: Phys. Lett. B163, 37 (1985)
14. G.S. Adkins, C.R. Nappi, and E. Witten: Nucl. Phys. B228, 552 (1983)
15. J. Goldstone and F. Wilczek: Phys. Rev. Lett. 47, 986 (1981)
16. G. Ripka and S. Kahana: Phys. Lett. B155, 327(1985)
17. G.S. Adkins, C.R. Nappi, and E. Witten: Nucl. Phys. B228, 552 (1983)
18. D.E. Kahana, A.D. Jackson, and G. Ripka: Nucl. Phys. A459, 663 (1986)
19. S. Kahana: In Proceedings of Bates Users' Workshop, MIT, August 1985; and In Hadronic Probes and Interactions, AIP Conference Proceedings 133, 222-240 (1985)
20. G. Ripka and S. Kahana: Saclay preprint

Strange Skyrmions

M. Praszałowicz

Institute of Physics, Jagellonian University,
Reymonta 4, PL-30-059 Kraków, Poland

The purpose of this talk is to discuss some difficulties encountered in extending an SU(2) Skyrme model [1,2] to three flavors [3]. We shall take a phenomenological point of view, being mostly interested in quantitative predictions. Viewed from this perspective the three flavor model does not reproduce the data [4,5].

The link between QCD and the model is the chiral symmetry group G (in our case G = SU(3)×SU(3)) broken by the QCD vacuum to SU(3)$_V$. The effective lagrangian can be only guessed. In our discussion we restrict ourselves to the minimal lagrangian proposed by Skyrme [1] a long time ago:

$$L = \frac{F_\pi^2}{16}\int d^3x \; \text{Tr}(\partial_\mu U \partial^\mu U^+) + \frac{1}{32e^2}\int d^3x \; \text{Tr}([\partial_\mu U U^+, \partial_\nu U U^+]^2) + N_c \Gamma_{wz}. \quad (1)$$

Here U is an SU(3) matrix, N_c number of colors. F_π (186 MeV) and e (4-6) are treated as free parameters. The second term in (1) (Skyrme term) was added by Skyrme [1] in order to stabilize the soliton, the third one (Wess-Zumino term) originates from topology [6] and its coefficient is bound to be N_c. The form of the soliton solution takes a form [3-6]

$$U_o(x) = \text{diag}\,(\hat{U}_o, 1). \quad (2)$$

$\hat{U}_o$ is an SU(2) matrix:

$$\hat{U}_o = \exp\,\{i\,\vec{\tau}\cdot\hat{n}\,f(r)\}, \quad (3)$$

where $\hat{n}$ is a unit radial vector and f(r) is a function of r only. f(0)=π and f(∞)=0 [4] guarantee that suitably defined baryon number [1,2,6] is 1.

The global symmetry of the energy, which leaves vacuum (i.e. U_o = 1) invariant, is given by

$$U_o \Rightarrow A\,U_o A^+, \quad (4)$$

where A $\in$ SU(3)/U(1), since [λ_8, U_o] = 0.

In order to quantize the model, one has to construct a hamiltonian in dynamical variables, which are introduced by making A time dependent. The symmetry of such zero modes is given by [3,4]

$$g_L A(t) g_R^+, \text{ with } g_L \in SU(3) \text{ and } g_R \in SU(2) \times U(1). \tag{5}$$

The right U(1) factor results in a constraint [3]

$$Y_R = N_C/3. \tag{6}$$

The hamiltonian describing the baryon masses takes a form

$$H_0 = M_{CL} + \frac{C_2(SU(2)_L)}{2 I_B} + \frac{1}{2 I_A}\left[C_2(SU(3)_R) - \frac{N_C^2}{12} - C_2(SU(2)_L)\right], \tag{7}$$

where M_{CL}, I_A and I_B are known funtionals of f [3] and the wave function of a baryon state in (p,q) representation of SU(3) is given by

$$\psi = \langle (p,q), Y, I, I_3 | A | (p,q), Y_R, J, -J_3 \rangle, \tag{8}$$

where I and J denote isospin and spin respectively. The constraint (7) selects the representations of triality zero:

$$8, 10, \overline{10}, 27, 35, \overline{35}, 64, \ldots \tag{9}$$

for $N_C=3$. The success of the model is the prediction that the lowest baryonic states belong to the Octet and Decuplet representations of SU(3). However for large N_C only high representations appear and the lowest states do not belong to 8 or 10.

Let us observe that the first two terms in Eq.(7) coincide with the SU(2) formula (see Ref.[2]), so the term in square brackets comes entirely from the strange zero modes.

Supplementing the lagrangian of Eq.(1) by the symmetry-breaking terms [3]

$$L_{BR} = a \int d^3x \, Tr(U+U^+ - 2) + b \int d^3x \, Tr(\lambda_8(U+U^+ - 2)), \tag{10}$$

and parametrizing A in the following way [7,8]:

$$A = e^{-i\alpha_1 \frac{\lambda_3}{2}} e^{-i\alpha_2 \frac{\lambda_2}{2}} e^{-i\alpha_3 \frac{\lambda_3}{2}} e^{-i\beta\lambda_4} e^{-i\gamma_1 \frac{\lambda_3}{2}} e^{-i\gamma_2 \frac{\lambda_2}{2}} e^{-i\gamma_3 \frac{\lambda_3}{2}}, \tag{11}$$

one gets the following hamiltonian [4,7]:

$$H_1 = g \, (1 - \langle (1,1), 0,0,0 | A | (1,1), 0,0,0 \rangle) = \frac{3}{4} g \sin^2 \beta, \tag{12}$$

where $g \propto \frac{m_K^2}{F_\pi^2 e^2}$ (for $m_\pi = 0$). For small g the energy splittings are proportional to the averages of H_1 between the baryonic states [3,4]. Fitting F_π and e to baryon masses [4] we obtain the spectrum of Fig.1a (light line) with $F_\pi = 46$ MeV and e=5.1. We can of course do better by treating all constants in (7) and (12) as free parameters. The resulting spectrum is depicted in Fig.1b (dashed line). Similarly one can calculate baryonic magnetic moments [9].

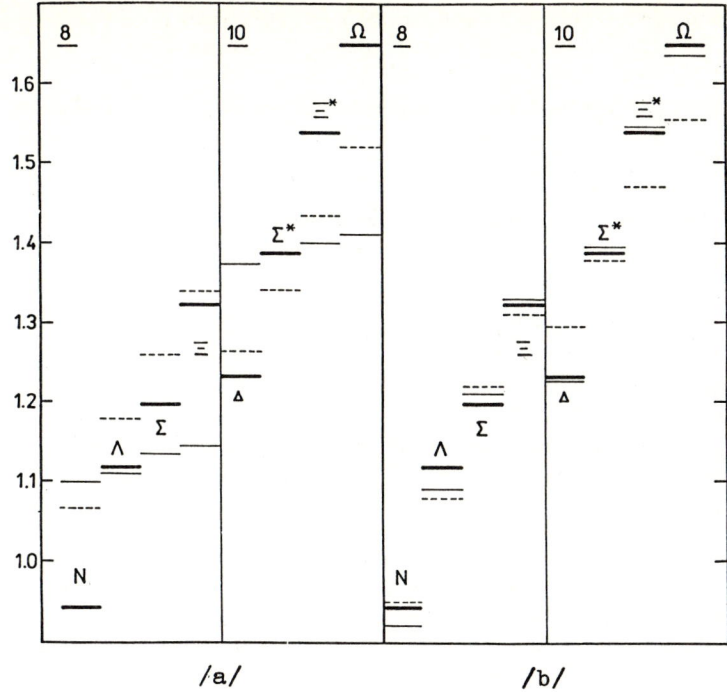

Fig.1 Baryon mass spectrum (solid lines) compared with the Skyrme model predictions:
/a/ perturbative fit - light lines - with F_π = 46 MeV and "exact" fit - dashed lines - with F_π = 86 MeV
/b/ 3-parameter fit - dashed lines and 4-parameter fit - light lines

The second order correction reads

$$E_2^B = \sum_{R'}{}' \frac{|\langle R',B| H_1 |R,B\rangle|^2}{M_B^{R'} - M_B^R} \approx \sum_{R'}{}' \frac{|\langle R',B| H_1 |R,B\rangle|^2}{M^{R'} - M^R}, \quad (13)$$

where R' denotes the SU(3) representations which mix with R. The approximation used in the last part of Eq.(13) consists in replacing the mass difference of the states which actually mix by a difference of the average masses of the representations R and R'. This introduces a new parameter [10]. The results are plotted in Fig.1b (light line).

The new prediction of the model is the existence of exotic states belonging to $\overline{10}$, 27 and higher representations of SU(3). For example the lightest $\overline{10}$ state ($M^{\overline{10}}$=1706 MeV) of spin 1/2 and strangeness +1 has a mass of the order of 1530 MeV.

In fact the hamiltonian (12) breaks the left SU(3) of Eq.(5) to SU(2)×U(1). The correct quantization procedure would consist in treating the strange degrees of freedom as nonzero modes. This leads to the problem of diagonalizing the following equation:

$$\{C_2(SU(3)_L)+2I_A H_1\}\psi=\{C_2(SU(3)_L)+\omega^2\sin^2\beta\}\psi=\varepsilon\psi. \quad (14)$$

This can be done numerically [7,10].

Before evaluating the baryon masses let us examine the limit of infinitely heavy Kaon. Yabu and Ando [7] investigate a harmonic approximation to (14). Their result reads

$$\varepsilon=\omega(|s|+2)-\frac{3}{4}+\frac{1}{4}(3s-2|s|)+\frac{s^2}{8}+\frac{1}{2}[I(I+1)+J(J+1)]+O\left(\frac{1}{\omega}\right), \quad (15)$$

where s denotes strangeness. The energy ε_o of a state with all quantum numbers equal zero is still nonzero. If we subtract ε_o from the baryon energy (as a sort of vacuum fluctuation) the result is very impressive: the Nucleon and Delta energies have the following limit for large ω:

$$E_{N,\Delta} \xrightarrow[m_K\to\infty]{} M_{CL}+\frac{1}{2I_B} C_2(SU(2)). \quad (16)$$

So in the large m_K limit we recover the SU(2) mass formula (see Eq.(7)), i.e. the strange zero modes are frozen as one would expect. The baryon spectrum obtained in that way is plotted in Fig.1a (dashed line). Yabu and Ando claim that F_π=89 MeV and e=3.87. Their result, although much better than the perturbative one, is still not satisfactory.

There is still one more approach to the SU(3) Skyrmion, where the symmetry is broken by the Ansatz itself [11]. The picture which emerges is that the Hyperon is a bound state of a soliton and a Kaon. Although theoretically very appealing this picture seems to fail phenomenologically [12].

It is a pleasure to thank M.A. Nowak and M. Jeżabek for stimulating discussions and their help in preparing this talk. I wish to thank the organizers for the invitation to this very stimulating meeting.

References
1. T.H.R. Skyrme, Proc. Royal Soc. A260 (1961) 127, Nucl. Phys. 31 (1962) 556
2. G.S. Adkins, C.R. Nappi, E. Witten, Nucl. Phys. B228 (1983) 552
3. M.A. Nowak, P.O. Mazur, M. Praszałowicz, Phys. Lett. 147B (1984) 137;
   S. Jain, S.R. Wadia, Tata Inst. Preprint, TIFR-TH-84-7, 1984;
   E. Guadagnini, Nucl. Phys. B236 (1984) 35
4. M. Praszałowicz, Phys. Lett. 158B (1985) 264

5. M. Chemtob, Nucl. Phys. B256 (1885) 214
6. E. Witten, Nucl. Phys. B223 (1983) 422, 433
7. H. Yabu, K. Ando, Kyoto preprint, KUNS 851, 1987
8. D.F. Holland, J. Math. Phys. 10 (1969) 531;
   T.J. Nelson, J. Math. Phys. 8 (1967) 857
9. C.R. Nappi, Proceedings of the Symposium on Anomalies, Geometry and Topology, 1985;
   E. Guadagnini, Pisa preprint, IFUP TH 10/84
10. M. Praszałowicz, Proceedings of Kraków Workshop on Skyrmions and Anomalies, World Scientific, preprint TPJU 5/87
11. C.G. Callan, I. Klebanov, Nucl. Phys. B262 (1985) 365
12. M.A. Nowak, Proceedings of Kraków Workshop on Skyrmions and Anomalies, World Scientific 1987

Diquarks in Exclusive Reactions

P. Kroll

Department of Physics, University of Wuppertal,
P.O. Box 100127, D-5600 Wuppertal 1, Fed. Rep. of Germany

1. INTRODUCTION

Inclusive reactions at large transverse momenta are in general well understood in terms of hard scatterings between elementary constituents in the framework of perturbative QCD. When applying this hard scattering scheme to exclusive reactions at large momentum transfer one runs into many difficulties. In the case of hadron-hadron scattering $AB \to CD$ one can only account for two features [1]. The first is the power law dependence of the cross-sections at fixed c.m.s. angle θ

$$d\sigma/dt(AB \to CD) = f(\theta)[\alpha_S(s)/s]^{n_A+n_B+n_C+n_D-2}, \qquad (1.1)$$

where n_I is the minimum number of constituents in hadron I. $\alpha_S(Q^2) = 12\pi/(25ln(Q^2/\Lambda^2))$ is the QCD running coupling constant. This power law behaviour seems to be in fair agreement with data. The second result is the helicity sum rule

$$\lambda_A + \lambda_B = \lambda_C + \lambda_D, \qquad (1.2)$$

which, however, is violated by all available experimental data. The large single spin asymmetry in elastic proton-proton scattering observed at Brookhaven [2] is a particularly spectacular example of such a violation, which has been estimated to be of the order of 20-25% [3].

Since the helicity sum rule is a consequence of perturbative QCD, namely the interaction of vector gluons with (almost) massless spin-half quarks, one clearly needs some non-perturbative effects in order to allow for the required helicity flips. In a recent paper [4] ANSELMINO, PIRE and myself have proposed a model in which such higher twist contributions are due to diquarks which may exist inside baryons as quasi-elementary constituents taking part in the hard scattering at intermediate values of momentum transfer. Asymptotically, of course, the diquarks dissolve into quarks and the usual perturbative picture emerges again. In the following I am going to report on this work as well as on some further applications of the diquark model [5]. For details one is referred to these papers.

From different experimental and theoretical approaches there have been many indications suggesting the presence, inside baryons, of diquarks. They were first introduced a long time ago in hadron spectroscopy as an intermediate step in the building up of baryons out of three quarks. More recently they have been used in nuclear physics. Dynamical calculations starting from non-relativistic potentials have shown that indeed inside baryons two quarks seem to cluster in configuration space whereas the third one is rather well separated from them [6].

Even more important for our aim, diquarks have also been found to play a rôle in inclusive hard scattering reactions. The most obvious place to signal their presence is deep inelastic lepton-hadron scattering. Indeed the combined SLAC and EMC structure function data show a need for higher twist terms because QCD evolution alone cannot account for the observed scaling violation at $x \geq 0.4$ [7]. If this higher twist term is modelled by lepton-diquark elastic scattering the structure function may be written as

$$F_2(x, Q^2) = F_2^{LT}(x, Q^2) + B(x) \left(\frac{1}{1 + Q^2/Q_0^2} \right)^n, \qquad (1.3)$$

where the first term on the right hand side is the standard leading twist contribution with a logarithmic Q^2 dependence. Using the EMC parametrization of the leading twist term, one finds a value of 3 - 4 GeV^2 for the diquark form factor parameter Q_0^2. Other authors have advocated somewhat different diquark models. Thus, for example, DONNACHIE and LANDSHOFF [8] assumed the diquark to break off when it is hit by the virtual photon. The diquark form factor in (1.3) appears only linearly in this case and a fit to the structure function data provides $Q_0^2 \approx 1 GeV^2$. The Stockholm group [9] takes the rather extreme view that the scaling violations are entirely due to diquarks, a point of view we do not share.

We note in passing that the diquarks do not lead to contradictions in e^+e^- annihilation. For $\sqrt{s} \geq 10 GeV$ the form factors (appearing quadratically in the cross sections) suppress the diquark contributions to $R = \sigma(e^+e^- \to \text{hadrons})/\sigma(e^+e^- \to \mu^+\mu^-)$ to such a degree that the usual explanation of R in terms of quark contributions is preserved.

2. DIQUARKS

In order to build up $SU(6)$ baryon wave functions one may first couple two quarks together in a diquark and then couple the diquark to the third quark. With zero orbital angular momentum between the two quarks forming a diquark only spin $0(S)$ and spin $1(V)$ diquarks appear. In terms of colour one has to consider antitriplet states since only these can form together with a quark an ordinary baryon.

In terms of diquarks a baryon wave function reads

$$\psi_{B\lambda} = \frac{1}{\sqrt{3}} \sum_{ij\mu\mu'\alpha\bar{\alpha}} C_{ij}^{\mu\mu'}(B\lambda) \delta_{\alpha\bar{\alpha}} \quad | \quad D_{i\mu\bar{\alpha}} q_{j\mu'\alpha} >; \qquad (2.1)$$

λ is the 3-component of the baryon spin, $q_{j\mu\alpha}$ represents a quark of flavour j, spin μ and colour α and, correspondingly, $D_{i\mu\alpha}$ a diquark.

As an example the proton wave function is given explicitly (colour indices are omitted for convenience)

$$\psi_{p+} = \tfrac{1}{\sqrt{2}} \left\{ \tfrac{1}{3}[V_0(ud)u_+ - \sqrt{2}V_0(uu)d_+ \quad -\sqrt{2}V_1(ud)u_- + 2V_1(uu)d_-]\frac{\phi_V(x)}{\sqrt{x}} \right.$$

$$\left. + S(ud)u_+ \frac{\phi_S(x)}{\sqrt{x}} \right\}, \qquad (2.2)$$

where the longitudinal momentum wave function ϕ_V, for vector diquarks may differ from that of the scalar ones, ϕ_S. This is a particular way to introduce some $SU(6)$ breaking. For $\phi_V = \phi_S$, (2.2) represents the exact $SU(6)$ wave function of the proton. The new point is now that the diquarks are viewed as bound states of two quarks which act as quasi-elementary constituents of the baryons. This assumption cannot be proven because this would require QCD to be solved non-perturbatively. However, as was discussed in the introduction there is overwhelming phenomenological support for diquarks. Thus it seems worthwhile to investigate the rôle diquarks may play in exclusive reactions.

The coupling of photons and gluons into diquarks follows standard prescriptions. For gluons one has

$$SgS \quad -iG_S\lambda^\alpha/2(q_1+q_2)^\mu \qquad (2.3)$$

$$VgV \quad -i\lambda^\alpha/2[G_1(q_1+q_2)^\mu g^{\kappa\nu} - G_2(q_2^\kappa g^{\mu\nu} + q_1^\nu g^{\mu\kappa}) + G_3(q_1+q_2)^\mu q_1^\nu q_2^\kappa].$$

The λ's are the usual Gell-Mann colour matrices; with the omission of these, (2.3) holds also for photons. The four couplings G_S, G_1, G_2 and G_3 are in fact form factors depending on the momentum transfer Q^2. For intermediate values of Q^2 the diquarks should resemble proper elementary particles. This requirement together with that of gauge invariance for reactions like $Sg \to Sg$ or $Vg \to Vg$ leads to ($g_S = \sqrt{4\pi\alpha_S}$)

$$G_S = g_S F_S(Q^2) \qquad G_1 = G_2 = g_S F_V(Q^2) \gg Q^2 G_3. \qquad (2.4)$$

The asymptotic behaviour of the diquark form factors follows from perturbative QCD [1]

$$G_S \sim 1/Q^2 \qquad G_1, G_2 \sim 1/Q^4 \qquad G_3 \sim 1/Q^6. \qquad (2.5)$$

For most of the applications at intermediate values of Q^2 the following parametrization is a suitable combination of the idea of quasi-elementary diquarks, (2.4), with the asymptotic pure quark picture, (2.5),

$$F_S(Q^2) = \frac{\alpha_S(Q^2)Q_0^2}{Q_0^2+Q^2}; \qquad F_V(Q^2) = \frac{\alpha_S(Q^2)Q_1^2}{Q_1^2+Q^2} \qquad \lambda_1 = \lambda_2 = 0$$

$$\tilde{F}_V(Q^2) = \frac{Q_2^2}{Q_2^2+Q^2} F_V(Q^2) \quad \text{otherwise.}$$

(2.6)

This parametrization will be used for the coupling to photons as well as to gluons.

3. THE GENERALIZED HARD SCATTERING SCHEME

Let us look at baryon-baryon scattering. In the scheme of Ref.[1] one has to consider all diagrams in which the six valence quark lines are connected by 5 gluons. Without explicit calculation which is beyond feasibility at present because of the numerous diagrams contributing ($\approx 10^5$), the power law (1.1)

$$d\sigma/dt \sim (\alpha_s(s)/s)^{10} \qquad (3.1)$$

and the helicity sum rule (1.2) are obtained.

Treating a baryon as a bound state of a quark and a diquark essentially amounts to a reduction of the effective number of constituents; only 3 instead of 5 gluons are needed to connect the diagrams. Consequently, the power of 10 is altered to

$$d\sigma/dt \sim (\alpha_S(s)/s)^6 F^4(s). \tag{3.2}$$

$F(s)$ is generic for the diquark form factors. Due to their behaviour (2.6), (3.2) turns asymptotically into the pure quark result. Yet for intermediate values of s (3.2) gives quite a different effective power. Indeed, as can be seen from Fig.1, for $pp \to pp$ scattering the transition from (3.2) to (3.1) is compatible with the data.

Furthermore, the possibility of helicity flips at VgV vertices allows all single and double flip amplitudes to be non-zero. Hence, the helicity sum rule (1.2) is violated at intermediate values of s.

An explicit calculation of baryon-baryon scattering in the scheme of Ref.[1] generalized to diquarks seems feasible. The introduction of diquarks reduces the computational effort enormously, only about 100 diagrams have to be calculated. The idea of diquarks, of course, could also play a role in many other processes. The simplest place to look at is the electromagnetic form factor of the nucleon. This will be discussed in the next section. Also simple are the two photon processes, like $\gamma\gamma \to p\bar{p}$ or Compton scattering. Work is in progress [11].

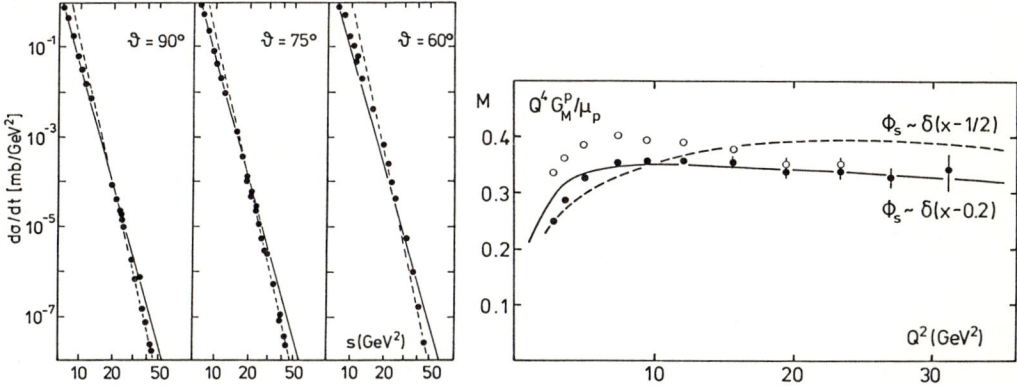

Fig.1 : (left) Differential cross-sections for elastic pp scattering vs s at fixed angles. Data from [10]. Solid (dashed) line: power laws $s^{-6}(s^{-10})$.

Fig.2 : (right) $Q^4 G_M/\mu_p$ as a function of Q^2. Data taken from [12] using $G_E = G_M(\bullet)$ and (4.4)(∘), respectively. The line is obtained with $Q_0^2 = 3.2 GeV^2$

4. ELECTROMAGNETIC FORM FACTORS OF THE NUCLEON

As usual the $\gamma^* p \to p$ vertex is analyzed in terms of the form factors G_M and F_2

$$I^\mu_{\lambda_2 \lambda_1} = ie_0 \bar{u}_p(q_2, \lambda_2)[\gamma^\mu G_M(Q^2) - \frac{\kappa}{2m_p}(q_1 + q_2)^\mu F_2(Q^2)] u_p(q_1, \lambda_1), \tag{4.1}$$

where κ is the anomalous magnetic moment of the proton and m_p the proton mass. According to Ref.[1] this vertex may be expressed by

$$I^\mu_{\lambda_2\lambda_1} = \sum_{\{\lambda\},\alpha,\beta} \int dx\, dy \psi^*_{\lambda_2}(y) T^{\alpha\beta\mu}_{\{\lambda\}}(x,y,Q^2)\psi_{\lambda_1}(x), \qquad (4.2)$$

where $\alpha\beta$ is either a quark or diquark, $T^{\alpha\beta\mu}$ is the amplitude for the elementary $\gamma^*\alpha\beta \to \alpha\beta$ subprocess (where the photon scatters off α) and the sum goes over all the helicities and kinds of quarks and diquarks contributing to I^μ. Internal transverse momenta in the proton are neglected. Calculating the elementary amplitudes and inserting them together with the wave functions (2.2) into (4.2) provides the magnetic form factor

$$G^p_M(Q^2) = \frac{8\pi C_F}{3Q^2}\Bigg\{\int dxdy \phi^*_S(y)\frac{\alpha_S(\hat{Q}^2)F_S(\hat{Q}^2)}{(1-x)(1-y)}\phi_S(x)$$

$$-(Q^2/8m^2)\, F_V(Q^2)\int dxdy \phi^*_V(y)\frac{\alpha_S(\tilde{Q}^2)(1-x)(1-y)F_V(\tilde{Q}^2)}{xy}\phi_V(x)\Bigg\} \qquad (4.3)$$

with $\hat{Q}^2 = (1-x)(1-y)Q^2$ and $\tilde{Q}^2 = xyQ^2$. m is the mass of the vector diquark for which the value 580 MeV is used. The colour factor C_F is 1/3.

The QCD evolution leads to a mild Q^2 dependence of the wave functions which in the restricted Q^2 region we are interested in does not significantly alter the Q^2 dependence contained in $T^{\alpha\beta}$ and is therefore neglected. Note the first term, originating from the T^{qS} amplitude, is dominant.

The comparison with experimental data requires some caution. The measurements of the elastic electron-proton cross-sections at relatively large Q^2 are customarily analyzed under the assumption

$$G^P_E(Q^2) = G^P_M(Q^2)/\mu_p. \qquad (4.4)$$

This scaling law, however, contradicts most of the current theoretical analyses including the diquark model in which the asymptotic result

$$G^P_E(Q^2)/G^P_M(Q^2) \longrightarrow 1 \qquad (4.5)$$

is quickly approached. Analyzing the most recent data [12] with the assumption (4.4) or with $G_E = G_M$ respectively leads to differences at intermediate values of Q^2 which may be considered as systematic errors. In Fig.2 the two data sets obtained that way are compared with the results of the diquark model. Obviously, a reasonable understanding of G_M is received with $Q^2_0 = 3.22 GeV^2$, a value in agreement with that needed to explain the scaling violations in the deep inelastic structure functions (cf Sect.1).

A similar calculation for the magnetic form factor of the neutron yields approximately

$$G^n_M \simeq -G^P_M/2 \qquad (4.6)$$

in fair agreement with the data.

5. A CONSTITUENT SCATTERING MODEL

Before carrying through the time consuming detailed computations in the framework of the generalized hard scattering scheme we have first exploited a constituent scattering model which may represent a reasonable approximation to reality in our kinematical region of interest. The model provides the power laws (1.1) and so far as only quarks are considered the helicity sum rule (1.2). The main idea of this model is that only one constituent, quark or diquark, is active i.e. undergoes a scattering with another constituent from a second hadron whereas the other constituents are considered as spectators (see Fig.3). A hadronic amplitude T for the reaction $AB \to CD$ is a convolution of wave functions with an elementary scattering amplitude $\hat{T}$ for the constituent reaction $ab \to cd$ which is approximated by the Born amplitude for on-shell constituents. Assuming large transverse momenta in the wave functions to be strongly suppressed by an exponential cut-off, one arrives after some calculation at

$$T(s,t) \sim \sum_{abcd} \int dx dy \psi_C^*(x) \psi_D^*(y) \hat{T}^{ab \to cd}(\hat{s}, \hat{t}) \psi_A(x) \psi_B(y) exp\left[\frac{t}{2a^2}((1-x)^2 + (1-y)^2)\right] \tag{5.1}$$

with a characterizing a soft process scale below which there is no suppression by the wave function; a being of the order of $300 MeV$.

At large momentum transfer $|t| \gg a^2$ the e-function renders the integral negligibly small unless the active constituents carry most of the hadron momenta. I.e. the integral becomes dominated by the region $x, y \approx 1$ hence by the end-point behaviour of the wave functions

$$T(s,t) \sim \sum_{abcd} \hat{T}^{ab \to cd}(\hat{s}, \hat{t}, x=y=1) \int_{-\sqrt{2}a/\sqrt{-t}}^{1} dx \psi_C^*(x) \psi_A(x) \int_{1-\sqrt{2}a/\sqrt{-t}}^{1} dy \psi_D^*(y) \psi_B(y). \tag{5.2}$$

This is the so-called end-point model advocated for by [13,14]. The two integrals in (5.2) are just the end-point contributions to the $AC(BD)$ transition form factors $F_{AC(BD)}(t)$.

Using again the $SU(6)$ wave functions (2.1) the hadronic amplitudes read at large $|t|$ (in the helicity basis)

$$T_{\{\lambda\}}(s,t) = \sum_{abcd} C_{ab \to cd} F_{AC}^{ac}(t) F_{BD}^{bd}(t) C_{\{\lambda\lambda'\}}^{ab \to cd}(AB \to CD) \hat{T}_{\{\lambda'\}}^{ab \to cd}(\hat{s}, \hat{t}, x=y=1); \tag{5.3}$$

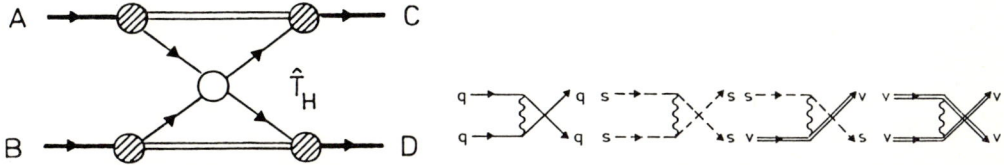

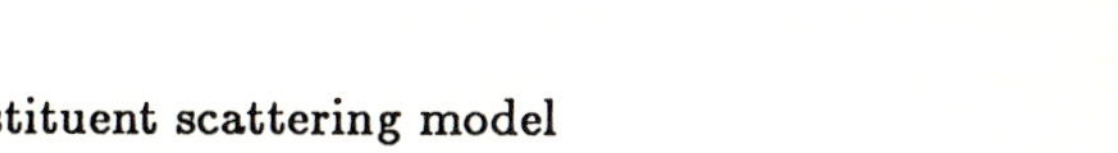

Fig.3 : (left) Basic diagram of the constituent scattering model

Fig.4 : (right) Elementary reactions to Born order contributing to $pp \to pp$

$\{\lambda\}, \{\lambda'\}$ are generic for sets of helicities. The coefficients $C(AB \to CD)$ are products of Clebsch-Gordan coefficients to be calculated from the appropriate $SU(6)$ wave functions. The overall normalization is not fixed in this type of model. Moreover, the wave function may behave differently when a quark or a diquark carries the full hadronic momentum. This possibility is taken into account by introducing the normalization constants $C_{ab \to cd}$.

6. NUCLEON-NUCLEON SCATTERING

Let us apply the model described in the preceding section to proton-proton scattering at large momentum transfer and study whether the subtle spin effects observed experimentally can be explained by diquarks. Quantum number considerations tell us that only the elementary reactions shown in Fig.4 can contribute to $pp \to pp$. The elementary amplitudes are computed to Born order according to the Feynman rules of QCD or to the diquark couplings described in Sect.2. Colour neutralization requires an interchange of the active constituents after the gluon exchange.

Inserting the elementary amplitudes into (5.3) and antisymmetrizing the hadronic amplitudes properly, one obtains the usual pp helicity amplitudes ϕ_i. Their pure quark component is SZWED's model [14]. Due to the diquarks the single (ϕ_5) and double (ϕ_2) helicity flip amplitudes are non-zero; the elementary reaction $VV \to VV$ contributes to the first one whereas to the double flip amplitude $SV \to SV$ contributes as well. As a consequence of the non-vanishing of these two amplitudes the helicity sum rule does not hold at intermediate values of momentum transfer, but by virtue of the diquark form factors it is recovered asymptotically. The model also shows the transition in the power law behaviour of the fixed angle cross-sections from s^{-6} to s^{-10} as seems to be required by the data (cf Fig.1).

The model has several free parameters: the overall normalization C_0 and the relative normalization between the diquark and quark contributions, $C_{SS} = C_{VV}$, the diquark masses and finally the diquark form factor parameters $Q_0^2 = Q_1^2$ and Q_2^2 (cf (2.6)). The mass of the scalar diquark turns out to be unimportant, any value below $700 MeV$ will do. The QCD scale parameter Λ is taken to be $100 MeV$.

The model has been fitted to the pp differential cross-section data [10] in the lab momentum region $11.75 GeV/c \leq p_L \leq 30 GeV/c$ and to spin correlation data. The $|t|$ region has been restricted to $\geq 5 (GeV/c)^2$. A good fit to the data has been obtained with the set of parameters

$$A: C_0 = 0.97 \quad C_{SS} = C_{VV} = 0.92 \quad m = 580 MeV$$
$$Q_0^2 = Q_1^2 = 3.22 GeV^2 \quad Q_2^2 = 15 GeV^2. \tag{6.1}$$

As an example of the results the spin correlation parameter A_{NN} is shown in Fig.5. For comparison also shown is the pure quark model and a fit B with a rather large value of $Q_0^2 (= 7.43 GeV^2)$, which is more in the spirit of the Stockholm model [9]. Fits A and B are equally good in terms of χ^2.

Having fitted pp scattering all parameters are fixed and one can predict $p\bar{p} \to p\bar{p}$ and $pn \to pn$ scattering, there is no free parameter left for these reactions. With $p\bar{p}$

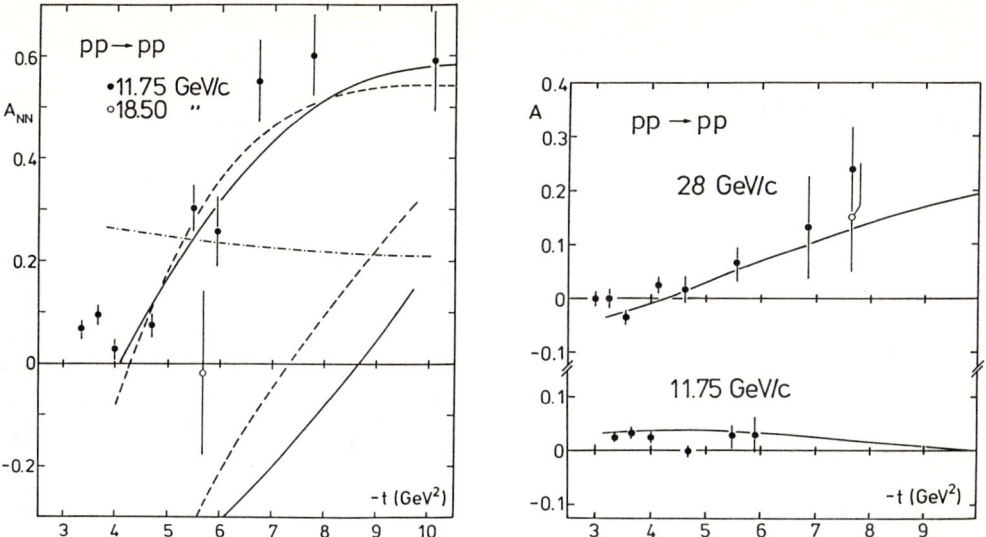

Fig.5 : (left) The parameter A_{NN}. Data from [10,15]. Solid, dashed and dash-dotted lines are the results of fit A, B and pure quark respectively.

Fig.6 : (right) The asymmetry A. Data from [2,10]. Solid lines are results of the diquark model

scattering the analytic structure of the model amplitudes is tested ($s - u$ crossing) whereas pn scattering checks their flavour dependence. In both cases good agreement between predictions and data is obtained.

It has been already mentioned that in contrast to pure quark models there is a non-zero single flip amplitude ϕ_5 in the diquark model. However, since the elementary amplitudes are calculated only to Born order, ϕ_5 is real as well as the other amplitudes. To demonstrate that this single flip contribution can easily account for the observed asymmetry [2] an angle independent phase δ between ϕ_5 and the other amplitudes has been introduced by hand. Taking the set A of parameters with $\delta(28 GeV/c) = 36.5^0$ (very similar to the value found in the empirical analysis carried out by FARRAR [3]) and $\delta(11.75 GeV/c) = 3.4°$ a good fit to the asymmetry data is obtained (see Fig.6). Theoretical sources for the phase are numerous but unfortunately any explicit calculation is hard to carry through.

7. OTHER APPLICATIONS

The constituent scattering model can easily be applied to many other reactions. As an example of meson-baryon reactions, one may consider $\pi^- p \to \rho^- p$ for which the helicity density matrix elements $\rho_{\lambda\lambda'}$ have been measured at $p_L = 9.9 GeV/c$ and at a c.m.s. angle of 90°[16]. Proceeding along the same lines as for pp scattering one can compute the amplitudes for this process. Using then the set A of parameters, the following values for the density matrix elements of the ρ^- at 90° are obtained:

	theory	experiment[16]
$\rho_{11} = \rho_{-1-1}$	0.39	0.44 ± 0.15
ρ_{00}	0.22	0.12 ± 0.30
ρ_{1-1}	0.23	0.32 ± 0.10
ρ_{10}	0.06	-0.01 ± 0.05

(7.1)

In the pure quark model, helicity conservation holds which provides $\rho_{1-1} = \rho_{10} = 0$. The introduction of diquarks definitely improves the theoretical results.

An interesting class of reactions is $p\bar{p}$ annihilation into a hyperon-antihyperon pair. These reactions go through successive annihilation and creation of constituent pairs, some of them, like $\Sigma^-\bar{\Sigma}^+$, only through diquark annihilation/creation. In Ref.[5] it is argued that for annihilation reactions the constituent scattering model may be extended to all values of t. The most important reason for this is that the argument of α_S is s. Therefore, at large energies, independent of the value of t, one-gluon exchange appears to be a reasonable approximation. The integral in (5.1) is not dominated by the end-point region in this case.

Results of the diquark model for integrated cross-sections are shown in Fig.7 (diquark parameters as in (6.1), normalizations fitted, broken $SU(6)$). Good agreement with the data is obtained. An extraordinary success of the diquark model is the prediction of

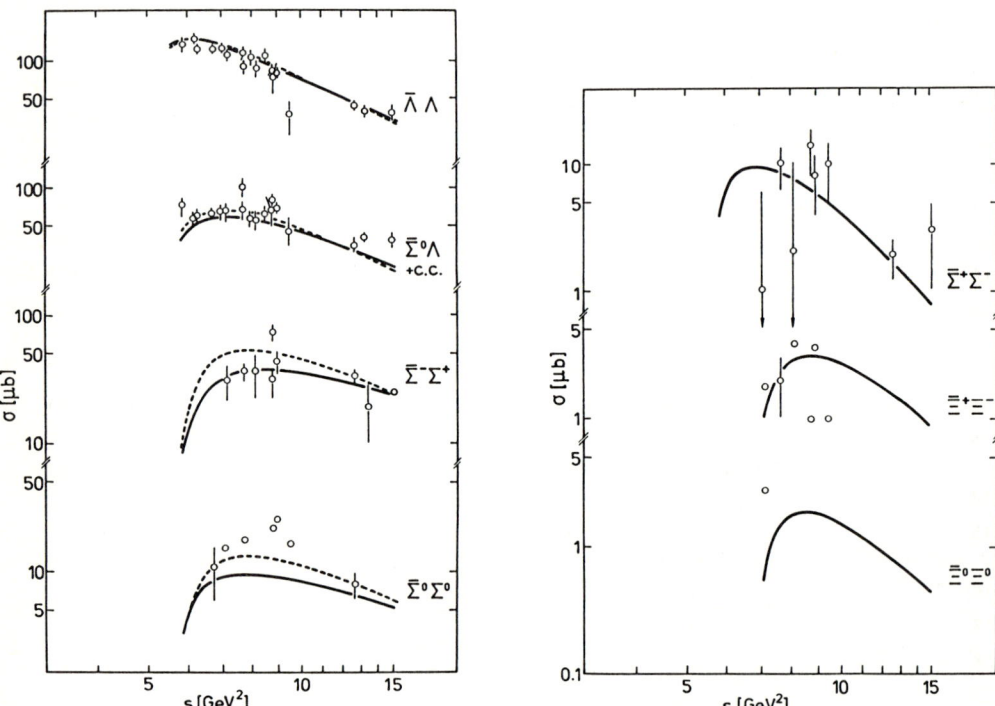

Fig.7 : Reaction cross-sections for hyperon channels. Data from [17]. Solid (dashed) line: diquark (pure quark)

the correct magnitude and energy dependence ($\sim s^{-6}$ instead of s^{-2}) of the cross-sections of reactions proceeding only through diquark annihilation.

Other models for $p\bar{p} \to Y\bar{Y}$ also based on constituent scattering take the non-relativistic limit of the elementary amplitudes (i.e. their active quarks have $x \approx 0$) and, of course, consider only quarks. These models do not predict the energy dependence of the cross-sections.

8. CONCLUSIONS

Spin phenomena in large angle exclusive reactions show many clear signs of non-perturbative effects. Guided by many experimental results and theoretical works, the active presence of diquarks is proposed, which may act as quasi-elementary constituents of baryons. The hard scattering scheme is generalized to diquarks. The main consequences of this are the change of the power laws and the violation of the helicity sum rule. Explicit studies are carried out in a constituent scattering model. Applications to pp, pn and $p\bar{p}$ elastic scattering, to $\pi p \to \rho p$ and to $p\bar{p} \to Y\bar{Y}$ reveal the diquark hypothesis to work very successfully in the intermediate energy region.

References

[1] S.J. Brodsky and G.R. Farrar, Phys.Rev.D11,1309(1975)
     G.P. Lepage and S.J. Brodsky, Phys.Rev.D22,2157(1980)
[2] P.R. Cameron et al, Phys.Rev.D32,3070(1985)
[3] G. Farrar, Phys. Rev. Lett. 56,1643(1986)
[4] M. Anselmino, P. Kroll and B.Pire, preprint WU-B-87-2, Wuppertal(1987)
[5] P. Kroll and W. Schweiger, preprint WU-B 87-9, Wuppertal(1987)
[6] S. Fleck, these proceedings
[7] J.J. Aubert et al.,Nucl. Phys. B259.189(1985)
[8] A. Donnachie and P.V. Landshoff, Phys. Lett. 95B,437(1980)
[9] S. Ekelin and S. Fredriksson, Phys. Lett. 162B,375(1985)
[10] J. Bystricky and F. Lehar, Nucleon-nucleon scattering data, Fachinformationszentrum Karlsruhe, Physics data Vol. 11-1(1985)
[11] F. Caruso, to be published
[12] R.G. Arnold et al.,Phys. Rev. Lett. 57,174(1986)
[13] S.D.Drell and T.M.Yan, Phys.Rev.Lett.24,181(1970); D.Horn and M.Moshe, Nucl.Phys.B48,557(1972)
[14] J.Szwed, Nucl.Phys.B229,53(1976)
[15] D.G. Crabb et al., Phys. Rev. Lett. 41,1257(1978); E.A. Crosbie et al., Phys. Rev. D23,600(1981); G.R. Court et al., Phys. Rev. Lett. 57,507(1986)
[16] S. Heppelmann et al.,Phys. Rev. Lett. 55,1824(1985)
[17] V.Flaminio et al.,CERN-HERA 84-01 (198
[18] H.Genz and S.Tatur, Phys.Rev.D30,63(1984)
[19] H.R.Rubinstein and H.Snellman, Phys.Lett.165B,187(1985)
[20] S.Furui and A.Faessler, preprint, Tübingen (1986)

Diquark Clustering in Baryons

S. Fleck[1,2], *B. Silvestre-Brac*[1], *C. Gignoux*[1], *and J.M. Richard*[1,3]

[1]Institut des Sciences Nucléaires, 53, av. des Martyrs,
  F-38026 Grenoble Cedex, France
[2]Laboratoire de Physique Nucléaire, T24–14, E5, Université Paris 7,
  2, place Jussieu, F-75251 Paris Cedex 05, France
[3]Laboratoire de Physique Théorique et Hautes Energies*, T16–E1,
  Université Pierre et Marie Curie, 4, place Jussieu,
  F-75252 Paris Cedex 05, France

We study to what extent baryons consist of a diquark surrounded by a quark, the diquark being a localized cluster made out of two quarks. Our quantitative analysis is performed in the framework of potential models.

I Introduction

The concept of diquark has been introduced and discussed very often in the study of the spectroscopy, production, structure or interaction of baryons [1]. In some extreme models, point-like diquarks are postulated almost on an equal footing with quarks, whereas, in more conservative pictures, the diquark denotes only the existence of some short-range two-body correlations in the baryon wave-function.

In baryon spectroscopy, there are several questions whose discussion leads to a comparison between the three-quark model and the quark-diquark model.

i) The three-quark model predicts obviously more states. In particular, if one introduces the standard Jacobi coordinates

$$\rho \propto r_2 - r_1 \qquad \lambda \propto 2 r_3 - r_1 - r_2 \qquad (1)$$

there are some positive parity resonances where both ρ and λ oscillators are orbitally excited. These states are absent from the quark-diquark model. If they exist, they should, however, be weakly coupled to the usual meson-nucleon entrance channels and thus hardly isolated without ambiguity in a phase-shift analysis.

ii) The leading Regge trajectory of baryons exhibits a behaviour

$$J = \alpha' M^2 \qquad (2)$$

with the *same* slope as in the meson case. This property which is automatically fulfilled in the quark-diquark model (the diquark has the same colour charge as an antiquark) was often used as a case against the three-quark models. MARTIN [2], however, showed that the property (2) emerges from a three-body picture with linear confinement (of two-body or three-body character). For the baryons lying on the leading Regge trajectory, there is a dynamical clustering which induces a quark-diquark structure.

iii) There have been empirical models describing simultaneously mesons as quark-antiquark systems and baryons as made of a quark and a diquark interacting through the same potential [3]. The question is whether the success of this approach is purely accidental or due to the existence of a strong two-body clustering inside baryons.

We will briefly report here the results of our investigations, referring to a forthcoming publication [4] for more details. Assuming that the quarks interact non-relativistically through a simple potential, we compare the exact three-body binding energy to an approximate quark-diquark energy and examine systematically the two-body correlations in the wave-function to detect any possible diquark clustering.

II Central Potential : Ground States

Let us first consider three quarks interacting through a flavour-independent central potential $v(r)$ with a confining tail. The exact three-body binding energy $E_3(q_1q_2q_3)$ is easily calculated by standard techniques : hyperspherical coordinates, harmonic oscillator expansion, Faddeev equations, etc. We compare it to the quasi two-body energy $E_2(Dq_3)$ obtained first by computing the diquark mass m_D out of

$$h_{12} = m_1 + m_2 + p_1^2/2m_1 + p_2^2/2m_2 + v(r_{12}) \qquad (3)$$

and inserting it into

$$h_{D3} = m_D + m_3 + p_D^2/2m_D + p_3^2/2m_3 + 2\,v(r_{D3}) \qquad (4)$$

Note that the approximations involved in replacing the central hamiltonian

$$h_{123} = \sum m_i + p_i^2/2m_i + \sum v(r_{ij}) \qquad (5)$$

by the sequence (3-4) are often antivariational, i.e. tend to underestimate the baryon mass. Some results are shown in Table 1, where an empirical power-law potential fitting the ground state baryons has been adopted, namely

$$v(r) = (A + B\,r^\beta)/2 \qquad (6)$$

with the parameters of ref. [5], i.e. (in GeV units) : $A = -8.337$, $B = 6.9923$, $\beta = 0.1$; $m_u = m_d = 0.300$, $m_s = 0.600$, $m_c = 1.895$.

Also shown in Table 1 are the r.m.s. distances $\langle r_{ij}^2 \rangle^{1/2}$ between quarks, computed out of the exact three-body wave-function. As expected, the QQq baryons exhibit the most pronounced clustering : in a flavour independent potential, heavy quarks experience more binding and remain close together at the bottom of the potential well.

Several diquark-quark approximations may be attempted for the same baryon, (1-2)3, (2-3)1 or (3-1)2. As seen in Table 1, the best result always corresponds to the pairing of the quarks with the shortest separation.

Table 1: Average quark separations and masses of ground state baryons calculated either with the three-body hamiltonian (5) or in the quark-diquark approximation (3-4), using the power-law potential (6).

	Three-body calculation			Quark-diquark approximation	
	$\langle r_{12}^2 \rangle^{1/2}$	$\langle r_{31}^2 \rangle^{1/2}$	Mass (123)	Mass (1-2)3	Mass 1(2-3)
uus	4.9336	4.4808	1.2679	1.3214	1.2337
uuc	4.7884	3.9826	2.4520	2.1522	2.3083
ssu	3.7821	4.3379	1.4300	1.3214	1.2337
ssc	3.4955	3.0380	2.7014	2.4929	2.5690
ccu	2.3263	3.6732	3.6848	3.6341	3.3964
ccs	2.2424	2.8815	3.7762	3.6990	3.5717

Beyond integrated quantities like $E_3(q_1q_2q_3)$ or $\langle r_{ij}^2 \rangle^{1/2}$, one may examine the wave-function $\Psi(\rho,\lambda)$ itself, for instance by plotting various sections of the density distribution

$$d(\rho,\lambda,\Theta_{\rho\lambda}) = \int \rho^2 \lambda^2 \, d\Omega_\rho \, d\varphi_{\rho\lambda} \, |\Psi(\rho,\lambda)|^2 \sin\Theta_{\rho\lambda} \qquad (7)$$

The detailed results will be displayed in ref [4]. They essentially confirm the above conclusions based on Table 1.

III Central Potential : The Leading Regge Trajectory

Since the arguments of ref. [2] on the slope of the baryon leading Regge trajectory were semi-classical, i.e. rigorous only at high angular momentum ℓ, it is interesting to check by explicit three-body calculations what occurs at moderate values of ℓ, say $\ell \approx 2\text{-}8$. We have done this study with the potential (6), and also with a more conventional coulomb-plus-linear model

$$v(r) = -k/r + \lambda r \qquad (8)$$

with the parameters of ref. [6], i.e. $m_q = .337$; $m_s = .600$; $m_c = 1.870$; $k = .5203$; $\lambda = .11857$ in GeV units.

The masses are shown in Table 2. The "exact" binding energy is obtained by expanding the wave-function in the harmonic-oscillator basis truncated at N = 8 quanta, and is compared to the quark-diquark approximation. Our hamiltonian is not valid for high ℓ, since the quarks become extremely relativistic, and, as a consequence, the resulting masses are no longer increasing functions of the constituent masses. Hopefully our qualitative conclusions concerning diquark clustering will be valid in more realistic pictures of excited baryons.

Table 2: Masses of $\ell=0$ and $\ell=8$ baryons, calculated either exactly or in the quark-diquark approximation, using the potential (8).

	$\ell = 0$			$\ell = 8$		
	exact	(1-2)3	1(2-3)	exact	(1-2)3	1(2-3)
uud	1.204	1.023	1.024	3.843	3.703	3.709
uus	1.354	1.141	1.203	3.735	3.554	3.841
uuc	2.496	2.225	2.352	4.592	4.340	4.887
ssu	1.493	1.375	1.311	3.728	3.985	3.669
ccu	3.686	3.629	3.424	5.340	6.129	5.285

An analysis of the results in Table 2 and of the density $d(\rho,\lambda,\Theta)$ of eq. (8) gives the following conclusions. For baryons with equal masses (qqq), the diquark clustering, which is elusive for $\ell = 0$, becomes more and more pronounced with increasing ℓ, in agreement with ref. [2]. The quark-diquark approximation, however, does not work too well for the binding energy (the use of m_D in the kinetic energy of eq. (4) dramatically reduces the orbital barrier if $m_D > 2m$).

For single flavoured baryons (qqQ), a clear (qq) clustering arises at large ℓ. The ground state with $\ell > 0$ is mostly a λ-type of excitation, with the (qq) pair almost always in s-wave, i.e. $\ell_\rho = 0$, $\ell_\lambda = \ell$. The orthogonal states with $\ell_\rho > 0$ are slightly higher in mass. The (qq)Q quark-diquark approximation to the binding energy, again, is not impressively good. The (qQ)q approximation should not be compared to the $\ell_\rho=0$ ground state but to some excitation.

For baryons with two heavy flavours (QQq), the (QQ) clustering observed at $\ell=0$ is lost when ℓ increases. The QQq system, indeed, finds it more economical to excite orbitally the ρ-oscillator rather than the λ one, so that the average QQ separation increases. At high ℓ, of course, one recovers a new clustering, of the (Qq)Q type.

Note that, here and in other sections, we consider quark-diquark configurations like (Qq)q or (Qq)Q in which the Pauli principle is explicitly broken. However, when the diquark clustering is well pronounced, the exchange kernels are very small so that proper symmetrization would not change the binding energy much.

IV The Role of Spin-Spin Forces

We have also studied the baryon wave-functions computed out of a potential model, where the central component (8) is supplemented by a spin-spin term.

$$V_{SS} = G \sum \sigma_i \sigma_j \, e^{-\Lambda r} / (m_i m_j r) \qquad (9)$$

as in ref. [6] or elsewhere [7]. The parameters of ref. [6] correspond to G = 0.09805 and Λ = 0.4341 in GeV units. The baryon spectrum has already been studied with the model [8].

Table 3: Masses of $\ell=0$ and $\ell=8$ baryons with spin 1/2, calculated either exactly or in the quark-diquark approximation, using the potential (8) and (9).

	$\ell = 0$			$\ell = 8$		
	exact	(1-2)3	1(2-3)	exact	(1-2)3	1(2-3)
uud	1.032	0.740	0.912	3.771	3.722	3.638
sdu	1.186	1.129	1.037	3.667	3.785	3.502
uus	1.264	0.952	1.128	3.754	3.569	3.783
cdu	2.327	2.324	2.136	4.531	4.862	4.318
uuc	2.493	2.155	2.323	4.609	4.348	4.859
ssu	1.373	1.173	1.239	3.694	3.997	3.618
ccu	3.636	3.558	3.397	5.326	6.132	5.260

The results corresponding to the spin 1/2 configurations are shown in Table 3. They have been analyzed in connection with a detailed inspection of the density distributions $d(\rho,\lambda,\Theta)$ and compared to the results of the previous section for a spin-independent potential. For the $\ell=0$ case, the conclusions on diquark clustering are essentially unchanged. We simply notice that the overall size of the baryons is slightly reduced by the short-range hyperfine interaction.

Let us consider now the proton trajectory. In the spin-independent case, the λ-excitation, where the two u quarks are clustered in a s-wave, was slightly below the ρ-excitation in which the d quark is alternatively clustered with each u quark. When the spin-spin potential is added, the lowest state of given ℓ is now the ρ-excitation, and the λ-excitation is shifted upwards by the chromomagnetic repulsion which makes also the u-u clustering less pronounced.

The Λ trajectory consists essentially of a λ-excitation in which the s quark carries the orbital momentum around a (ud) diquark. The same is true for the Σ^0, although, due to the chromomagnetic repulsion, the (ud) clustering is less effective.

V The Born-Oppenheimer Approximation

The quark-diquark model fails in describing the first excitations of the QQq baryons with two heavy flavours. While a clear quark-diquark structure q(QQ) is seen, for $\ell=0$, the diquark is broken for $\ell > 0$. In fact, these states are very efficiently described in the Born-Oppenheimer approximation. The hamiltonian (5) is first rewritten in reduced coordinates.

$$h_{123}^{red} = p_\rho^2/2M + p_\lambda^2/2\mu + v(\rho) + v(\lambda-\rho/2) + v(\lambda+\rho/2) \qquad (10)$$

where $1/\mu = 1/m + 1/2M$. One first computes, for fixed ρ the ground state of a single quark in a non-central potential, corresponding to

$$\{p_\lambda^2/2\mu + v(\lambda-\rho/2) + v(\lambda+\rho/2)\}\, \varphi(\lambda,\rho) = \varepsilon(\rho)\, \varphi(\lambda,\rho) \qquad (11)$$

The usual Born-Oppenheimer approximation consists of neglecting the ρ-dependence in $\varphi(\lambda,\rho)$, leading to

$$\{p_\rho^2/2M + v(\rho) + \varepsilon(\rho)\}\, \psi(\rho) = E_{BA}\, \psi(\rho) \qquad (12)$$

which overestimates the binding i.e. $E_{BA} < E_{exact}$.

An alternative is the so-called variational Born-Oppenheimer approximation, where the solution of (11) is incorporated in a trial wave-function $\tilde{\psi}(\rho)\, \varphi(\lambda,\rho)$. The effective potential is now $v(\rho) + \varepsilon(\rho) + \langle \varphi | \nabla_\rho^2 | \varphi \rangle$, leading to the upper bound $E_{VAB} > E_{exact}$. The results in Table 4 show clearly that the Born-Oppenheimer treatment of the (QQq) baryons is extremely accurate. It can also be used in situations where the light quark is treated relativistically, for instance in the bag model.

Table 4: Ground state, radial and orbital excitation of the (ccu) system, calculated from potential (6) exactly or in the Born-Oppenheimer approximations.

(ℓ,n)	B.A.	exact	V.B.A.
(0,0)	3.6840	3.6848	3.6852
(0,1)	4.1092	4.1096	4.1117
(1,0)	3.9689	3.9712	3.9709

VI Conclusion

We have analyzed the baryon wave-functions calculated in simple potential models. The occurrence of a quark-diquark structure depends on the flavour content and the orbital momentum which is considered. In this low-Q^2 physics, the clustering of quarks into diquarks is far from being a general pattern.

Acknowledgments :

We would like to thank P. Kroll and A. Martin for encouraging discussions.

References :

1 See, for instance, the contribution by P. Kroll in this volume
2 A. Martin, Zeit. Phys. C32, 359 (1986)
3 D.B. Lichtenberg, W. Namgung, E. Predazzi and J.G. Wills, Phys. Rev. Lett. 48, 1653 (1982)
  K.F. Liu and C.W. Wong, Phys. Rev. D28, 170 (1983)
4 B. Silvestre-Brac, C. Gignoux, S. Fleck, J.M. Richard, in preparation

5 J.M. Richard, P. Taxil, Phys Lett. 128B, 453 (1983)
  see also, A. Martin, Phys Lett. 100B, 511 (1981)
6 R.K. Bhaduri, L.E. Cohler, Y. Nogami, Nuovo Cimento 65A, 376 (1981)
7 S. Ono, F. Schöberl, Phys. Lett. 188B, 419 (1982)
8 C. Gignoux, B. Silvestre-Brac, Phys. Rev. D32, 743 (1985)

Hadron Wave Functions with Condensate Induced Running Masses

M. Lavelle[1],*, E. Werner[1], and St. Glazek[2]

[1]Institute of Theoretical Physics, University of Regensburg,
D-8400 Regensburg, Fed. Rep. of Germany
[2]Institute of Theoretical Physics, Warsaw University,
PL-Warsaw, Poland

1. Introduction

In this paper we try to combine the high-energy perturbative features of QCD with the low-energy nonperturbative properties, the manifestation of the latter being the interaction with quark, gluon and mixed condensates. The main motivation for our approach in the nonperturbative domain lies in the success of QCD sum rules [1,2] which make abundant use of the concept of slow quarks and gluons interacting with the nonperturbative vacuum. Here we use this interaction to generate running quark masses which interpolate between the current quark masses in the large q domain and the constituent masses which find their empirical justification in many successful constituent quark models. A preliminary account of these ideas was given in a recent paper [3] in which deep inelastic and static properties of nucleons were successfully described with a wave function employing a gross simplification of running masses; namely fixed current masses for the deep inelastic properties and fixed constituent masses for the static properties.

We use light front coordinates (LFC) since in this scheme the necessary decoupling of the centre of mass and relative motion is achieved automatically, so that arbitrary motions of the hadron constituents can be described in a relativistic framework. In addition, the absence of Wigner rotations for light front spinors reduces the effect of boosts to changes of the momenta of the constituents (and of the invariants derived from them, e.g. running masses) without changing spins.

The essential assumption in the dynamics is that the complicated many-body nature of the hadron wave function can be accounted for by the concept of dressed quarks, which obtain their dynamical masses and effective vertices from interactions with the QCD vacuum condensates.

2. Light Cone Kinematics for 2- and 3-Particle Systems

LC coordinates in position and momentum space are defined as in [4]

$$x = (x^-, x^1, x^2, x^+) \quad ; \quad x^\pm = x^0 \pm x^3 \quad ; \quad p = (p^-, p^1, p^2, p^+) \quad ; \quad p^\pm = p^0 \pm p^3 \quad .$$

The on-shell condition for a momentum four vector reads: $p^- = (m^2 + p_\perp^2)/(p^+)$.

For a <u>two-particle system</u> one defines the total momentum components by $P^{+,\perp} = p_1^{+,\perp} + p_2^{+,\perp}$ and the longitudinal momentum fractions by $x_i = p_i^+/P^+$; $x_1 + x_2 = 1$. The relative momentum of particles 1 and 2 is $q = x_2 p_1 - x_1 p_2$, with $q^+ = 0$.

We define a quantity D_0 which is a measure of the off-shellness of a configuration with LC momenta $p_i^+, p_i^\perp, i = 1, 2$ [4]:

---

*Supported by BMFT (MEP 0233 REB 4)

$$D_0 = P^2 - (p_1 + p_2)^2 = P^+ \left(P^- - \sum_{i=1,2} p_i^- \right) = M^2 - \left(\frac{q^{\perp 2}}{x_1 x_2} + \frac{m_1^2}{x_1} + \frac{m_2^2}{x_2} \right) . \quad (1)$$

D_0 is boost invariant; furthermore the contributions from CM-motion (M^2) and relative motion (the term in the last set of parentheses) decouple.

For a <u>three-particle system</u> the corresponding definitions are

$$P^{+,\perp} = \sum_{i=1}^{3} p_i^{+,\perp} \quad ; \quad x_i = \frac{p_i^+}{P^+}$$

$$q_k = x_j p_i - x_i p_j \quad ; \quad Q_k = (x_i + x_j) p_k - x_k (p_i + p_j) \quad ; \quad \text{i,j,k cyclic.}$$

From the definition it follows that $q_k^+ = 0$, $Q_k^+ = 0$. In analogy to the two-particle case we have

$$\begin{aligned} D_0 &= P^+ (P^- - \sum_{i=1}^{3} p_i^-) \\ &= M^2 - \left(q_k^{\perp 2} \frac{x_i + x_j}{x_i x_j} + Q_k^{\perp 2} \frac{1}{x_k (x_i + x_j)} + \sum_{i=1}^{3} \frac{m_i^2}{x_i} \right). \end{aligned} \quad (2)$$

For the discussion of running masses we have to introduce the concept of single-particle off-shellnesses for the constituent particles; we define this off-shellness by

$$\hat{p}_i^2 = (P - \sum_{j \neq i} p_j)^2 - p_i^2.$$

Making use of momentum conservation for the $+$ and $\perp$ components, one obtains $\hat{p}_i^2 = x_i D_0$. According to their definition, the $\hat{p}_i^2$ for different particles are simply related by $\hat{p}_i^2 x_j = \hat{p}_j^2 x_i$ and we have $\sum_i \hat{p}_i^2 = 1$.

3. Self-Consistency Equation for $\hat{p}_i^2$.

Assuming a definite functional relationship for running masses of the form $m_i = m_i(\hat{p}_i^2)$, one obtains from (1) and (2) for the 2- and 3-particle cases the following equations for $\hat{p}_i^2$:

$$\hat{p}_i^2 = x_i M^2 - \left[\frac{q_\perp^2}{1 - x_i} + m_i^2 (\hat{p}_i^2) + \frac{x_i}{1 - x_i} m_j^2 \left(\frac{1 - x_i}{x_i} \hat{p}_i^2 \right) \right], \quad (3)$$

$$\hat{p}_i^2 = x_i M^2 - \left[q_k^{\perp 2} \frac{x_i + x_j}{x_j} + \frac{x_i Q_k^{\perp 2}}{x_k (x_i + x_j)} + m_i^2 (\hat{p}_i^2) + \frac{x_i}{x_j} m_j^2 \left(\frac{x_j}{x_i} \hat{p}_i^2 \right) + \frac{x_i}{x_k} m_k^2 \left(\frac{x_k}{x_i} \hat{p}_i^2 \right) \right]. \quad (4)$$

For given relative and longitudinal momenta these are transcendental equations for the off-shellness $\hat{p}_i^2$. Once one off-shellness is known the other(s) follow immediately.

4. Condensate-Induced Running Masses

The lowest-order contributions (in the sense of Wilson's OPE) come, for light quarks, from the quark condensate $<: \bar{\psi}\psi :>$, and, for gluons, from the gluon condensate $<: \alpha_s G^2 :>$. They lead to running masses of the form [5]

(1) Quarks: $m_q(p^2) = C_q/p^2$, $C_q = -\frac{4\pi}{3}\alpha <: \bar{\psi}\psi :>$,
(2) Gluons: $m_g^2(p^2) = C_g/p^2$, $C_g = \frac{\pi}{2} <: \alpha_s G^2 :>$,
The empirical values are $C_q = (0.24 GeV)^3$; $C_g = \frac{\pi^2}{2}(0.33 GeV)^4$.
One could suspect that higher-order terms in $1/\hat{p}^2$, i.e. higher-dimensional condensates, should become important at small $\hat{p}^2$. However, due to the properties of the solutions of (3) and (4) this is not the case, as will now be discussed.

5. Qualitative Features of the Solution of the Self-Consistency Equations

With the form given above for the running quark mass, (3) and (4) reduce to cubic equations for $\hat{p}_i^2$. Their solution yields immediately the running masses $m(\hat{p}_i^2)$ and the total D_0. The following qualitative features are to be noticed:

    a) $\hat{p}_i^2$ is always negative (in contrast to the case of constant masses) and is bounded from above: $-\infty < \hat{p}_i^2 < -\eta^2$, with $\eta^2 \sim 0.3$ GeV. This means that the behaviour of $m(\hat{p}^2)$ for $|\hat{p}| \leq 0.3$ GeV is irrelevant for the resulting off-shellness. Higher-dimensional condensates, which contribute mainly in the region B, are therefore suppressed in the solution of the consistency equation (Fig.1).

    b) The running masses always remain finite, even if transverse momenta go to zero.

The results of the numerical solution for the two-particle case are shown in Fig.2. In the upper part the running masses $m_1(\hat{p}_1^2)$ and $m_2(\hat{p}_2^2)$ are shown together with their sum $m = m_1 + m_2$ for the indicated values of $q^\perp$. The lower part of the figure shows $-D_0$. The result for $-D_0$ for the constant mass case ($m = 0.33$ GeV) is given for comparison.

For the 3-particle case the corresponding results are shown in Table 1 as a function of $q_3^\perp$, $Q_3^\perp$, x_1, x_2.

For given D_0 the proton wave function is constructed in the following way (for details see [3]):

$$|\psi\rangle_P = \sum_{k_j,\lambda_j} f(D_0) \left(\sum_i a_i I_i\right) \left(u^\dagger_{k_1\lambda_1} u^\dagger_{k_2\lambda_2} d^\dagger_{k_3\lambda_3}\right)_c |0\rangle.$$

Here $f(D_0)$ is an empirical function which decreases monotonically with increasing $-D_0$ with a width determined by the r.m.s. radius of the hadron. In [3] it was chosen as an exponential function. In this case it reduces to the form of the Isgur-Karl wave function in the nonrelativistic limit. The sum over i is a sum over Ioffe currents in which the three quark spinors are coupled to the proton spinor. The remaining factor contains the properly coupled creation operators for u- and d-quarks.

Assuming a typical half-width of ~ 0.7 GeV for $f(D_0)$ one notices the following interesting features:

For a given $q^\perp$ or $(q_3^\perp, Q_3^\perp)$, the largest probability amplitudes are obtained for $x_1 = x_2 = 0.5$ or $x_1 = x_2 = x_3 = 1/3$ respectively (equal running masses). However, there are rather

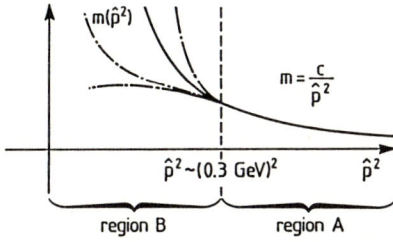

Fig.1: Regions of off-shellness which are relevant (A) /not relevant (B) for the solution of (3) and (4)

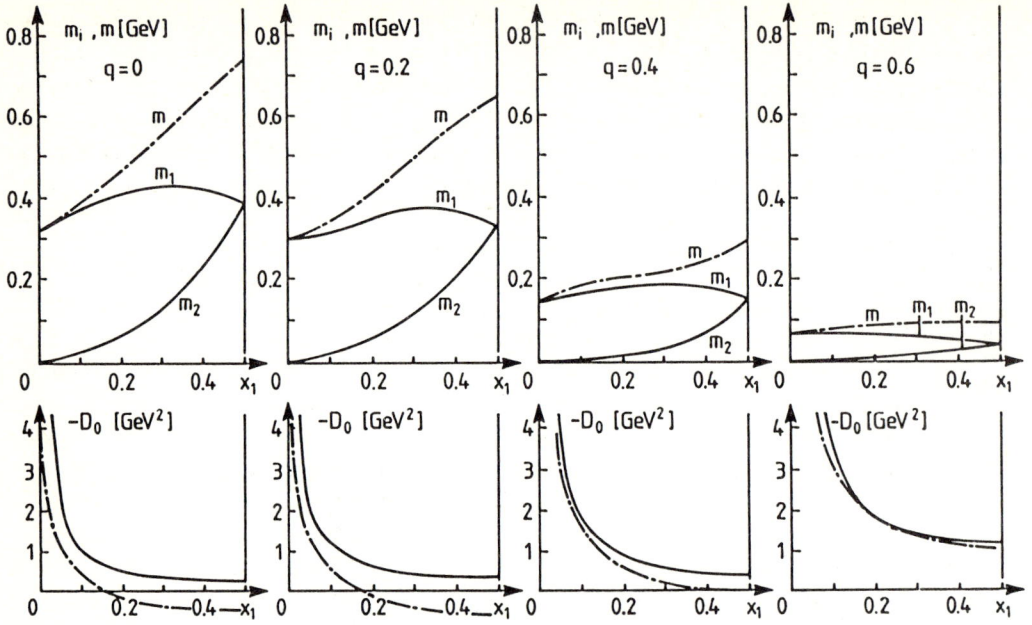

Fig.2: Running masses and off-shellness D_0 for 2 particles

Table 1: Running masses and off-shellness D_0 for 3 particles.

$q_3^\perp = Q_3^\perp = 0$ GeV/c

x_1	x_2	$-D_0$	m_1	m_2	m_3
0,1	0,1	1,08	0,32	0,32	0,02
0,1	0,2	0,91	0,40	0,15	0,03
0,1	0,3	0,89	0,42	0,09	0,04
0,2	0,2	0,43	0,38	0,38	0,29
0,2	0,3	0,41	0,47	0,26	0,21
0,2	0,4	0,40	0,49	0,18	0,18
0,3	0,3	0,30	0,40	0,40	0,23
1/3	1/3	0,29	0,34	0,34	0,34

$q_3^\perp = 0$; $Q_3^\perp = 1/3$ GeV/c

x_1	x_2	$-D_0$	m_1	m_2	m_3
0,1	0,1	1,19	0,27	0,27	0,02
0,1	0,2	1,01	0,35	0,13	0,03
0,1	0,3	0,96	0,37	0,29	0,05
0,2	0,2	0,53	0,33	0,33	0,07
0,2	0,3	0,48	0,37	0,21	0,11
0,2	0,4	0,46	0,40	0,15	0,15
0,3	0,3	0,35	0,33	0,33	0,22
1/3	1/3	0,33	0,30	0,30	0,30

$q_3^\perp = 1/3$; $Q_3^\perp = 0$ GeV/c

x_1	x_2	$-D_0$	m_1	m_2	m_3
0,1	0,1	1,76	0,16	0,16	0,01
0,1	0,2	1,34	0,23	0,09	0,02
0,1	0,3	1,24	0,26	0,06	0,03
0,2	0,2	0,67	0,23	0,23	0,05
0,2	0,3	0,55	0,31	0,18	0,09
0,2	0,4	0,54	0,32	0,12	0,12
0,3	0,3	0,39	0,28	0,28	0,11
1/3	1/3	0,35	0,28	0,28	0,28

$q_3^\perp = 1/3$; $Q_3^\perp = 1/3$ GeV/c

x_1	x_2	$-D_0$	m_1	m_2	m_3
0,1	0,1	2,20	0,12	0,12	0,01
0,1	0,2	1,64	0,18	0,07	0,01
0,1	0,3	1,42	0,22	0,05	0,02
0,2	0,2	0,85	0,17	0,17	0,04
0,2	0,3	0,64	0,13	0,03	0,04
0,2	0,4	0,66	0,13	0,11	0,11
0,3	0,3	0,51	0,18	0,18	0,12
1/3	1/3	0,47	0,19	0,19	0,19

large regions with $x_1 \neq x_2$ or $x_1 \neq x_2 \neq x_3$ where the masses can be quite different, but where the wave functions are still large. ($-D_0$ rises rather slowly in this region.) Evidently, the wave function falls off with increasing relative momenta. It is seen that the constituent quark model with $m \sim 330$ MeV is approached for $q_3^\perp \to 0, Q_3^\perp \to 0, x_1 = x_2 = x_3 = 1/3$. There is, however, a large region in x_i-space where the dynamics of the running masses due to their relative motion is important with large deviations from the constituent quark model. This is most clearly seen when one compares the off-shellness $-D_0$ for the case of running masses and for constant masses (see Fig.2, lower part). For any wave function having a reasonable range in momentum space this will lead, for example, to different elastic form factors. Of course, these will also be influenced by running vertices which will appear on an equal footing with running masses and result from the internal structure of the dressed quarks.

References

[1] M.A. Shifman, A.I. Vainstein & V.I. Zakharov, Nucl.Phys. B 147 (1979) 385; 448; 519

[2] B.I. Ioffe, Nucl.Phys. B 188 (1981) 317

[3] St. Głazek, J. Namysłowski & E. Werner, Phys. Lett. 158 B (1985) 150

[4] J.M. Namysłowski, Progr.in Part. and Nucl. Phys. 14 (1984) 1

[5] J.M. Namysłowski, Preprint Univ. of Warsaw IFT/12/87

Vector Meson Interactions in the Effective Lagrangian

B. Moussallam

Division de Physique Théorique, IPN,
Université Paris-Sud, F-91406 Orsay Cedex, France

The part of the effective Lagrangian containing vector mesons has rarely been used in the past. We study how well it works in describing recent data on the two-photon production of rho meson pairs. It appears necessary to go beyond the tree amplitudes and to include the effect of tensor mesons.

At low energies the quark degrees of freedom are frozen and one may believe that the interactions between mesons are governed by an "effective Lagrangian"(EL)[1]. This idea has attracted increased interest recently, based on the hope of explaining the spectrum and the properties of the baryons in this way as well [2]. In the literature, the interactions of pseudoscalars is vastly documented, mainly at very low energies [3] but also in the 1-2 GeV range [4]. In the following, we would like to inquire how well one can understand the vector mesons in the EL framework.

It was recognized for a long time that there are strong ties between the electromagnetic interactions of hadrons and the vector mesons, through the concept of vector dominance [5]. It states that the hadrons do not couple directly to the photon, the coupling occurs via an intermediate neutral vector boson. This is an indirect (and not very clearly understood) by-product of compositeness and it works reasonably well, for instance, in the case of the pion or kaon form factors. It looks interesting then, to consider from this point of view reactions like $\gamma\gamma \to V_1 V_2$, for which experimental results have recently become available [6]. Here, we will concentrate on the case where V_1 and V_2 are ρ mesons, which has the following peculiar characteristics. The cross-section of the neutral channel, $\rho^0\rho^0$, has a large enhancement, $\sigma \simeq 130$ nb, centered slightly below the threshold (recall that the ρ has a significant width). By contrast, the charged channel is strongly suppressed, only an upper bound is available: $\sigma(\rho^+\rho^-) < 30$ nb. How can we understand such a difference?

If we have a Lagrangian for, say, pseudoscalars and we want to have vector dominance, then the ρ meson has to be introduced as if it were the gauge boson associated with isospin, except for a mass term. The result is as follows [7]:

$$\mathcal{L}_{(1)} = -\frac{1}{4}(\partial_\mu \vec{\rho}_\nu - \partial_\nu \vec{\rho}_\mu + g\vec{\rho}_\mu \wedge \vec{\rho}_\nu)^2 + \frac{1}{2}m_\rho^2 (\rho_\mu^0 - \frac{e}{g}A_\mu)^2$$
$$+ \frac{1}{2}m_\rho^2(\rho_\mu^{+2} + \rho_\nu^{-2}) - \frac{1}{4}(\partial_\mu A_\nu - \partial_\nu A_\mu)^2 \,. \qquad (1)$$

Other light, nonstrange particles that one may believe to play a role in the reaction are the pion and the omega meson. The necessary couplings are summarized by

$$\mathcal{L}_{(2)} = \frac{1}{2}(\partial_\mu \vec{\pi} + g\vec{\rho}_\mu \wedge \vec{\pi})^2 - \frac{1}{4}(\partial_\mu \omega_\nu - \partial_\nu \omega_\mu)^2$$
$$+ \frac{1}{2}m_\omega^2 (\omega_\mu - \frac{e}{3g}A_\mu)^2 - \frac{g'}{2m_\rho}\varepsilon^{\mu\nu\alpha\beta}(\partial_\mu \omega_\nu \partial_\alpha \vec{\rho}_\beta \cdot \vec{\pi}). \quad (2)$$

The coupling constant g can be estimated to be about 5 from $\rho^0 \to e^+ e^-$ decay rate and g' can be expressed in terms of g, m_ρ and f_π from the explicit expression of the Wess-Zumino action [8].

The Lagrangian (1) fixes the form of the triple and quartic vertices of the ρ meson. If one starts from $\rho^0 \rho^0$ (as required by vector dominance from the two-photon channel), it is easily seen that at the tree level, all the self-interaction diagrams lead to the $\rho^+ \rho^-$ state, in complete contradiction with what experiment indicates. The contribution from the one-pion exchange terms (which are multiplied by λ in (3)) turns out to be rather weak. At threshold ($\sqrt{s} \simeq 1540$ MeV) the resulting amplitudes have a simple expression:

$$A(\gamma\gamma \to \rho^0\rho^0) = \lambda e^2(-2I_1 + I_2 + I_3 - 2I_4), \quad (3a)$$
$$A(\gamma\gamma \to \rho^+\rho^-) = e^2(-2(1+\lambda)I_1 + (4+\lambda)(I_2+I_3) - (4+2\lambda)I_4), \quad (3b)$$

where $\lambda \simeq 0.25$ and I_i are four independent kinematical invariants which do not vanish at threshold ($I_1 = \varepsilon_1.\varepsilon_2\, \varepsilon_3^*.\varepsilon_4^*$, $I_2 = \varepsilon_1.\varepsilon_3^*\, \varepsilon_2.\varepsilon_4^*$, $I_3 = \varepsilon_1.\varepsilon_4^*\, \varepsilon_2.\varepsilon_3^*$ and $I_4 = \varepsilon_1.\varepsilon_2\, p_1.\varepsilon_3^*\, p_1.\varepsilon_4^*/m_\rho^2$). Clearly, the production of charged ρ pairs is much larger than that of neutral ones at this stage.

For one thing, the use of Born diagrams at this energy is too naive. Past experience with two-body pion and kaon interactions shows that much better results are obtained by computing the amplitudes at the one-loop level and Padé approximating the perturbation expansion in every partial wave [4]. One may also think of introducing heavier bosons in the Lagrangian. Let us examine this, first. Candidates to play a role are the scalar isoscalar meson ε (other scalars being known to couple very weakly to the two-photon channel [9]) and the tensors f and A_2. The following Lagrangian can express the couplings to the vectors:

$$\mathcal{L}_{(3)} = -\frac{g_s}{4m_\rho}\phi_\varepsilon(\omega_{\mu\nu}\omega^{\mu\nu} + \vec{\rho}_{\mu\nu}\vec{\rho}^{\mu\nu})$$
$$- \frac{g_T}{m_\rho}\left[f^{\mu\nu}(\omega_{\mu\alpha}\omega_\nu^\alpha + \vec{\rho}_{\mu\alpha}.\vec{\rho}_\nu^\alpha) + \vec{A}_2^{\mu\nu}.\vec{\rho}_{\mu\alpha}\omega_\nu^\alpha\right]. \quad (4)$$

The electromagnetic widths of f and A_2 are well known, which allows us to determine g_T. The case of ε is not as clear. Recently, it was suggested that the $\gamma\gamma \to \pi^+\pi^-$ data near threshold implies a very large two-photon width ($\Gamma \simeq 10$ keV)[10], but this interpretation depends on a particular model and needs to be confirmed. From a theoretical point of view, a large width is expected if the ε is considered as a q$\bar{\text{q}}$ p-wave state [11], but a mixture of 2q2$\bar{\text{q}}$ strongly depletes the prediction [12]. A recent estimate, based on the linear σ-model shows that there is a cancellation between the fermionic and pionic contributions which results in a rather small value(< 1 keV)[13]. One problem in disentangling the scalar contribution in two-photon reactions is that the tensor mesons behave partly as scalars when they are off-shell.

The main part comes from exchanges of ε and f in the s-channel. At threshold, we obtain the following amplitudes (which are the same for both the neutral and charged channels):

$$A_T = \frac{10}{9} \left(\frac{g_T}{g}\right)^2 \frac{1}{4 - \left(\frac{m_f}{m_\rho}\right)^2} e^2 (I_1 - 4I_2 - 4I_3 + 4I_4), \tag{5a}$$

$$A_S = \frac{10}{9} \left(\frac{g_S}{g}\right)^2 \frac{1}{4 - \left(\frac{m_\varepsilon}{m_\rho}\right)^2} e^2 (2I_1). \tag{5b}$$

A comparison with (3) shows that these two contributions go in the right direction: they cancel the production of charged ρ pairs and enhance the neutral ones. Using the experimental value of the f width, the constant multiplying e^2 in (5a) turns out to be about 0.4. The cancellation provided by the tensor is therefore incomplete. Ignoring the scalar, at this point we would have about equal neutral and charged cross-sections.

Now we estimate (crudely) the effect of higher order diagrams, considering only vectors and pseudoscalars. The theory is not renormalizable strictly speaking, because the vectors are massive and self-interacting. It is possible, however, to make sense of the amplitude at order one-loop. Here, we will be content with a simpler K-matrix estimate. Recall that it can be considered as a [1,1] Padé approximant to the one-loop amplitude where the real part is set to zero, but the imaginary part is correctly given by a unitarity relation. In order to implement this, we calculate the tree amplitudes from (1) and (2) for all the helicity states of the reactions $\gamma\gamma \to \gamma\gamma$, $\gamma\gamma \to 2\rho^0$, $\gamma\gamma \to \rho^+\rho^-$, $\rho^0\rho^0 \to \rho^+\rho^-$, $\rho^0\rho^0 \to 2\rho^0$, and $\rho^+\rho^- \to \rho^+\rho^-$, and perform (numerically) a partial wave projection. One should also consider the $\omega\omega$ channel, but we neglect it in view of the smallness of the one-pion exchange term in (3). We thereby obtain a 22x22 matrix, K^j. The T matrix is related to it by

$$T^j = K^j \left(1 - \frac{i}{16\pi\sqrt{s}} P K^j\right)^{-1}, \tag{6}$$

where P is a diagonal matrix containing the CMS momenta of the different channels involved and takes the Bose symmetry into account. Doing this, we obtain $\sigma(\rho^+\rho^-) \simeq \sigma(\rho^0\rho^0) \simeq 55$ nb, which is a significant improvement over the tree level result. One gets the feeling that combining the effect of K-matrix unitarization and of the tensor (and perhaps scalar) resonances, one may eventually succeed in explaining the experimental results.

Finally, let us mention other interpretations of these data. Alexander et al.[14] have pointed out that in terms of Regge trajectories the reaction for the neutral pair is dominated by the Pomeron, and the cross-section should therefore (asymptotically) be larger than the charged one. They are able to construct an amplitude which fits the $\rho^0\rho^0$ data even at low energies in this way. More speculative arguments were put forward in refs.15, where the results are explained in terms of interference effects between several new tensor resonances. We have shown in the present work that there is an important background contribution which should be taken into account.

References:

[1] G. t'Hooft, Nucl. Phys.B75, 461 (1974); S. Weinberg, Physica 96A , 327 (1979)

[2] I. Zahed and G.E. Brown, Phys. Reports 142, 1 (1986)

[3] S. Weinberg, 18, 188 (1967); J.A. Cronin, Phys. Rev. 161, 1483 (1967); P. di Vecchia et al., Nucl. Phys. B181, 318 (1981)

[4] J.L. Basdevant and. B.W. Lee, Phys. Lett. 29B, 437 (1969); L.H. Chan and R.W. Haymaker, Phys. Rev. D10, 4143, (1974)

[5] J.J. Sakurai, in "Currents and mesons",Chicago Press (1969)

[6] M. Poppe, Int. Jour. of Mod. Phys. A1, 545 (1987)

[7] N.M. Kroll, T.D. Lee and B. Zumino, Phys. Rev. 157, 1376 (1967)

[8] O. Kaymackallan, S. Rajeev and J. Schechter, Phys. Rev. D30, 594 (1984)

[9] D. Antreasyan et al., Phys. Rev. D33, 1847 (1986)

[10] C. Couraud et al., Nucl. Phys. B271, 1 (1986)

[11] V.M. Budnev and A.E. Kaloshin, Phys. Lett. 86B, 351 (1979)

[12] N.N. Achasov, S.A. Devyanin and G.N. Shestakov, Zeit. Phys. C16, 351 (1982)

[13] S. Hadjitheodoridis and B. Moussallam, IPN report (1987), to be published in Phys. Rev. D.

[14] G. Alexander, A. Levy and U. Maor, Zeit. Phys. C30, 65 (1986)

[15] N.N. Achasov, S.A. Devyanin and G.N. Shestakov, Zeit. Phys. C16, 55 (1982); B.A. Li and K.F. Liu, Phys. Rev. D30, 613 (1984)

Infrared Aspects of QCD

H.M. Fried

Physics Department, Brown University,
Providence, RI 02912, USA

These very brief remarks sketch a fairly new approach for the estimation of strong-coupling (SC) effects in continuum QCD, which have been devised by F. Guérin, T. Grandou, H.-T. Cho and myself. They have been applied, with reasonable success, so far only to two-dimensional systems; but applications to $(QCD)_4$, as well as to other non-linear problems in mathematical physics, are underway.

The essential motivation stems from the qualitative success of an array of eikonal models, for scattering and production, developed during the past two decades. If one imagines a two-particle scattering amplitude given as the sum over an infinite number of virtual meson exchanges, with some of those mesons virtually separating into closed loops and towers, etc., then the eikonal limit corresponds to that situation where the momentum transfer q is much less than the magnitude of the CM spatial momentum p of the incident particles, $(q/p) \ll 1$. But for such processes the order of magnitude of the momentum k of any virtual meson exchanged between the scattering particles is typically on the order of q, and hence the eikonal limit corresponds to the infrared (IR) restriction: $(k/p) \ll 1$.

The resulting amplitude can be expressed in closed form — although phenomenology must be used for cross channel exchanges of objects more complicated than towers — and one finds a function containing all powers of the appropriate coupling constant, which result is then applicable to SC situations. Thus the real lesson of eikonal physics is that SC approximations may be constructed, directly in the continuum, by retaining the IR, or "soft" part of all relevant virtual processes, and neglecting — as the first step of a systematic approximation scheme — all virtual effects containing non-IR, or "hard" momentum dependence.

How can one use these ideas for non-scattering situations, such as the construction of a correlation length, where there are no asymptotic momenta (and where the Bloch-Nordsieck techniques of scattering theory are not applicable)? This problem was treated in connection with the known chiral symmetry-breaking of $(QED)_2$ [1,2] and then generalized to $(QCD)_2$ [3,4]; these are studies of how $<\bar{\psi}\psi>$ maintains a non-zero value in the limit of vanishing fermion mass, m. The problem is simplest in two dimensions, for there gluonic renormalization effects need not be considered. It is relevant to SC because both m and the coupling g have the same dimensions, so that the true, dimensionless coupling is g/m; hence the SC and the chiral limits are the same. There are surely simpler ways of calculating $<\bar{\psi}\psi>$ in $(QED)_2$, and our interest here was only in the comparison of the results of these continuum, SC calculations with the known, exact answer. In addition, it is possible to extend these techniques to the calculation of $<\bar{\psi}\psi>$ in $SU(N) - (QCD)_2$, for arbitrary N. The IR method is in the process of being developed for use in other, more complicated, four-dimensional problems, but these will not be discussed here.

An exact form for $<\bar{\psi}\psi>$ can most simply be obtained by an Action Principle argument, which yields (in Euclidean space)

$$<\bar{\psi}\psi> = -\frac{\partial}{\partial m} \ln <0+|0-> / \int d^2x \ ,$$

where the vacuum-to-vacuum amplitude possesses a cluster expansion,

$$\ln <0+|0-> = \sum_{n=1}^{\infty} \frac{Q_n}{n!} \ .$$

Each connected Q_n describes the linkage of n closed fermion loops by photons (or gluons); for simplicity we here discuss only the "quenched" approximation involving Q_1, a single loop with arbitrary many photons (or gluons) exchanged across that loop. In our method there are no divergences, UV or IR, in quenched approximation, as explained in Ref. 1.

The essential calculation involves an IR approximation to the log of the fermion determinant, $L[A]$, and we here write this using two exact forms, the first that given by SCHWINGER [5] in terms of a quadrature over proper time,

$$L[A] = -\frac{1}{2} \int_0^\infty \frac{ds}{s} \ e^{-ism^2} \left\{ Tr\left[e^{-is(\gamma \cdot [i\partial + gA])^2}\right] - (g \to 0) \right\} \ ,$$

and the second form obtained by use of the exact FRADKIN [6] representation

$$L[A] = -\frac{1}{2} \int_0^\infty \frac{ds}{s} \ e^{-ism^2} \int d^2x \cdot N(s) \cdot \int d(\phi_\mu) e^{\frac{i}{4} \int_0^s ds' \phi^2} \delta\left(\int_0^s ds' \phi(s') \right) \cdot trU(s) \ ,$$

$$U(s) = \left(exp\left[-ig \int_0^s ds' \left(\phi_\mu(s') A_\mu(x - \int_0^{s'} ds'' \phi) - i\sigma_{\mu\nu} F_{\mu\nu}(x - \int_0^{s'} ds'' \phi) \right) \right] \right)_+ \ ,$$

where $N^{-1}(s) = \int d(\phi_\mu) e^{\frac{i}{4} \int_0^s \phi^2}$, and $\sigma_{\mu\nu} = \frac{1}{4}[\gamma_\mu, \gamma_\nu]$.

In QED, $F_{\mu\nu} = \partial_\mu A_\nu - \partial_\nu A_\mu$, while in QCD $A_\mu = \lambda_a A_\mu^a, F_{\mu\nu} = \lambda_a F_{\mu\nu}^a$, and $F_{\mu\nu}^a = \partial_\mu A_\nu^a - \partial_\nu A_\mu^a + g f_{abc} A_\mu^b A_\nu^c$, with λ_a the SU(N) Gell Mann matrices. One notes that if $\phi_\mu(s')$ is parametrized as a 4-velocity, $\phi_\mu = dx_\mu/ds'$, then Fradkin's representation takes the form of a Feynman path integral. It will be convenient subsequently to perform the continuation $s = -i\tau$, and to refer to τ as the proper time.

The IR method defined here is composed of two distinct approximations, each of which may be systematically corrected. These steps are (i) the "soft" approximation, where every $\tilde{A}_\mu(k)$ is rewritten as

$$\tilde{A}_\mu(k) = \tilde{A}_\mu(k) e^{-k^2/\mu_c^2} + \tilde{A}_\mu(k) \left[1 - e^{-k^2/\mu_c^2}\right] = (\tilde{A}_\mu)_{soft} + (\tilde{A}_\mu)_{hard} \ ,$$

and all dependence on A, in $L[A]$ is replaced by A_{soft}. This leading approximation will then contain all powers of g^2, and is therefore a candidate for a SC calculation.

Even with this restriction to soft frequencies, these forms are too complicated to allow a full computation, and we must have recourse to a further approximation. This is where the choice of μ_c enters, for if it can be arranged that $0(k_\nu \int_0^{s'} ds'' \phi_\nu(s'')) < 1$, then (ii) a "multipole" expansion can be defined whose leading term is obtained by discarding the $\int_0^{s'} ds'' \phi$ inside the F of $L[A]$, and by keeping the first non-vanishing term in the similar expansion of the argument of the A dependence of $L[A]$. Since $k \leq \mu_c$, and $\phi \sim (\tau)^{-1/2}$, one may choose $\mu_c \sim 1/\sqrt{\tau}$, or $\mu_c = c/\sqrt{\tau}$, with c a real constant on the order of unity. (Since τ itself scales as m^{-2}, and all subsequent integrals converge, in effect this means $\mu_c \sim m$.) The $\int d[\phi]$ of the Fradkin representation is now Gaussian — although τ-dependent — and all steps of the computation can be performed.

One notes that the result of approximations (i) and (ii) will depend on the constant c, in contrast to the exact result which must be independent of c. Rather like a constant of integration, or a subtraction constant in a dispersion relation, c must be determined by a comparison external to this computation. Its precise value depends on the accuracy of the approximation, rather like what one would find in a finite renormalization scheme. But since $c \sim 0(1)$, even without fixing a precise value of c, our calculations become estimates; and since these will be estimates of SC quantities directly in the continuum, they are not without a certain interest. It is also amusing to see just how, in this two-dimensional computation, a basically non-perturbative effect is obtained: each closed fermion loop containing more than a single photon individually vanishes as $m \to 0$; but the limit of the sum of all such terms does not vanish, and provides our non-perturbative result.

There is one most important question that must be resolved before one can express confidence in our results, and that is the matter of gauge invariance. In QED, $L[A]$ can be shown to be a functional of $F_{\mu\nu}$, so that our IR approximations are automatically independent of gauge. In QCD, however, $F_{\mu\nu}$ is not invariant; rather, the direction of its color vector is rotated by any gauge transformation, while its magnitude is left unchanged. The problem can quite generally be phrased in the following way. Non-Abelian gauge invariance here represents invariance of $L[A]$ under local gauge transformations in configuration space. Our soft approximation, however, corresponds to a local restriction in momentum space, and hence to a non-local restriction in configuration space. How can gauge invariance under local coordinate transformations be preserved under non-local restrictions? In answer to this question, we point to two explicitly gauge invariant constructions of the IR method (Ref. 4), and to a detailed computation which is invariant but not manifestly so (Ref. 3). The interesting point is that each of these three different methods for setting up the IR approximation yields the same answer for $<\bar\psi\psi>$, except that the constants c appearing in each result differ by factors of the $0(1)$.

Finally, we state our results, and begin by writing the exact, well-known answer [7] for $<\bar\psi\psi>$ in $(QED)_2$,

$$<\bar\psi\psi>^{exact}|_{\frac{m}{g}\to 0} = -\frac{ge^\gamma}{2\cdot\pi^{3/2}},$$

where γ is Euler's constant. The IR analysis of Ref. 1 generates for this quantity, in quenched approximation, the value

$$<\bar\psi\psi>^Q_{IR}|_{\frac{m}{g}\to 0} = -\frac{gc}{4\cdot\pi^{3/2}},$$

while that of Ref. 2, which goes beyond the quenched approximation to include an estimate of all "chain graphs" in the cluster expansion, yields

$$<\bar\psi\psi>^{ch.gr.}_{IR}|_{\frac{m}{g}\to 0} = -\frac{gc}{4\cdot\pi^{3/2}}[1-\frac{1}{4}],$$

showing an expected diminution of the magnitude of the quenched result [8]. It is amusing to see that the peculiar "phase space" of the exact answer, proportional to $(\pi)^{-3/2}$, is reproduced by the IR method.

In $(QCD)_2$, the quenched computation of Ref. 3 yields

$$<\bar{\psi}\psi>^Q_{IR}|_{\frac{m}{g}\to 0} = -gc\xi(N) ,$$

where one can easily show the bounds: $aN \leq \xi(N) \leq bN^{3/2}$, with $\underline{a}$ and $\underline{b}$ numerical constants, independent of N. In the large N limit, $\xi(N) \to bN^{3/2}$, reproducing the form of ZHITNITSKI's result [9] in the planar limit, $g^2 N = constant$. In the IR estimate there is no restriction on g; but there is no available way of determining c except by comparison with Zhitnitski in the large N, planar limit.

In summary, we have shown that an IR method can be devised for use in estimating at least these simplest of SC problems. Attempts at an IR description of $(QCD)_4$ are presently underway.

References

1. F. Guérin and H. M. Fried, Phys. Rev. D33 (1986) 3039.

2. H. M. Fried and T. Grandou, Phys. Rev. D33 (1986) 1151.

3. T. Grandou, H.-T. Cho and H. M. Fried, "Chiral Symmetry Breaking in Continuum $(QCD)_2$ by an Infrared Method," Nice preprint TH86/13.

4. H.-T. Cho, H. M. Fried and T. Grandou, "Non-Abelian Gauge Invariance and the Infrared Approximation," Nice preprint TH86/14.

5. H. Schwinger, Phys. Rev. 82 (1951) 664.

6. E. S. Fradkin, Nucl. Phys. 76 (1966) 588.

7. S. Coleman, Ann. Phys. (NY) 101 (1976) 239; S. Coleman, R. Jackiw and L. Susskind, *ibid.* 93 (1975) 267.

8. E. Marinari, G. Parisi, and C. Rebbi, Nucl. Phys. B190 (1981) 734.

9. A. R. Zhitnitski, Phys. Letters 165B (1985) 405.

Many-Body Techniques Applied to QCD and Aspects of Confinement*

D. Schütte

Institut für Theoretische Kernphysik der Universität Bonn,
Nussallee 14–16, D-5300 Bonn, Fed. Rep. of Germany

1. INTRODUCTION

The numerical complexity of lattice Monte-Carlo calculations for the spectrum of QCD makes it desirable to investigate alternative formulations of gauge field theories. A natural alternative is the treatment of the field theoretical eigenvalue problem by suitable many-body approximation techniques. (Replace the Feynman path-integral by the Schrödinger equation).

It is the purpose of this talk to give a survey on the structure of such a field theoretical many-body problem, especially with respect to renormalization (regularization and check of scaling in the continuum limit) and with respect to the different choices of regularization for gauge theories (lattice regularization and complete gauge fixing).

A comparison of the structure of the many-particle problems within these two regularization schemes yields some new aspects of confinement.

2. GENERAL STRUCTURES OF QUANTUM FIELD THEORIES

A Poincaré invariant Lagrangian $L(\phi, \partial_\mu \phi)$ (for simplicity, we discuss in this section only the scalar case) yields, via canonical quantization, operators

$$P^\mu(\phi,\pi), \quad L^{\mu\nu}(\phi,\pi) \quad , \quad [\phi(\vec{x}), \pi(\vec{y})] = -i\delta(\vec{x}-\vec{y}) \tag{1}$$

of energy-momentum and angular momentum which <u>formally</u> obey the commutation relation of the Poincaré algebra. In principle, the decomposition of this reducible, unitary representation of the Poincaré group should yield bound states (elementary particles) and scattering states as eigenstates of the Hamiltonian $P^0 = H$. However, the pathological character of H for a quantum field theory with interaction does not allow to argue in such a direct way and renormalization has to be invoked to make the theory (hopefully) well defined. Since non-perturbative methods must be used for the computation of bound states, a convenient strategy [1] for renormalization is a <u>regularization</u> of the (t=0) field operators:

$$\phi(\vec{x}) \rightarrow \phi_N(\vec{x}) = \sum_{\alpha=1}^{N} f_\alpha(\vec{x}) q_\alpha \tag{2}$$

and a <u>check of scaling</u> in the continuum limit. In (2), $\{f_\alpha(\vec{x}), \alpha = 1,2...\}$ is a suitable single particle basis and $N = N(\Omega, M)$,

---
*Supported by the Deutsche Forschungsgemeinschaft

Ω = volume, M = momentum cutoff. This yields the eigenvalue problem of the regularized Hamiltonian $H(g,\Omega,M) = H(p_1...p_n, q_1...q_n)$, $p_\alpha = -i\partial/\partial q_\alpha$, as that of a well-defined "standard" many-body system (The wave functions are $\Psi = \Psi(q_1...q_N)$ and $\langle\Psi,\Psi\rangle = \int d^N q \Psi^*\Psi$). In principle, the solution of this eigenvalue problem yields then observables $A(g,M,\Omega)$ whose "thermodynamical" limit $A(g,M) = \lim_{\Omega\to\infty} A(g,M,\Omega)$ will exist [2]. (Note, however, that the reliable determination of $A(g,M)$ is, in practise, a major computational problem). The function $A(g,M)$ behaves pathologically for $M\to\infty$. However, for a renormalizable field theory, a universal function $g(M)$, the "running coupling constant" [3] should exist such that

$$\lim_{M\to\infty} A(g(M),M) = \text{well-behaved for all observables A} \quad (3)$$

For a Yang-Mills theory or massless QCD, the M-dependence of $A(g,M)$ is trivial:

$$A(g,M) = M^D A(g,1) \quad \begin{aligned} D &= \text{dimension of A} \\ &= 1 \text{ for masses} \\ &= 2 \text{ for string tension} \end{aligned} \quad (4)$$

and (3) reduces to a predetermined "scaling" behaviour of $A(g,1)$

$$A(g,1) \simeq \exp(-D/\beta_0 g^2) \quad \text{for} \quad g\to 0 \quad . \quad (5)$$

This structure of $A(g,1)$ is observed (within errors) in lattice Monte-Carlo calculations [4] which supports the confidence that the continuum limit (3) should be in order.

3. REGULARIZATION OF GAUGE FIELD THEORIES

In the case of gauge field theory, the construction of a regularized theory yielding $A(g,\Omega,M)$ is more involved since now the physical variables are the "orbits" of gauge fields, i.e. the equivalence classes of fields $A_\mu(x) \in$ Lie SU(n) ($x = (t,\vec{x})$) where

$$A'_\mu \simeq A_\mu \iff \text{there exists } U(x) \in SU(n) \text{ with}$$
$$A'_\mu = U(A_\mu - i\partial_\mu)U^{-1} \quad . \quad (6)$$

The convenient way to work with such equivalence classes is to choose a representative (fix the gauge) and to check that observables $A(g,M)$ are independent from the choice of this representative. In the framework of canonical quantization, a convenient starting point is a partial gauge fixing by setting $A_0=0$. (This leaves open time-independent gauge transformations). Quantization of the gauge field Lagrangian yields then a formally covariant theory only on the "physical" states $\Psi(\vec{A})$ obeying the Gauss law [5,6]

$$\Psi(A) = \Psi(A') \quad \text{for any } \vec{A}' \text{ with}$$
$$\vec{A}'(\vec{x}) = U(\vec{x})(\vec{A}-i\vec{\nabla})U^{-1}(\vec{x}) \quad . \quad (7)$$

This reduces the variables of the theory to the gauge orbits. There exist two regularizations of the gauge field theory which are consistent with the Gauss law, the lattice regularization and the regularization of equivalence classes using complete gauge fixing. (Hereby, the Coulomb gauge appears to be the most consistent choice).

3.1 Lattice Regularization

This regularization is described in detail in ref. [1]. The basic idea is to use gauge invariant wave functions $\Psi(A)$ related to the parallel transport along a closed loop κ

$$\chi_\kappa(A) = \text{tr } P \text{ expi} \int_\kappa \vec{A} d\vec{x} . \tag{8}$$

$\vec{A}(\vec{x})$ is regularized with the characteristic functions related to a (3-dimensional, finite) lattice (with links $\kappa_1 \ldots \kappa_N, \kappa_\alpha$ goes $\vec{x}_\alpha$ to $\vec{y}_\alpha$)

$$\vec{A}(\vec{x}) = \sum_{\alpha=1}^{N} \sum_{a=1}^{R} \vec{f}_\alpha(\vec{x}) \lambda_a q_{\alpha a} \tag{9}$$

$\{\lambda_1 \ldots \lambda_R\}$ = Basis Lie SU(n).

$\chi_\kappa(A)$ is then computable for lattice loops and is just the trace of the product of link-parallel transports (we assume $\epsilon = 1$, ϵ = lattice constant)

$$U_\alpha = \text{Pexp i} \int_{\kappa_\alpha} \vec{A} d\vec{x} = \exp i\lambda_a q_{\alpha a} . \tag{10}$$

Introducing U_α as new variables leads then to the Kogut-Susskind Hamiltonian form of standard lattice QCD [7,8] where (without quarks)

$$H_{ks} = \sum_\alpha \frac{g^2}{2} \text{tr } E_\alpha^2 - \frac{1}{2g^2} \sum_\square \text{tr}(P_\square + P_\square^{-1} - 2) \tag{11}$$

(E_α is canonically conjugate to U_α, $P_\square$ is the parallel transport (8) along a plaquette). The wave functions (describing a quantum many-top problem) must obey the Gauss law on the lattice ($g(\vec{x}) \in SU(n)$, arbitrary)

$$\Psi(U_1' \ldots U_N') = \Psi(U_1 \ldots U_n); \quad U_\alpha' = g(\vec{x}_\alpha) U_\alpha g^{-1}(\vec{y}_\alpha) . \tag{12}$$

The computation of observables related to the eigenfunctions of H_{ks}, but framed in the language of Feynman path integrals, leads to the standard lattice Monte-Carlo QCD [4]. Attempts to solve the eigenvalue problem of H_{ks} directly using many-body techniques are discussed in ref. [9].

3.2 Regularization in the Coulomb gauge

In the framework of complete gauge fixing, a specific description of the orbit space of gauge fields is used by choosing representatives. Herefore, one relates any $\vec{A}(\vec{x})$ to a (unique) representative $\vec{F}_Q(\vec{x})$ ($Q \in I$ = suitable index set) such that a (modulo the center of SU(n) unique) gauge field $U(\vec{x})$ exists with [1]

$$\vec{A} = U(\vec{F}_Q - i\vec{\nabla})U^{-1} . \tag{13}$$

This allows us to change coordinates in the functional space $\vec{A} \Leftrightarrow (U,Q)$ and the Gauss law may be trivially fulfilled for $\Psi = \Psi(Q)$ (independent of U). The set $\vec{F}_Q, Q \in I$ is conveniently determined via a gauge fixing condition. Translational and rotational invariance favours

$$\vec{\nabla}\vec{F}_Q = 0 \text{ , the Coulomb gauge .} \tag{14}$$

The regularization is then introduced on these representatives:

$$\vec{F}_Q(\vec{x}) = \sum_{\alpha=1}^{N} \vec{f}_\alpha(x) Q_\alpha , \quad (Q_1 \ldots Q_N) \ R^N . \tag{15}$$

The condition that the set $F_Q, Q \in I \subset R^N$ should be a set of <u>unique</u> representatives for the fields $\vec{A}(\vec{x})$ leads to the condition that $Q = (Q_1 \ldots Q_N)$ should be restricted to a "Gribov-domain" I which is bounded by the Gribov horizon [10]. Quantitative estimates of Gribov [10] and Cutkosky [11] yield that this horizon is well approximated by an ellipsoid in R^N, i.e. I is to a good approximation probably given by

$$I = \{Q \in R^N \mid \sum c_\alpha^2 Q_\alpha^2 < 1\} . \tag{16}$$

The precise definition of I is given in refs. [1,10].

The regularized Coulomb gauge Hamiltonian $H_C(Q, \partial/\partial Q)$ of a Yang-Mills theory emerging from the non-linear coordinate transformation $\vec{A} \leftrightarrow (U,Q)$ is rather complicated [1,5,6]. The important feature of H_C is that it is singular at the Gribov horizon. Attempts to cope with the eigenvalue problem of H_C have been undertaken by several groups [11,12]. Like in the case of the lattice regularization, no many-body approximation has been found up to now which yields observables fulfilling the rather stringent scaling condition (5).
E.g., standard non-perturbative many-body techniques (exp S, cluster expansions) were applied to compute glue-ball spectra [12], but these techniques turned out to be suitable only for large momenta, the small momentum behaviour of the wave functions, where the Gribov-horizon is expected to play an important role, has still essentially to be improved.

4. ASPECTS OF CONFINEMENT

The singularity of the Coulomb gauge Hamiltonian H_C at the Gribov horizon suggests a special confinement scenario [1]. Herefore, one introduces static quarks and studies the structure for the strong coupling limit and its transition to the continuum limit. Within the lattice regularization, confinement is trivial in the strong coupling limit (due to the Gauss law), the weak coupling case is not accessible analytically. Lattice Monte-Carlo calculations, however, strongly suggest that confinement is still valid: the string tension is different from zero for $g \to 0$ and scales (at leasts up to 20%) [4].

Within the Coulomb gauge, confinement is non-analytic even in the strong coupling case. However, certain necessary conditions [1] (induction of non-trivial vacuum polarization effect as an indication of a string formation) hold because of the singularity of H_C at the Gribov horizon.

Assuming that confinement is true in the strong coupling limit, some possible conclusions for the nature of confinement for $g \to 0$ can be drawn. It is suggested that the eigenfunctions $\Psi(Q_1 \ldots Q_N)$ for $g \to 0$ behave with respect to the variable Q_α ($\alpha = (\vec{k},r,a)$ = (momentum, polarization, colour) [2]) for <u>small</u> $\vec{k}_\alpha$ ($|\vec{k}_\alpha| < M_0$) just like in the strong coupling case $g \to \infty$, whereas one should have perturbative structures for $|\vec{k}_\alpha| \gg M_0$. The reason, herefore, is that the Gribov horizon shrinks for small k_α (e.g. $C_\alpha^{-1} \simeq |k_\alpha|^m$, $m > 1$).

It is expected that $M_0 = M_0(g)$ scales (i.e. $M_0 \simeq \exp(-1/\beta_0 g^2)$), but since it remains finite for any finite g, the strong coupling nature of $\Psi(Q)$ for Q_α with $|\vec{k}_\alpha| < M_0$ is sufficient to keep confinements in the continuum limit. (For details see ref. [1]). Numerical investigations aiming at supporting these suggestions are in progress.

4.1 References

1. D. Schütte, The many-body Structure of QCD within different Regularization Schemes, Proceedings of the International Workshop on Quarks, Gluons and Hadronic Matter, Cape Town 1987, to be published
2. D. Schütte, Phys. Rev. D31 (1984) 810
3. J.B. Kogut, Rev. Mod. Phys. 55 (1983) 775
4. B.J. Pendleton, Non-Perturbative QCD, contribution to the same proceedings of the Les Houches Workshop 1987
5. N.H. Christ and T.D. Lee, Phys. Rev. D22 (1980) 939
6. H. Cheng and E.C. Tsai, Canonical Quantization of Non-Abelian Gauge Field Theories and Feynman Rules, MIT preprint, to appear in Nucl. Phys. B
7. J. Kogut, L. Susskind, Phys. Rev. D11 (1975) 395
8. M. Creutz, Quarks, Gluons and Lattices (Cambridge University Press, 1983)
9. D. Horn et al, Phys. Rev. D31 (1985) 2589
   W. Furmanski, A. Kalowa, Yang-Mills Vacuum - an Attempt of Lattice Loop Calculus, (Preprint 1986) CALT-68-1330
   S.A. Chin, O.S. van Roosmalen, E.A. Umland and S.E. Koonin, Phys. Rev. D31 (1985) 3201
10. V.N. Gribov, Nucl. Phys. B139 (1978) 1
11. R.E. Cutkosky, Phys. Rev. D30 (1984) 447
12. B. Faber, H. Nguyen-Quang and D. Schütte, Phys. Rev. D34 (1986) 1157
    H. Nguyen-Quang, Dissertation Bonn (1986)

Hadronic Reactions at Large Momentum Transfers

J. Soffer

Centre de Physique Théorique*, CNRS, Luminy, Case 907,
F-13288 Marseille, Cedex 9, France
*Laboratoire propre LP-7061, Centre Nationale de la Recherche Scientifique

1. INTRODUCTION

Hadronic collisions at large momentum transfers are expected to provide significant tests of our knowledge of strong interactions dynamics at short distances. This physics is quite interesting because the description of the hard part of these processes contains most valuable information on the deep structure of hadrons. However, one should be prepared to deal with small event rates and small cross sections (e.g. a few nanobarns or less for exclusive reactions) because indeed, only a tiny part of the hadronic wave functions is involved at high energy and large scattering angles. It is in precisely this kinematic domain that new experiments have shown a strong dependence of scattering processes on the initial or/and final hadron spin state. We will see that these spin effects are puzzling and must play a major role for a comprehensive understanding of the underlying dynamics. The outline of this report is as follows. In the next section we will review the subject of nucleon - nucleon elastic scattering at large angles, the experimental situation and the implications for various theoretical models. Section 3 is devoted to exclusive meson - nucleon scattering. In Section 4 we will consider the large Q^2 behaviour of the nucleon and the pion elastic form factors and the interesting possibility of having a zero in the space-like region. High $p_\perp$ inclusive hyperon production where one observes a large transverse polarization which remains a challenge for theorists will be discussed in Section 5.

2. NUCLEON - NUCLEON ELASTIC SCATTERING

Let us recall that nucleon - nucleon elastic scattering is described in terms of five independent helicity amplitudes ϕ_i (i = 1,...,5) usually denoted as follows : $\phi_1 = <++|++>$, $\phi_2 = <++|-->$, $\phi_3 = <+-|+->$, $\phi_4 = <+-|-+>$ and $\phi_5 = <++|+->$. We will be concerned with the following set of observables defined in [1] whose expressions in terms of the ϕ_i's are

$$\sigma_0 = s^2 \, d\sigma/dt = 1/2 \, (|\phi_1|^2 + |\phi_2|^2 + |\phi_3|^2 + |\phi_4|^2 + 4|\phi_5|^2)$$
$$A = \text{Im} \, [\, (\phi_1 + \phi_2 + \phi_3 - \phi_4) \, \phi_5^* \,] / \sigma_0$$
$$A_{NN} = \text{Re} \, (\, \phi_1 \phi_2^* - \phi_3 \phi_4^* + 2|\phi_5|^2 \,) / \sigma_0$$
$$D_{NN} = \text{Re} \, (\, \phi_1 \phi_3^* - \phi_2 \phi_4^* + 2|\phi_5|^2 \,) / \sigma_0 \tag{1}$$
$$K_{NN} = \text{Re} \, (\, \phi_3 \phi_2^* - \phi_1 \phi_4^* + 2|\phi_5|^2 \,) / \sigma_0$$
$$A_{LL} = - (|\phi_1|^2 + |\phi_2|^2 - |\phi_3|^2 - |\phi_4|^2) / 2\sigma_0$$
$$A_{SS} = \text{Re} \, (\phi_1 \phi_2^* + \phi_3 \phi_4^*) / \sigma_0$$
$$A_{LS} = \text{Re} \, [\, (\phi_1 + \phi_2 - \phi_3 - \phi_4) \, \phi_5^* \,] / \sigma_0 .$$

$d\sigma/dt$ is the unpolarized cross section for which a fair amount of large angle data is known as well as for A called the analyzing power and for A_{NN} the transverse spin - spin correlation parameter. Let us take a brief look at this data first at fixed energy. At p_{lab} = 24 GeV/c (see Fig. 1a), after the characteristic narrow diffraction peak, $d\sigma/dt$ has a slower falloff in t up to $\theta_{c.m.}$ = 45°. At 6 GeV/c the analyzing power for np elastic scattering is negative at large angles

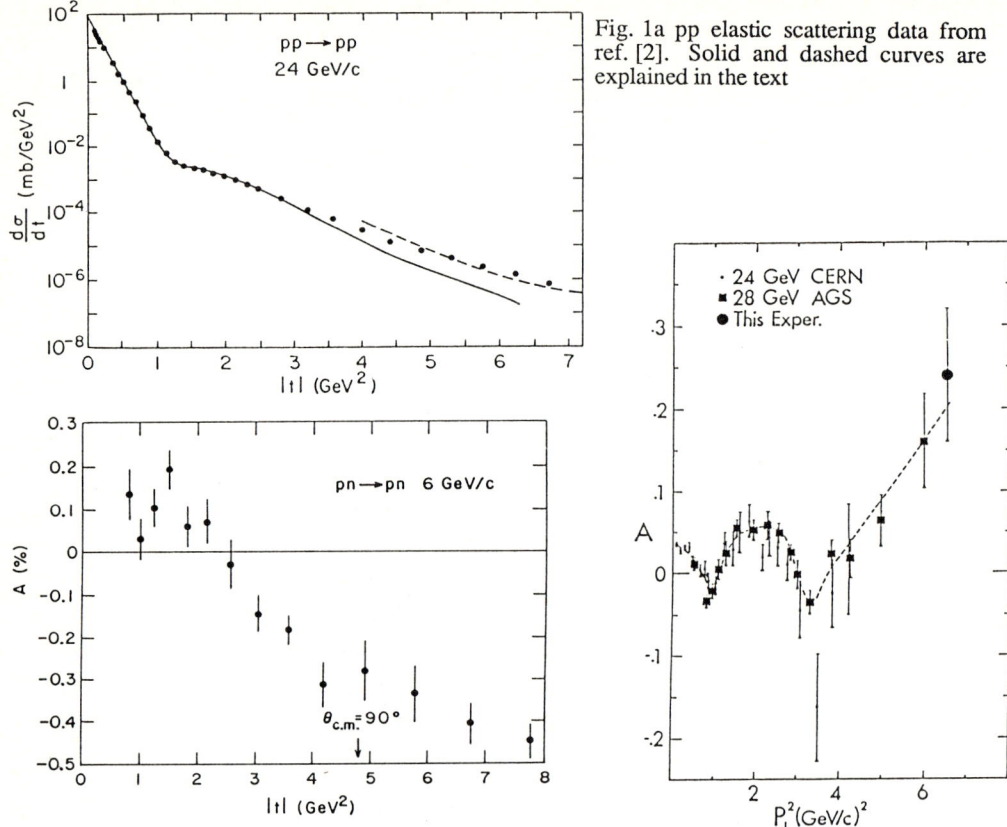

Fig. 1a pp elastic scattering data from ref. [2]. Solid and dashed curves are explained in the text

Fig. 1b The analyzing power A for np elastic scattering at 6 GeV/c from ref. [3]

Fig. 1c The analyzing power A for pp elastic scattering versus $p_\perp^2$ taken from ref. [4]. The curve is hand-drawn to guide the eye

and of the order of -40 % near 90° (see Fig. 1b). For pp elastic scattering at 28 GeV/c, A is of the order of 5 % or so in the small angle region but increases to a much higher positive value for $\theta_{c.m.} \sim 45°$ as shown in Fig. 1c. At fixed angle $d\sigma/dt$ near 90° has a very fast drop off with energy up to 20 GeV/c or so (see Fig. 2a) and A_{NN} has a dramatic rise above 8 GeV/c up to 60 % at the highest energy ever measured as shown in Fig. 2b. At $\theta_{c.m.} = 50°$ (see Fig. 2c) the behaviour of A suggests the appearance of a new regime for $p_{lab} > 20$ GeV/c, which remains to be confirmed by the measurement of A_{NN} at higher energy. Is there a simple interpretation of these experimental facts ? pp and $\bar{p}$p elastic scattering at small angles are well described by a model based on a new impact picture for low and high energy [8] whose predictions given in 1979 in the TeV energy range, are in excellent agreement with recent CERN $\bar{p}$p collider data [9]. The scattering amplitudes constructed in the framework of this model provide an accurate representation of the soft part of the scattering process which allows a fair description of $d\sigma/dt$ at the PS energy, up to $|t| \simeq 4$ GeV^2 as shown by the solid curve in Fig. 1a. Beyond this value the diffractive contribution drops faster than the data, suggesting that it cannot be the "whole story" since the hard part of the scattering becomes important at large angles and should be included to complete the physical picture. Before we discuss different possibilities let us make some remarks about the special case of pp elastic scattering at 90°. As in the forward direction, life is simpler because only three amplitudes survive since $\phi_3 = -\phi_4$ and $\phi_5 = 0$.

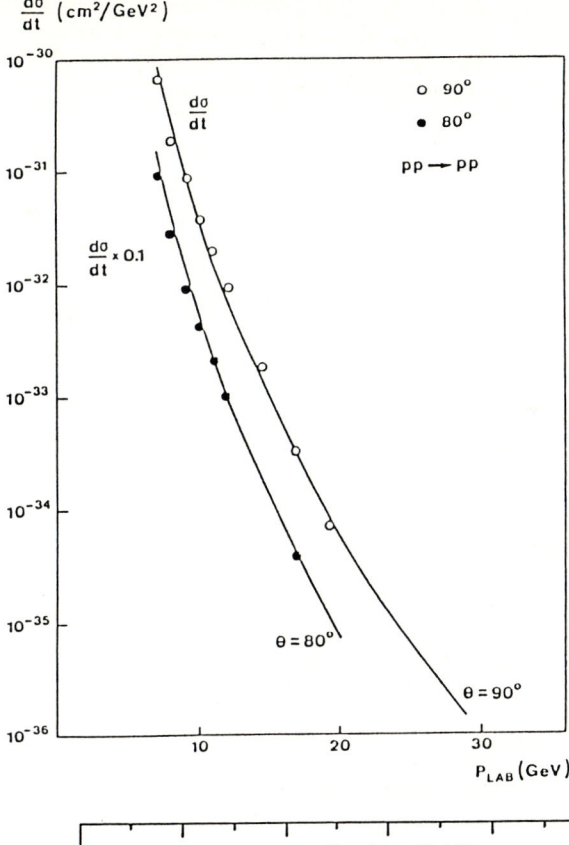

Fig. 2a pp elastic scattering data at large angles from ref. [5]. The curves are explained in the text

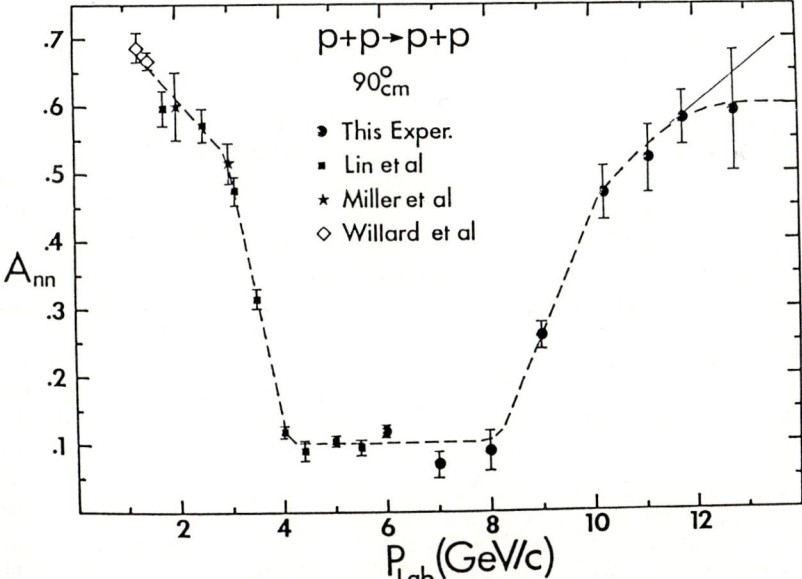

Fig. 2b The A_{NN} parameter for pp elastic scattering at $\theta_{c.m.} = 90°$ versus p_{lab} from ref. [6]. The curve is hand-drawn to guide the eye

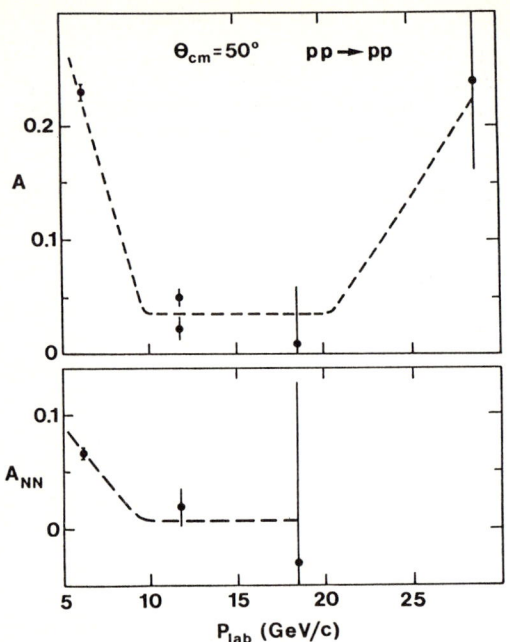

Fig. 2c pp polarization parameters A and A_{NN} versus p_{lab} for $\theta_{c.m.} = 50°$ (data from [4] and [7]). The curves are hand-drawn to guide the eye

So instead of twenty-five observables resulting from the initial five amplitudes, we have only nine observables; eight of them vanish (e.g. A = 0) and there are eight linear relations e.g.

$$D_{NN} = K_{NN} \quad \text{and} \quad 1 + A_{LL} = A_{NN} - A_{SS}. \tag{2}$$

In addition there exist special positivity relations, generalizations of the obvious constraint $|A| \leq 1$, like for example

$$|D_{NN}| \leq \frac{1 + A_{NN}}{2}. \tag{3}$$

Knowing A_{NN} and A_{LL}, (2) and (3) can be used to predict A_{SS} and to put limits on $|D_{NN}|$. These relations which are a guide for experimentalists should be respected by any dynamical assumptions. Clearly they do not hold for np elastic scattering because of non-identical particles.

One possible theoretical framework is perturbative QCD (PQCD) predicting for the cross section of any exclusive process $a + b \rightarrow c + d$, the asymptotic scaling law

$$\frac{d\sigma}{dt} = \frac{1}{s^n} F(\theta_{c.m.}) \tag{4}$$

with n = 10 for baryon-baryon scattering. This rapid energy dependence is in reasonable agreement with the data shown in Fig. 2a. However the exact expression for $F(\theta_{c.m.})$ providing both the absolute normalization and the angular dependence of $d\sigma/dt$ involves in PQCD an impressive number of diagrams [10] and is not yet available. There is another simple feature of PQCD that one ought to test with the spin-dependent observables, namely, *the helicity conservation rule* [11]. The sum of the initial helicities must be equal to the sum of the final ones,

$$\lambda_a + \lambda_b = \lambda_c + \lambda_d, \tag{5}$$

as a consequence of the vectorial nature of the gluon and of the near masslessness of light quarks. For both pp and np elastic scattering this rule implies that the amplitudes ϕ_2 and ϕ_5 must be zero at large angles. Let us examine several consequences of this restrictive situation at large angles in addition to the fact that the helicity amplitudes resulting from PQCD are expected to be essentially real, although this last assumption can be relaxed, as we will see later.

i) The analyzing power A must vanish because it is linear in ϕ_5, but this is obviously not satisfied by the data presented earlier.

ii) There are new constraints between the double spin correlation parameters, for example,

$$A_{SS} = - A_{NN}$$

and

$$\left|\begin{matrix} D_{NN} \\ K_{NN} \end{matrix}\right| = \sqrt{\tfrac{1}{2}(1-A_{LL})(1+A_{LL} \pm \sqrt{(1+A_{LL})^2 - 4A_{NN}^2})}, \qquad (6)$$

so by measuring A_{NN} and A_{LL} it is possible to predict A_{SS}, $|D_{NN}|$ (or $|K_{NN}|$), which are consistency checks. At 11.75 GeV/c, the data does not satisfy this because in particular for $\theta_{c.m} > 65°$ one has [12]

$$(1 + A_{LL} - 2 A_{NN})/4 = (-0.072 \pm 0.02), \qquad (7)$$

which, according to (6), should be a positive quantity !

iii) A_{NN} now has a much simpler form and reads

$$A_{NN} = \frac{- 2 \phi_3 \phi_4}{\phi_1^2 + \phi_3^2 + \phi_4^2} . \qquad (8)$$

At this stage one may ask the following question : is it easy to understand the rapid variation seen in the data, namely, $A_{NN} \sim 0$ at 50° and $A_{NN} \sim 60\%$ at 90° ? From (8) one needs either $\phi_3/\phi_1 \sim 0$ or $\phi_4/\phi_1 \sim 0$ at 50° and $\phi_3/\phi_1 = - \phi_4/\phi_1 = \sqrt{3}/2$ at 90°. In most models $\phi_3/\phi_1 \sim 1$ *at all angles*, so the ratio of the double flip amplitude ϕ_4 to the non-flip amplitude ϕ_1 should vary quickly between 50° and 90° as already noted before [13], in contrast with the statement made in [14]. Another consequence of $\phi_1 \sim \phi_3$, provided ϕ_5 is not zero, is the fact that A_{LS} defined in (1) is expected to be very small for all angles, in agreement with experimental observation [12]. The constituent-interchange model [15] which satisfies the helicity conservation rule is even more restrictive because one has

$$\phi_1 = \phi_3 - \phi_4 ,$$
$$\phi_3 = 56 \, f(\theta_{c.m}) + 68 \, f(\pi - \theta_{c.m}) , \qquad (9)$$
$$\phi_4 = - 68 \, f(\theta_{c.m}) - 56 \, f(\pi - \theta_{c.m}) ,$$

with

$$f(\theta_{c.m}) = 0.5 / (1 - \cos \theta_{c.m})^4.$$

It leads to $A_{NN}(90°) = 1/3$ and almost the same value at 50°, in complete contradiction with the 12 GeV/c data [6].

We regard this situation as evidence for a serious need of non-perturbative effects in a kinematic region where PQCD is believed to be relevant. The difficulties mentionned above do not appear in the framework of Quark Geometrodynamics (QGD) and its more refined version Anisotropic Chromodynamics (ACD) whose applications to large angle scattering have been discussed elsewhere [16]. This approach displays many features of the naive quark model and in addition it incorporates quark confinement explicitly. The two-body large angle amplitudes are obtained by folding the hadron vertex functions with the elementary quark-quark scattering amplitudes. The vertex functions are constructed with the assumption that during the scattering process the initial and final baryons have in common two spectator quarks which conserve spin, momentum and the internal degrees of freedom. The energy dependence of the vertex functions is consistent with the large Q^2 behaviour of the hadron form factors which will be specially discussed in Section 4. At the level of the basic quark-quark amplitudes it is assumed that the short-distance force (which is color neutral) is generated by the exchange of several infinite towers of mesons, which are not only transverse vector states (analogous to gluons in PQCD), but also pseudoscalar and longitudinal vector states. As a consequence, the spin structure of the amplitudes is not as restrictive as that of PQCD and does not obey the helicity conservation rule. Therefore one obtains a set of five real amplitudes given in Table 1 of [16] which allows a good description of $d\sigma/dt$ near 90° as shown by the solid curves in Fig. 2a. The model does not provide the absolute normalization which was fixed at 10 GeV/c and 90°. By simply adding these contributions to the impact picture amplitude one gets the dashed line shown in Fig. 1a in better agreement with the data for $|t| > 5$ GeV^2 or so. For $p_{lab} = 28.7$ GeV/c at the highest available t-value, $|t| = 10$ GeV^2, we find $d\sigma/dt = 3.7 \times 10^{-8}$ mb/GeV^2 which can be compared with the data $d\sigma/dt = (3.5 \pm .7) 10^{-8}$ mb/GeV^2 reported in [17]. If we now calculate the analyzing power A at 28 GeV/c we find a rapid increase near $\theta_{c.m} = 45°$ up to 30 % or so in agreement with the data as shown in Fig. 3. This large effect is due to a maximum interference between the hard scattering amplitudes, which are real, and the dominantly imaginary impact picture contribution. The arrow indicates that below $p_T^2 = 5$ GeV^2, the solid line turns to a broken line which is not a reliable prediction. In Fig. 3 we also show our prediction for A at 50 GeV/c, which is reliable down to $p_T^2 = 2$ GeV^2 or so because the cross section is also well described down to smaller p_T^2 values. In Fig. 4 the predictions for A_{NN} exhibit a rise to about 100 % near 90°. Many other interesting predictions for pp and np elastic scattering can be found in [18]. Let us emphasize that if the predictions for A and A_{NN} turn out to be verified, they would confirm the correctness of the phase of the impact picture and of the magnitude of the hard scattering single flip amplitude ϕ_5. Higher-order twist effects have been invoked to avoid the difficulties of PQCD, and by heuristic arguments [19] one chooses $\phi_5 \sim 1/5\, \phi_1$ with an arbitrary phase difference $\eta \sim 45°$. There is also an arbitrary phase in [20] where a nice description of large angle exclusive reactions is given, within a consistent picture assuming the existence of diquarks inside baryons.

Finally we briefly mention an attempt to produce a phase in PQCD. It is based on the consideration of subtle Sudakhov effects which generate the so-called Chromo-Coulomb Phase [21] analogous to the QED Coulomb Phase. At fixed angle one gets a factor

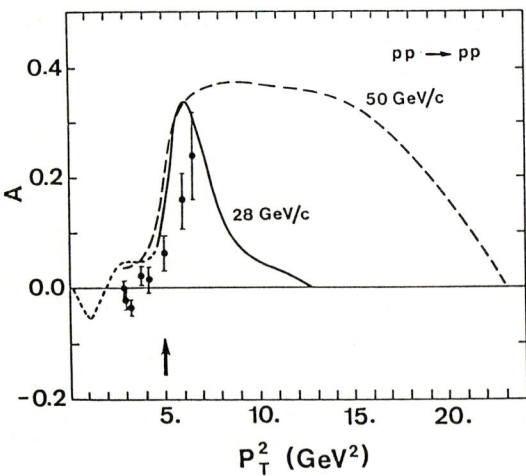

Fig. 3 Model predictions for the pp analyzing power at 28 GeV/c and 50 GeV/c taken from ref. [18] (data at 28 GeV/c from ref. [7])

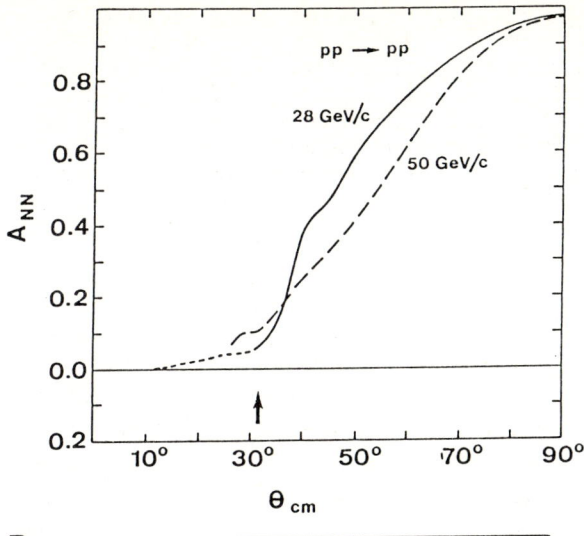

Fig. 4 Model predictions for A_{NN} versus $\theta_{c.m}$ in pp elastic scattering at 28 GeV/c and 50 GeV/c taken from ref. [18]

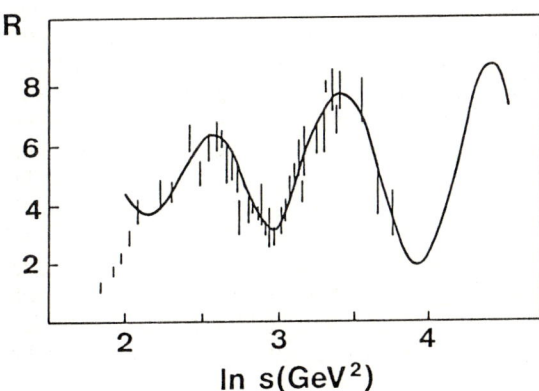

Fig. 5 The quantity $R = S^{10} d\sigma/dt$ for pp elastic scattering at 90° taken from ref. [21]

$$e^{-\ln^2(-s)} = e^{-\ln^2 s} e^{i\pi \ln s}, \qquad (10)$$

introducing an energy dependent phase in the scattering amplitudes. There is some evidence for an oscillatory pattern with energy of the pp elastic cross section at 90° as shown in Fig. 5. It is also conceivable that some spin observables at fixed angle oscillate as possibly suggested from Fig. 2c, but clearly more data is needed.

3. MESON - BARYON SCATTERING

Meson-baryon scattering at large angles is also a good testing ground for the theoretical ideas discussed above. As before we will see that the simplified picture of gluon exchange does not seem to emerge from the data. Most of this large angle data is available for unpolarized differential cross sections but in general they are poorly measured near 90° due to the lack of statistics for $p_{lab} \geq 10$ GeV/c. The QGD-ACD approach which has been applied to this class of reactions [16] allows prediction of the angular distribution of several elastic and inelastic reactions as well as their relative normalization [22]. The predicted cross sections for $\pi^- p \to \pi^- p$ and $\pi^- p \to \rho^- p$ are in fair agreement with the 10 GeV/c data from [23] and the model can also accommodate the ρ^--spin alignment effect that has been recently observed in the reaction $\pi^- p \to \rho^- p$ [24]. In the c.m. ρ^- helicity frame the angular distribution in terms of the density matrix elements ρ_{ij} of the ρ is

$$\frac{4\pi}{3} W(\theta,\phi) = \rho_{00} \cos^2\theta + \rho_{11} \sin^2\theta - \rho_{1-1} \sin^2\theta \cos 2\phi$$

$$- \sqrt{2} \operatorname{Re} \rho_{10} \sin 2\theta \cos\phi, \tag{11}$$

where θ is the polar angle and ϕ the azimuth of the charged pion produced in the decay. We give below the experimental values and in parenthesis the theoretical expectations

$$\begin{aligned} \rho_{00} &= 0.12 \pm 0.30 & (0.096) \\ \rho_{11} &= 0.44 \pm 0.15 & (0.451) \\ \rho_{1-1} &= 0.32 \pm 0.10 & (0.355) \\ \rho_{10} &= -0.10 \pm 0.05 & (-0.01) \end{aligned} \tag{12}$$

which shows a remarkable agreement between theory and experiment. On the other hand, in PQCD as a consequence of (5) one finds $\rho_{10} = \rho_{1-1} = 0$, corresponding to a flat ϕ dependence. For reactions of the type $0^- 1/2^+ \to 1^- 1/2^+$, according to QGD - ACD all vector mesons are expected to be produced with a strong spin alignment and this characteristic feature should be tested in other reactions, for example ρ^+, ρ^0 or K^{*0} production. However the model has some problems concerning the relative normalization of the type $0^- 1/2^+ \to 0^- 1/2^+$. At 90° one expects $d\sigma/dt\,(\pi^+ p \to \pi^+ p) = d\sigma/dt\,(\pi^- p \to \pi^- p)$ while the inelastic channels $\pi^- p \to K^0 \Lambda$ and $\pi^- p \to K^+ \Sigma^-$ should be at least one order of magnitude smaller. The inelastic reactions are indeed suppressed compared to the elastic ones but at 10 GeV/c $\pi^+ p \to \pi^+ p$ is about three times $\pi^- p \to \pi^- p$ [25]. There is an indication that one gets larger cross sections when quark interchange is present and *not* gluon exchange. This would imply a large $\pi^- p \to \pi^0 n$ cross section, say at least 0.5 nb/GeV^2 at 10 GeV/c and similarly for $K^- p \to \pi^0 \Lambda$. These are crucial experimental tests which remain to be checked and a detailed theoretical evaluation of these charge and hypercharge exchange processes is certainly worth investigating. As we will see in Section 5 the polarization of the hyperons produced inclusively is large and very intriguing, so we believe it should also be measured in large angle exclusive reactions.

4. STRUCTURES IN HADRON FORM FACTORS

As we have seen above, the large Q^2 behaviour of hadron form factors is related to the energy dependence of large angle exclusive hadronic reactions. Nevertheless in the standard picture where hadrons are assumed to be made of quarks (antiquarks), which may have a structure, it is plausible to have some zeros in hadron form factors which need not have a definite sign. There are many such examples in nuclear physics where one observes that the position of the zeros occurs at larger Q^2 values for lighter nuclei. For the nucleon, such zeros have been predicted from PQCD [26] on the basis of an antisymmetric wave function in which the valence quarks do not share the nucleon momentum equally. The proton form factor is expected to be negative asymptotically and given the fact that it is positive at low Q^2, it should change sign, at least once, for some space-like Q^2 value which is not precisely known. New measurements of elastic electron scattering from protons [27] have been used to extract precise values of the proton magnetic form factor $G_M^P(Q^2)$ up to $Q^2 = 31.3$ GeV^2 as shown in Fig. 6. We see clearly that $Q^4 G_M^P$ decreases with increasing Q^2 in agreement with some PQCD calculations, which, however, are unable to provide a reliable normalization of the form factor. In a phenomenological approach [30] it was conjectured that the form factor could be described in terms of a function which has the fastest allowed decrease by analyticity with a zero around $Q^2 = 45$ GeV^2. Clearly a zero for much lower Q^2 values is absent in the data and excluded in a simple quark model [31]. A comparison of the data with the result of the fit of [30] is shown by the shaded area in Fig. 6, in better agreement with the new data. If such a zero exists, the

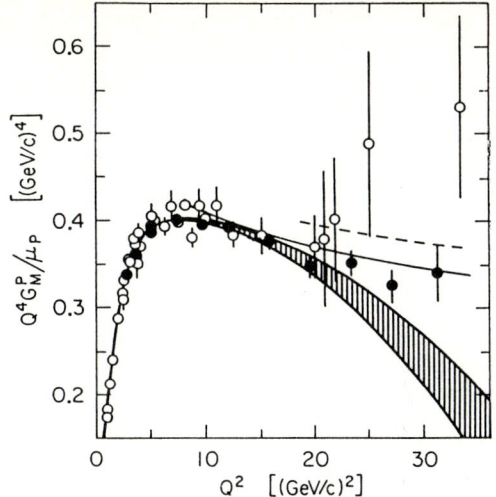
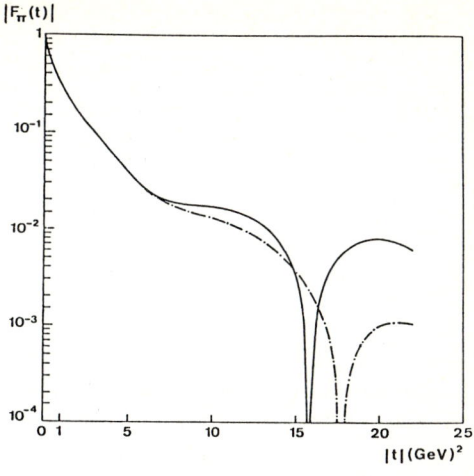

Fig. 6 The proton magnetic form factor versus Q^2. Open circles show old data as given in ref. [28]. Solid circles show recent data from ref. [27]. Solid (ref. [29]) and dashed (ref. [26]) curves are PQCD calculations. The shaded area corresponds to the uncertainty of the solution of ref. [30]

Fig. 7 The pion form factor extracted from π^-p elastic scattering. The two curves from ref. [32] reflect the uncertainty in the large-t data

proton charge density would have a depression near the origin, which gives a strong constraint for the proton wave function. Similarly, from a Chou-Yang analysis of π^-p elastic scattering from FNAL, it was found that the pion form factor should have a zero for Q^2 between 15 and 20 GeV^2 as shown in Fig. 7. The confirmation of the existence of these zeros and their exact location represent crucial information for our understanding of the composite nature of hadrons.

5. INCLUSIVE HYPERON PRODUCTION

Let us now consider inclusive hyperon production, where remarkable spin effects have been continuously discovered over the last ten years. An inclusive cross section is obtained by summing over many inelastic channels. Since any single transverse spin asymmetry A requires some phase difference between two amplitudes (see A in (1) as an example), a sizeable polarization effect can result only from a remarkable coherence between the many different inelastic channels involved. For large transverse momentum p_T, A is strongly suppressed in PQCD as a consequence of (5) holding for each inelastic channel. The experimental situation contradicts badly these expectations, as shown in Fig. 8, which gives a comparison of the A observed for different hyperons produced inclusively by unpolarized protons. The Λ polarization has the following features :

- it is *negative* with respect to the direction $\vec{p}_{inc} \times \vec{p}_\Lambda$
- for p_T below 0.8 GeV/c, its magnitude is almost linear on p_T with a slope increasing with x
- for p_T above 1 GeV/c, its magnitude is independent of p_T up to $p_T \sim 3.5$ GeV/c and almost linear on x
- it is almost independent of energy from 12 GeV/c to 2000 GeV/c.

This last property shows that a European Hadron Facility with an intense polarized beam could be used to improve our understanding of this polarization effect by measuring for example the spin transfer D_{NN} and the double helicity asymmetry A_{LL}. In the FNAL energy range the measured polarizations of Ξ^0 and Ξ^- are the same as that of Λ and opposite in sign

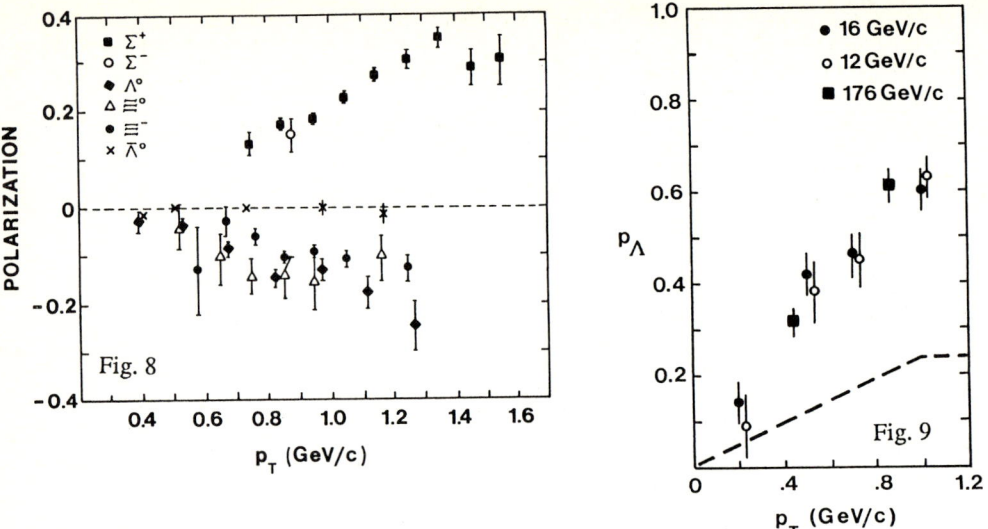

Fig. 8 Polarization of different hyperons produced by 400 GeV/c protons on Be versus p_T from [33] and references therein

Fig. 9 Λ polarization in $K^-p \to \Lambda X$ versus p_T at 12 GeV/c and 16 GeV/c from ref. [34] and at 176 GeV/c from [35]

for Σ^+ and Σ^- as shown in Fig. 8. In addition one finds that in the proton beam fragmentation region p and $\bar{\Lambda}$ are produced unpolarized. In meson-induced reactions one finds a large *positive* polarization for the hyperons produced in $K^-p \to \Lambda X$, $K^-p \to \Xi^- X$ and $K^+p \to \bar{\Lambda} X$. For illustration we show in Fig. 9 the case of $K^-p \to \Lambda X$ with again a remarkable energy independence. The dashed line shows for comparison the magnitude of P_Λ in proton-induced reactions. In the SU(6) limit, the (ud) system of the lambda is in a singlet state so the spin of the lambda is that of the s quark whereas for the sigma it is opposite to it, since the (ud) system is in a triplet state. This is the simple argument for the sign difference between $\Lambda(\Xi)$ and Σ polarizations. The data also tell us that for the proton-induced reactions the s quark contained in the Λ is a *slow* quark coming from the sea with a spin *down*, whereas in meson-induced reactions it is a *fast* quark with a spin *up* contained in the initial kaon beam. Some qualitative features of these data are consistent with simple theoretical ideas [36, 37], but we still lack a serious understanding of the subtle dynamical origin of this important phenomena where experiment is far ahead of theory.

It is a pleasure to thank Jean-Marc Richard for organizing this stimulating workshop in such a friendly atmosphere.

References :

1. C. Bourrely, E. Leader and J. Soffer, Phys. Rep. <u>59</u>, 95 (1980)
2. J.V. Allaby et al., Nucl. Phys. <u>B52</u>, 316 (1973)
3. Y. Makdisi et al., Phys. Rev. Lett. <u>45</u>, 1529 (1980)
4. P.R. Cameron et al., Phys. Rev. <u>D32</u>, 3070 (1985)
5. Landolt-Börnstein, Group 1, vol. 7 (1973), vol. 9a (1980) and vol. 9b1 (1982) edited by H. Schopper (Springer-Verlag, Berlin)

6. A.C. Fernow and A.D. Krisch, Ann. Rev. Nucl. Sci. $\underline{31}$, 107 (1981)
7. H. Miettinen et al., Phys. Rev. $\underline{D16}$, 549 (1977)
   J. O'Fallon et al., Phys. Rev. Lett. $\underline{39}$, 733 (1977)
   S.L. Lin et al., Phys. Rev. $\underline{D26}$, 550 (1982)
   D. Peasle et al., Phys. Rev. Lett. $\underline{50}$, 802 (1983) ; $\underline{51}$, 2359 (1983)
   G. Court et al., Phys. Rev. Lett. $\underline{57}$, 507 (1986)
8. C. Bourrely, J. Soffer and T.T. Wu, Phys. Rev. $\underline{D19}$, 3249 (1979)
9. C. Bourrely, J. Soffer and T.T. Wu, Nucl. Phys. $\underline{B247}$, 15 (1984) and Phys. Rev. Lett. $\underline{54}$, 757 (1985)
10. G. Farrar, Phys. Rev. Lett. $\underline{53}$, 28 (1984)
11. S.J. Brodsky and G.P. Lepage, Phys. Rev. $\underline{D24}$, 2848 (1981)
12. I.P. Auer et al., Phys. Rev. $\underline{D34}$, 1 (1986)
13. J. Soffer, preprint BNL 38691 (August 1986)
14. H.J. Lipkin, Phys. Lett. $\underline{181B}$, 164 (1986)
15. G. Farrar, S. Gottlieb, D. Sivers and G. Thomas, Phys. Rev. $\underline{D20}$, 202 (1979)
    S. Brodsky, C. Carlson and H.J. Lipkin, ibid. $\underline{20}$, 2278 (1979)
16. G. Nardulli, G. Preparata and J. Soffer, Nuovo Cimento $\underline{83A}$, 361 (1984) and references therein
17. G. Cocconi et al., Phys. Rev. $\underline{138B}$, 165 (1965)
18. C. Bourrely and J. Soffer, Phys. Rev. Lett. $\underline{54}$, 760 (1985) and Phys. Rev. $\underline{D35}$, 145 (1987)
19. G. Farrar, Phys. Rev. Lett. $\underline{56}$, 1643 (1986)
20. M. Anselmino, P. Kroll and B. Pire, preprint WU B 87-2 and P. Kroll these proceedings
21. B. Pire and J. Ralston, Phys. Lett. $\underline{117B}$, 233 (1982)
22. G. Nardulli, G. Preparata and J. Soffer, Phys. Rev. $\underline{D31}$, 626 (1985)
    G. Preparata and J. Soffer, Phys. Lett. $\underline{180B}$, 281 (1986)
23. G.S. Blazey et al., Phys. Rev. Lett. $\underline{55}$, 1820 (1985)
24. S. Heppelmann et al., Phys. Rev. Lett. $\underline{55}$, 1824 (1985)
25. G. Bunce et al., to be published
26. V.L. Chernyak and I.R. Zhitnitsky, Nucl. Phys. $\underline{B246}$, 52 (1984)
27. R.G. Arnold et al., Phys. Rev. Lett. $\underline{57}$, 174 (1986)
28. M.D. Mestayer, SLAC - 214, Ph. D. Thesis, Stanford University (1978)
29. S.J. Brodsky and G.P. Lepage, Phys. Scripta $\underline{23}$, 945 (1981)
30. C. Bourrely, J. Soffer and T.T. Wu, Z. Phys. C, Particles and Fields $\underline{5}$, 159 (1980)
31. N. Isgur, G. Karl and J. Soffer, Phys. Rev. $\underline{D35}$, 1665 (1987)
32. C. Bourrely et al., Phys. Lett. $\underline{132B}$, 191 (1983)
33. C. Wilkinson et al., Phys. Rev. Lett. $\underline{58}$, 855 (1987)
34. T.A. Armstrong et al., Nucl. Phys. $\underline{B262}$, 356 (1985)
35. S.A. Gourlay et al., Phys. Rev. Lett. $\underline{56}$, 2244 (1986)
36. B. Andersson, G. Gustafson, G. Ingelman and T. Sjostrand, Phys. Rep. $\underline{97}$, 31 (1983)
37. T. Degrand, J. Markkanen and H. Miettinen, Phys. Rev. $\underline{D32}$, 2445 (1985)

Polarized Parton Distributions and the Magnitude of Spin Effects at Very High Energies

P. Taxil

Centre de Physique Théorique*, CNRS, Luminy, Case 907,
F-13288 Marseille, Cedex 9, France
*Laboratoire propre LP-7061, Centre Nationale de la Recherche Scientifique

Parton distributions are a basic ingredient of any calculation in hard scattering hadronic processes. They contain some crucial information on the real nature of the proton beam made of quarks, gluons and quark-antiquark pairs from the sea : they tell us in which way each constituent carries a fraction x of the proton momentum. As we are interested in spin effects, we need additional information : how is the proton spin distributed among the various partons ? This information is contained in the polarized parton distributions.

For a given parton we denote by $f_{\pm}(x,Q^2)$ the parton distributions in a polarized nucleon with helicity either parallel (+) or antiparallel (-) to the parent nucleon helicity. Polarized distributions are defined as usual as the difference : $\Delta f = f_+ - f_-$, the unpolarized distributions being given by the sum $f = f_+ + f_-$. In the simple picture of a proton made of valence quarks only, one gets from the simple SU(6) wave function (conservative SU(6) model [1]) : $\Delta u_v(x) = 2/3\ u_v(x)$ and $\Delta d_v(x) = -1/3\ d_v(x)$, a choice which does not satisfy the Bjorken sum rule [2], as emphasized in [3], and is not in good agreement with data from SLAC experiments with a polarized electron beam on a polarized target [4].

A model [5] in which the valence quarks carry most of the helicity only at large x is in better agreement with data and has been extensively used as a starting point before turning on the Q^2 evolution due to scaling violations (see [3], [6] and references therein). Conversely at smaller x there is a dilution effect due to the interaction of valence quarks with the other constituents and the parametrization of $\Delta u(x)$ and $\Delta d(x)$ is modified accordingly (see [3] for an extensive discussion). At the same time, polarization for the gluons and the quark-antiquark pairs from the sea can be obtained in a model of gluon bremsstrahlung and quark pair creation. As the vector gluon character of QCD leads to helicity-conserving subprocesses (this important property has already been emphasized in this workshop [7]) we know the way helicity is transmitted along quark and gluon lines. It is then possible to parametrize $\Delta G(x)$ and $\Delta q_{sca}(x)$, taking into account various constraints (angular momentum conservation, sum rules...). One gets (at low Q^2) that valence quarks, gluons and the sea carry respectively around 75%, 20% and 5% of the proton spin [3].

On the experimental side note that the EMC collaboration has recently studied the deep inelastic scattering of polarized muons on a polarized target [8] at higher values of Q^2 (20 -100 GeV^2) and this will allow one to get more information on the polarization of the quarks for $0.1 \leq x \leq 0.6$. It has also to be stressed that an experiment on a polarized deuterium target would be useful to separate $\Delta u(x)$ and $\Delta d(x)$.

At higher energies, one can reach higher values of Q^2 : it can reach up to 10^4 GeV^2 at the HERA ep collider (30 GeV electron beam against 800 GeV proton beam) where polarization would allow some insights on polarized quark distributions in this range of Q^2 to be obtained and also give a unique opportunity to perform accurate tests of Standard Model physics.

On the theoretical side, the Q^2 evolution of the polarized distributions has to be calculated, according to QCD, by solving Altarelli - Parisi equations. Numerical [9] as well as analytical [10] studies have been performed. Concerning the unpolarized distributions, it is well know (see e.g. [11,12]) that they grow for small x and diminish for large x, gluons and quark-antiquark pairs playing an increasing role at very small x. Since the evolution can be described as due to gluon emission from quark lines and subsequent pair creation, helicity conservation leads to the same type of behaviour for $\Delta q(x,Q^2)$, $\Delta G(x,Q^2)$ and $\Delta q_{sea}(x,Q^2)$. This increase at low x (for illustration we refer to figures in [3,9,10]), which is a direct property of perturbative QCD, will have some important consequences in very high energy measurements with polarized beams.

The interest of spin effects at present hadronic collider energies has already been stressed for testing Standard Model physics [6,13,14] as well as new theoretical models [15].We have chosen to illustrate the importance of spin-dependent distributions by giving some predictions [16] for the next generation of supercolliders : the U.S.project SSC at $\sqrt{s}$ = 40 TeV or the CERN LHC at $\sqrt{s} \approx 20$ TeV. The interest of these machines for physics in the multi-TeV regime has been reviewed recently [11] in the unpolarized case and the feasibility of polarized proton beams is the object of careful studies [17].

The kinematic region of interest is 10^2 GeV^2 < Q^2 < 10^8 GeV^2 and 10^{-5} < x < 10^{-1}. After an accurate resolution of the evolution equations, starting from an improved version of the CARLITZ-KAUR [5] model, one can derive simple analytic parametrizations for the various polarized distributions $\Delta f(x,Q^2)$ (displayed in [18] along with a parametrization of the unpolarized distributions obtained by following [12]).

From this point one can estimate the magnitude of cross sections by introducing various differential parton-parton luminosities [11] defined as

$$\tau \frac{dL_{ij}}{d\tau} = \frac{\tau}{1+\delta_{ij}} \int_\tau^1 \frac{dx}{x} [f_i^{(a)}(x) f_j^{(b)}(\tau/x) + f_j^{(a)}(x) f_i^{(b)}(\tau/x)] , \quad (1)$$

where $f_i^{(a)}(x)$ is the distribution (evaluated at some value of Q^2) of partons of type i carrying the fraction momentum x of hadron a. Equation (1) represents the number of parton i - parton j collisions per unit of the scaled energy square $\tau = Q^2/s$, s being the total energy square.

Extending to the case of a hadronic reaction with *one* polarized beam, a single helicity hadron asymmetry

$$A_L = \frac{d\sigma^{(-)}/d\tau - d\sigma^{(+)}/d\tau}{d\sigma^{(-)}/d\tau + d\sigma^{(+)}/d\tau} \quad (2)$$

due to a parity violating interaction, can be written as

$$A_L = \sum_{ij} \frac{\tau(dL_{\Delta ij}/d\tau)}{\tau(dL_{ij}/d\tau)} \hat{a}_L^{ij} \quad (3)$$

in obvious notation, $\hat{a}_L^{ij}$ being the various subprocess asymmetries.

For example, in the case of the production of a single intermediate boson $W^\pm$ in pp collisions occurring via the simple Drell-Yan mechanism, A_L is given immediately by (3), summing over the appropriate combinations of quark-antiquark pairs.As can be seen from Fig.1, in such a process where $\hat{a}_L^{ij}$ are 100% (maximal parity violation), A_L is substantial even at extremely high energies.

If we turn now to an inclusive reaction of the type $a + b \rightarrow c$ (or jet) + X, where parity is conserved, we need two polarized beams to define a double helicity asymmetry A_{LL}

$$A_{LL} = \frac{d\sigma_{a(+)b(+)} - d\sigma_{a(+)b(-)}}{d\sigma_{a(+)b(+)} + d\sigma_{a(+)b(-)}} \quad (4)$$

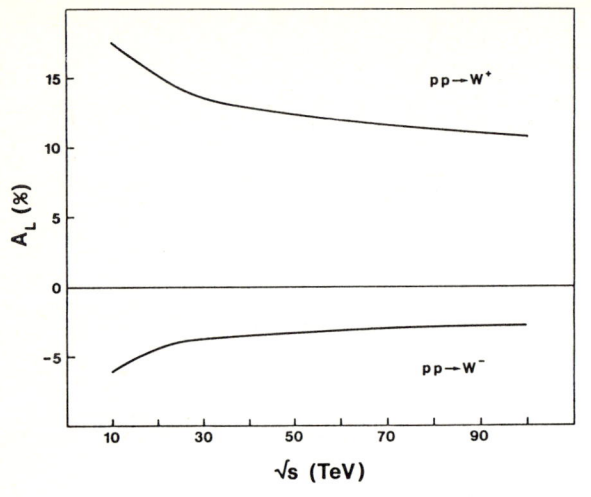

Fig. 1 Predictions for the single helicity hadron asymmetry A_L in $pp \to W^+ + X$ and $pp \to W^- + X$, versus energy

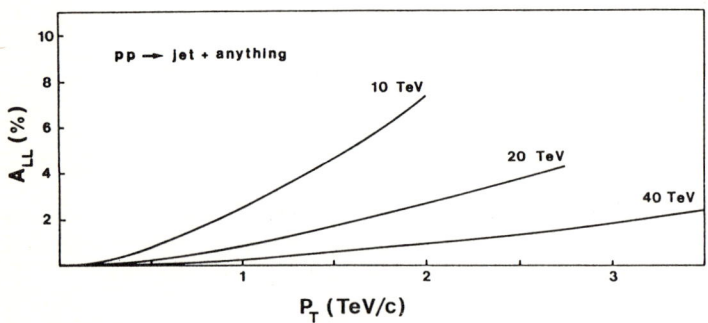

Fig. 2 Predictions for the double helicity hadron asymmetry A_{LL} in $pp \to$ jet $+ X$ at $y = 0$ versus p_T for three different energies

given by

$$A_{LL} d\sigma = \sum_{ij} \int dx_a dx_b \left[\Delta f_i^{(a)}(x_a, Q^2) \Delta f_j^{(b)}(x_b, Q^2) + (i \to j) \right] \hat{a}_{LL}^{ij}. \quad (5)$$

The subprocess asymmetries $\hat{a}_{LL}^{ij}$ can be found in [6] for QCD jet production. A_{LL} as shown in Fig.2 is small due mainly to the presence of two polarized distributions in computing (5), and the rise in p_T reflects the Q^2 evolution of the polarized parton distributions.

For physics beyond the Standard Model, the observation of a non-zero A_L in a purely hadronic process could reveal the presence of new parity violating interactions of subconstituents at a scale of a few TeV. Also $\hat{a}_{LL}^{ij}$ in the production of supersymmetric partons are -100% so they lead to large and negative A_{LL} in high p_T jet production with missing energy.

In conclusion, the notion that constituent asymmetries are completely diluted by the rapid growth in parton multiplicities at high Q^2 and low x is certainly too naive a picture. Some effects, large enough to be measured, can be present in the multi TeV regime and reveal the success of the Standard Model or the presence of new physics. The polarized distributions at low Q^2 are a basic ingredient of any calculation but, for the moment, the experimental information is very poor. We hope that the possible presence of large spin effects at very high energies will push experimentalists to make an effort for a very precise determination of polarized structure functions.

References :

1. J. KUTI and W. WEISSKOPF, Phys. Rev. $\underline{D4}$, 3418 (1971)
2. J.D. BJORKEN, Phys. Rev.$\underline{148}$, 1467 (1966), $\underline{D1}$ 1376 (1970)
3. P. CHIAPPETTA and J. SOFFER, Phys. Rev.$\underline{D31}$, 1019 (1985)
   P. CHIAPPETTA, J.Ph. GUILLET and J. SOFFER, Nucl. Phys. $\underline{B262}$, 187 (1985)
4. G. BAUM et al., Phys. Rev.Lett. $\underline{51}$,1135 (1985)
5. R. CARLITZ and J. KAUR, Phys. Rev. Lett. $\underline{38}$, 1116 (1977)
   J. KAUR, Nucl. Phys. $\underline{B128}$, 219 (1977)
6. N.S. CRAIGIE, K. HIDAKA, M. JACOB and F.M. RENARD, Phys. Reports, $\underline{99}$,69 (1983)
7. J. SOFFER, these proceedings
8. R. GAMET : In proceedings of the Int. Eur.Conf. on High Energy Physics, Bari 1985 ed. by L. NITTI and G. PREPARATA (Laterza,Bari 1985) p. 565
9. P. CHIAPPETTA, J.Ph. GUILLET and J. SOFFER, Phys. Lett. $\underline{183B}$, 213 (1987).
10. M.B. EINHORN and J. SOFFER, Nucl. Phys. $\underline{B274}$, 714 (1986)
11. E. EICHTEN, I. HINCHLIFFE, K. LANE and C. QUIGG, Reviews of Modern Physics, $\underline{56}$, 579 (1984) and references therein
12. J.P. RALSTON, Phys. Lett., $\underline{172B}$, 430 (1986).
13. P. AURENCHE and J. LINDFORS, Nucl. Phys. $\underline{B185}$, 301 (1981)
    P. CHIAPPETTA and J. SOFFER, Phys. Lett. $\underline{152B}$, 126 (1985)
14. M. JACOB, these proceedings
15. P. CHIAPPETTA , J. SOFFER and P. TAXIL, Phys. Lett. $\underline{162B}$, 192 (1985)
16. C. BOURRELY, F.M. RENARD, J. SOFFER and P. TAXIL, Spin Effects at Supercollider Energies, Physics Reports (in preparation)
17. see AIP Conference Proceedings n° 145, (1986) (Ed. A.D. KRISCH, A.M.T. LIN and O. CHAMBERLAIN).
18. C. BOURRELY, J. SOFFER and P. TAXIL, Phys. Rev. $\underline{D36}$, 3373 (1987)

Part II

Annihilation

$N\bar{N}$ Annihilation into Two Mesons

A.M. Green

Research Institute for Theoretical Physics, University of Helsinki,
Siltavuorenpenger 20 C, SF-00170 Helsinki, Finland

1. Introduction

The interest in $N\bar{N}$ annihilation into two mesons is twofold. Firstly, on the experimental side there has been considerable progress from recent LEAR experiments which have determined several new $\bar{p}p \to M_1 M_2$ branchings for stopped $\bar{p}$'s – see the contributions by C.AMSLER and G. SMITH at this workshop. However, it should be emphasized that this progress in only partial and, hopefully, the Crystal Barrel and Obelisk proposals for the next series of LEAR experiments will clarify the situation to a much greater extent. On the theoretical side, for a quark description of $N\bar{N}$ annihilation into two mesons it is necessary to introduce some mechanism for quark-antiquark ($q\bar{q}$) annihilation. This has resulted in many theoretical papers resembling the menu of a Chinese restaurant. From column A (see Fig. 1) a choice (or combination) is made between the two basic $q\bar{q}$ vertices:
   a) The $^3P_0$-model, in which the $q\bar{q}$ annihilate into the vacuum (often called the Pair-Creation model) – Fig. 1a).
   b) The $^3S_1$-model, in which the $q\bar{q}$ annihilate into a single gluon (or effective gluon) – Fig. 1b).

Having made this choice, the number of orders from column B has to be decided – 'one' as in Fig. 1c) or 'three' as in Fig. 1d). Frequently it is then found that one combination can explain some experimental features and another can explain other features. It is the purpose of the present paper to convince the reader that the single combination Figs. 1a) plus 1c) is dominant.

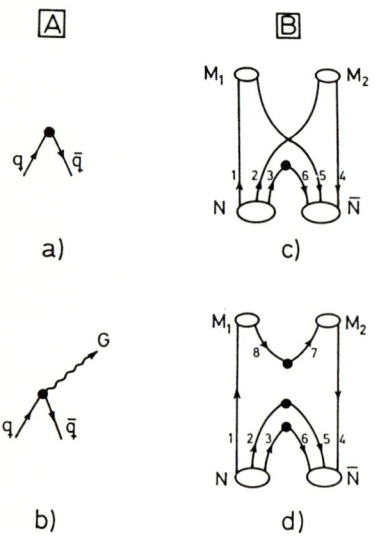

Fig. 1. The menu

2. The $^3P_0$-model versus the $^3S_1$-model

Before discussing $N\bar{N}$ annihilation it is useful to see how these two models for $q\bar{q}$ annihilation fare in their description of the <u>meson</u> decays $M_1 \rightarrow M_2 M_3$, since this is presumably a simpler situation. For the $^3P_0$-model the most comprehensive study has been made by KOKOSKI and ISGUR /1/, who show that this approach is very successful for essentially all meson decays. Furthermore, they give a model based on the flux tube breaking of ISGUR and PATON /2/ that can explain this success. Later, this flux tube model will be discussed again in connection with $N\bar{N}$ annihilation. For the $^3S_1$-model there, apparently, has never been such a comprehensive study. However, as shown by GREEN and NISKANEN /3/ the $^3S_1$-vertex can be expressed in terms of the $^3P_0$-vertex as

$$\Theta(^3S_1) = F\Theta(^3P_0), \quad F = \left[1 + 3/2(s_1(s_1+1) + 2 - s(s+1))\right] \quad (2.1)$$

where s_1 is the spin of the decaying meson (e.g. 1 and 0 for ρ and B) and s is the total spin of the decay products (e.g. 0 and 1 for $\pi\pi$ and $\omega\pi$). For the decays $\rho \rightarrow \pi\pi$ and $B \rightarrow \omega\pi$ the value of F becomes 7 and 1 respectively. Therefore, since the $\Theta(^3P_0)$ alone gives a good description of both of these decays <u>the $^3S_1$-model will fail miserably</u>.

Since the $^3S_1$-model is not the dominant mechanism for meson decays, why should it be so for $N\bar{N}$ annihilation? Of course, the effective gluon involved is different in the two cases. However, it seems unreasonable that the differences are so large that the $^3S_1$-model could be enhanced to such an extent that it now dominates over the $^3P_0$-vertex. In the view of the author, the main reason why the $^3S_1$-vertex is used in $N\bar{N}$ annihilation is because it is able to give, in <u>lowest order</u>, $N\bar{N}$ (S-wave) $\longrightarrow$ Two s-wave mesons ($M_1 M_2$) [e.g. $N\bar{N}(^{13}S_1) \rightarrow \pi^{\mp}\rho^{\pm}$] i.e. Figs. 1b) plus c). On the other hand, the usual treatment of the $^3P_0$-model in lowest order gives <u>zero</u> for Figs. 1a) plus c) - contrary to experiment. The $^3S_1$-advocates then say "Since the $^3S_1$-model gives a non-zero result, it must be better than the $^3P_0$-model" - an example of the "Anything" is better than "Nothing" philosophy.

3. The $^3P_0$-model in $N\bar{N}$ annihilation

Of course, advocates of the $^3P_0$-model have not been perturbed by the "conclusion" of the last subsection, since the $^3P_0$-vertex to third order [i.e. Figs. 1a) plus d)] readily gives $N\bar{N}$(S-wave)$\rightarrow M_1^s M_2^s$. Unfortunately, some of the branching ratios predicted do not agree with experiment. In particular, the model gives

$$R_s = \frac{N\bar{N}(^{31}S_0) \rightarrow \pi\rho}{N\bar{N}(^{13}S_1) \rightarrow \pi\rho} \approx 1, \quad (3.1)$$

whereas, experimentally $R_s \ll 1$. Also, the $^3P_0$-model in lowest order (i.e. Figs. 1a) plus c)) gives

$$R_p = \frac{N\bar{N}(^{33}P_{12}) \rightarrow \pi\rho}{N\bar{N}(^{11}P_1) \rightarrow \pi\rho} = 18, \quad (3.2)$$

whereas, experimentally $R_p \ll 1$. This observation of I=1 suppression in $N\bar{N}$ (S- and P-waves) is known as the $\pi\rho$ - puzzle. It has been suggested by DOVER et al. /4/ that $R_s \ll 1$ is not indicating a selection rule but is simply a reflection that in the $N\bar{N}(^{31}S_0)$ state $\pi^{\pm}\rho^{\mp}$ face a competition with annihilation into $\pi^{\circ}f$ resulting in a quenching of the $\pi^{\pm}\rho^{\mp}$ channel.

The above resolution of the S-wave $\pi\rho$ -puzzle leaves much to be desired. Firstly, the need to introduce three $^3P_0$-vertices is difficult to justify on any more basic approach such as the flux tube model /1,2/. Secondly, the need to employ a cancellation by competition with the π^*f channel is not very "natural". However, as recently pointed out by GREEN /5/ there is an alternative where $R_* \ll 1$ can be achieved by simply Figs. 1a) plus c) i.e. the $^3P_0$-vertex to lowest order. As a first step in convincing the reader of this statement, it is necessary to show that Figs. 1c) and d) are in a certain sense "equivalent" to each other. This can be done by the chain of events

$$N\bar{N}(S) \xrightarrow{1} M^s{}_{14}\epsilon_{25}\epsilon_{36} \xrightarrow{2} M^s{}_{14}\epsilon_{25}(l=0) \xrightarrow{3} M^s{}_{15}M^s{}_{24}(l=1), \qquad (3.3)$$

where the first step is to recouple the 3 quarks and 3 antiquarks of the $N\bar{N}(S)$ system into <u>one s-wave meson and two ϵ-mesons</u> (i.e. mesons with quantum numbers of the vacuum — $^{13}P_0$). These need not be considered as real mesons, but as objects with the quantum numbers of these mesons. In other words, this first step is simply a mathematical recoupling. The second step involves the annihilation of $\epsilon_{36}(=q_3\bar{q}_6)$ by the usual $^3P_0$-vertex to leave an s-wave meson $M^s{}_{14}$ and the ϵ-meson (ϵ_{25}) in a <u>relative s-wave</u>. The final step can then be considered as yet another mathematical recoupling in which antiquarks $\bar{q}_4$ and $\bar{q}_5$ interchange to give two s-wave mesons $M_{15}M_{24}$ in a relative p-wave — see ref./5/ for details. The extent to which Figs 1c) and d) are equivalent can be checked by comparing different branching ratios. The best compilation of ratios using Figs. 1a) plus d) is given by MARUYAMA et al./6/. For the 11 possible branchings $N\bar{N}(S) \to M^s{}_1 M^s{}_2$ [e.g. $N\bar{N}(^{13}S_1) \to \pi^{\pm}\rho^{\mp}$ etc.] the ratio

$$R = \frac{\text{Branching ratio ref. /5/}}{\text{Branching ratio ref. /6/}} = \frac{16}{81},$$

whereas for the 48 possible branchings $N\bar{N}(S) \to M^s{}_1 M^s{}_2 (l=0,2)$ [e.g. $N\bar{N}(^{31}S_0) \to \eta\delta^0$, $N\bar{N}(^{11}S_0) \to \omega H$ etc] the value of R is 1/3. Presumably, the fact that R differs in the two cases (16/81 versus 1/3) is simply a matter of definition — but at present the actual source of the difference has not been located. Of course, it may be said that the above "equivalence" is obvious, since in Fig. 1d) the annihilation of $q_2\bar{q}_5$ and the creation of $q_7\bar{q}_8$ is represented in Fig. 1c) by a continuous $q\bar{q}$ state with the quantum numbers of the vacuum $(q_2\bar{q}_5)^{0^+}$. However, between the two interpretations there is a fundamental difference which involves the question of strangeness. In Fig. 1d) the creation of $q_7\bar{q}_8$ from an SU(3) vacuum gives the combination $1/\sqrt{3}(u\bar{u} + d\bar{d} + s\bar{s})$. On the other hand Fig. 1c), since it is simply a recoupling of $N\bar{N}$ wavefunctions, never involves strange quarks. This offers an instant explanation of the observed K^+K^--suppression, with the actual decay $N\bar{N}(^3S_1) \to K^+K^-$ being mediated by some higher order mechanism such as an effective $^3S_1$ vertex — see RUBINSTEIN and SNELLMAN /7/ and FURUI /8/. As shown in Ref. /9/ such a $^3S_1$ vertex need not have anything to do directly with gluons, but can emerge by simply forming an overlap of 3 quarks and 3 antiquarks each described by four component spinors. This is a natural extension of the usual $^3P_0$-vertex considered as the overlap of a single quark and a single antiquark — each described as four component spinors. There the leading term arises through the product of the "large" component in the q or $\bar{q}$ wavefunction with the "small" component of the $\bar{q}$ or q wavefunction.

<u>The conclusion of the above is that $N\bar{N}$(S-wave) can annihilate into two s-wave mesons using the $^3P_0$-vertex to only first order i.e. Figs. 1a) plus c).</u>

At this stage it is convenient to point out a difference between the diagrams drawn in Figs. 1c) and d) and those drawn by GENZ elsewhere in this workshop. For example, a possible configuration in Fig. 1c) after the $^3P_0$-vertex could be $[\rho_{14}\ \epsilon_{23}]^{J=1\ I=1}$, which is represented by <u>four</u> flavour combinations. Under the recoupling of step 3 in Eq. (3.3) some combinations of $M^a{}_{15}M^a{}_{24}$ could involve matrix elements such as

$$\langle d_1\bar{d}_4|\rho_{14}\rangle\ \langle \rho_{14}|u_1\bar{u}_4\rangle,$$

which in the quark line rules of GENZ /10/ is called an annihilation diagram and is drawn in a way similar to Fig. 1d). The present interpretation of Fig. 1c) in terms of meson-like configurations of definite spin and isospin is therefore a specific combination of the rearrangement and annihilation diagrams of GENZ.

4. The $\pi\rho$-puzzle

Having convinced the reader that an alternative interpretation of Fig. 1d) is a recoupling of the configuration $M^a{}_{14}\epsilon_{23}(\ell=0)$ as described in Eq. (3.3), the natural question to raise is the role of other configurations involving p-wave meson states. Such states $M^a{}_{14}M^a{}_{25}$ enter on the <u>same</u> footing as $M^a{}_{14}\epsilon_{23}$. For example, the $N\bar{N}(^{11}S_0)$ state can be rewritten as

$$N\bar{N}(^{11}S_0) \rightarrow \sqrt{6}(\eta\epsilon) - \sqrt{2}(\pi\delta) + \sqrt{6}(\omega H) - \sqrt{2}(\rho B) \qquad (4.1)$$

- others being given in ref. /5/. Equation (3.3) is now replaced by

$$N\bar{N}(S) \xrightarrow{1} \Sigma_\rho A(p) M^a{}_{14}M^a{}_{25}\epsilon_{36} \xrightarrow{2} \Sigma_\rho A(p) M^a{}_{14}M^a{}_{25}(\ell=0)$$

$$\xrightarrow{3} \Sigma_\rho A(p) C\ M^a{}_{15}M^a{}_{24}(\ell=1). \qquad (4.2)$$

This now leads to the selection rule

$$N\bar{N}(S,\ \text{SPIN SINGLET}) \not\rightarrow M^a{}_1 M^a{}_2 (\ell=1). \qquad (4.3)$$

At present this is only a numerical observation — but not a particularly suprising one when the amplitudes in Eq. (4.1) are seen to come in pairs that eventually cancel each other. <u>This immediately explains the $\pi\rho$-puzzle in S-waves</u> referred to in Eq. (3.1) without resorting to any competition with the $\pi°f$ channel. Furthermore, as seen in table 1 some of the other singlet branchings are also reasonably consistent with zero.

Table 1. Spin singlet branchings

Branching	Expt. (see C. AMSLER at this workshop)
$N\bar{N}(^{11}S_0) \rightarrow \omega\omega$	1.4 ± 0.6 %
$N\bar{N}(^{11}S_0) \rightarrow \rho°\rho°$	0.12 ± 0.12
$N\bar{N}(^{31}S_0) \rightarrow \rho°\omega$	$\begin{cases} 2.1 \pm 0.2 \\ 0.7 \pm 0.3 \\ 3.9 \pm 0.6 \end{cases}$

For comparison the branching $N\bar{N}(^{13}S_1) \rightarrow \pi\rho$ is 4.6 ± 0.4 %. Of course, it must be remembered that a more meaningful comparison would require the inclusion of phase space and form factor effects — see GREEN /11/.

In table 2 a comparison is made between the rather meagre experimental results and the predictions of the present model for spin triplet states. In parentheses are given the predictions of Fig. 1d) i.e. simply using the $M^B{}_{14}\epsilon_{25}$ configuration. For $^{13}S_1$ and $^{33}S_1$ the theories are separately normalized to $\pi\rho$ and $\eta\rho$ respectively, since initial state interactions rule out any direct comparison between the two $N\bar{N}$ states.

Table 2. Spin triplet branchings

Branching	Theory x 10^2	Expt. (C.AMSLER at this workshop)
$N\bar{N}(^{13}S_1) \longrightarrow \rho\pi$	4.6 (4.6)	4.6 ± 0.4 %
$\longrightarrow \eta\omega$	6.9 (0.77)	1.3 ± 0.2
$N\bar{N}(^{33}S_1) \longrightarrow \eta\rho$	0.65 (0.65)	0.65 ± 0.16
$\longrightarrow \pi\omega$	12 (1.3)	< 1 %
$\longrightarrow \pi^{\pm}\pi^{\mp}$	2.6 (0.65)	0.31 ± 0.03
$\longrightarrow \rho\rho$	5.2 (4.6)	< 9 ± 3

Again it should be emphasized that a meaningful comparison would require the inclusion of phase space and form factor effects – see GREEN /11/. The only definite statement is that the addition of p-wave mesons other than the ϵ can result in massive changes in the predicted branchings e.g. $N\bar{N}(^{33}S_1) \longrightarrow \pi\omega$ increases by an order of magnitude.

So far only the $\pi\rho$-puzzle in $N\bar{N}$ S-waves has been discussed. For the P-wave case mentioned after Eq. (3.2) it is necessary to consider the chain

$$N\bar{N}(P) \xrightarrow{1} \Sigma_\rho \ A(p_1p_2) \ M^B{}_{14}M^B{}_{25}\epsilon_{36} \xrightarrow{2} \Sigma_\rho \ A(p_1p_2) M^B{}_{14}M^B{}_{25}(\ell=0) \quad (4.4)$$

$$\xrightarrow{3} \Sigma_\rho \ A(p_1p_2) \ C \ M^B{}_{15}M^B{}_{24}(\ell=0).$$

At present the first step – the evaluation of the $A(p_1p_2)$ – is not complete /11/. Hopefully this mechanism will overwhelm the usual rearrangement

$$N\bar{N}(P) \longrightarrow \Sigma_s \ A(s_1s_2) \ M^B{}_{14}M^B{}_{25}\epsilon_{36} , \quad (4.5)$$

which resulted in Eq. (3.2) and is in gross disagreement with experiment.

It should be added that it is <u>not</u> unreasonable that p-wave mesons (or more precisely – p-wave meson-like configurations) are so important, when it is remembered that in the $p\bar{p}$ annihilation cross section P-waves already dominate over S-waves for $p(Lab) \geq 200$ MeV/c – see Fig. 15 in GREEN and NISKANEN /12/.

A potential problem, when introducing two p-waves into $N\bar{N}(S)$ is that <u>nodes</u> are unavoidably generated in the $N\bar{N}$ relative coordinate. These arise when the fourier transform of $k^2Y_0(k)$ is made and come at ≈ 0.5 fm. However, it is not clear to what extent this is a real effect. Similar problems arose earlier in meson decays. There they were seen to be unrealistic and so various prescriptions were invoked to remove the nodes – see refs. /13/.

The inhibition $N\bar{N}(^{31}S_0) \not\rightarrow \pi\rho$ is not absolute since the branching is seen in $\bar{p}n \rightarrow \pi^-\rho^0$. However, this should presumably be considered as arising from a higher order effect of which there are many possibilities:

a) An effective $^3S_1$-operator could be induced as a higher order correction to the $^3P_0$-vertex /9/ as already mentioned at the end of section 3 in connection with K-meson production.

b) By one-pion-exchange the $N\bar{N}$(S-wave) state could become $N^*\bar{N}$(P-wave) where N^* is an <u>odd parity baryon</u> – as discussed in section 3.3.4 of GREEN and NISKANEN /12/.

c) In Eq. (4.1) the meson configurations could be considered more than simply a mathematical rearrangement of $N\bar{N}$. If these became true mesons – albeit in virtual configurations – then the mass differences (e.g. $M(\delta\pi) = 1.12$ GeV versus $M(B\rho) = 2$ GeV) could prevent an exact cancellation after the recoupling into two s-wave mesons.

5. In Search of a Theory for $N\bar{N}$ Annihilation

So far in $N\bar{N}$ annihilation no explicit mention has been made of <u>gluons</u> in the $^3P_0$-model – their presence being only implicit with the use of oscillator wavefunctions for the N, $\bar{N}$ and mesons. Of course, at the energies of interest it is the non-perturbative regime of QCD that enters, and so "complete solutions" cannot be expected. However, a step in the right direction could be the use of the flux tube model of ISGUR and PATON /2/ that has proved so successful for the description of meson decays /1/. A possible scenario would then be the one depicted in Fig. 2. The first step in Figs. 2a) to 2b) is similar to the meson decay problem, since it simply involves the breaking of the flux tube between junctions J_1 and J_2 by means of the $^3P_0$ operator. The difficult step is that between Figs. 2c) and 2d). However, since the operation of the plaquette does not directly involve the two quarks and antiquarks, these remain in the same states of spin and flavour. Therefore, the effect of P is dominated by an effective interaction between the quarks and antiquarks that is <u>spin and isospin independent</u>. Schematically, the scenario of Fig. 2 can be written as

$$M = \langle M_1 M_2 | P | B \rangle \frac{1}{\Delta E(B)} \langle B | \Omega | N\bar{N} \rangle \tag{5.1}$$

$$\approx \langle M_1 M_2 | F\Omega | N\bar{N} \rangle , \tag{5.2}$$

where F is a spin-isospin independent factor simulating the role of the Baryonium and Plaquette. In other words, the model presented earlier in sections 3 and 4 is still valid from the point of spin

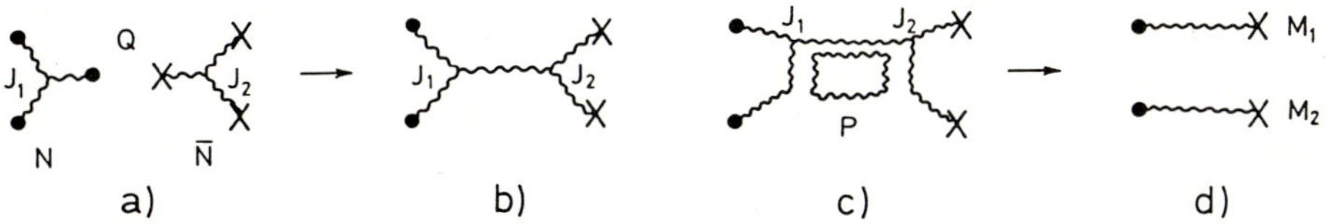

Fig. 2. Scenario for $N\bar{N}$ annihilation into two mesons $M_1 M_2$
  a) Action of the $^3P_0$ operator (Ω) to give:
  b) Baryonium (B)
  c) In baryonium the junctions J_1, J_2 can approach each other to be annihilated by the Plaquette operator (P) to give:
  d) Two mesons $M_1 M_2$.

and isospin. However, now because of F in Eq. (5.2) it should not be expected that the <u>strength</u> of the $^3P_0$-vertex in Figs. 1a) plus c) is the same as in meson decay. The above also shows that probably a better interpretation of Figs. 1a) and 1c) and Eq. (4.2) is that the recoupling of quarks (14)(25) into (15)(24) takes place at the step in Fig. 2 where the two junctions are annihilated. Needless to say, this approach has a very long way to go before it is on the same footing as the meson decay works of ref. /1/.

6. Conclusions

a) For the main contributions to $N\bar{N}$ annihilation there is no justification or need for the use of the $^3S_1$-model, i.e. the vertex in Fig. 1a) dominates.

b) With the $^3P_0$-model there is no need to go to third order as in Fig. 1d), since this is essentially equivalent to the first order contribution of Fig. 1c). However, a <u>recoupling</u> of quarks and antiquarks is necessary.

c) As a consequence of b) it is necessary to include all p-wave meson-like configurations (M^P_2) in the rearrangement $N\bar{N}(S) \longrightarrow M^P_1 M^P_2 \in_3$.

d) At a more microscopic level involving explicit gluons, the removal of junctions is predominantly spin and isospin independent. Therefore, their effect can to some extent be simulated by modifying the strength of the $^3P_0$-vertex extracted from meson decays.

<u>$N\bar{N}$ annihilation is dominated by the $^3P_0$-model in first order</u> i.e. Figs. 1a) plus c).

References

1. R. Kokoski and N. Isgur, Phys.Rev. <u>D35</u>(1987) 907
2. N. Isgur and J. Paton, Phys.Rev. <u>D31</u>(1985) 2910
3. A.M. Green and J.A. Niskanen, Mod.Phys.Lett. <u>A1</u>(1986) 441
4. C. Dover, Proc. 2nd Conf. on the Intersection between Particle and Nuclear Physics, Lake Louise, Canada ed. D.F. Geesaman (AIP conference proceedings No. 150) p.272
   C. Dover, P. Fishbane and S. Furui, Phys.Rev.Lett. <u>57</u>(1986) 1538
5. A.M. Green, Helsinki preprint HU-TFT-87-8
6. M. Maruyama, S. Furui and A. Faessler, Tübingen preprint January 1987
7. H. Rubinstein and H. Snellmann, Phys.Lett. <u>B165</u>(1985) 187
8. S. Furui, Z.Phys. <u>325</u>(1986) 375
9. A.M. Green, J.A. Niskanen and S. Wycech, Phys.Lett. <u>B172</u> (1986) 171
10. H. Genz, Phys.Rev. <u>D28</u>(1983) 1094 and at this workshop
11. A.M. Green, in preparation
12. A.M. Green and J.A. Niskanen, Prog. in Particle and Nuclear Physics, ed. A. Faessler (Pergamon Press 1986), Helsinki preprint HU-TFT-85-60
13. R. Koniuk and N. Isgur, Phys.Rev. <u>D21</u>(1980) 1868
    N. Törnqvist and P. Zenczykowski, Z.Phys. <u>C30</u> (1986) 83

Charged Two-Meson Production from $\bar{N}N$ Annihilation at Rest *

G.A. Smith

Laboratory for Elementary Particle Science, Department of Physics,
The Pennsylvania State University, University Park, PA 16802, USA
PS183 Collaboration (LEAR),
Athens[1]-Irvine[2]-Karlsruhe[3]-New Mexico[2]-Penn State[4]

1. Introduction

We study $\bar{N}N$ annihilation at rest for several important reasons: (1) it allows a test of quark models in non-perturbative QCD; (2) initial states of known angular momentum (S,P) can be prepared (liquid, gaseous H_2 - see talk of C. AMSLER in these proceedings), thus facilitating theoretical considerations; (3) quantum number conservation filters initial $\bar{N}N$ states when specific final states are considered ($^{2I+1,2J+1}L_J(\bar{N}N) = {}^{33}S_1$ only for $\pi^+\pi^-$, $^{11}S_0$ only for $\pi^\pm\delta^\mp$, $^{31}S_0$ only for $\rho^\circ\omega^\circ$, etc.), again facilitating theory; and (4) experiments, with (2) and (3) in mind, may uncover dynamic selection rules which can be traced back to fundamental quark processes. Also, $\bar{N}N$ annihilation at rest has been used by our collaboration (and others) to search for exotic, narrow 4-quark states [1,2]. No states have been found, with limits set at $\sim$ a few $\times 10^{-4}$ per annihilation.

Theoretical quark model predictions for $\bar{N}N$ annihilation are numerous. In general, they work within the framework of annihilation and rearrangement diagrams, leading to 2,3 meson final states. The fundamental $q\bar{q}$ coupling is either $^3P_0$ (vacuum), or $^3S_1$ (gluon). The reader is referred to the contribution to this conference by A.M. GREEN, and references cited therein, for further details. It is probably fair to say that no one version of this theory has emerged strongly over another. It is our objective to challenge the theory with new measurements of 2-body branching ratios, hopefully leading to a better theoretical understanding of the annihilation process.

2. The PS183 Detector

The PS183 detector took data at LEAR in 1983-1986. A schematic plan of the detector is shown in Fig. 1. Beams of antiprotons from 200 to 450 MeV/c were stopped in a liquid hydrogen (deuterium) target. The annihilation vertex was located using the reconstructed beam track, and tracks observed in the drift chambers (R,N) on either side of the target. In Fig. 2(a) we show the three orthogonal profiles (horizontal, vertical, transverse) of reconstructed vertices. The solid line indicates the target flask boundaries. One sees that all interactions are inside the target and confined to a region of $\sim \pm 1$ cm around the stopping point.

For this study, particles which penetrated the magnet (operated at 3.5 or 7.5 kG field) to the far set of drift chambers (P) have been momentum analyzed. In Fig. 2(b) we show the results of a careful study of the reliability of momentum reconstruction of such tracks, using known reference lines from $\bar{p}p \to \pi^+\pi^-$, K^+K^- and K^+ (stop) $\to \mu^+\nu, \pi^+\pi^\circ$ at the two fields. For each process we plot the percent deviation of the measured momentum versus the known value momentum, when the field was up (+), down (-) and combined (±). One sees that typically all measurements are

---
* Based on a Ph.D. thesis submitted by M.J. Soulliere to The Pennsylvania State University, 1987
[1] Work supported in part by the Greek Ministry of Research and Technology.
[2] Work supported in part by the U.S. Department of Energy.
[3] Work supported in part by the Federal Ministry for Research and Technology, FRG.
[4] Work supported in part by the U.S. National Science Foundation.

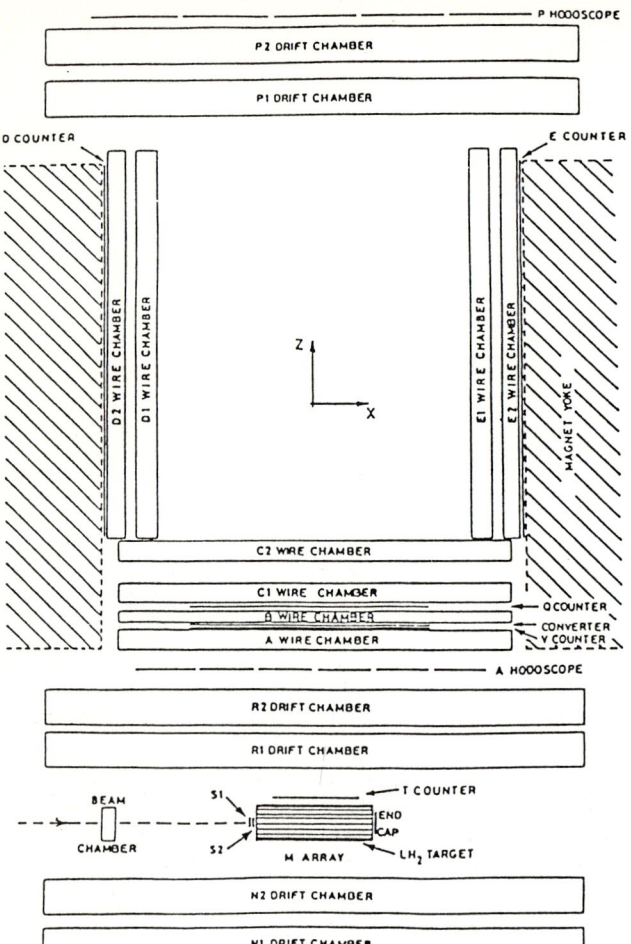

Fig. 1: Plan view of PS183 spectrometer

consistent with known values to better than 2×10^{-3}, or $\gtrsim 2$ MeV/c. The corresponding resolution (width) measurements of these lines are plotted in Fig. 3(a). At 7.5 kG, where the resolution is superior, the rms width is always < 1.5%. As we will see, the high degree of precision indicated by these measurements is essential to extracting reliable signals from the spectra observed in this experiment.

Particle masses are measured by time-of-flight (TOF) between the A and P hodoscope scintillation counters. Results are given in Fig. 3(b), where we plot momentum versus mass squared for (+),(-) charges separately. Pions, kaons (and protons) are cleanly separated. The protons result from scatters of pions (near 180°) within the target, or materials close to the target.

3. Results

Inclusive $\pi^{\pm}$ spectra taken in 1984 are shown in Fig. 4(a). The pion and rho peaks are clearly evident. Fits have been made to these spectra using direct resonances, reflection of resonances and phase space background. These results are shown in Fig. 4(b). The $\pi\pi$ peak is not shown, as it is off scale. The $\rho^{\circ}\omega^{\circ}$ and low momentum tail to the $\pi\rho$ peak are reflections of events where the ρ° sends a particle through the spectrometer. Peaks which were not evident in Fig. 4(a) are found consistently

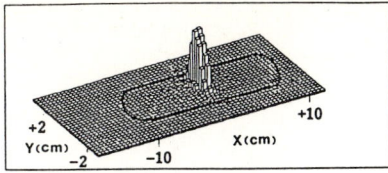
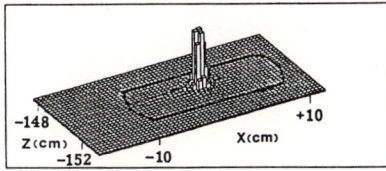
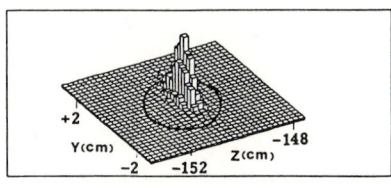

Fig. 2(a): Pion vertex distributions at B = 7.5 kG

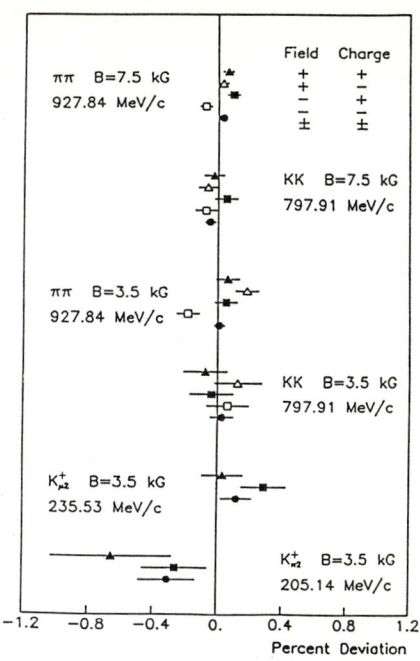

Fig. 2(b): Percent deviation of fitted calibration lines

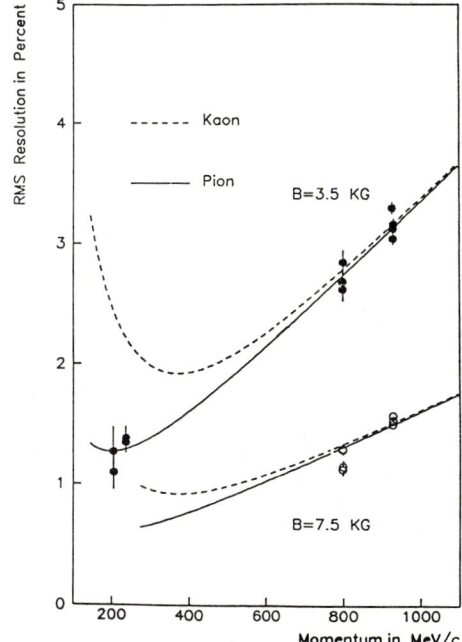

Fig. 3(a): RMS momentum resolution of pions and kaons in percent

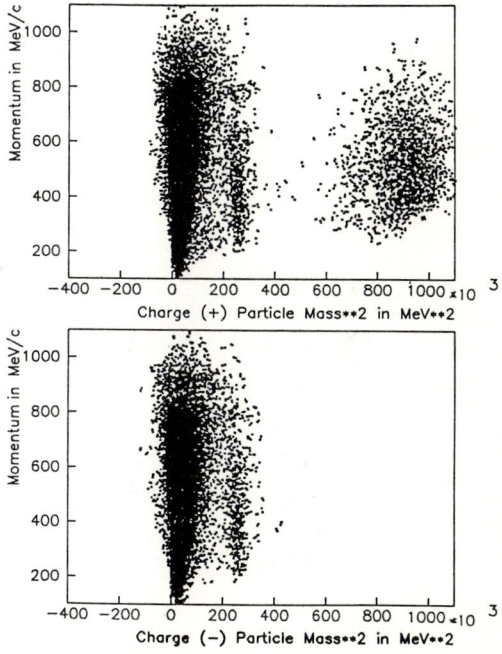

Fig. 3(b): Momentum versus mass squared at B = 3.5 kG for particles identified by TOF

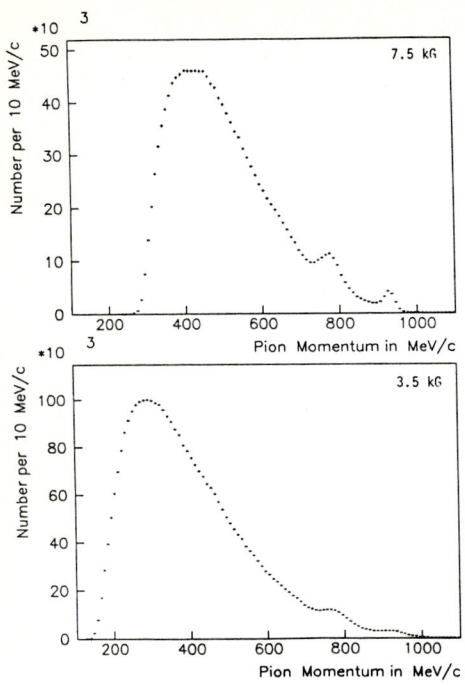
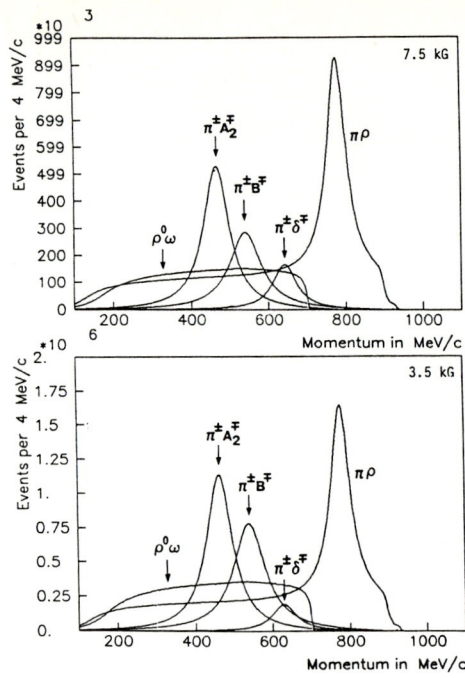

Fig. 4(a): Pion momentum spectra at B = 7.5, 3.5 kG

Fig. 4(b): Pion momentum structures from 2-meson final states at B = 7.5, 3.5 kG

Table 1: Percent yields of 2-meson final states

Meson Final State	PS183 B = 7.5 kG	PS183 B = 3.5 kG	CERN[3]	Columbia[3]
$\pi^{\pm}\pi^{\mp}$	0.3093 ±0.0033±0.031	0.3164 ±0.0069±0.032	0.37 ±0.03	0.32 ±0.03
$\pi^{\pm}\rho^{\mp}$	3.014 ±0.041±0.30	3.008 ±0.095±0.30	3.87 ±0.21	2.9 ±0.4
$\pi^{\pm}\delta^{\mp}$	0.76 ±0.12±0.08	0.50 ±0.19±0.05	< 0.16	—
$\pi^{\pm}B^{\mp}$	1.76 ±0.20±0.18	2.71 ±0.39±0.27	0.79 ±0.11	0.75 ±0.25
$\pi^{\pm}A_2^{\mp}$	2.64 ±0.20±0.26	3.19 ±0.27±0.32	4.65 ±0.51	—
$\rho^0\omega$	3.49 ±0.56±0.35	4.67 ±0.71±0.47	2.26 ±0.23	0.7 ±0.3
$K^{\pm}K^{\mp}$	0.0993 ±0.0024±0.0099	0.0977 ±0.0029±0.0098	0.096 ±0.008	0.11 ±0.01
$K^{\pm}K^{*\mp}$	0.1457 ±0.0087±0.015	0.1394 ±0.0082±0.014	0.104 ±0.010	0.086 ±0.018

in both spectra. These are due to the $\pi^{\pm}A_2^{\mp}$, $\pi^{\pm}B^{\mp}$ and $\pi^{\pm}\delta^{\mp}$ reactions. The kaon spectra have also been fitted in a similar manner, revealing the $K^{\pm}K^{\mp}$ and $K^{\pm}K^{*\mp}$ reactions.

Yields for the two sets of fits are given in Table 1. The two sets of PS183 measurements are consistent, with one exception ($\pi^{\pm}B^{\mp}$ - only 3% confidence for agreement). Previous bubble chamber measurements [3] are provided for comparison. In this regard, we note that the CERN, Columbia groups did not observe $\pi^{\pm}\delta^{\mp}$, and had values for $\pi^{\pm}\rho^{\mp}$ and $\rho^{\circ}\omega$ which are not in agreement (3% and 0.1% confidence respectively). As the PS183 statistical errors are generally quite superior to the bubble chamber errors (two exceptions are $\pi^{\pm}B^{\mp},\rho^{\circ}\omega$), and the $\pi^{\pm}\delta^{\mp}$ reaction has been observed for the first time, a clear step forward has been made in our knowledge of annihilation at rest.

4. The $\pi^{\pm}\delta^{\mp}$ Reactions

The observation of the $\pi^{\pm}\delta^{\mp}$ reaction by PS183 is a new result. At this conference the Asterix group also discussed their recent observation of $\delta^{\pm} \rightarrow \eta^{\circ}\pi^{\pm}$ in $p\bar{p}$ annihilation at rest (see contribution to this conference by C. AMSLER). An interesting systematic appears to emerge when one compares the results of these 2 experiments (see Table 2). Namely, PS183 and Asterix find a mass (~ 1030 MeV) and width (90 MeV - PS183 only), both of which are larger than the conventional δ values reported in the Review of Particle Properties [4]. Since the δ is a controversial object (see H. LIPKIN, this conference), this clear discrepancy in the mass and width may be a signature of new physics. A good measurement of $\delta \rightarrow K\bar{K}$ would be extremely useful in this regard.

Table 2: A comparison of δ masses and widths from PS183, Asterix and Rev. Part. Prop.

Groups	Mass (MeV)	Width (MeV)
PS183 (ave. of 1984 data)	1042±5	130±50
PS183 (1986 data-preliminary)	1035±3	90±16
PS183 (overall ave.)	1037±4	90±16
Asterix (this conference)	1025±10	54 (fixed)
Rev. Part. Prop. [4]	983±2	54±7

References

1. A. Angelopoulos et al, Phys. Lett. 159B, 210 (1985).
2. A. Angelopoulos et al, Phys. Lett. 178B, 441 (1986).
3. R. Armenteros and B. French in High Energy Physics, Vol. IV, ed. E.H.S. Burhop, Academic Press, New York, 1969, p. 237 and M.J. Soulliere, Ph.D. thesis, The Pennsylvania State University, 1987.
4. M. Aguilar-Benitez, Phys. Lett. 170B, 1 (1986).

The S-, P-Wave Problem in $\bar{N}N \to \pi\pi$ at Rest

G.A. Smith

Laboratory for Elementary Particle Science, Department of Physics,
The Pennsylvania State University, University Park, PA 16802, USA
PS183 Collaboration (LEAR),
Athens[1]-Irvine[2]-Karlsruhe[3]-New Mexico[2]-Penn State[4]

1. Introduction

There is a long standing impression that the P-wave contribution to $\bar{N}N \to \pi\pi$ at rest is large [1]. The agrument for this proceeds as follows. By parity and total angular momentum conservation, $\bar{p}p \to \pi^+\pi^-$ (I=1) proceeds from L=0, C=odd $\bar{N}N$, whereas $\bar{p}p \to \pi^+\pi^-$ (I=0) and $\pi^\circ\pi^\circ$ proceed from L=1, C=even $\bar{N}N$. Furthermore, for $\bar{p}d$, charge independence predicts $f(\pi^+\pi^-n) = \frac{1}{2}f(\pi^-\pi^\circ p) + 2f(\pi^\circ\pi^\circ n)$ [2]. Thus, for S-wave where $f(\pi^\circ\pi^\circ n) = 0$, we have the prediction $f(\pi^-\pi^\circ p)/f(\pi^+\pi^-n) = 2$.

To date, several measurements which bear directly on this problem have been reported. It is well established that the branching ratio (BR) for $\bar{p}p \to \pi^+\pi^-$ is $(32\pm3) \times 10^{-4}$ [3]. A value for the BR for $\bar{p}p \to \pi^\circ\pi^\circ$ of $(4.8\pm1.0) \times 10^{-4}$ has been reported [4], although two other measurements which will be discussed later are substantially smaller. Two measurements for the ratio $\frac{\bar{p}d \to p\pi^-\pi^\circ}{\bar{p}d \to n\pi^+\pi^-}$ have also been reported: 0.68 ± 0.07 [5] and 0.70 ± 0.05 (or 0.55 ± 0.05 depending on whether π^+ or π^- inclusive spectra are analyzed for $n\pi^+\pi^-$ [6]) where the former value carries with it the condition that the "spectator" nucleon in the deuteron has a momentum less than 300 MeV/c. When the above numbers are taken in combination, one finds that the ratio $(\pi\pi)L=1, I=0/(\pi\pi)TOTAL = 0.40\pm0.08$ (H_2) and 0.75 ± 0.08 (D_2). The difference between H_2 and D_2 has been associated with the Fermi motion in the deuteron [1].

The authors of Ref. [6] also observe that the widths of their pion lines from $\pi^+\pi^-n$ are larger than from $\pi^-\pi^\circ p$, and that this cannot be explained by resolution and Fermi motion in the deuteron. This leads them to conclude that the enlarged width is due to Doppler broadening resulting from a bound $\bar{p}p$ state with a binding energy and width of ~ 20 MeV each.

2. The PS183 Detector

The detector has been described in detail in two other articles which appear in these proceedings [3,7]. For the data reported here the target was filled with liquid deuterium. Especially important to the present work is the excellent resolution of the spectrometer and the multiplicity measurement provided by the cylindrical wire chamber (CWC) around the target.

3. Results

In Fig. 1 we show the π^- spectrum for CWC=1. A very prominent peak appears at 924 MeV/c with a width (FWHM) of 60 MeV/c. This we associate with $\bar{p}d \to \pi^-\pi^\circ p$, where the proton ranges in the target ($p \gtrsim 300$ MeV/c). The width is compatible with resolution (32 MeV/c [3]) smeared by the Fermi motion of the neutron target (~ 50 MeV/c). The background has been calculated using phase space and known resonances (η, ω, ϕ),

---

[1] Work supported in part by the Greek Ministry of Research and Technology.
[2] Work supported in part by the U.S. Department of Energy.
[3] Work supported in part by the Federal Ministry for Research and Technology, FRG.
[4] Work supported in part by the U.S. National Science Foundation.

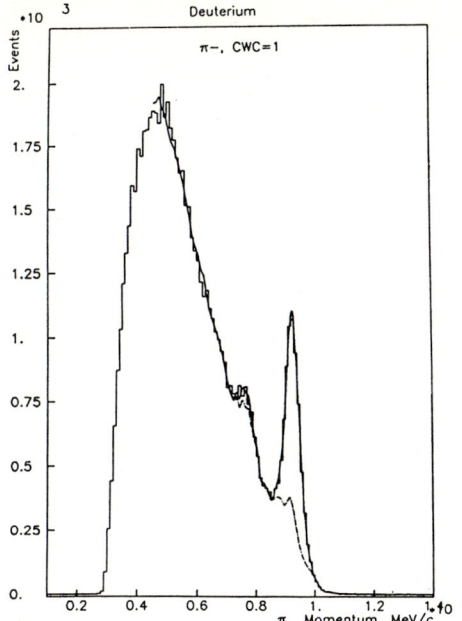
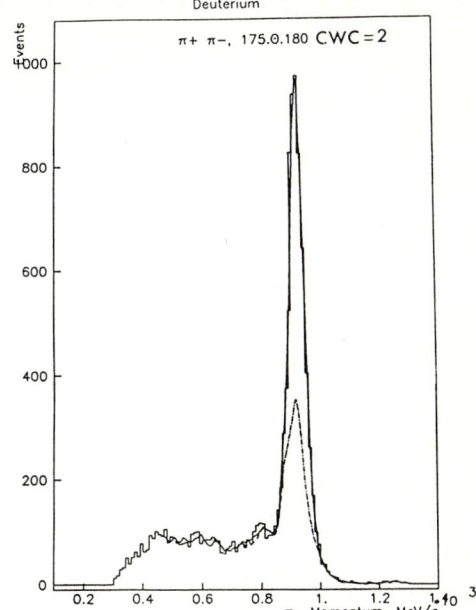

Fig. 1: Pi-minus momentum spectrum in deuterium for CWC=1.

Fig. 2: Pi-plus(minus) momentum spectrum in deuterium for CWC=2 and the angle between the $\pi^\pm$ and the away-side track in the range 175°-180°.

smeared by the deuteron wavefunction. The single largest background contribution is due to $\rho^-\pi^0$, where the π^- from $\rho^- \to \pi^-\pi^0$ "reflects" into the spectrometer. This accounts for the small enhancement of the background below the $\pi^-\pi^0$ peak. The fitted yield of $\pi^-\pi^0$ events is $2361\pm274/10^6$ annihilations.

In Fig. 2 we show the $\pi^\pm$ spectrum for CWC=2 and the angle between the $\pi^\pm$ and the away-side particle in the range 175-180°. Again, a very prominent peak which we associate with $\bar{p}d \to \pi^+\pi^-n$ at 923 MeV/c with a width (FWHM) of 60 MeV/c is seen. As before, fits to the spectrum have been made, allowing for the angle cut. The prominent background peak under the $\pi^+\pi^-$ peak is due to the strong $\pi^\pm\rho^\mp$ process, where the $\pi^\mp$ from $\rho^\mp \to \pi^\mp\pi^0$ "reflects" into the spectrometer. The fits give 1973 ± 202 $\pi^+\pi^-n$ events/10^6 annihilations.

From these two measurements, we find the ratio $\frac{\bar{p}d \to \pi^-\pi^0 p}{\bar{p}d \to \pi^+\pi^-n} = 2.39\pm0.37$, where one must halve the number of $\pi^+\pi^-n$ events to avoid double-counting problems. This value is barely 1σ from the prediction of pure S-wave $\bar{N}N$ interaction. We believe that the clear conflict of this measurement with previous ones [5,6] is explained by the vastly superior resolution of our detector and by taking into account the $\rho\pi$ reflection problem in our fits. When combined with the more recent measurements of the branching ratio for $\bar{p}p \to \pi^0\pi^0$ of $(1.4\pm0.3)\times10^{-4}$ [8] and $(2.06\pm0.14)\times10^{-4}$ [9], one must conclude that the evidence for a large P-wave $\bar{N}N$ interaction at rest is weak.

References

1. See R. Bizzarri, Physics at LEAR With Low-Energy Cooled Antiprotons, ed. U. Gastaldi and R. Klapisch, Plenum Pub., New York, 1984, p. 193.
2. H. Lipkin and M. Peshkin, Phys. Rev. Lett. 28, 862 (1972).

3. A more precise value for this number has been reported. See G.A. Smith, "Charged Two-Meson Production from $\overline{N}N$ Annihilation at Rest," these proceedings.
4. S. Devons et al, Phys. Rev. Lett. 27, 1614 (1971).
5. L. Gray et al, Phys. Rev. Lett. 30, 1091 (1973).
6. D. Bridges et al, Phys. Lett. 180B, 313 (1986).
7. G.A. Smith, "Search for Unusual Behavior in $\bar{p}$-Nucleus Annihilation at Rest," these proceedings.
8. G. Bassompierre et al, Proc. 4th European Antiproton Symposium, Barr, 1978, Vol. 1, p. 139.
9. L. Adiels et al, contribution to the VIII European Symposium on Nucleon-Antinucleon Interactions, 1-5 Sept. 1986, Thessaloniki, Greece.

Spin-Dependent Observables in p̄p Elastic Scattering at Low Energy

R. Bertini[3], H. Catz[3], A. Chaumeaux[3], B. Fabbro[3], J.-C. Faivre[3],
H. Fanet[3], J. Pain[3], F. Perrot[3], E. Vercellin[3], E. Boschitz[1], W. Gyles[1],
C.R. Ottermann[1], T. Tacik[1], E. Descroix[2], R. Harountunian[2],
J.Y. Grossiord[2], A. Guichard[2], J. Arvieux[4], J.M. Durand[4], J. Yonnet[4],
S. Mango[5], J. Konter[5], and B. Van den Brandt[5]

[1] Kernforschungszentrum und Universität Karlsruhe,
  D-7500 Karlsruhe, Fed. Rep. of Germany
[2] I.P.N. Lyon, F-69622 Villeurbanne Cedex, France
[3] Service de Physique Nucléaire-Moyenne Energie,
  CEN Saclay, F-91191 Gif-sur-Yvette Cedex, France
[4] L.N.S., CEN Saclay, F-91191 Gif-sur-Yvette Cedex, France
[5] S.I.N., CH-5234 Villigen, Switzerland

I. INTRODUCTION

At the end of the first period of the LEAR era our knowledge of the N̄N system is still very incomplete. This situation is related to the larger complexity of the N̄N scattering compared to the rather well known NN system. In order to get a feeling of this complexity, let us observe that in the N̄N scattering there is no generalized Pauli principle that excludes in NN for each isospin some partial waves. Moreover the phase shifts become complex due to the presence of the strong annihilation. These two features each double the number of required parameters to describe the N̄N interaction.

New data have been provided by LEAR experiments on the total p̄p cross section /1-3/, angular distributions of the differential cross sections of p̄p elastic scattering /2-3/ and charge exchange p̄p → n̄n [ref./3/]. However data on spin dependent observables, that is analysing power, spin rotation parameters etc, are almost inexistent.

An additional difficulty to provide data on spin observables is due to the lack of p̄ polarized beams and to the fact that the very low value of p̄C analysing power /4/ prevents the measurement of the polarization of the scattered p̄. This excludes two tools that have been essential to determine spin-dependent observables in the NN system. A different approach is therefore needed in order to provide the required data. It is the aim of this contribution to present a method to measure spin observables in the p̄N elastic scattering utilizing a polarized target, a spin rotation device and a proton recoil polarimeter. This will be discussed in section 2, but first let us resumé the present status of the theoretical understanding.

The theoretical approach to N̄N scattering is mainly based on potential models. The first ingredient of this theoretical description is a form of theoretical NN potential based on meson exchange which is G-parity transformed to an N̄N potential. This G-parity transformation reverses the signs of the potential contributions of the odd G-parity meson exchanges. In the NN potentials large cancellations occur between the contributions of different mesons, in the N̄N potentials these cancellations no longer occur /5/ and these potentials are generally more attractive than in NN. Therefore, more partial waves are contributing significantly to the cross sections at low energies. The second ingredient in N̄N models is some kind of annihilation mechanism. The annihilation cross section is large ($\sigma_{an}/\sigma_{el} \gtrsim 2$), and is responsible for the large imaginary part of the potential. Several different approaches exist for describing the annihilation : one may apply a suitable boundary condition /6/, use an optical potential /7,8/ or do an actual coupled-channel calculation /9/. All these approaches fit reasonably well the existing data on the spin-integrated cross sections. For the spin-dependent observables the predictions depend consistently on the theoretical inputs. For these observables we adopt the definition given in ref./10/, where a cartesian frame is adopted in which the x and z axis are in the scattering plane, perpendicular and parallel to the momentum of the incoming beam, the y axis is normal to the scattering plane.

Our purpose is to provide with data the observables P, D, A, A', R and R'. Predictions for the polarization P(θ) in $\bar{p}p$ elastic scattering exist :

a) in the optical potential approach with two sets of parameters (model I and II) for the annihilation potential, assumed to be spin, isospin and energy independent, in ref./11/ ;
b) in the optical potential approach but with an annihilation potential dependent on the state (spin and isospin) and the energy /8/ and with a different cut-off radius ;
c) in the coupled channel approach /9/

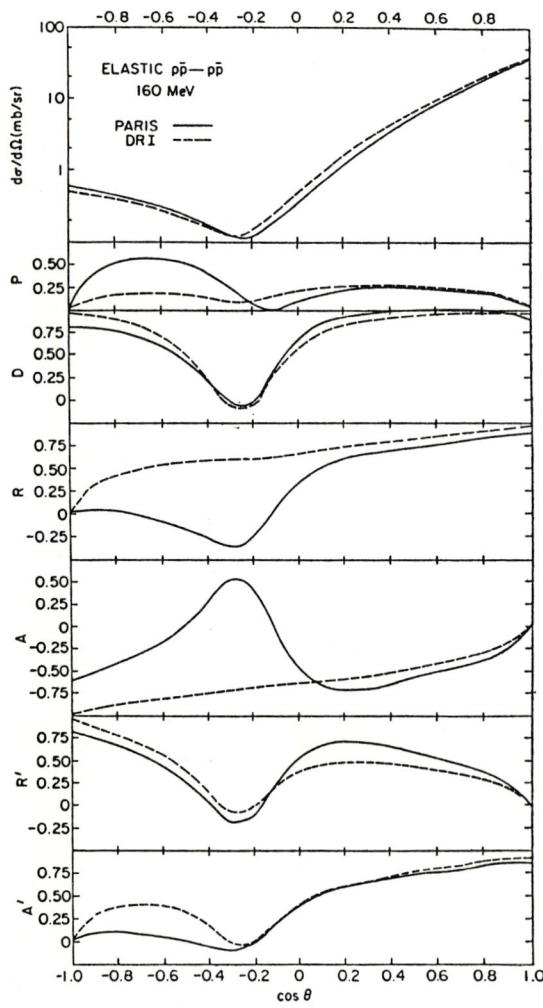

Fig. 1. Differential cross section dσ/dΩ and spin observables P, R, A, R', A' for elastic $p\bar{p}$ scattering at E = 160 MeV. The predictions of the PARIS and DRI models are shown as solid and dashed lines.

All these calculations predict sizeable amounts of polarization and a strong angular dependence, that could be roughly related to the potential in the following way : forward angles are more sensitive to the long-range part of the potential, backward angles to the short range part.

The energy dependence of $P(\theta)$ has also been computed as a function of the energy in ref./11/. A strong energy dependence for $P(\theta)$ is predicted.

Predictions for the observables $d\sigma/d\Omega$, P, A, A', R and R' have been given, at several energies, in ref. /12/. They have been made by the use of the PARIS potential /8/ and the model DR1 (refs./5-11/). Angular distributions of these observables are shown in fig. 1, for the energy at 160 MeV. It appears that, whereas the dependence of $d\sigma/d\Omega$ on different theoretical inputs is quite similar, strong differences are shown, mainly at backward angles, by the spin-dependent observables.

2. EXPERIMENTAL METHOD

Before describing the proposed experimental set-up let us first develop some arguments that have motivated our choices. As already mentioned, some symmetries, valid in the pp elastic scattering, no more occur in the $\bar{p}p$ elastic scattering. Therefore the full angular range $0 \leq \theta_{cm} \leq 180°$ has to be studied. An inspection of the kinematics shows that for each laboratory angle there are two solutions : one for each kind of particles ($\bar{p}$ or p). This implies particle identification. Moreover, as we are in the low energy region, energy loss and straggling, Coulomb multiple scattering and annihilation in the target, strongly limit the maximum target thickness. The situation is even more involved if one needs a polarized target. For example if the polarized material is propanediol $[C_3H_6(OH)_2]$ or butanol $[C_4H_9(OH)]$ the $\bar{p}p$ elastic scattering has to be selected from the reaction channels on the nuclear content of the target. All these factors strongly limitate, unlikely in the pp case, the angular range obtainable in coincidence experiments and even make hopeless, in the low energy domain, the measurement if thick polarized targets are utilized (3 cm or so).

The best way to identify antiprotons (protons) and to select the proper reaction channel is to measure accurately their angle and their momentum with a magnetic field. This can be done with the proposed set-up illustrated in fig. 2. It consists of :

a) The SPES II spectrometer,
b) a frozen spin polarized target and
c) a proton recoil polarimeter.

a) <u>The SPES II spectrometer</u>

The SPES II spectrometer has operated successfully at CERN since many years and produced large amount of data both on hypernuclei production and $\bar{p}p$ and $\bar{p}d$ reactions at the P.S. [ref./13/] and in $\bar{p}$-nucleus scattering /14/ at LEAR. The major modifications that will be introduced to the set-up utilized in ref./14/ will be : new M.W.P. chambers ⑥ with a more performant encoding system to increase the data-taking speed, the polarized target ④ and a monitor counter ③ to check continuously the thickness of the target content (fig. 2). The antiproton flux will be measured with counters ①. In the $\bar{p}p$ elastic scattering for the measurement of $P(\theta)$ we will detect $\bar{p}(p)$ in the angular domain $0 \leq \theta_L \leq 45°$, corresponding to the forward (backward) $\theta_{C.M.}$ domain. In such a way we will detect always the particle that has the highest energy and limit therefore energy straggling and multiple scattering effects. The target thickness (3 mm at 300 MeV/c has been so chosen that the energy resolution on the missing mass spectrum will always be lower than 2 MeV. Energy resolution is an important parameter to select good events ($\bar{p}p$) from target contaminations ($\bar{p}$ nucleus reactions). Assuming an energy resolution of 2 MeV we have estimated the spectrum contamination from $\bar{p}A$ events taking the cross section values of ref./14/ and found them to be always less than 1 % when $\bar{p}$ are detected. When protons are detected the ratio peak over

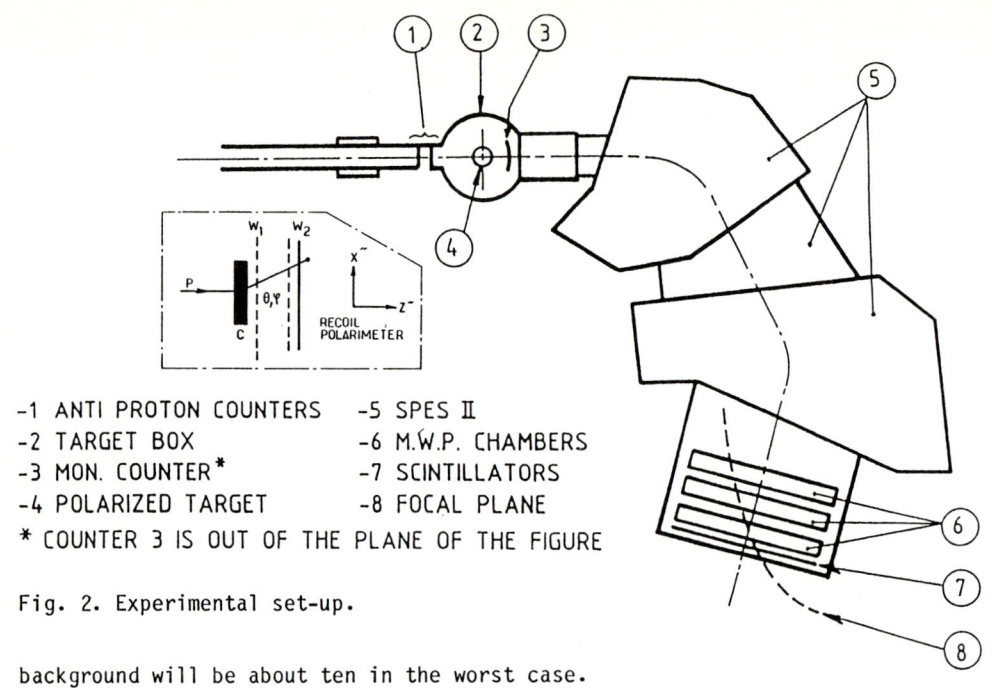

-1 ANTI PROTON COUNTERS -5 SPES II
-2 TARGET BOX   -6 M.W.P. CHAMBERS
-3 MON. COUNTER*   -7 SCINTILLATORS
-4 POLARIZED TARGET   -8 FOCAL PLANE
* COUNTER 3 IS OUT OF THE PLANE OF THE FIGURE

Fig. 2. Experimental set-up.

background will be about ten in the worst case.

For the measurement of D, A, A', R and R' only protons will be detected (angular domain accessible $0 < \theta_L < 50°$) and the recoil polarimeter, sketched in fig. 2, added behind the M.W.P. chambers ⑥.

b) The polarized proton target

For the polarized proton target we intend to use a frozen spin configuration. The development of this set-up is based on the long experience with polarized targets at SIN /15-17/. The 2.5 T magnetic field needed to polarize the protons is produced by a superconducting split coil magnet. Due to the conduction cooling of the magnetic coils it is possible to supply a vertical or horizontal field. This fact enables us to get all proton spin directions needed for the proposed experiment. For the set-up with a vertical field there are no restrictions on the angles of the scattered or recoil particles. The restrictions in the case of the horizontal field configuration are small due to the optimized geometry of the coils. The accessible θ region is 272° over 360°, $\Delta\phi = \pm 20°$. The homogeneity of the field is 1 part in 10^4 over a spherical volume with a diameter of 2.5 cm. Therefore we can vary the thickness of the target between 3 mm and 20 mm. For the target material we use butanol or propanediol. The proton polarization is determined by comparing the dynamic polarization signal with the natural polarization signal at thermal equilibrium /17/. Using this set-up we can obtain proton polarization of at least 80 %. To decrease the influence of the magnetic field on the trajectories of the incoming and outgoing particles we shall lower the magnetic field of the target to 0.35 T. This is achieved with the target in the frozen spin mode. With this set-up it is also possible to polarize deuterons.

c) The recoil polarimeter

This device will be utilized for the measurement of D, A, A', R, R'. When protons are detected, we measure their polarization putting on the focal plane of SPES II the polarimeter, sketched on the left side of the fig. 2. Then, the protons, scattered by the carbon analyser C, are detected by the multiwire proportional chambers W_1 and W_2. By track reconstruction we can obtain the polar (θ) and azimuthal (ϕ) angles. The intensity distribution of the scattered events $W(\theta,\phi)$ is related to the polarization components of the proton along the axis $x"(P_{x"})$ and $y"(P_{y"})$ through the relation /18/.

$$W(\theta,\phi) \propto \frac{d\sigma}{d\Omega} [\ 1 + P_{y"}\ A_c(\theta)\ \cos\phi - P_{x"}\ A_c(\theta)\ \sin\phi\ ] D(\theta,\phi)$$

where $D(\theta,\phi)$ is the polarimeter efficiency and $A_c(\theta)$ the carbon analysing power. Unlike the $\bar{p}C$ scattering, in the pC scattering $A_c(\theta)$ is large enough to measure the proton polarization as it has been shown in many experiments /19-21/.

The spin of the scattered proton (from $\bar{p}p$), if oriented in the scattering plane, is rotated by the magnetic field of SPES II by an angle δ relative to the particle momentum given by

$$\delta = \gamma\ (g/2 - 1)\ \alpha$$

Where γ is the Lorentz factor, $g/2$ is the magnetic moment of the proton and α is the bending angle of SPES II (about 104°). This is a very interesting feature as it allows the measurement of spin components, oriented longitudinally (z axis) before precession.

Knowledge of A and R require measurements of the spin transverse components ; R' and A' require the measurement of the spin components originally oriented along the scattered particle ($\bar{p}p$) direction of motion.

Because of the recoil of the focal plane with kinematics and because the proton energy decreases to too low values, only lab. angles $0 \leq \theta_L \leq 50$ can be measured corresponding to the backward $\theta_{c.m.}$ region ($\bar{p}p$).

A part of this program is the subject of the PS. 198 experiment /22/, which has its first data taking scheduled for the beginning of November 1987.

REFERENCES

1. A.S. Clough et al. : Phys. Lett. 146B 299 (1984).
2. C.I. Beard et al. : in Third LEAR Workshop, p.225. eds. U. Gastaldi, R. Klapisch, J.M. Richard and J. Tran Thanh Van (Tignes, 1985).
3. W. Bruckner et al. : Phys Lett. 158B, 180 (1985).
   W. Bruckner et al. : Phys Lett. 166B, 113 (1986).
   W. Bruckner et al. : Phys Lett. 169B, 302 (1986).
4. C.I. Beard et al. : in Third LEAR Workshop, 239, eds. U. Gastaldi, R. Klapisch, J.M. Richard and J. Tran Thanh Van (Tignes, 1985).
5) W.W. Buek et al. : Ann. Phys. (N.Y.) 121 ,47 (1979) ; C.B. Dover et al., 121, 70 (1979).
6) A. Delville et al., Am. : J. Phys. 46 (1978) 907 ; O.D. Dalkarov et al., Nuovo Cimento 40A, 152 (1977).
7) R.A. Bryan et al.: Nucl. Phys. B5 201 (1968) ; C.B. Dover et al.: Phys. Rev. C21 1466 (1980) ; T. Ueda in Progr. Theor. Phys. 62 1670 (1979).
8) J. Cote et al.: Phys. Rev. Lett. 48 1319 (1982).
9) P.H. Timmers et al.: Phys. Rev. D29 1928 (1984).
10) N. Hoshizaki in Suppl. Progr. Theor. Phys. 42, 107 (1968).
11) C.B. Dover et al.: Phys. Rev. C25, 1952 (1982).
12) C.B. Dover : in Low Energy $\bar{p}p$ strong interactions ; theoretical perspective, invited talk at the workshop on the design of a low energy antimatter facility in USA, Madison, Wisconsin ; October 3-5 (1985) and private communication.

13) W. Brückner et al.: Phys. Lett. 79B, 157 (1978); R. Bertini et al.:
    Phys. Lett. 83B, 306 (1979); Phys. Lett. 90B, 375 (1980); Nucl. Phys.
    A360, 315 (1981); Nucl. Phys. A368, 365 (1981); Nucl. Phys. B209, 269 (1982) ;
    Phys. Lett. 136B, 29 (1984); Phys. Lett. 158B, 19 (1985).
14) D. Garreta et al.: Phys. Lett. 135B, 266 (1984); Phys. Lett. 149B, 29 (1984);
    Phys. Lett. 150B, 95 (1985).
15) C. Amsler et al.: Phys. Lett. 57B, 289 (1975).
16) G.R. Smith et al.: Phys. Rev. C29, 2206 (1984).
17) B. Van den Brandt et al.: Verhandlungen of the German Physical Society,
    München (1985), Physik Verlag, Weinheim, 317.
18) D. Besset et al.: Nucl. Instr. Meth. 166, 515 (1979).
19) E. Aprile et al.: Nucl. Instr. Meth. 215, 147 (1983).
20) R.D. Ransome et al.: Nucl. Instr. Meth. 201, 315 (1982).
21) J.B. Roberts : A.I.P. Conf. Proc. 42, 67 (1978).
22) R. Bertini et al.: in Measurement of spin-dependent observables in the p̄N
    elastic scattering from 300 MeV/c to 700 MeV/v LEAR up PS 198 and ref. DPhN
    Saclay 2306 bis (1985).

Antiproton-Nucleus Annihilation

J. Cugnon

Institut de Physique B5, Université de Liège,
Sart Tilman, B-4000 Liège 1, Belgium

1. INTRODUCTION

Just before the opening of the LEAR era, annihilations of antiprotons on nuclei were considered of considerable interest [1], especially after a paper by J. RAFELSKI [2], who speculated about exotic processes triggered by the annihilation. The bulk of the experimental results obtained so far in the LEAR regime is largely consistent with the following simple scenario : the incident $\bar{p}$ annihilates on a single nucleon, mainly at the surface of the nucleus, creating a few pions which cascade through the nucleus, eject some fast nucleons and leaves the residual nucleus with a moderate excitation, which is evacuated by evaporation. (We analyse this scenario in Section 2). However, $\bar{p}$-annihilation on nuclei presents some special interest in two respects : (i) it may provide a very useful tool for studying the break-up of the nucleus under the injection of increasing excitation energy. This process could show some typical features of critical dynamics. (ii) The very presence of other nearby nucleons could alter some characteristics of the annihilation process. We successively discuss these possibilities in Sections 3 and 4. For the last case, we analyze the present indications for unusual annihilations.

2. THE DYNAMICS OF THE $\bar{p}$-NUCLEUS ANNIHILATION

2.1 The Simplest Dynamical Scheme

The simplest dynamical scheme one can think of is composed of three stages :
  (1) The $\bar{p}$ annihilates on a single nucleon at the surface of the nucleus, after an electromagnetic cascade for the annihilation at rest and after some possible distortion of its motion for the annihilation in flight.
  (2) The created pions cascade through the nucleus, interacting with the nucleons, ejecting a few of them as a result of two-body interactions.
  (3) After this fast ejection process is over, the residual nucleus, still with some excitation energy (due to the nucleons which are involved in step (2) but which have not enough energy to escape), evaporates a few particles, mainly neutrons.

It can be furthermore assumed that the annihilation is a well-defined event in space-time (with the same properties as in free space) and that the multiple scattering process proceeds through collisions occurring successively in space-time. The whole scheme can be made quantitative by describing the complicated multiple scattering process (and even stage (1)) with the intranuclear cascade (INC) model. This has been done by at least four groups in refs. [3-6] where the model is described. In short, the cascade is viewed as a succession of binary collisions (and decays) which occur according to their free-space cross-sections (except for Pauli blocking effects). The following collisions are usually included : $\pi N \to \pi N$, $\pi N \to 2\pi N$, $\pi N \leftrightarrow \Delta$, $NN \to NN$, $NN \leftrightarrow N\Delta$, leading to a complicated propagation of the pions travelling through the nucleus and to their possible absorption.

The INC has been shown to successfully reproduce the bulk of available $\bar{p}$-nucleus data, and especially the inclusive measurements of ref. [7], as shown in Fig. 1 for one particular case.

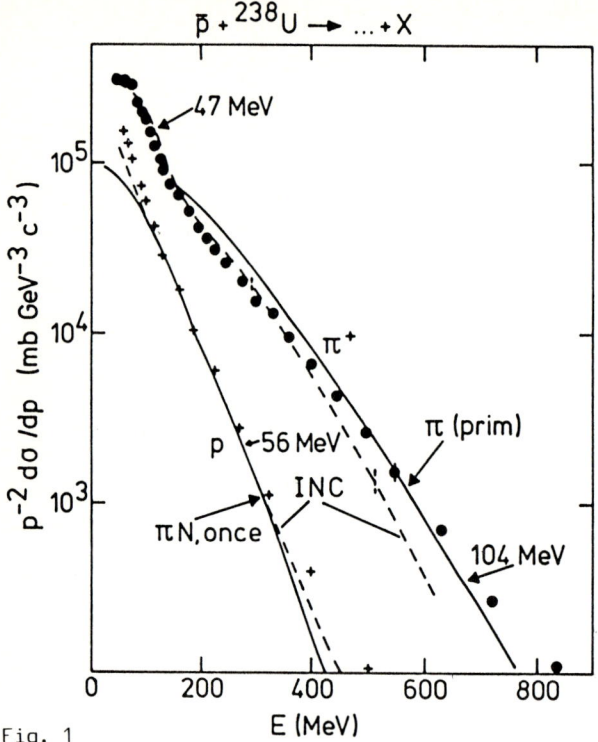

Fig. 1

π^+ and proton spectra for $\bar{p} + {}^{238}U$ annihilations (E is the kinetic energy). Dots and crosses : experimental data [7]. Dashed curves : INC calculations [7]. The full curves correspond to the spectra of the primordial pions and of the nucleons which have been hit once by a pion, respectively. The "temperatures" are indicated by the numbers (see text).

2.2 The INC Dynamics

One of the striking features of Fig. 1 is the Maxwell-Boltzmann tails in the π^+ and p spectra with large corresponding temperatures. This should not be interpreted as the evidence of the heating of a piece of nuclear matter. We eludicate this point in describing the main features of the INC dynamics for the annihilation at rest. The antiproton annihilates at the edge of the nucleus, giving birth to about 5 pions according to the available phase space. These pions display a thermal-like spectrum akin to the one of the observed high energy π^+ (see Fig. 1). Therefore this part of the spectrum is due to noninteracting pions (about half of the original pions). The pions which penetrate the nucleus make a few collisions, hitting the nucleons. Since they are quite energetic a single hit is sufficient to tranfer a large amount of energy to the nucleons. This is shown also in Fig. 1 where the spectrum of the ejected nucleons which have been hit once and once only by a pion is displayed. Secondary collisions give rise to the low energy pions and protons. The average number of collisions per pion is the order of 3-4 [8,9].

While travelling through the nucleus, pions are transforming back and forth into deltas. At the most one delta is present inside the nucleus. About one sixth of the pions is ultimately absorbed. The interaction between the pions and the baryon system transfers energy to the latter. This is shown in Fig. 2, which fixes the time scale by the same token. The large amount of energy transferred to the nucleus is evacuated

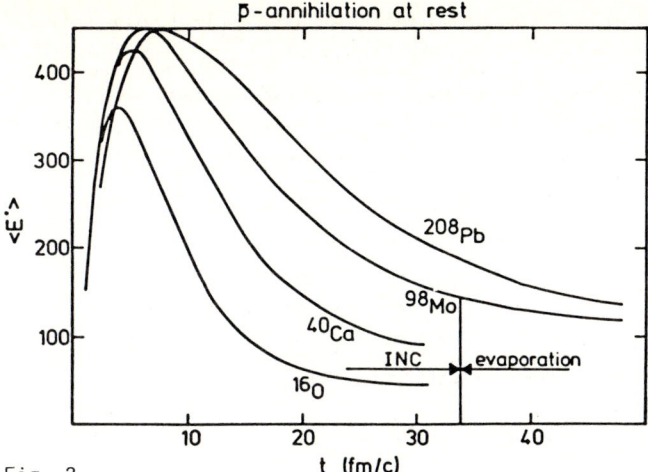

Fig. 2

Time evolution of the average excitation energy E* in the target. The separation between the INC and the evaporation phases is indicated for the ^{98}Mo nucleus. Adapted from ref. [11].

on two time scales. First, the largest part is released very quickly by the ejection of fast nucleons (about 4 for the ^{98}Mo case). After this time, one may consider that the remaining excitation energy is more or less randomized, and is released by evaporation. The separation between the standard INC cascade involving hard collisions and the evaporation is shown in Fig. 2. However this separation is admittedly loosely defined.

2.3 Residue Distribution

The shape of the residual mass distribution has recently been measured by radiochemical techniques [10] for the $\bar{p}$-Mo case. On the average, about 15 nucleons are removed from the A = 98 nucleus. This means that about 10 nucleons are evaporated. This agrees with the calculations of ref. [11] (see Fig. 3) in which an evaporation calculation is switched on at the end of the INC.

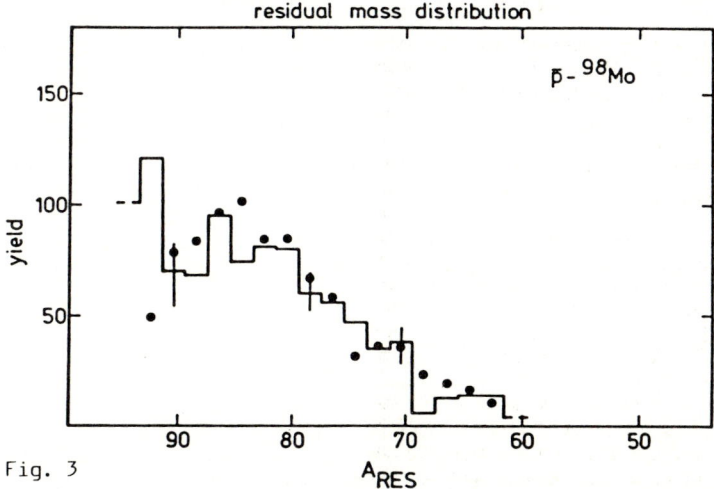

Fig. 3

Residual mass distribution for $\bar{p}$-Mo annihilation. Data (dots) from ref. [10], INC results [histograms]. Adapted from ref. [11]

In conclusion, the following picture seems to emerge at low energy : the annihilation produces an extremely localized thermalised system which decays into pions. These pions produce some streaks in the nucleons, ejecting a few fast nucleons. The nucleus does not undergo a very strong disturbance (except for the very small ones). It keeps its cohesion and de-excites by evaporating a few nucleons.

3. NUCLEAR FRAGMENTATION

3.1 Introduction

There is currently a great interest among nuclear physicists in the way the nucleus of mass A behaves after it receives excitation energy E^*. The answer to this question depends, of course, on the way the excitation energy is injected. If the excitation energy is rapidly thermalized, the answer is known in at least two limiting regimes. If $E^*/A \lesssim 2-3$ MeV, the nucleus loses its energy by evaporating particles, a few neutrons essentially. If $E^*/A \simeq 20-100$ MeV, the nucleus basically disintegrates into its constituents [12] : p,n plus a few created pions. For the intermediate regime, it is believed (and there are already experimental indications of this phenomena) that as E^*/A increases, the nucleus first loses more and more nucleons, then fragments into many large pieces (of the order of C or O nuclei), a feature referred to as the multifragmentation, and for still larger E^*/A, fragments in many more smaller fragments (see ref. [13] for a review). There are some speculations on the fact that the onset of multifragmentation could be rapid, showing some features of a phase transition : the condensation transition, studied by Fisher [14], which corresponds to the transition from a liquid phase to an assembly of droplets. The properties of this transition should bear some relationship with the saturating properties of nuclear forces and the nuclear surface tension.

3.2 Percolation Models

As we have seen, in a $\bar{p}$-nucleus system, the excitation energy is, very likely, not rapidly randomized. Rather, most of the excitation energy is used to eject rapidly fast particles. For such a case, the current ideas are that the evolution of the nucleus will depend very much on the geometrical properties of the "damage" caused to the nucleus by this fast process. More precisely, some authors think that the evolution is very much akin to cluster formation in percolation models [15,16]. The latter tell that, if the percentage p of voids created in a medium is small, a few small clusters will break loose from a very large one (of size comparable with the size of the original system). If p is larger than a critical value p_c, the system breaks into many clusters of small size. The behaviour close to p_c is very similar to a phase transition : it is referred to as the percolation transition and can be characterized by critical exponent entering in several observables like the distribution of the largest cluster (see Fig. 4), of the multiplicity... This figure is just for illustrative purpose, since it is believed that p_c is larger for the nuclear percolation phenomena.

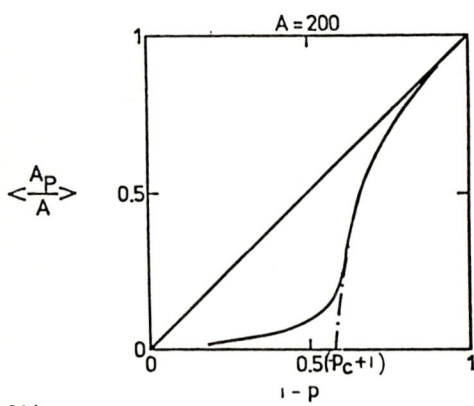

Fig. 4

Ratio of the largest cluster size to the remaining size after removal of p % sites of an infinite 2D lattice (dot-and-dashed line). The full line is the result of a percolation model for a A = 200 nucleus [34].

3.3 Specific Features of the $\bar{p}$-Nucleus System

The $\bar{p}$ annihilation appears as an alternative way to study the break-up of a nucleus, campared to the most common tools, namely high energy proton-induced reactions and intermediate energy heavy ion reactions. High energy protons and antiprotons are very similar in the sense that they first induce a rapid knock-out (spallation) of nucleons, whereas intermediate heavy ions give rise apparently to a more complicated dynamical path [8]. Furthermore, both high energy protons and heavy ions are subject to another difficulty, namely the fact that many impact parameters, pertaining to different values of p in the analysis above, are mixed up in the observation. This is definitely not the case for $\bar{p}$-annihilation at rest, where the initial system is always "prepared" in the same way.

From Section 3.2, it seems however that $\bar{p}$-annihilations at rest are subcritical in the sense of the current nuclear percolation models (see Section 3.2), although, according to some models [16] the energy deposited seems close to the required value for the onset of multifragmentation. Thence, the important question is how to make $\bar{p}$-nucleus overcritical ? Two answers at least are possible : (1) Study $\bar{p}$-annihilation in flight. Rough estimates indicate that $\bar{p}$-nucleus will become overcritical for energies above 1 GeV to 5 GeV, depending upon the models [17]. As we mentioned, one would thus lose the simplicity of the analysis, typical of at-rest annihilation. (2) Go to antideuterium (or heavier antiparticle) annihilations.

4. UNUSUAL ANNIHILATIONS

4.1 Introduction

Are there "unusual" annihilations, whose properties are different from those implied by the scenario of Section 2.1 ? The minimal departure would be a $\bar{p}$-annihilation on two nucleons. This possibility, first proposed by S. KAHANA [18], seems plausible in the light of the arguments of ref. [19], which tell that the ($\bar{p}p$) fireball lives long enough to interact with another nucleon before decaying inside a nucleus.

In ref. [19], the properties of these annihilations are studied using the simplest assumption, namely that the decay of B = 1 ($\bar{p}$NN) system is governed by phase space. In other words, the decay is described in the frame of a microcanonical ensemble. The rate for a given multiplicity n is

$$f_n(\sqrt{s}\,;\,m_1,\ldots,m_n) = C^n\, R_n(\sqrt{s}\,;\,m_1,\ldots,m_n) \quad , \tag{1}$$

where R_n is the invariant phase space integral [20]. The only free parameter C is fitted to reproduce the pion multiplicity distribution in $\bar{p}p$. The most important prediction of ref. [19] is a considerable enhancement of strange particle production in B = 1 annihilations, compared to B = 0. This is a pure consequence of phase space, since the threshold for the ΛK channel in B = 1 is well below the $\bar{K}K$ channel in B = 0. This conclusion is illustrated in Table 1, which gives relative yields for several channels. Comparison is also done (for the quark content) with quark-gluon predictions. It has been stressed in refs. [21,22] that strangeness producing reactions are very fast inside the plasma. However, it is rather unprobable that $\bar{p}$-annihilation leads to plasma formation. In ref. [23], it is shown that if $\bar{p}p$ annihilation generates a plasma, the strangeness saturation degree of the latter should be only of 10 % at most, to have sensible results. Therefore, if strangeness enhancement production is observed in $\bar{p}$-nucleus, it could be due to either unsaturated plasma, although not probable, or to B $\neq$ 0 annihilations, saturating phase space in the hadronic phase.

4.2 Analysis of Various Systems

(a) $\bar{p} + d$. This system has been studied in ref. [24] for the $\bar{p}d \to p + \pi^-$ channel, in ref. [25] for the $\bar{p}d \to \bar{K}Kp$ channel and recently by another group (see ref. [26]). The very observation of the exclusive $p\pi^-$ channel [25] is already an evidence for

B = 1 annihilation. In the $\bar{p}d \to \bar{K}Kp$ reaction, the protons are observed with a very high momentum tail, totally unconsistent with a "spectator" proton, which would have only the average momentum inside the deuteron. The importance of this tail indicates that B = 1 annihilations occur about 10 % of the time [19]. With this frequency and using ref. [19], one can make definite predictions for the branching ratios for $\bar{p}d \to p\pi^-$ and $\bar{p}d \to \Lambda + x$: they turn out to be 3×10^{-5} and 4.7×10^{-3} respectively. They compare rather well with experiment, which yields 0.9×10^{-5} and 3.6×10^{-3} respectively.

(b) $\bar{p}$-nucleus. The theoretical problem here is to determine the frequency of B = 1 annihilations. Only crude estimates are made in ref. [19], based on geometrical considerations only. Anyway, we can look whether pure B = 0 annihilations are sufficient to reproduce some observables involving strangeness. The latter can be :

(1) K^+ yield : the K^+ abundance is practically unchanged by the cascade process since K^+ scatters only elastically with the nucleons. At rest, the K^+ abundance in $\bar{p}p$ is 0.025 per annihilation. An enhancement in $\bar{p}$ nucleus would indicate the presence of B = 1 annihilation. Unfortunately, there is no measurement available up to now.

(2) K^+/π^- ratio : this ratio may be easier to measure since it does not require a measurement over a large angular range. However, the situation is a little bit complicated here since pions can be absorbed and since large nuclei (N > Z) favour π^-'s. This could perhaps explain the observations of ref. [26]. The last problem can be minimized by using N = Z nuclei and/or considering the $K^+/(\pi^- + \pi^+)$ ratio.

(3) Λ^0 production : this has been measured by Condo et al. [27] with a poor statistics. On the average, they observe $\sim (1.9 \pm 0.4) \times 10^{-2}$ Λ^0 per annihilation. This figure could be obtained with B = 0 annihilations only if all the $\bar{K}$'s transform into Λ by scattering through the nucleus. This obviously is not plausible in view of the small $\bar{K}N \to \Lambda\pi$ cross-section (see below).

(4) <u>Hypernuclei formation</u> : in a remarkable experiment [28], a group at LEAR has observed delayed fissions (with lifetime $\sim 10^{-10}$ sec) following $\bar{p}$ annihilation in flight on ^{235}U and on ^{209}Bi nuclei. The lifetime is so long that, according to the authors, the only possible explanation is the formation of a Λ-hypernucleus. When the Λ decays, the released energy leads to fission. The mechanisms which produce the hypernucleus can be either (a) $\bar{p}N \to K\bar{K}$, $\bar{K}N \to \Lambda\pi^-$, followed by the fixation of the hyperon on the nucleus or (b) $\bar{p}NN \to \Lambda K$, followed by the fixation of the Λ. The first mechanism is plausible, since the $\bar{K}$ issued from the annihilation has just the momentum ($\sim$ 700 MeV/c), which favours the substitutional fixation of the Λ created by $\bar{K}N \to \Lambda\pi^-$ [29]. Let P_0 and P_1 be the relative probability of the B = 0 and B = 1 annihilation, P_r the probability for a $\bar{K}$ to make a $\bar{K}N \to \Lambda\pi^-$ reaction inside the nucleus, and $P_f^{(0)}$, $P_f^{(1)}$ the fixation probability of the Λ in B = 0 and B = 1 cases. The yield of hypernuclei per annihilation is then given by

$$Y = P_0 B_{\bar{K}}^{(0)} P_r P_f^{(0)} + P_1 B_\Lambda^{(1)} P_f^{(1)} \quad , \qquad (2)$$

where $B_{\bar{K}}^{(0)}$ and $B_\Lambda^{(1)}$ are the branching ratios for $\bar{K}$ production in B = 0 and Λ production in B = 1 respectively. Assuming $P_0 = P_f^{(0)} = 1$, $B_{\bar{K}}^{(0)} = 0.06$, and evaluating P_r with $\sigma(\bar{K}N \to \Lambda\pi^-) = 1.6$ mb, one obtains $Y = 10^{-4}$ at the most [30]. The observed yields are 3×10^{-4} and 9×10^{-4} for Bi and U respectively. The latter can be explained only if the B = 1 annihilations are possible. With $P_1 \approx 0.1$, the experimental value is obtained, if $P_f^{(1)} \approx 10^{-1}$ for Bi, which seems quite reasonable.

4.3 Other Possible Signals for Unusual Annihilations

The understanding of the dynamics of the B = 1 annihilation is very difficult, since the description of the $\bar{p}p$ annihilation at the quark level is still in its infancy [31,32]. Nevertheless, it is very reasonable to believe that the proximity of other nucleons can disturb considerably the complicated rearrangement of quarks occurring in the course of annihilation. It is quite possible that the distribution of the number of emitted pions can be altered compared to free space annihilation. Therefore, the charged pions multiplicity distributions and if possible, the total pion multiplicity distributions should be considered as a possible way for looking to unusual annihilations. Of course, pion multiplicity are changed by the cascade, but this effect could in principle be handled satisfactorily by the INC model.

5. CONCLUSION

Although the bulk of the experimental data on $\bar{p}$-nucleus annihilation is consistent with a conventional view of the process, we have pointed out two important aspects of the annihilations. First, it may provide a useful tool for studying multifragmentation of nucleus. For this, measurements of the fragment mass yield is needed, as well as exclusive measurements of the fragmentation. We stress the theoretical importance of the nuclear multifragmentation. It is as important to understand how a nucleus loses its cohesion as to understand the origin of its self-boundedness. In this respect, this question is closely related to the general title of this School, since most of the matter in the Universe is organized in nuclei. From the phase transition theory, the problem is interesting since it deals with (unknown) transition from a Fermi liquid to a droplet (fog) phase.

The second important aspect is the unusual annihilations. We have indicated the available evidence for B = 1 annihilations. The mechanism for such annihilations is far from being understood, but their experimental study should be pursued. More generally, the modification of the annihilation process due to the presence of surrounding nucleons has not been investigated. Experimentally, this might be done by studying pion and charged particle multiplicity distributions in $\bar{p}$-nucleus annihilations.

Table 1. Branching ratios (in percent)

	π's only	$\langle n_\pi \rangle$	$\bar{K}K$	ΛK	ΣK	$\langle K^+ \rangle$	$\langle K^+/\pi \rangle$	$\bar{3s}/q$
$\bar{p}p$, at rest exp.	95	5.01	5	-	-	~2.5	0.5	-
$\bar{p}p$, microcan. ref. [19]	95.5	5.05	4.5	-	-	-	-	-
$\bar{p}NN$, microcan. ref. [19]	88.5	4.73	2.71	2.86	5.52	5.5	1.4	11
$\Phi(3N)$, microcan. ref. [19]	85.4	4.36	1.77	4.32	8.07	-	0.019	-
quark-gluon T = 200 MeV ref. [21]	-	-	-	-	-	-	25	-
quark-gluon, $\bar{p}p$ ref. [23]	-	4.5	-	-	-	-	0.5 if F=0.1	-
$\bar{p}A$, cannon. ref. [33]	-	-	-	-	-	-	-	10-15

REFERENCES

1. U. Gastaldi, K. Kilian : Physics Possibilities with LEAR, CERN/EP/79-94 (1979)
2. J. Rafelski : Phys. Lett. 91B, 281 (1980)
3. A.S. Iljinov, V.I. Nazaruk, S.E. Chigrinov : Nucl. Phys. A382, 378 (1982)
4. M. Cahay, J. Cugnon, P. Jasselette, J. Vandermeulen : Phys. Lett. 115B, 7 (1982)
5. M.R. Clover et al. : Phys. Rev. C26, 2138 (1982)
6. D. Strottmann, W.R. Gibbs : Phys. Lett. 149B, 288 (1984)
7. P.L. McGaughey : Phys. Rev. Lett. 56, 2156 (1986)
8. J. Cugnon, J. Vandermeulen : In Proceedings of the Antiproton 86 Conference, Thessaloniki (to be published)
9. P.L. McGaughey, M.R. Clover, N.J. DiGiacomo : Phys. Lett. 166B, 264 (1986)
10. E.F. Moser et al. : Phys. Lett. 179B, 25 (1986)
11. J. Cugnon, P. Jasselette, J. Vandermeulen : submitted to Nucl. Phys.
12. S. Nagamiya, M. Gyulassy : Adv. Nucl. Phys. 13, 201 (1984)
13. J. Hüfner : Phys. Rep. 125, 129 (1985)
14. M.E. Fisher : Physics 3, 225 (1967)
15. X. Campi : J. Phys. A19, L917 (1986)
16. J. Desbois, O. Granier, C. Ngô : Z. Physik A325, 245 (1986)
17. W. Gibbs : In Proceedings of Lake Louise Conference, Alberta, Canada, 1986
18. S. Kahana : In Proceedings of the Workshop on Physics at LEAR, ed. by U. Gastaldi, R. Klapisch (Plenum Press, New York 1984) p.485
19. J. Cugnon, J. Vandermeulen : Phys. Lett. 146B, 16 (1984)
20. R. Hagedorn : Relativistic Kinematics (Benjamin, Reading Mass. 1963) ch.7
21. P. Koch, B. Müller, J. Rafelski : Phys. Rep. 142, 167 (1986)
22. J. Rafelski : this Workshop
23. S.C. Phatak, N. Sarma : Phys. Rev. C31, 2113 (1985)
24. R. Bizarri et al. : Lett. Nuovo Cimento 2, 431 (1969)
25. B.Y. Oh et al. : Nucl. Phys. B51, 57 (1973)
26. G. Smith : this Workshop
27. G.T. Condo, T. Handler, H.O. Cohn : Phys. Rev. C29, 1531 (1984)
28. J.P. Bocquet et al. : Phys. Lett. B182, 146 (1986)
29. R. Bertini : this Workshop
30. M. Epherre-Rey Campagnolle : In INS Symposium on Hypernuclear Physics, Tokyo 1986
31. A.M. Green : this Workshop
32. H. Genz : this Workshop
33. C. Derreth et al. : Phys. Rev. C31, 1360 (1985)
34. X. Campi, J. Desbois : In Phase Space Approach to Nuclear Dynamics (World Scientific Publishing Company, Singapore 1985) p.238

Search for Unusual Behavior in $\bar{p}$-Nucleus Annihilation at Rest

G.A. Smith

Laboratory for Elementary Particle Science, Department of Physics,
The Pennsylvania State University, University Park, PA 16802, USA
PS183 Collaboration (LEAR),
Athens[1]-Irvine[2]-Karlsruhe[3]-New Mexico[2]-Penn State[4]

1. Introduction

In 1986 the LH_2/D_2 target of the PS183 spectrometer at LEAR was briefly replaced with C/U targets. This was motivated by theoretical models on $\bar{p}$ annihilation in nuclei (discussed later), as well as the apparent lack of precise momentum and mass-identified spectra as could be readily provided by PS183.

Annihilation of antiprotons in nuclei at rest is a complicated process which has been considered in the framework of several models. One of these, the intranuclear cascade model (INC), has been refined by several groups [1-3] to the point of providing reasonable descriptions of existing data [4]. The capture of the $\bar{p}$ is into an antiprotonic atomic state, leading to a surface annihilation. The antiproton then initiates a cascade of pions and nucleons through the nucleus. Another treatment of the problem follows from the relativistic fluid dynamic model (RFDM) [5], which provides, for example, predictions for energy-angle correlated emission of nucleons.

Both of these models require elaborate and lengthy Monte Carlo computations. Furthermore, these models do not point to any features of the annihilation which might signal more fundamental physics, such as quark dynamics. For this reason, our attention was drawn to the $\overline{N}NN$ fireball model (B=1 model) discussed by KAHANA [6] and developed by CUGNON and VANDERMEULEN [7], where rather dramatic effects involving strangeness and high momentum nucleons and pions have been suggested. These predictions are illustrated in Table 1, taken directly from Ref. [7]. In summary, 11.5% of B=1 events involve strangeness, compared to 5% in $\overline{N}N$ annihilation. Of the 11.5%, 8.8% are hyperon events, which are not produced at all in $\overline{N}N$ annihilation at rest. This enhancement in strange particle production would be signaled by an increase of the K^+/π ratio from $\sim$ 5% ($\overline{N}N$) to $\sim$ 15% ($\overline{N}NN$) [7]. An independent prediction of enhanced strange particle production has also been made recently by DERRETH et al. [8] in a similar model. An even more dramatic prediction for this ratio ($\sim$ 25%) is provided if a quark-gluon plasma is formed in $\bar{p}$-nucleus annihilation [9]. In addition, one might expect to find high momentum proton-pion pairs at 180°, signaling the decay of the $\overline{N}NN$ fireball into $N\pi$.

The simplest system in which one might expect to find $\overline{N}NN$ effects is the deuteron. Surprisingly, only one experiment has published $\bar{p}$ data at rest which bear on this problem. Based on 6 events, BIZZARRI et al. [10] reported a branching ratio for the B=1 reaction $\bar{p}d \to \pi^- p$ of $(0.9\pm0.4) \times 10^{-5}$. The theoretical prediction is $\sim 3 \times 10^{-4}$ [7], which suggests perhaps at most a 3% B=1 interaction rate. They also reported a rate for $\Lambda^\circ(\Sigma^\circ) K^{\circ+}\pi$'s of 3.6×10^{-3}, compared to a theoretical prediction of $\sim 4.7 \times 10^{-2}$ [7], again suggesting a small (7%) B=1 rate. No such data exist for annihilation at rest on heavier nuclei.

---

[1] Work supported in part by the Greek Ministry of Research and Technology.
[2] Work supported in part by the U.S. Department of Energy.
[3] Work supported in part by the Federal Ministry for Research and Technology, FRG.
[4] Work supported in part by the U.S. National Science Foundation.

Table 1: Predicted branching ratios for $\overline{N}NN$ annihilation ($\sqrt{s} = 3M_n$) [from Ref. 5]

Channel type	Percentage	(n_π)
$N\pi$'s	88.5	4.73
$N\pi$	5.2×10^{-2}	
$NK\overline{K}\pi$'s	2.7	1.16
$\Lambda K\pi$'s	2.9	2.51
$\Sigma K\pi$'s	5.5	2.32
$\Xi KK\pi$'s	0.4	0.39
all	100	4.42

2. The PS183 Detector

The detector has been described in detail in another article appearing in these proceedings [11]. For the 1986 runs, two changes were made to the detector. First, a cylindrical proportional chamber (CWC) with a solid angle of $\sim 95\%$ of 4π steradians was installed around the target. Second, four neutron counters were stacked immediately behind the N drift chamber, covering an area of 100 x 20 cm² and a depth of 4x10 = 40 cm. The measurements presented in this paper utilized the detector as described previously [11], with the C/U targets and the CWC which provided the charge multiplicity of an event. Neutron spectra which bear on the question of the existence of high momentum nucleons are presently under analysis. As before, charged particles were momentum analyzed in the magnet and their masses were measured by time-of-flight (TOF), as illustrated in Fig. 1. A clean separation among π, K, p, d (and possibly even t) particles is seen. Further refinements in the analysis, including the detection of nuclear fragments by TOF and energy loss methods in the A scintillation hodoscope between the target and the magnet, are foreseen.

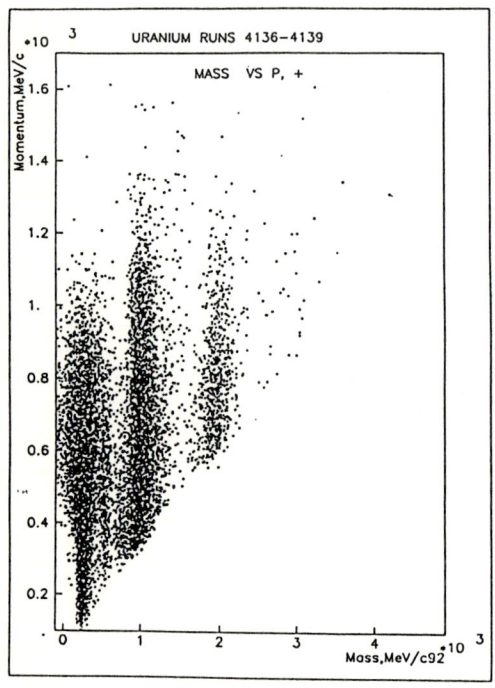

Fig. 1: Momentum versus mass for particles identified by TOF (B = 3.5 kG)

3. Results

(a) $\bar{p}d \rightarrow \pi^-p$ (B=1)

In Fig. 2 we show plots of momentum versus angle between the particle detected in the magnet and a second away-side particle seen in the N drift chamber (CWC=2) for π^- tracks (Fig. 2(a)) and p tracks (Fig. 2(b)). Very clear clusters of events (~ 40 π^-, ~ 40 p) are seen at 1250 MeV/c and 180°, exactly where $\bar{p}d \rightarrow \pi^-p$ at-rest events are expected. These ~ 80 events give a branching ratio of $(28\pm3)\times10^{-6}$ per annihilation, which is a factor of 3 larger than the value quoted by BIZZARRI et al. [10]. This suggests a B=1 rate of nearly 10%.

(b) $\bar{p}d \rightarrow \Sigma^-K^+$ (B=1)

We have carried out a search for this reaction in CWC=2 events, as it would provide further evidence for the B=1 interaction. Experimentally, this is more difficult than reaction (a), as the colinearity of such events is destroyed by the Σ^- decay and the expected location of the K^+ line (1089 MeV/c) is much closer to background from $\overline{NN}$ (B=0) $\rightarrow K^+K^-$ events. At the moment we have no signal, corresponding to a 90% confidence limit of ~ 8×10^{-6} per annihilation.

(c) $\bar{p} \rightarrow$ High Momentum Protons (B=1)

In Fig. 3 we present proton (and π^- for contrast) spectra from hydrogen (A=1), deuterium (A=2), carbon (A=12) and uranium (A=238). One notices the very rapid increase in the p/π^- ratio for momenta above ~ 400 MeV/c (where the acceptance of the spectrometer is constant) which takes place between A=2 and A=12,238. There is clearly an abundance of high momentum protons, nearly equaling the number of π^-, for large

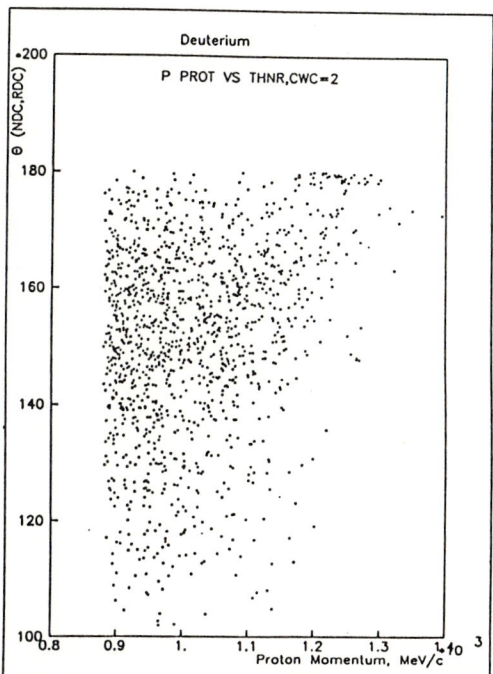
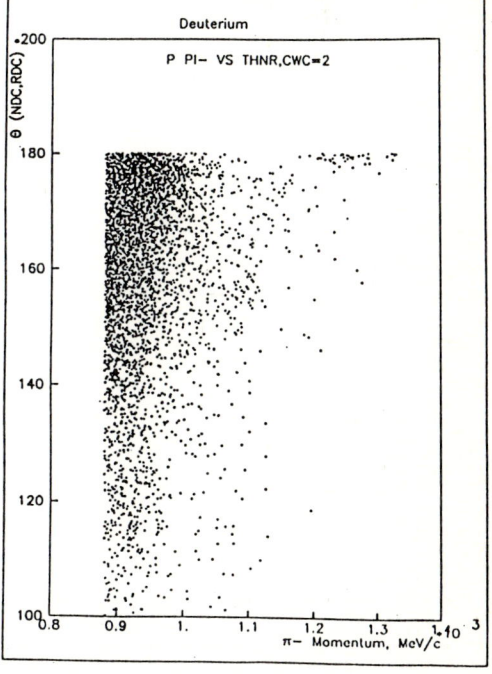

Fig. 2(a): Pi minus momentum versus the angle between the pi minus and an away-side particle (unidentified) for CWC=2

Fig. 2(b): Proton momentum versus the angle between the proton and an away-side particle (unidentified) for CWC=2.

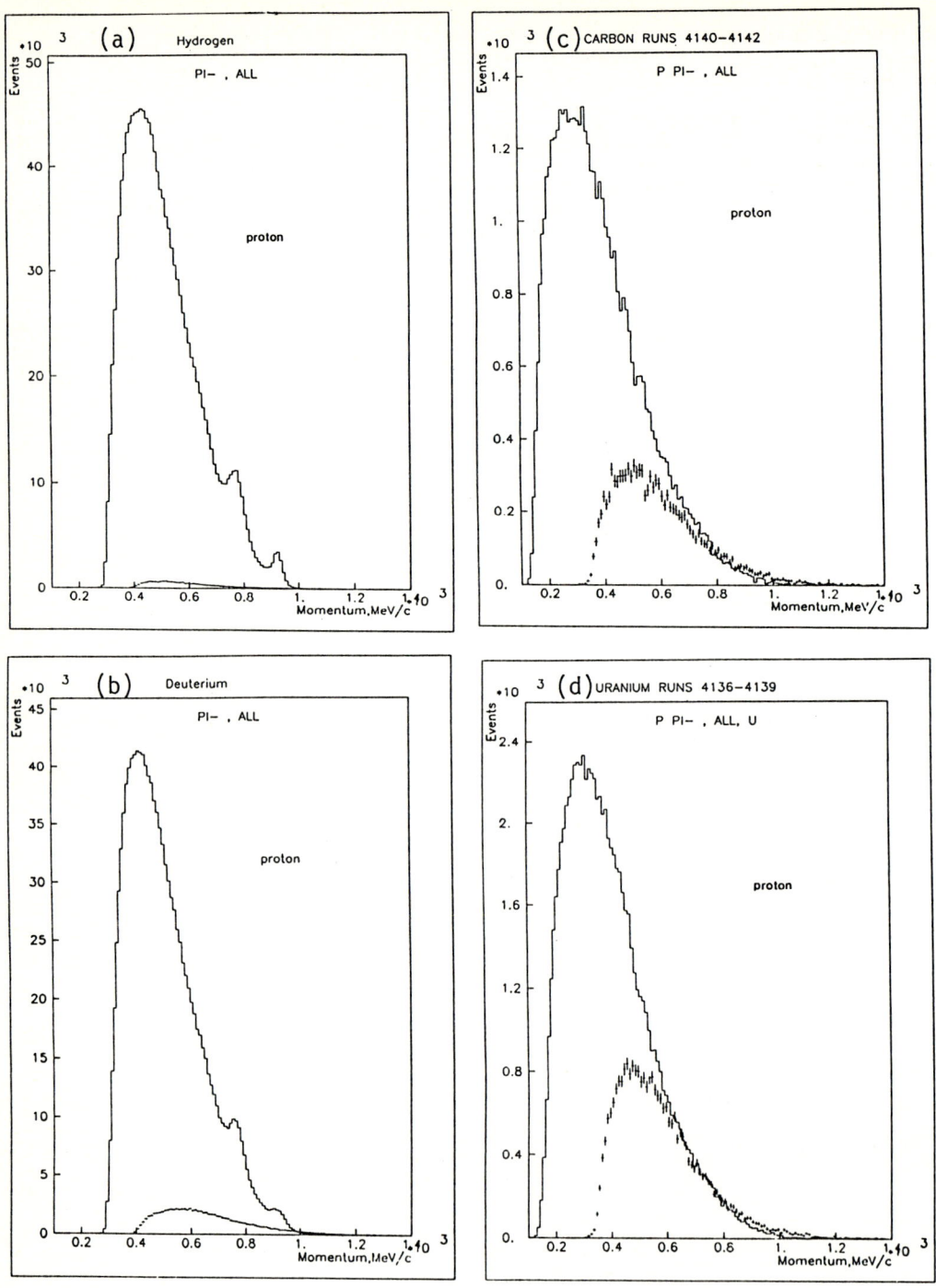

Fig. 3: Pi minus and proton momentum spectra for (a) hydrogen (b) deuterium (c) carbon and (d) uranium

A. The question arises as to whether an INC calculation could account for such protons. This, as well as a systematic search for exclusive reactions involving such protons (i.e. $\pi^-\Delta^+, p\rho^\circ, p\omega^\circ$, etc.) will be the subject of future studies.

(d) Inclusive $K^\pm$ Production

We now turn to strangeness as a possible signature of new physics in $\bar{p}A$ annihilation. In Fig. 4 we show K^+ and π^- spectra from the four targets. Apparent features include narrow ρ, π, K^* and K enhancements in the spectra for hydrogen and deuterium. A careful examination of these spectra shows that the integrated K^+/π^- ratio above 500 MeV/c is 2.2%, 2.8%, 3.0% and 3.0% in hydrogen, deuterium, carbon and uranium

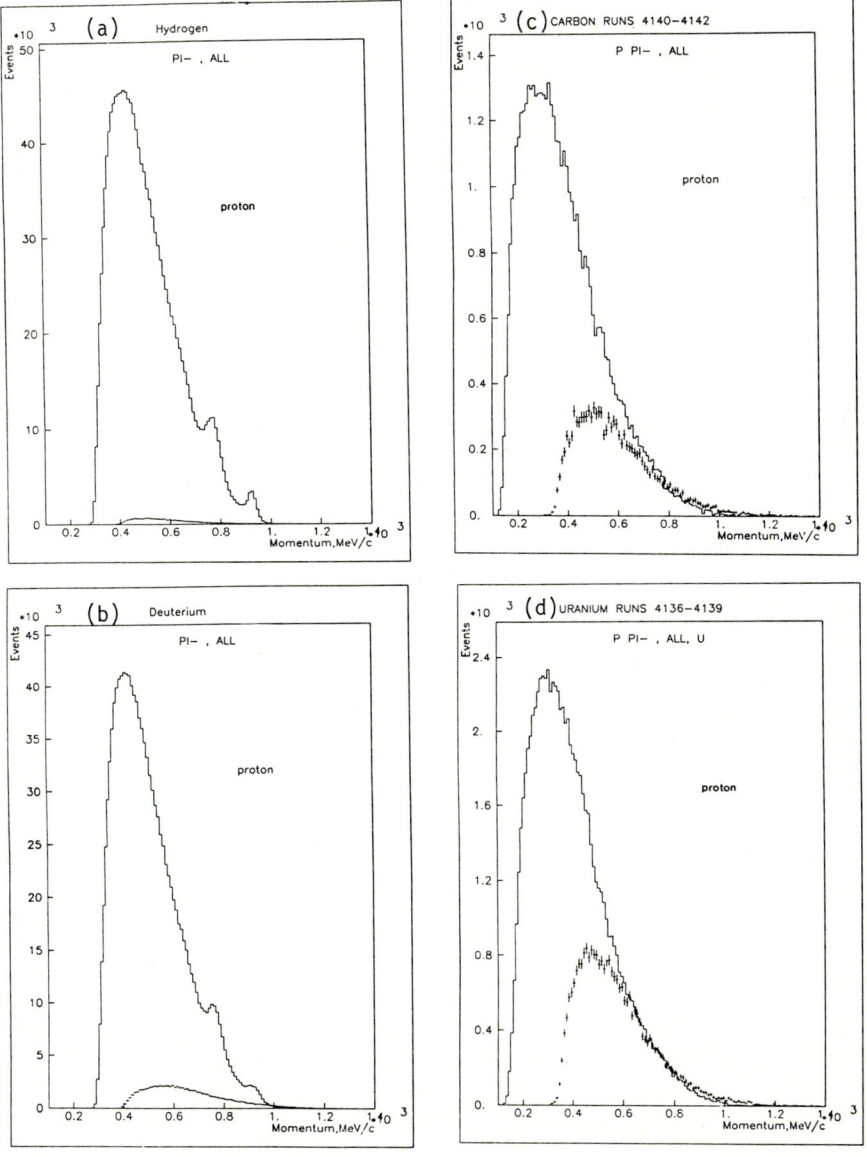

Fig. 4: K^+/π^- spectra for (a) hydrogen (b) deuterium (c) carbon and (d) uranium

respectively. A similar trend (2.2%, 3.0%, 3.2% and 3.2%) is seen in the K^-/π^- data. We conclude that some growth in the K/π ratio is observed above 500 MeV/c with increasing A, but that it is not confined to K^+ as the B=1 model would prefer.

(e) $\bar{p}A \to K\bar{K}$ and $\pi\pi + (A-1)$

We have detected a signal for the exclusive reaction $\bar{p}A \to K\bar{K} + (A-1)$ for A=238. In Fig. 5(a) we show the total energy of the K^- for events with CWC=1 ($\bar{p}n \to K^-K^0$ like events). A signal of 10±3 events is fitted over background. The average energy of the peak, 904±8 MeV, is 36±8 MeV below that expected from an unbound nucleon. We interpret the missing energy (2×38±8 = 76±16 MeV) as being accounted for in terms of binding and kinetic energy of released nucleons. The width of the peak (9±7 MeV rms) is compatible with that expected from resolution [11], namely 15 MeV.

A similar search has been made for $\bar{p}A \to \pi\pi + (A-1)$. In Fig. 5(b) we show for A=12 the total energy of the $\pi^\pm$ with CWC=2 and the angle between the $\pi^\pm$ and the away-side particle between 175-180° ($\bar{p}p \to \pi^+\pi^-$ like events). A peak of 16±5 events at 887±8 MeV is observed with a width (rms) of 27±8 MeV. This peak is 53±8 MeV below that expected from an unbound nucleon. These data suggest that the $\pi^+\pi^-$ peak in carbon appears at somewhat lower energy and with somewhat larger width than the K^-K^0 peak in uranium. Studies are underway to determine absolute yields for these lines.

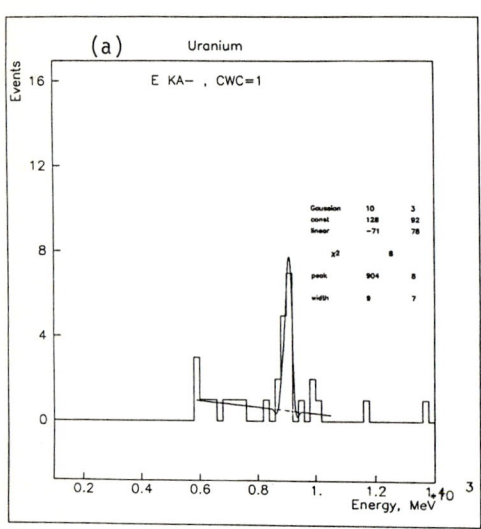
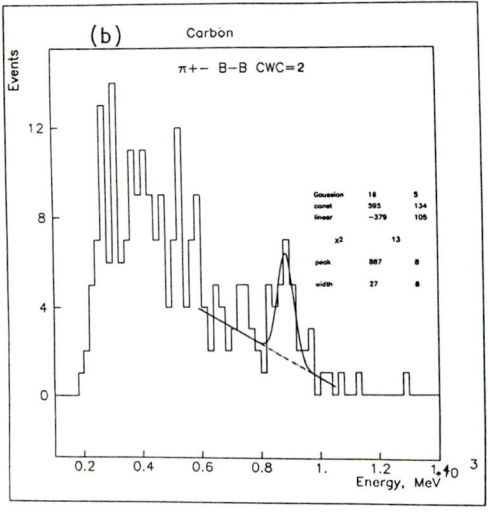

Fig. 5(a): Total energy of K^- from uranium for events with CWC=1

Fig. 5(b): Total energy of $\pi^\pm$ from carbon for events with CWC=2 and the angle between the $\pi^\pm$ and the away-side particle at 175°-180°

References

1. J. Cugnon and J. Vandermeulen, Nucl. Phys. A445, 717 (1985) and references cited therein.
2. M.R. Clover et al., Phys. Rev. C, 26, 2138 (1982).
3. A.S. Iljinov et al., Nucl. Phys. A382, 378 (1982).
4. P.L. McGaughey et al., Phys. Rev. Lett. 56, 2156 (1986).
5. D. Strottman, Phys. Lett. 119B, 39 (1982).
6. S. Kahana, Physics at LEAR with Low-Energy Cooled Antiprotons, ed. U. Gastaldi and R. Klapisch, Plenum Pub., New York, 1984, p. 485.
7. J. Cugnon and J. Vandermeulen, Phys. Lett. 146B, 16 (1984) and LEAR Workshop, Tignes, 1985, p. 559.
8. C. Derreth et al., Phys. Rev. C, 31, 1360 (1985).
9. J. Rafelski, Phys. Rep. 88, 331 (1982).
10. R. Bizzarri et al., Lett. Nuovo Cimento, Vol. II, N.9, 431 (1969).
11. G.A. Smith, "Charged Two-Meson Production from $\bar{N}N$ Annihilation at Rest," these proceedings.

Part III

Structure Functions

Nucleons in Nuclei from Quasi-Elastic Electron Scattering

A. Gérard

DPhN/HE, C.E.N. Saclay,
F-91191 Gif-sur-Yvette Cedex, France

1. INTRODUCTION

One challenging problem in modern nuclear physics is to understand how the internal structure of the nucleon interferes with the dynamics of nucleons in a nucleus. The point of view of traditional nuclear physics is to separate completely the motion of nucleons in the nucleus, described by a *wave function* and the internal structure of the nucleons, described through *form factors*. In the standard approach, free nucleon form factors are used together with some minimal prescription to take into account off-shell effects.

This conventional wisdom is nowadays questioned. On the one hand, a deformation of nucleons in the nucleus has been proposed as an explanation of the EMC effect and to anomalies observed in the quasi-elastic response functions. On the other hand, possible effects of quark antisymmetrization have been investigated in recent theoretical works.

The purpose of this paper is to review the present status of data in quasi-elastic electron scattering, to connect them with recent theoretical developments and to outline some future directions of research not accessible to present electron facilities.

2. QUASI-ELASTIC ELECTRON SCATTERING. INCLUSIVE MEASUREMENTS.

Notation for the kinematics is depicted in Fig. 1. In the kinematical region $\omega \simeq Q^2/2M_N$ electron scattering from nuclei is known as essentially a *quasi-free* process. This conviction is grounded on a vast amount of experimental data covering a momentum transfer range from 0 to 12 fm^{-1}.

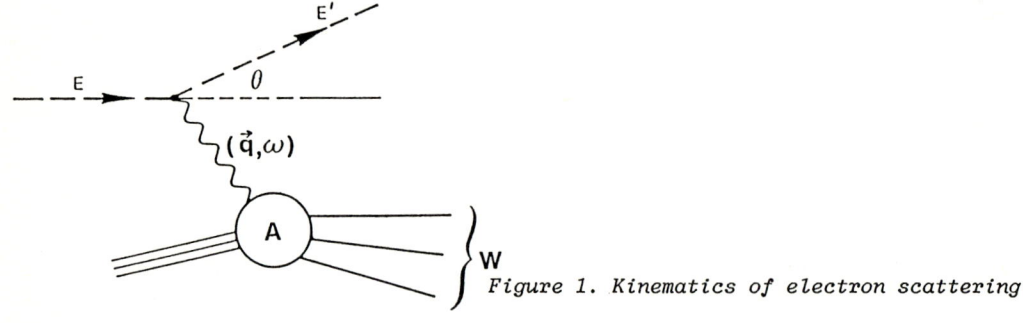

Figure 1. Kinematics of electron scattering

2.1 Low Q^2 inclusive data

Fifteen years ago things appeared clear. The simple Fermi gas model by E. MONIZ et al./1/ accounted for the gross features of the 1971 Stanford data. One of the major achievements of the eighties was the experimental *separation of the transverse and longitudinal response functions* R_T and R_L *by the Rosenbluth technique* in inclusive (e,e') measurements up to $\simeq$ 0.3 GeV^2. Various nuclei: ^{3}He , ^{3}H, ^{12}C, ^{40}Ca, ^{48}Ca, ^{56}Fe, ^{238}U have been studied at Saclay and Bates /2-4/. Except in the case of A=3 nuclei, conventional calculations (independent particle models + Schrödinger formalism + free nucleon form factors) *overestimate the longitudinal structure function* by 20 to 40%. As a consequence the Coulomb Sum Rule (CSR) is not saturated even at the highest momenta reached in these experiments (Fig. 2). This missing strength is a puzzling problem and several explanations have been proposed to explain it. Some of them are reviewed in this article. A popular one, first given by J. NOBLE /5/ is that the radius of the proton would get larger when embedded in the nuclear medium, thus quenching its electromagnetic form factors.

2.2 High Q^2 data. Y-scaling

In 1975, G. WEST /6/ suggested that a scaling variable, called y, could be used in quasi-elastic electron scattering. This y-scaling phenomenon was experimentally demonstrated in 1979 from data on ^{3}He measured at SLAC up to 100 fm^{-2}/7/ and their analysis by I. SICK et al./8/. Basically, the y-scaling is expressed by the relation

$$\sigma(q,\omega)/\sigma_{eN}(q)\ d\omega = F(y)dy$$

where σ_{eN} is the elementary electron-nucleon cross-section. The scaling variable y is somehow linked to the momentum component of the nucleon parallel to the virtual photon momentum q but there is still discussion of the best choice for this variable and of the domain of validity of the y-scaling.

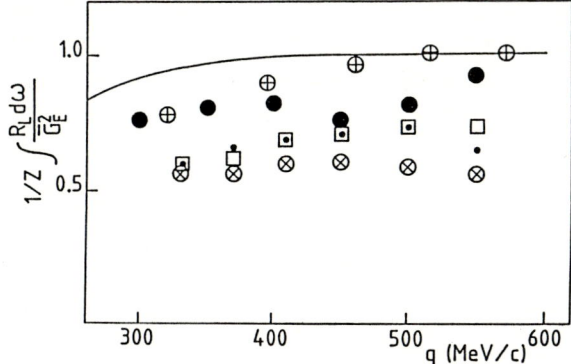

Figure 2. The Coulomb Sum Rule measured in Saclay experiments /3/ for ^{3}He (⊕), ^{12}C (●), ^{40}Ca (⊗), ^{48}Ca (□) and ^{56}Fe (•) compared to DE FOREST /37/.

In a recent SLAC experiment, the onset of scaling and its dependence on A were studied in more detail /9/. The y-scaling is now a well-established phenomenon in the *low energy side* of the quasi-elastic peak for momenta much above the nucleus Fermi momentum. It is interpreted as reflecting the quasi-free nature of the scattering process in that kinematical region. Moreover, it has been pointed out by I. SICK /10/ that using in σ_{eN} either electromagnetic form factors or a mass of the nucleon *rescaled* by more than 2 or 3% destroys the scaling. This statement, reinforced by the new SLAC data, appears to be a strong indication that, at least *in the large q domain*, the nucleon electromagnetic form factors are unlikely to be modified by large factors as advocated to explain the quenching of the longitudinal response function at *low momenta*.

3. QUENCHING OF LONGITUDINAL STRUCTURE FUNCTION: PROPOSED EXPLANATIONS

Quasi-elastic electron scattering is usually described in the framework of the Impulse Approximation. This approximation assumes that the nucleon electromagnetic current J_μ is the superposition of individual one-body currents:

$$J_\mu = \sum_i j_\mu^N(i)$$

The target nucleus is described as a sum of independent nucleons. The elementary nucleon current j_μ is the free nucleon one with some prescription to take into account off-shell effects. It is commonly believed that the kind of prescriptions given by T. DE FOREST /11/, which obey *gauge invariance*, allow little ambiguity in the choice of the current operator.

Several explanations have been proposed to explain a possible breakdown of the Impulse Approximation in quasi-elastic scattering.

3.1 Collective effects

When the wavelength of the virtual photon is larger than the internucleon distance, collective effects are expected to occur. They have been treated in the framework of semiclassical RPA theory. Using a semi-phenomenological particle-hole interaction, W.M. ALBERICO et al./12/ account fairly well for the rather *flat shape* of the longitudinal response at small momenta. This flatness is due to the splitting of the ph force into two isospin channels: the isovector (T=1) part is essentially repulsive whereas the isoscalar part (T=0) becomes attractive at the surface of the nucleus.

However, they do not account for the *magnitude* of the response unless the radius of the proton be *increased by 13% in* ^{12}C *and 20% in* ^{40}Ca *and* ^{56}Fe !

3.2 Ground state NN correlations

Going *beyond the independent particle picture* by using more realistic ground state wave functions is a difficult undertaking. Models are available for ^3H and

^{3}He from Faddeev equations /13/, and for A = 3,4 nuclei and infinite nuclear matter from variational calculations /14/. Such solutions have been used by R.SCHIAVILLA et al./15/ to study the pair distributions and structure functions in these systems. In order to learn about the dependence of such quantities on the number of particles, they calculated them for droplets of *atomic helium liquids*, containing 4 to 240 atoms and found a rather smooth behaviour.

^{3}He data are found in good agreement with these calculations. For heavier nuclei, the general tendency is a *spreading out* of the longitudinal strength out of the quasi-free peak arising from a coupling of 2p-2h states to 1p-1h states. This effect tends to explain why part of the strength is missing in the longitudinal sum rule *when the integration is restricted* to the quasi-free peak. The agreement with the data at low ω does not suggest any necessity to modify the nucleon form factor but rather to refine the treatment of 2p-2h contributions which affect the large ω part of the structure function.

3.3 Distortion of the electromagnetic wave

Distortion of the incoming and outcoming electron wave in the nuclear Coulomb potential is usually accounted for by using an *effective momentum transfer*. A second effect is the *focusing* of the electron wave. Calculations adapted from phase-shift codes are not quite reliable for quasi-elastic scattering because of the large number of multipoles involved. More refined ones are in progress /16/ and there are propositions to check them by experiments comparing e^+ and e^- quasi-free scattering.

3.4 Mesonic Exchange Currents and N^* excitation

The contribution of Mesonic Exchange Currents to the longitudinal response has been recently discussed by B.DESPLANQUES /17/. Using some constraints given by the *charge form factors* of ^{3}H and ^{3}He, he produces estimates around $Q^2 = 6$ fm^{-2} of three possible contributions to a quenching of R_L:
- The one-pion exchange "pair term": 4%
- The two-pion exchange - correlated (σ-exchange): 3%
                        - uncorrelated   : 4%
- The N^* Roper resonance: 5%. This number tends to indicate that the Roper resonance, believed to be the first *radial excitation of the nucleon* is likely to play a significant role in nuclear charge properties.

These preliminary results and their connection with the swelling nucleon picture certainly deserve further investigation.

3.5 Relativistic effects. Dirac phenomenology

Field theoretic descriptions of nucleonic spinor fields interacting with mesonic fields, basically a scalar field σ and a vector field ω, originate in WALECKA's work /18/, further developed by SEROT /19/ who incorporated in this picture the π and ρ mesons. This *Dirac phenomenology* has achieved many successes in accounting for a large variety of nuclear phenomena, especially those involving spin observables. Schematically:

1. The nucleon motion is described by the Dirac Equation.

2. The mean field is built from the strongly *attractive scalar σ component* ($V_s \cong$ -400 MeV) and the strongly *repulsive vector ω component* ($V_0 \cong$ 300 MeV). Terms in $V_s - V_0$ which appear in the small components of the Dirac spinor tend to double their free space value, thus increasing the coupling to negative energy states. Roughly speaking, the nucleon is *dressed with nucleon-antinucleon pairs* which modify its electromagnetic properties. In quark language however, it is not clear why such a 3p-3h excitation should play a more important role than the mesonic 1p-1h excitations of the nucleon.

In electron scattering this relativistic effect could be manifested at two levels:

- The nuclear wave function, which should come out of relativistic mean-field calculations.

- The nucleon current, for which a new prescription must be defined.

Several attempts to calculate inclusive response functions on ^{12}C and ^{40}Ca have been developed /20,21,38/ in a relativistic Hartree approximation. They do not seem to fully solve the problem of accounting simultaneously for transverse and longitudinal structure functions.

3.6 Change in nucleon electromagnetic structure

3.6.1 Size of the nucleon, QCD and quantum mechanics

Changing the size of the nucleon in the nuclear medium has become a popular exercise for theorists. The change of a single length or mass scale appeared an elegant way to bring a unified explanation to high energy (EMC effect) and low energy (quenching of quasi-free structure function) phenomena. A solution to this problem from QCD first principles is presently far out of reach and some amount of phenomenology is unavoidable.

On the other hand the concept of *size* of a complex hadronic object like the nucleon must be handled carefully and is now currently believed to be *probe dependent*. In several models the baryonic size (distribution of baryon number) of the nucleon appears different from its electromagnetic size (distribution of charge density). See for instance WEISE's paper in these proceedings. The quantity studied through electron scattering characterizes the coupling of a virtual photon to an extended charged object.

A general quantum mechanics argument developed recently by M. OKA and R.D. AMADO /22/ helps to connect a large class of dynamical models of the nucleon. They show that the swelling of the nucleon is a direct consequence of the *attractive* nature of the nuclear potential and must emerge from any model which accounts for this attraction.

3.6.2 The pionic model

A change in the *peripheral structure* of the nucleon, namely its pion cloud, appears a natural interpretation of its size increase. A physical origin of such an effect has been proposed by M. ERICSON and M. ROSA-CLOT /23/. The occurrence in a nucleus of low energy electric dipole excitations would result in a large increase of the *electric polarisability* of the nucleon (by a factor of 3) and a significant increase of its *charge radius* ($\Delta \langle R_c^2 \rangle \cong .5 fm^2$).However, they say nothing about higher momenta of the charge form factor q^2 expansion so that the magnitude of such an effect is difficult to estimate in the kinematical conditions of the experiments. Whether such a distortion of the pion cloud affects the magnetic radius is not known.

3.6.3 Models of the nucleon

A wide class of models addressed the problem of a change of the confinement size of quarks in the nuclear environment. Most of them /24-27/ succeed in producing a size increase consistent with the rescaling argument proposed by F. CLOSE et al./28/ for the EMC effect. However, some calculations within the Skyrme soliton framework /29/ or the quark cluster model /30/ produce different predictions.

At this point it is clear that further experimental work is needed. A refined study of the coupling of photons to bound nucleons would bring additional information. This is the goal of a new generation of exclusive measurements in the quasi-elastic region.

4. SINGLE OUT ONE-NUCLEON KNOCKOUT. NEW (e,e'p) MEASUREMENTS

4.1 Basic principle

A specific way to study the coupling of a photon with a nucleon bound in a nucleus is to isolate in the nuclear structure functions the one-nucleon knockout mechanism. Detection in coincidence (e,e'p) experiments of the scattered electron and a knocked-out proton allows the missing momentum p_m and missing energy E_m to be determined. In PWIA, $p_m = -p$ is the momentum of the initial proton on its shell and the six-folded cross-section factorizes:

$$d^6\sigma/d\Omega_e d\Omega_p dEdp = K\sigma_{ep} S(E_m,p_m)$$

where

- K is a phase-space factor.
- $S(E_m,p_m)$ is the spectral function of the nucleus. Selecting E_m and p_m in a given range allows one to sample protons with a *given binding energy* and to select regions with *different nuclear densities*.
- σ_{ep} is the off-shell electron-proton cross-section.

Interference terms in the (e,e'p) cross-section vanish in the so-called "parallel kinematics"(p'⫽q) and there remains a Rosenbluth-like expression for σ_{ep}:

$$\sigma_{ep} \propto \sigma_{Mott} \ (W_T(q^2) + \epsilon \ W_L(q^2))$$

The transverse and longitudinal part of the structure function are extracted, as in inclusive experiments, by performing two measurements: one at a forward and one at a backward angle. Since the spectral function is not known, information on σ_{ep} has to be extracted from *ratios*, such as $\sigma_L(q^2)/\sigma_T(q^2)$, $\sigma_L(q_1^2)/\sigma_L(q_2^2)$, $\sigma_T(q_1^2)/\sigma_T(q_2^2)$, where the spectral function is eliminated.

4.2 Main difficulties

There are (at least) two difficulties in extracting reliable information on the bound nucleon current from these (e,e'p) experiments:

- The first one is experimental. A typical 30% effect in the L/T ratio is manifested as a 5 to 10% effect in the ratio backward/forward. To be sure that such an effect shows up requires control of the experimental parameters (efficiencies, phase space, target thickness,...) at the percent level!
- The second one is theoretical. It is known that the distortion of the outgoing in the nuclear potential destroys the simple PWIA factorization. However, measuring *ratios* helps: calculations in DWIA /31/ show that uncertainties on the Final State Interaction are much lower on ratios of distorted cross-sections than on the cross-sections themselves.

4.3 Results

Such measurements have been undertaken at Saclay, NIKHEF and Bates. Two classes of results have been reported:

- *Studies of the ratio σ_L/σ_T*. In NIKHEF data /32,33/, the 1p knockout in ^{12}C and the 1p and 1s knockout in ^{6}Li have been singled out. The Saclay data on ^{40}Ca /34/ have been averaged over a 60 MeV range of missing energy but they go up to higher momenta (10 fm^{-2}). All these results are consistent with the inclusive measurements: the ratio σ_L/σ_T measured in the exclusive (e,e'p) channel is *smaller by 20 to 30% than the Impulse Approximation prediction*. These results by themselves appear compatible with the hypothesis of a swollen nucleon. To get more specific information on the *form factors* of the bound nucleon it is necessary to study the Q^2 behaviour of the cross-section.

- *Q^2 behaviour of the cross-section*. This kind of study has been carried out at Saclay on ^{40}Ca, up to 15 fm^{-2} for the transverse (magnetic) part and up to 10 fm^{-2} for the longitudinal (Coulomb) part /34/. These data do not suggest a large modification of the electromagnetic form factors of the bound proton. In terms of magnetic radius, an upper bound of 5% increase looks reasonable. Data for the Coulomb part are not accurate enough to definitely rule out an increase of the electric radius.

5. CONCLUSIONS. UNSOLVED PROBLEMS

Obviously we are not at the end of the story!

From quasi-elastic scattering there is now increasing evidence against a nucleon swollen by more than a few percent. The nucleon is probably *too soft* in those models which predict a 15-20% radius increase. Following OKA and AMADO /22/ a constraint for the models could be the position of the *first monopole excitation* of the nucleon, namely the Roper N*(1440) resonance. Some increase of the charge radius cannot be ruled out, but it appears a low q phenomenon difficult to disentangle from collective effects.

On the other hand, the quenching of the longitudinal structure function is now confirmed by exclusive measurements and remains a puzzling problem.

It is clear that a deeper understanding of the *two-body effects* is badly needed. The observation of the spreading of the s-shell in the transverse part of the $^{12}C(e,e'p)^{11}B$ /39/ is an indication that such two-body mechanisms could play an important role. The nature of the two nucleon subsystem in nuclear matter is actually not known, especially when the nucleon are separated by distances smaller than 1 fm. Short range correlations, quark antisymmetrization, deformation of the nucleon, are likely to be intertwined problems and address the fundamental question of the degrees of freedom relevant to describe strong interacting matter, especially at short distances /35/.

To answer this question, a powerful method would be to single out all the components of the nuclear structure functions in *exclusive experiments*, covering a much larger part of the (q^2,ω) plane than present experiments. The two-body density could be measured through (e,e'NN) coincidence experiments. On the other hand the role of the resonances of the nucleon beyond the Δ in nuclear dynamics is unknown. For instance data on the excitation of the Roper resonance in nuclei would be of primary interest.

Such experiments are out of reach of present electron facilities. They require both energy and duty cycle. A multi-GeV continuous wave electron facility has been recognised as a powerful tool to deepen our understanding of nuclear short range dynamics and to explore the transition between the "classical" description of nuclear matter in terms of baryons and mesons, and the high energy domain where fundamental constituents, quarks and gluons, are expected to play an explicit role. But this question is addressed somewhere else in these proceedings /36/.

REFERENCES

1. E.Moniz et al.: Phys. Rev. Lett. **26**, 445 (1971)
2. M.Deady et al.: Phys. Rev. **C28**, 631 (1983) and **C33**, 1897 (1985)
3. P.Barreau et al.: Nucl. Phys. **A402**, 515 (1983)
   Z.E.Meziani et al.: Phys. Rev. Lett. **52**, 2130 (1984) and **54**, 1233 (1985)
   C.Marchand et al.: Phys. Lett. **153B**, 29 (1985)

4. C.C.Blatchley et al.: Phys. Rev. C34, 4 (1986)
5. J.V.Noble: Phys. Rev. Lett. 46, 412 (1981)
6. G.B.West: Phys. Rep. 18C, 264 (1975)
7. D.Day et al.: Phys. Rev. Lett. 43, 1143 (1979)
8. I.Sick et al.: Phys. Rev. Lett. 45, 871 (1980)
9. D.Day et al.: to be published
10. I.Sick: Phys. Lett. 157B, 13 (1985)
11. T. de Forest: Nucl. Phys. A392, 232 (1983)
12. W.Alberico et al.: Nucl. Phys. A462, 269 (1986)
13. C.R.Chen et al.: Phys. Rev. C33, 1740 (1986)
14. R.Schiavilla et al.: Nucl. Phys. A449, 219 (1986)
15. R.Schiavilla et al.: Nucl. Phys. A473, 267 (1987)
16. M.Traini and S.Turck-Chieze: in preparation
    G.Co' and J.Heisenberg; preprint 1987
17. B.Desplanques: Proceedings of "9° session d'études biennale de Physique Nucléaire", Aussois, March 9-13 (1987)
18. J.D.Walecka: Ann. of Phys. 83, 491 (1974)
19. B.D.Serot: Phys. Lett. 86B, 146 (1979)
20. G.Do Dang and Nguyen Van Giai: Phys. Rev. C30, 731 (1984)
    G.Do Dang, private communication
21. S.Nishizaki, H.Kurasawa and T.Suzuki: Phys. Lett. B171, 1 (1986)
22. M.Oka and R.D.Amado: Phys. Rev. C35, 1586 (1987)
23. M.Ericson and M.Rosa-Clot: Z. Phys. A324, 373 (1986)
24. M.Jändel and G.Peters: Phys. Rev. D30, 1117 (1984)
25. L.S.Celenza, A.Rosenthal and C.M.Shakin: Phys. Rev. C31, 232 (1985)
26. G.Chanfray and H.J.Pirner: preprint LYCEN/8610 (1986)
27. C.W.Wong: Nucl. Phys. A435, 669 (1985)
28. F.E.Close, R.G.Roberts and G.G.Ross, Phys. Lett. 129B, 346 (1983)
29. M.Oka, K.F.Liu and H.Yu: Phys. Rev. D34, 1575 (1986)
30. M.Oka: Phys. Lett. 165B, 1 (1985)
31. S.Boffi: Nuovo Cimento 76A, 186 (1983)
32. G.van der Steenhoven et al.: Phys. Rev. Lett. 57, 182 (1986)
33. G.van der Steenhoven et al.: Phys. Rev. Lett. 58, 1727 (1987)
34. D.Reffay-Pikeroen et al.: Submitted to Phys. Rev. Lett.
35. P.J.Mulders: Nucl. Phys. A449, 525 (1986)
    T. de Forest and P.J.Mulders: Phys. Rev. D35, 2849 (1987)
36. A.Gérard: contribution to the Round Table on Future Accelerators, these proceedings
37. T.de Forest: Nucl. Phys. A414, 347 (1984)
38. K.Wehrberger and F.Beck: Phys. Rev. C35, 298 (1987)
39. P.E.Ulmer et al.: Phys. Rev. Lett. 59, 2259 (1987)

New Results on the EMC Effect

G. D'Agostini

Centre de Physique des Particules de Marseille, Faculté de Luminy,
Case 907, F-13288 Marseille Cedex 9, France

In 1982 the European Muon Collaboration (EMC) saw an unexpected difference between the proton structure functions F_2 measured in iron and deuterium. This difference could not be explained by either the nuclear Fermi motion effects or the systematic errors. It was published in 1983 [1] and is now called "the EMC effect". The great interest in this discovery comes mainly from its nuclear interpretation i.e.: the modification of the nucleon structure by the surrounding nucleons in the nucleus. It is not my intention to cover the theoretical developments in the subject for which I prefer to refer to the G.Miller talk at this meeting. I shall concentrate on the experimental measurements with particular emphasis on new results. After a short historical review, I shall introduce the formalism needed to follow the discussions on the EMC effect. Then, results will be presented, mainly from the EMC and Bologna CERN Dubna Munich Saclay (BCDMS) collaboration at CERN, with a few new results from neutrino scattering (BEBC) and a few from SLAC and FERMILAB.

A quick confirmation of the first observation made by the EMC-NA2, was provided by the reanalysis of empty target data [2] taken 10 years before at SLAC. These first results established the X-dependence of the F_2 structure functions ratio. In the following years many experiments were carried out and brought in new information on this effect. Let us note the A-dependence from the SLAC-E139 experiment [3] and the Q^2-dependence from the BCDMS-NA4 experiment [4]. Neutrino scattering data are characterized by small statistics and are more appropriate for the sea quark determination. More generally we can say that all the experiments were in reasonable agreement with the exception of the large enhancement (15%) of the structure function ratio, F_2^{Fe}/F_2^D, observed by the EMC at low X (<0.15) which was not confirmed by the other experiments.

The deep inelastic lepton scattering provides one of the cleanest ways to probe the matter. The basic process is illustrated in Fig. 1.

In the laboratory frame, we measure the energy and the momentum of the incident and scattered leptons. Then we can define $\nu = E_i - E_f$, the energy of the probe in the laboratory frame, $Q^2 = 4 E_i E_f \sin^2(\theta/2)$, the probe mass squared, $Y = \nu/E_i$, the fraction of energy of the probe related to the incident energy and $X = Q^2/2M_N\nu$, the Bjorken

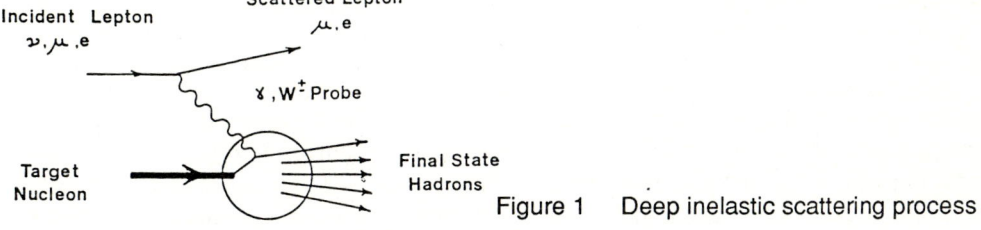

Figure 1   Deep inelastic scattering process

variable, dimensionless. M_N is the free nucleon mass. This one-boson exchange process can be expressed in terms of a leptonic tensor, perfectly known thanks to QED, times a hadronic tensor not given by the theory and containing the unknown structure of the nucleon. The cross section can then be expressed as a function of unknown structure functions F_1, F_2 (and, in the neutrino case only, F_3). We prefer, generally, to replace F_1 by $R=\sigma_L/\sigma_T$ the ratio of longitudinal and transversal probe polarisation cross sections.

Experimentally the structure function F_2 can be extracted from the X, Q^2-dependence of the differential cross sections provided R is known. To measure R we must vary a third independent variable as, for instance, the incident lepton energy E_i.

The first measurement of F_2 in 1968 showed that it depends essentially on one dimensionless variable X. This is called scale invariance or scaling. The parton model of R.P.Feynman gives a beautiful interpretation of scaling. In this model the nucleons are considered as constituted of free pointlike particles: the partons. In the simplest model, the masses and the transverse momenta are set equal to zero. The charged partons have spin 1/2. We define $q_i(X)$ as the probability density that the struck quark "i" takes the fraction X of the total nucleon momentum (M_N in the laboratory frame). In this model X is identical to the measured Bjorken variable $X=Q^2/2M_N\nu$ and F_2 is equal to $\Sigma e_i^2 X q_i(X)$, e_i being the charge of the parton "i". This gives a simple interpretation of F_2 ; F_2 is closely related to the momentum distribution of the nucleon constituents weighted by their charges squared. Moreover, in this model and neglecting $1/Q^2$ terms, R is equal to zero (this is due to the 1/2 spins). The measurement of the number of partons gives an infinite result interpreted as an infinite number of q-$\bar{q}$ pairs, called the sea. The remaining partons (3 for a nucleon) are called the valence partons. The momentum fraction taken by the charged partons is found to be around 50% which means that the remaining 50% is taken off by neutral partons.

In 1974, the Q^2-dependence of F_2 was observed in μ-scattering at Fermilab. This scaling violation is small and permanent with Q^2 (i.e.: logarithmic) and well explained by QCD. One can identify the charged partons to the quarks and the neutral partons to the gluons. At the leading order, using the Altarelli-Parisi equations, one can use a formalism very close to the parton model one (probabilistic interpretation of the quark distributions). One must add the Q^2-dependence in the q_i. The resulting expression for F_2 is as in the parton model: $F_2(X,Q^2) = \Sigma e_i^2 X q_i(X,Q^2)$. $R=\sigma_L/\sigma_T$ is no longer zero but proportional to the strong coupling constant $\alpha_S(Q^2)$. The Altarelli-Parisi equations allow us to know the Q^2 evolution of F_2 but not the X-dependence. This is related to the confinement problem. The X-dependence has to be parametrized for a given Q_0^2. The Q^2 evolution of F_2 is in fact related to the Q^2 evolution of the gluon distribution. Therefore, the X shape of the glue has also to be parametrized. This is the standard way to get information on the gluon distribution from electron, muon and neutrino scattering.

In the case of lepton-nucleus (A>1) deep inelastic scattering one could, in principle, repeat the procedure, changing only a few points so that one has more valence quarks ($3 \to 3 \cdot A$), a different scaling variable $X \to X_A = Q^2/2M_A\nu$ (the maximum momentum is now the nucleus mass) and a new name for the structure functions, for instance $F_2^N \to F_2^A$. In fact, the experimental results are usually expressed in terms of the *measured* X Bjorken variable and not X_A. This means that X can in principle reach values greater than 1 and up to A. The corresponding probability is small but non-zero. It is also well known that the nuclear cross section is, experimentally, roughly equal to A times the nucleon cross section σ_N and can be written as $\sigma_A = A \cdot \sigma_N + A \cdot \varepsilon$ (ε being small). For these reasons we prefer to define F_2^A from the nuclear cross section normalized to one nucleon, σ_A/A, with the same nucleon kinematical factors, as a function of the same variable X, Q^2. From now on the quantities will be considered normalized to one nucleon. The experimental results are presented in general either as the ratio F_2^A/F_2^N or as the cross section ratio σ_A/σ_N. These two quantities are not equal if $R=\sigma_L/\sigma_T$ is A-dependent.

The physical interpretation of the EMC effect is not straightforward within the QCD framework, mainly due to the confinement problem. The currently admitted picture is based on several models, each having a limited X domain of validity. Figure 2 shows a rather widely accepted set of models. Among them, the shadowing region (small X) and the Fermi motion region (large X) meet a large consensus.

The published iron over deuterium ratio of the EMC was done with a preliminary version of the deuterium data at 280GeV. These data are now fully analyzed and will be published soon [5] with the final version of this ratio. There is no significant change but the systematical errors are now included (Fig. 3). The result is dominated by the systematics of the experiment. This is mainly due to the fact that the EMC effect was unforeseen and no special care was taken to reduce the differences between the iron and the deuterium data. The main problems were due to the different target size and the resulting geometrical acceptances for the two targets; the iron target was a 1 meter long calorimeter and the deuterium target 6 meters long. Also, the deuterium data were taken several months after the iron (including the Christmas shutdown!) so that the apparatus status (efficiency, dead parts,...) and beam fluxes were completely different. In the new EMC experiment these points have been cured.

Data were taken in 1984-1985 from small parasitic targets mounted on a "chariot" during the running of the polarised target phase of NA2. Three targets (Cu,D_2,C) were used during each data-taking period, each distributed over the same volume, being cycled into the beam every few hours. Systematic differences in acceptance were thus minimized, effects due to changes in the apparatus also cancelled to negligible levels.The results are still preliminary but should be released soon. Figure 4 shows the F_2 ratio for copper over deuterium, compared with the initial EMC result for iron over deuterium. Such a comparison is valid because we know [3] that the A-dependence of this ratio is logarithmic or at least very weak. We see that this new EMC result is now dominated by the statistics and no longer by the systematics. The difference in the first two X bins seems significant and could hardly be explained by the A-dependence. It probably reflects a systematic effect in the first EMC experiment.

The new EMC result can be compared with a beautiful new BCDMS result [4] in Fig. 5, in which the old BCDMS result is also shown. Both the systematic (given by the lines) and the statistical errors are small. They are typically of the same order of magnitude.

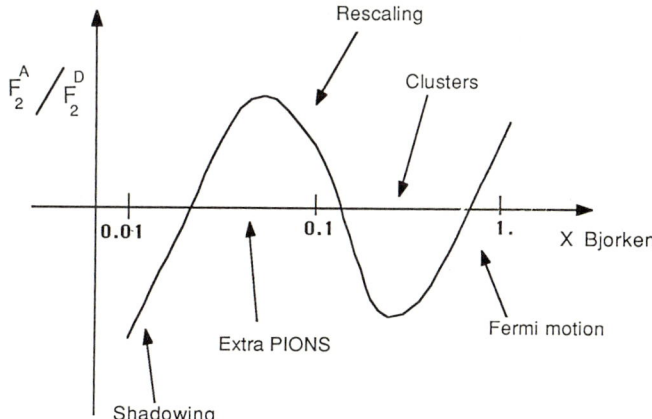

Figure 2

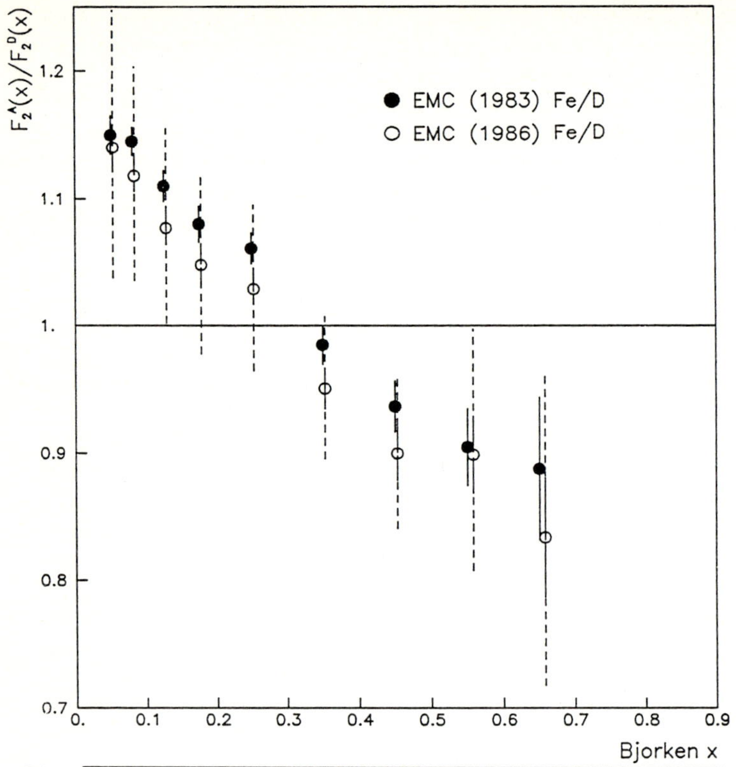

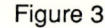

Figure 3

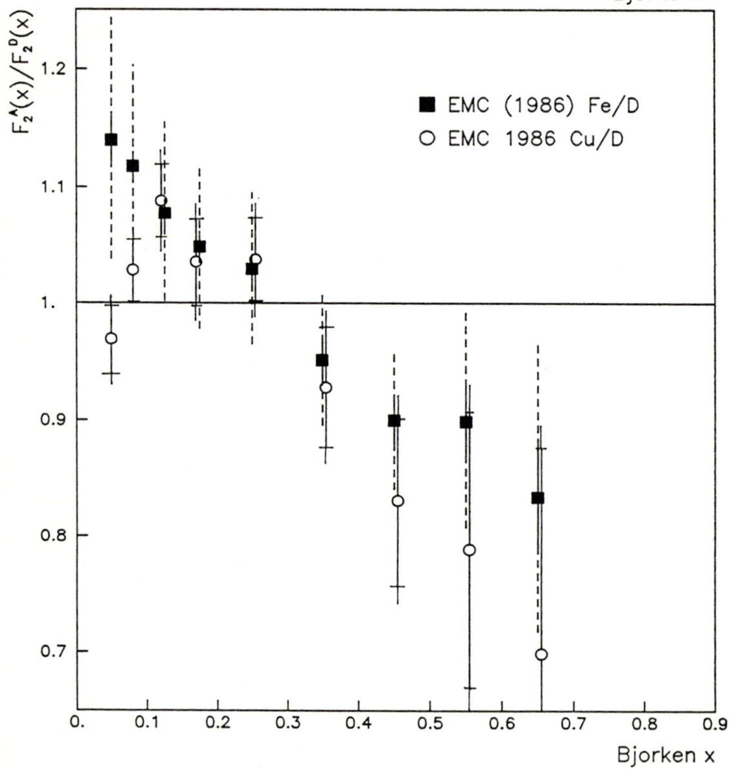

Figure 4

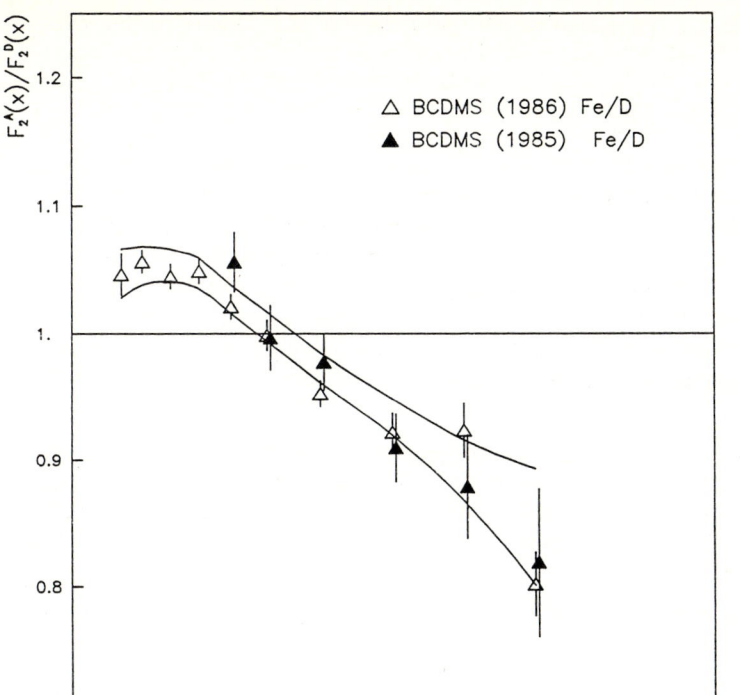

Figure 5

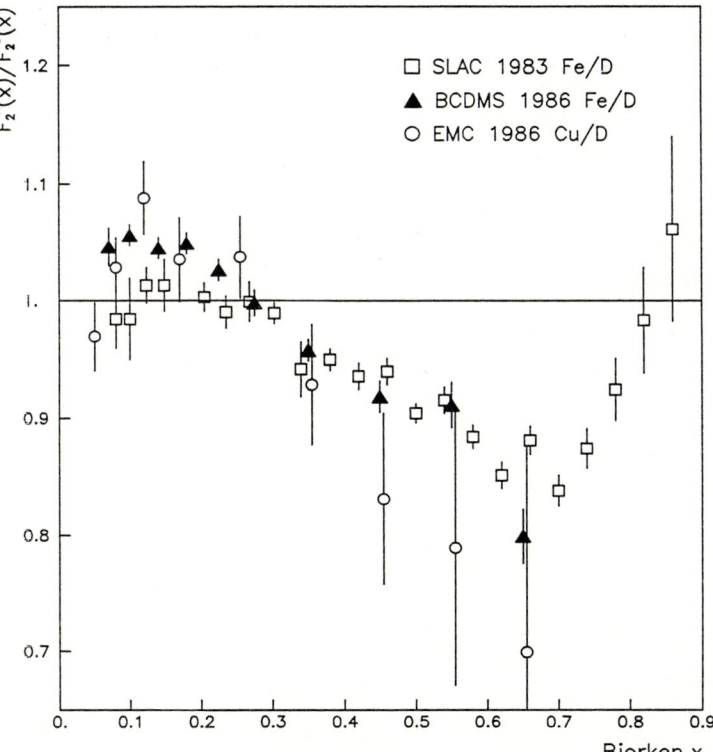

Figure 6

This new result improves the statistics and extends the acceptance to the low X region. There is now a definite enhancement of (4.5±0.5)% at low X. The comparison of both the new EMC, the new BCDMS and the old SLAC data [3], in Fig. 6, shows a good agreement except in the low X region.

The enhancement of the high Q^2 experiment at low X is not yet understood. Assuming the probability of an experimental error to be negligible, we can formulate two different hypotheses: a Q^2-dependence of the ratios or a A-dependence of R at low Q^2 ($\Delta R \approx 0.15$ would be needed). Concerning the Q^2-dependence of the ratio, the most precise measurement from BCDMS[4] does not show any clear trend for it but the data do not cover a large Q^2 range at low X. Concerning the A-dependence of R, Fig. 7 shows the old measurement performed at SLAC [6] and a new one [7] also from SLAC. Both are compatible with no dependence and neither bring information in the low X region, which is of interest.

Figure 8 shows the new EMC carbon over deuterium result compared with both the SLAC [3] and BCDMS (nitrogen over deuterium) [8] results. They are in very good agreement. In the low X region the enhancement, if any, is weaker than for the higher mass target Fe,Cu.... No comparisons between low and high Q^2 are possible in the low X region due to the lack of low Q^2 data.

Figure 9 presents the last neutrino result from WA25/WA59 [9] at CERN compared with the BCDMS result. The general trend of the ratio is observed but no final conclusion can be drawn.

Neutrino and anti-neutrino results are more useful in the sea distribution studies because they can measure F_3, which represents another constraint for the interpretations.

The only measured quantities are "clearly" the structure functions F_1, F_2 and F_3, but, assuming the quark-parton model decomposition, we can write the neutrino and the anti-neutrino cross sections as a function of the various quark and anti-quark distributions

$$\frac{d^2\sigma^{\nu N}}{dXdY} \propto [\,(u+d+2s) + (1-Y)^2(\bar{u}+\bar{d})\,]$$

and

$$\frac{d^2\sigma^{\bar{\nu} N}}{dXdY} \propto [\,(\bar{u}+\bar{d}+2\bar{s}) + (1-Y)^2(u+d)\,]\,.$$

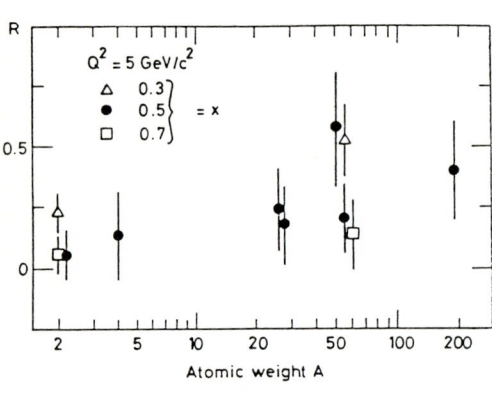

Figure 7 a

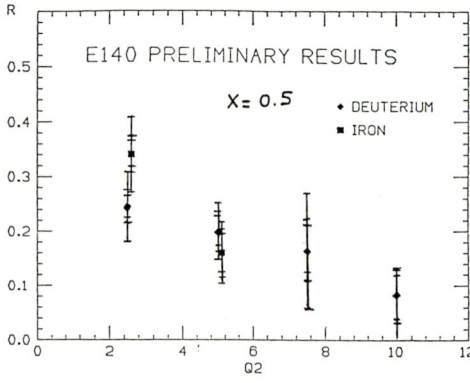

Figure 7 b

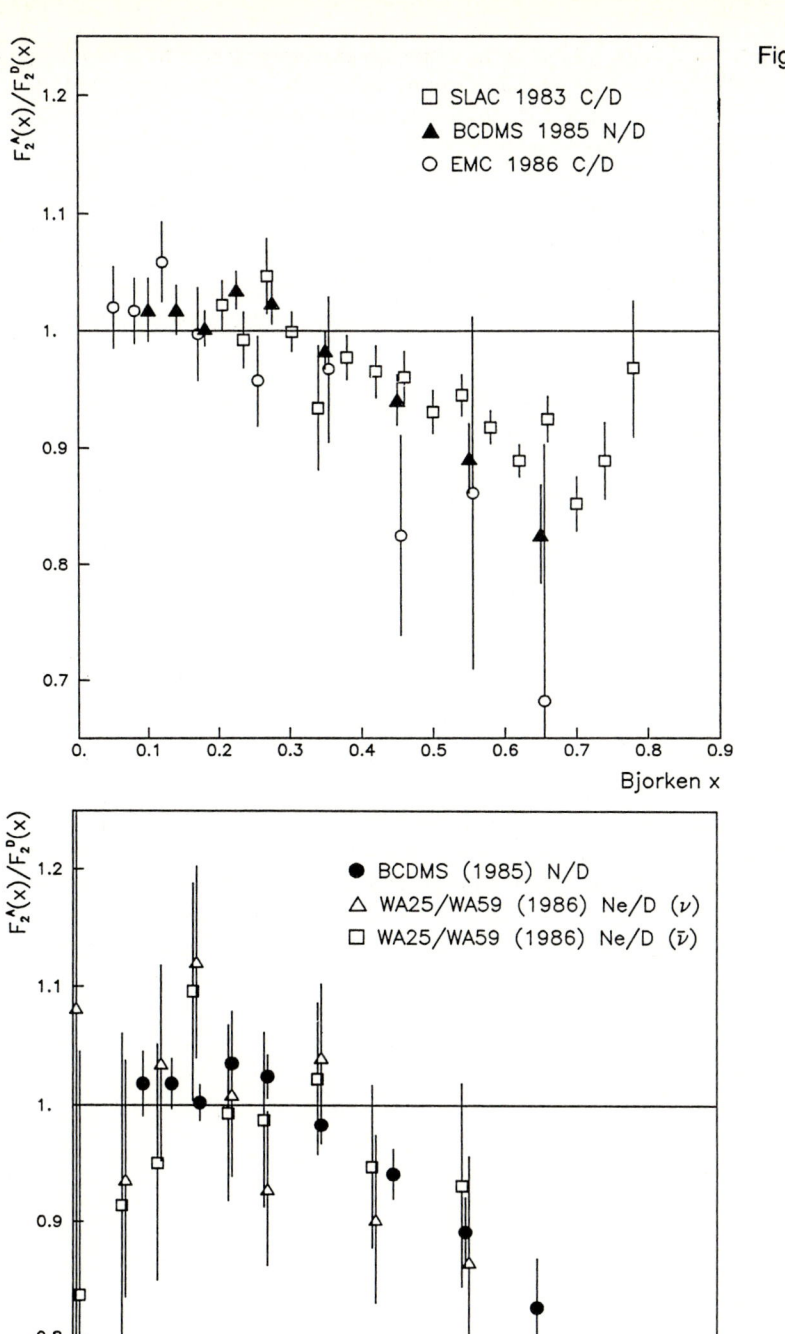

Figure 8

Figure 9

Using either linear combinations or the large Y trend we can have some information about the antiquark distributions. This has been done by the CDHS collaboration [10] with iron and hydrogen targets and their published $\bar{q}$ ratio is shown in Fig. 10.

From this result we see no significant difference, but the error bars are too large and the X range too small to discriminate between various models. They, however, exclude a large increase of the sea in the low X region. I should remark that the F_2 structure function published by the CDHS collaboration (in fact only the drawing was published) has been changed considerably [11] (up to 30% in low X region). Part of these changes come from the F_2 extraction procedure and not from a modification of the cross section. It is not clear to me what the implications of these changes in the anti-quark distribution ratio could be.

The WA25/WA59 collaborations made a similar analysis [9] with neon and deuterium data. They fitted the cross sections, parametrizing the various distributions. Using a standard shape for the sea

$$q_s \propto (1-X)^\gamma,$$

they found that the shape of the sea in neon is not significantly different from that in deuterium ($d\gamma \equiv (\gamma Ne - \gamma D) = 1 \pm 1$)) and that the fraction of sea in neon is $(7\pm6)\%$ or $(18\pm10)\%$ less than in deuterium, depending on the model for $R = \sigma_L/\sigma_T$. This result also excludes a large increase of the sea in a heavy nucleus.

In 1985 the EMC [12] published the ratio of the energy distribution of the J/Ψ ($c-\bar{c}$) bound state) shown in Fig. 11. Using reasonable assumptions and the photon-gluon fusion model (Fig. 12) this ratio can be interpreted as the gluon distribution ratio of the various targets, for $Q^2 \sim 10 GeV^2$.

This ratio can also be related to the ratio of the incoherent J/Ψ photo-production cross sections. If the A-dependence of the cross section is taken as $\sigma(Q^2=0) = \sigma_0 A^\alpha$, then, from the above ratio one can get $\alpha = 1.10 \pm 0.03 \pm 0.04$. A recent SLAC experiment [13] studied such J/Ψ cross sections for several targets using real photons of energy $\nu \sim 70 GeV$ similar to the virtual photon energy in the EMC experiment. This experiment gives $\alpha = 0.94 \pm 0.02 \pm 0.02$ which is inconsistent with the EMC result. The origin of this

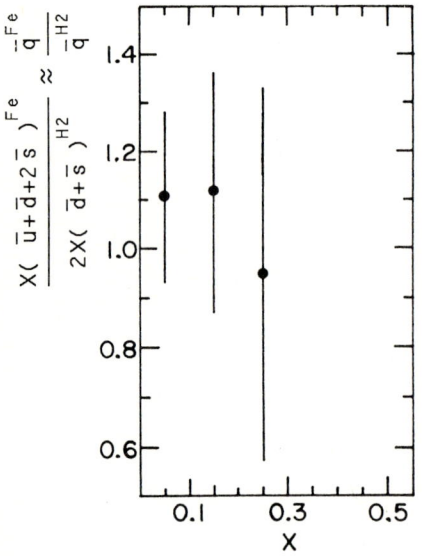

Figure 10

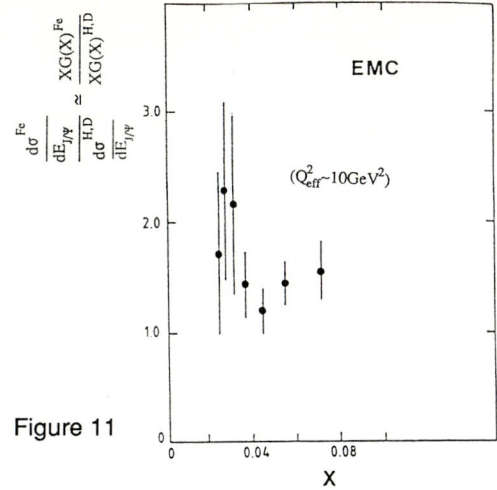

Figure 11

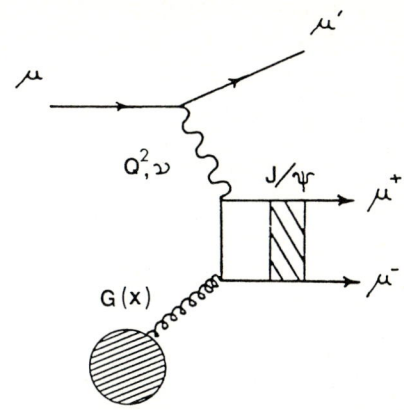

Figure 12 Photon-gluon fusion Model

difference is still unknown. It may be an experimental bias or related to the way of extending the result from virtual to real photons. It is not the photon-gluon fusion interpretation that is doubtful. We must emphasize that, if the simplest way to understand $\alpha > 1$ is to assume more gluons in heavy targets, the simplest way of interpreting $\alpha < 1$ is to assume that the J/Ψ is absorbed by the nuclear matter.

Another way to get information on the glue is to study the Q^2 evolution of the structure functions. Such an analysis was done [14] using the EMC iron and hydrogen structure functions. In order to discuss the experimental results on F_2, one is forced to do approximations since one is not able to solve the QCD problem for heavy targets. Here we are assuming that all the (small) perturbations introduced by the many (A) nucleons can be taken into account using *effective* quark distributions and keeping the QCD evolution equations. As the EMC effect is small this is a possible way to parametrize the EMC effect. I think that, following this assumption, one can get, at least qualitatively, information on the real quark distributions. Figure 13 shows the distributions we get for the valence part X(uv+dv)(X), the sea 2.X($\bar{u}+\bar{d}+\bar{s}+\bar{c}$)(X) and the glue XG(X) for both targets. We can say that the effective nucleus valence distributions are lower than those for the nucleon and that the sea and the glue are higher. This is true in the X range 0.1<X<0.6 where we have enough data to constrain the fits. The difference shown for the sea is inconsistent with the neutrino results, see for instance [10].

To conclude, I can say that the structure function ratios F_2^A/F_2^D for a heavy target (A~56) show now a definite enhancement of (4.5±0.5)% for the high Q^2 (~20GeV^2) experiments in the low X region (~0.08). There is still a disagreement in this region with the low Q^2 (~2GeV^2) SLAC experiment. The A-dependence of R and the Q^2-dependence of the F_2 ratio are, at least, very weak. The neutrino results do not show an increase of the sea in the nucleus; this seems, to me, inconsistent with the QCD analysis of the charged lepton structure functions. Such an analysis within the various parametrizations needs less valence, more sea and more glue for the nucleus distributions. The EMC results on the J/Ψ photo-production cross sections do not agree with a recent result from SLAC.

In almost all cases we need more data in the low X region and if I had to decide for the future I would require more data in this region but also more different types of measurements. Concerning the near future I will refer to the talk of K.Rith in this meeting.

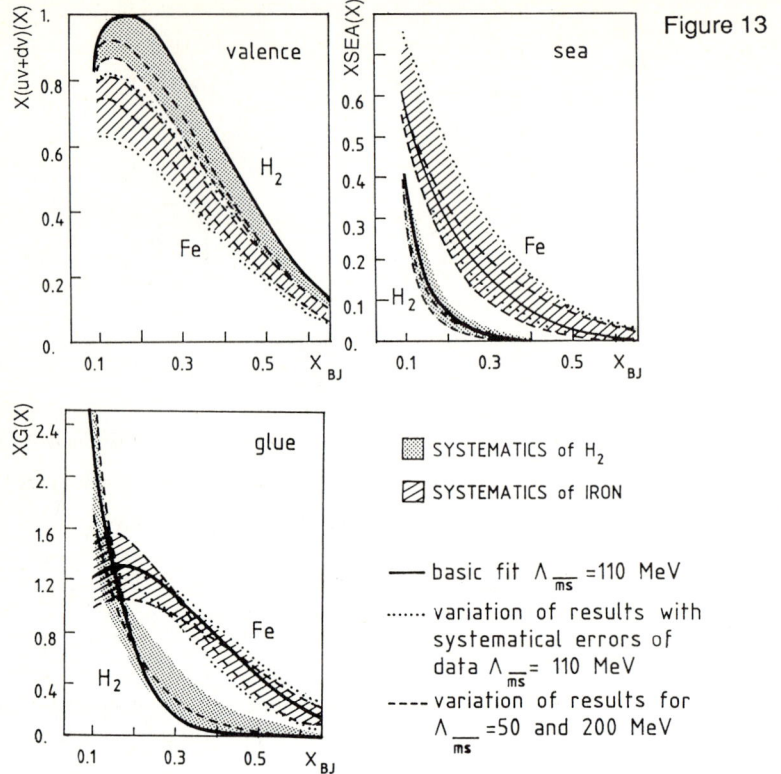

Figure 13

It is a pleasure to thank E.Aslanides and J.M.Richard for their invitation to give this talk. The friendly atmosphere of the Workshop was very stimulating for me.

REFERENCES
1) EMC Phys.Lett. 123B (1983) 275
2) A.Bodek et al. Phys.Rev.Lett. 50 (1983) 1431
3) A.Bodek et al. Phys.Rev.Lett. 52 (1984) 727
4) BCDMS coll. CERN-EP 87-13 Subm. to Phys.Lett. B
5) EMC F_2 structure function of Deuterium, Subm. to Nucl.Phys. B
6) J.Gomez SLAC Pub. 3552 (1985)
7) S.Dasu et al. Univ. of Rochester preprint UR958 (1986)
8) BCDMS coll. Phys.Lett. 163B (1985) 282
9) WA25/WA59 coll. CERN-EP 86-217 (1986) Subm. to Z.Phys. C
10) CDHS coll. Z.Phys. C25 (1984) 29
11) P.Buchholz Proc. International Europhysics Conference Bari 1985
    ed. L.Nitti & G.Preparata
12) EMC Phys.Lett. 152B (1985) 433
13) M.D.Sokoloff et al. Phys.Rev.Lett. 57 (1986) 3003
14) G.D'Agostini Thèse de Doctorat d'état Université d'Aix-Marseille II

Nuclear Effects in Quark and Gluon Distributions – Experimental Perspectives

K. Rith

Max-Planck-Institut für Kernphysik,
D-6900 Heidelberg, Fed. Rep. of Germany

I Introduction

The experimental status of the EMC effect and the different approaches for its theoretical interpretations have been presented already in detail at this conference by G. D'Agostini [1] and G. Miller [2]. In my contribution I will therefore concentrate on the discussion of future experiments at high momentum transfers and outline what we might learn from them.

Obviously the existing experimental information is still too scarce to decide which of the many models proposed to explain the EMC effect is the adequate one. (For a more detailed discussion see for instance reference [3]). The decrease of the nuclear structure function at medium x ($0.3 < x < 0.8$) measured by the SLAC experiment E 139 [4], for instance, can qualitatively be described by nearly any class of models like convolution approaches with contributions of "swollen" nucleons, excess pions, Δ-isobars or multiquark clusters, quark exchange calculations or x-rescaling and Q^2-rescaling models. This just tells us that inclusive data from this x region alone are insufficient to discriminate between the different explanations. We need in addition much more specific experimental information and have to clarify, for example:

1) The behaviour of the nuclear structure function F_2^A at small x ($0.001 < x < 0.2$) compared to the "free" nucleon one F_2^D:
   a) the x dependence,
   b) the A dependence and
   c) the Q^2 dependence.
2) The A dependence of
   a) the seaquark distribution,
   b) the gluon distribution.
3) Nuclear effects in $R = \sigma_L/\sigma_T$.
4) The Q^2 dependence of F_2^A/F_2^D at $x > 0.3$.
5) The behaviour of $F_2^{A_1}/F_2^{A_2}$ at $x > 1$.
6) F_2^n/F_2^p for nuclei.
7) The influence of the nuclear medium on hadron distributions.

I will discuss these topics in some detail and will point out by which measurements the required information could be obtained.

II Future measurements
II 1 The behaviour of the nuclear structure functions at low x

In my opinion further experimental information about the detailed behaviour of nuclear structure functions at low x is most essential to improve our understanding of nuclear effects in deep inelastic lepton nucleus scattering. It is this region where most probably we can learn something about the role quarks and gluons play for nuclear forces since it is dominated by the seaquark and gluon distribution. It is also the region where the theoretical expectations differ most drastically: pion model advocates for instance expect an <u>enhancement</u> of the structure function ratio for bound and free nucleons down to very low x, the x-rescaling model produces only a very <u>small deviation</u> from one (since the effect is proportional to $x \cdot dF_2/dx$), if no further ingredients are added, and from shadowing models one gets a rather <u>big reduction</u> of the nuclear cross section. And it is the region where the experimental situation is most confusing: many electron and muon experiments [5] have been performed to find out whether shadowing, which has been undoubtedly observed for real photons ($x=Q^2=0$) [6] also occurs for virtual photons. But unfortunately the results differ so much (mainly due to large uncertainties caused by enormous radiative corrections) that our knowledge about this region at low Q^2 is essentially zero. The high Q^2 muon data of the original EMC experiment [7] show a continuous rise of F_2^{Fe}/F_2^D towards small x (up to a value of ~1.15 at x=0.05), the SLAC E139 data [4] are consistent with unity for x<0.3 and neutrino data [8] suggest a bump around x=0.15 and a drop below one for x <0.1 similar to the older SLAC data of E 61 [9]. The BCDMS Fe/D data [10] show a clear enhancement of about 5-6% at low x and the new EMC results [1] for C/D and Cu/D indicate that at x~0.05 the ratio is falling below one even at Q^2 values around 5 GeV^2. The experimental situation is really unsatisfactory!

The theoretical confusion might be connected with one complication which occurs in the discussion of nuclear effects in quark and gluon distributions: nuclear physicists are used to working in the rest frame of the nucleus while the "infinite momentum frame" (or better the light cone) is preferred for the interpretation of deep inelastic scattering experiments. We might ask: what happens if we look at a nucleus in a frame where it is moving very fast? This has been discussed by several authors [11-13]. Let me briefly repeat their arguments to see what kind of effects might show up.
In the "infinite momentum frame" the virtual photon probes the quark and gluon distributions of a nucleon (rest mass m, rest frame diameter D, longitudinal momentum p) with a <u>transverse</u> resolution $\Delta D=1/\sqrt{Q^2}$. Longitudinally the nucleon is Lorentz contracted to a size $D'=D \cdot m/p$, but the <u>"longitudinal size"</u> of its constituents is given by $\Delta z=1/xp$. At low x, Δz can be much larger than D'. Now let us consider a nucleus (rest frame diameter D_A, mean nucleon distance d) with several nucleons along the direction of the virtual photon. If $\Delta z>d'=d \cdot m/p$, or equiva-

lently x<1/dm, then seaquarks and gluons of different nucleons start to overlap spatially. The smaller x, the more nucleons share their seaquark and gluon content. At x~1/D_Am all nucleons with the same impact parameter contribute and the effect saturates. Therefore in a nucleus the gluon and seaquark number density at small x, taken at the position of one nucleon, might be much bigger than for a free nucleon. Due to this "overcrowding" there might be a bigger chance for gluons and seaquarks to interact with each other and to reduce the number density again by annihilation [12]. As a consequence one finds fewer low momentum gluons and seaquarks per nucleon in a nucleus than in a free nucleon and one obtains a decrease of the nuclear cross section at low x: <u>shadowing</u>. It is important to point out that in this picture shadowing is caused by a modification of quark and gluon distributions in a nucleus, while in the vector meson dominance model (VDM) [14] these are unaffected and the decrease of cross section is caused by a modification of the interaction mechanism: the hadronic component of the photon which is absorbed already at the surface of the nucleus.

The magnitude and x dependence of this effect, as estimated by Qiu and Müller [12], is sketched in the bottom part of Fig. 1. They predict a maximum decrease of about 10% for carbon and about 20% for lead.

In my opinion their calculation is not complete, since a decrease of the number density of seaquarks and gluons must be compensated at some higher value of x by momentum conservation. This has been taken into account in the early model of Nikolaev and Zakharov [11], who at that time had only considered seaquarks. Their prediction is shown in Fig. 2. <u>Shadowing</u> at low x is compensated by an enhancement (<u>"antishadowing"</u>) at larger x.

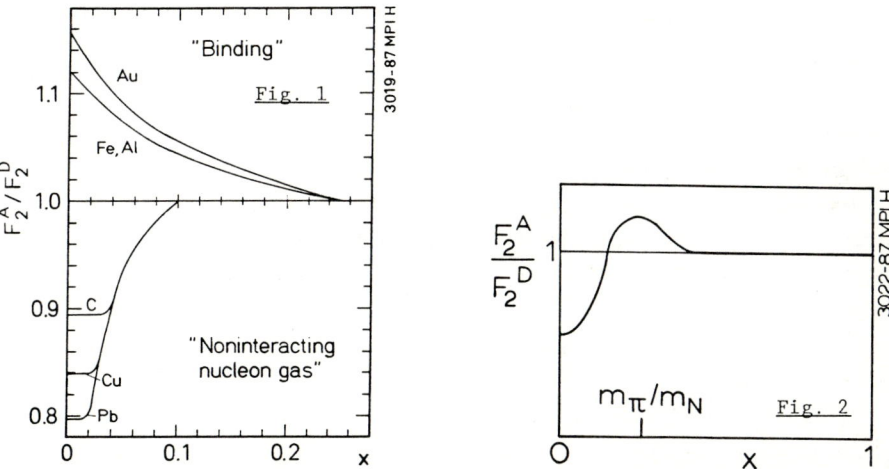

Fig. 1: Predicted behaviour of F_2^D at low x. Bottom: Shadowing model of Müller and Qiu [12]; top: pion model of Berger and Coester [15]

Fig. 2: Shadowing and antishadowing as predicted by Nikolaev and Zakharov [11]

One might ask: how does one take into account nuclear binding in this approach? Experts tell me that there is no unique prescription for how to boost a wave function of a bound system to the infinite momentum frame. Indeed, in this approach the nucleus in its rest frame is treated as a system of "noninteracting" nucleons. The interaction enters, however, indirectly via the mean nucleon distance d and the overall size D_A of the nucleus and shows up as an explicit interaction between gluons and seaquarks of different nucleons in the fast moving system.

Now let us consider that in addition to these indirect manifestations of binding some energy is taken away from the nucleons by the binding field. This process is considered at present by many people to be responsible for the decrease of the nuclear structure function at medium x, although it does not reproduce the measured A dependence. This field must reappear somewhere when we probe the nucleus on the quark gluon level, since quarks and gluons are the building blocks of all possible hadronic exchange particles.

As indicated in Fig. 3 one could expect that long range correlations (pions?) would be visible at small x while at large x we would be sensitive to short range correlations (multiquark clusters?). As an example of an additional binding effect the modification of the nuclear structure function due to excess pions, as calculated by Berger and Coester [15], is shown in the top curve of Fig. 1, labelled "binding". They predict an enhancement below x~0.2 which continuously increases to a maximum value of about 12-15% at very low x.

It is interesting to look also at the predicted Q^2 dependence, which is shown in Fig. 4 for x=0.02. In both the "pure pion model" and the "pure shadowing model" only a very small logarithmic Q^2 dependence, governed by perturbative QCD, is ex-

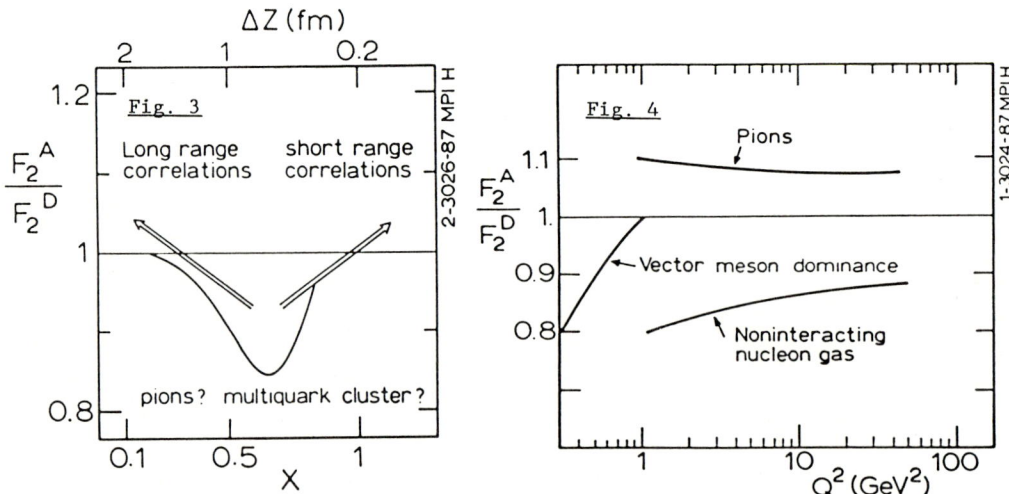

Fig. 3: Possible influence of binding effects on nuclear structure functions

Fig. 4: Model predictions for the Q^2 dependence of F_2^A/F_2^D at low x

pected. Also shown in this figure is the expectation of the VDM model where shadowing dies away very quickly with Q^2.

Which model or combination of models gives the correct description at low x? If the new EMC data, which indicate shadowing even at Q^2 values around 5 GeV^2 turn out to be correct, then the "pure pion model" as well as the VDM model, can be excluded.

Future experiments:

The existing data have much too low statistics to allow further conclusions. Obviously very high statistics data for several nuclei, which cover the whole x range below 0.001<x<0.2 and a Q^2 range as large as possible, are required to measure the magnitude of the decrease at very low x and of the enhancement at larger x and their A and Q^2 dependence.

These will be obtained by the NMC experiment at CERN [16], which has been taking data since summer 1986 (see Sect. III). A special low angle trigger allows this kinematic region to be explored in detail. Targets of relatively low weight (100g/cm^2) are being used. Foreseen are measurements on D, He, Li, Be, C, Al, Si, Ca, Nb and Ho. Ho is of special interest, since it is a deformed nucleus (major axis 7 fm, minor axis 5.1 fm) which can be easily polarized. Measurements with the virtual photon direction along and perpendicular to the polarization axis will allow a clean test of shadowing models without the uncertainties of different radiative corrections which occur when nuclei of different Z are compared. They can show whether the EMC effect at low x is a density effect or a size effect.

Furthermore the muon experiment E 665 at FNAL [17], which starts data taking this summer, is intended to study nuclear effects in the low x region. The Q^2 range covered by this experiment will be about twice that of the CERN experiments due to the higher beam energy, which at the beginning will be around 500 GeV.

II. 2 The A dependence of the seaquark and gluon distributions

The question whether the seaquark and gluon distributions in a nucleus are decreased at low x or whether there is an additional component due to the nuclear force could be studied in the cleanest way by looking at seaquarks and gluons directly. Unfortunately, charged lepton scattering experiments do not allow the seaquark distribution to be measured; this is only possible with neutrino/antineutrino experiments, and since there is no direct probe for the gluon distribution, one has to extract it indirectly.

There are two results for nuclear effects in seaquark distributions: the CDHS experiment [18] obtained for the ratio of the antiquark distributions in iron, 1/2 $(u+d+2s)_{\nu\ Fe}$, and in hydrogen, $(d+s)_{\nu\ H}$, a mean value of 1.10±0.11 (stat)±0.07 (syst), compatible with a small enhancement or no effect. An analysis of approximately 40000 neutrino/antineutrino events taken by the WA 25/59 collaboration with the BEBC bubble chamber with neon and deuterium filling [19] suggests a (16±8)% decrease of the sea in neon.

The CDHS result is shown in Fig. 5, together with model calculations for the Q^2 rescaling model and two pion models [20]. Obviously more accurate data would be very helpful to discriminate between these models or even exclude them.

Future experiments:

There is little hope of improving the situation by further neutrino experiments since they will always suffer from too low statistics from hydrogen or deuterium targets.

But there is a chance to get very accurate data for the anti-quark distribution at low x from the new Drell-Yan experiment E772 at FNAL. This experiment [21] will start data taking in June 1987 with a 800 GeV proton beam incident on a deuterium target and several nuclear targets. (For a detailed discussion and an extensive list of references see also Berger [22], Bickerstaff et al.[20] and Garvey [23]). One way to obtain information on the gluon distribution is from the cross section of J/ψ production (studied via the $\mu^+\mu^-$ decay of the J/ψ). They can be related in the framework of the photon-gluon-fusion model [24]. The result of the EMC is shown in Fig. 6, together with the predictions of the Q^2-rescaling, the x-rescaling and the shadowing model. Only the Q^2 rescaling model predicts an enhancement. The measured mean value of 1.45±0.12 (stat)±0.22 (syst) is much larger than expected from any model, but these data have too low statistics to allow any firm conclusion about the magnitude and the x and A dependence of the change of the

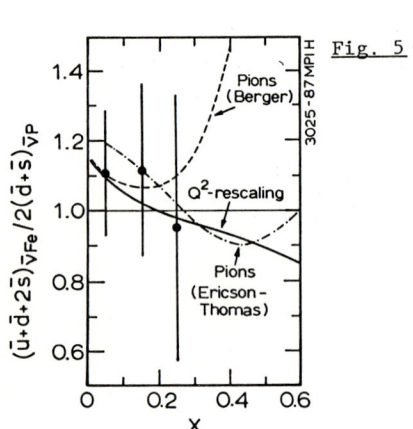

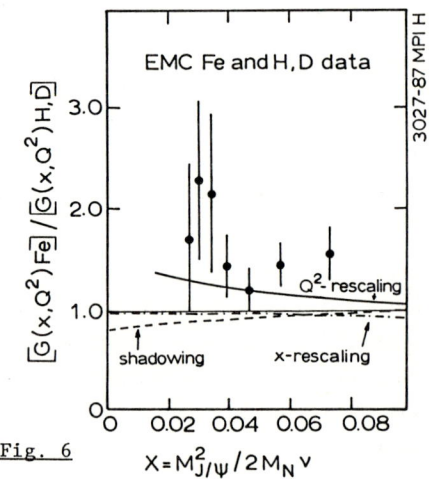

Fig. 5: Ratio of the seaquark distribution in iron and hydrogen as measured by CDHS

Fig. 6: Ratio of the gluon distribution in iron and hydrogen deuterium, extracted by EMC from J/ψ production

gluon distribution. A different result has been obtained by an experiment with
real photons at Fermilab [25]. The incoherent cross section for J/ψ production,
measured by this experiment, can be parametrized as $\sigma = \sigma_0 A^\alpha$ with $\alpha = 0.94 \pm 0.02 \pm 0.03$.
This corresponds to a 20% decrease of the cross section per nucleon at x=0 from
H/D to Fe, compatible with the shadowing model.

How much of this difference is due to the fact that the Fermilab experiment is at
x=0 while the EMC experiment covers the range 0.026<x<0.086, and how much is due
to experimental problems (subtraction of the coherent cross section) is at present
unclear. Obviously more and better data are needed.

Future experiments:

There will be a new measurement by the NMC on two nuclear targets (one low A,
one large A nucleus) where about 7000 J/ψ events are expected from each target.
Rather thick targets of 600g/cm^2 will be used in this "high luminosity run" These
statistics will allow a rather precise determination of the <u>change</u> of the gluon
distribution with A even when one restricts the analysis to the 15-20% inelastic
events where the theoretical uncertainties for the interrelation of J/ψ cross section and gluon distribution are assumed to be much smaller than for the elastic
ones.

II. 3 Nuclear effects on $R = \sigma_L/\sigma_T$

It is an interesting question whether the two structure functions F_1 and F_2,
related by

$$R = \sigma_L/\sigma_T = \frac{F_2 \left(1 + \frac{4M^2 x^2}{Q^2}\right) - 2xF_1}{2xF_1},$$

show the same nuclear dependence or whether R varies with A.

Small differences are predicted from the x-rescaling model [26]. R would also
change if higher twist contributions are different for bound and for free nucleons
[27]. Furthermore one expects a sizeable change in R with A (at low x) if the
gluon distribution is A dependent, since the magnitude of R, as predicted by perturbative QCD [28], depends very much on the shape of the quark and gluon distributions [29].

The data of E 139 suggested (within rather large error bars) a substantial
variation of R with A, which could explain the apparent difference between the E
139 cross section ratio and the EMC structure function ratio for iron and deuterium. Preliminary data of the dedicated experiment E 140 [30], however, indicate
that the A dependence at x>0.2, if it exists at all, might be very small [31].

Future experiments:

The NMC plans to measure the A dependence of R during the "high luminosity run"
in 1988. The expected data will allow the measurement of an absolute change in R

with A with an accuracy of 0.02 for several kinematic bins in the region
$0.05 < x < 0.3$ and $7 \text{ GeV}^2 < Q^2 < 50 \text{ GeV}^2$.

II.4 The Q^2 dependence of $F_2^{A_1}/F_2^{A_2}$ at x>0.3

The Q^2 dependence of the structure function ratios at x>0.3 can be easily calculated in the framework of perturbative QCD. The expected variations of $d(F_2^{A_1}/F_2^{A_2})/d\log Q^2$ are only a few permille or less except for very large x where they could reach several percent [32]. This is, however, only true if the Q^2 dependence of the individual structure functions themselves is completely governed by QCD. But it has been shown that additional higher twist contributions, which go like inverse powers of $\log Q^2$, are needed to fit the free proton structure function F_2^P over the whole Q^2 range covered by the low energy electron data from SLAC and the high energy EMC muon data. The magnitude of the higher twist contribution is given by the difference between the full curve (QCD + higher twist) and the dashed curve (QCD only) in Fig. 7.

If higher twist contributions (which are a measure of quark-quark correlations or the final state interactions between the struck quark and the debris of the nucleon/nucleus, respectively) were different for bound and for free nucleons, then a much stronger Q^2 dependence of the structure function ratios could be possible at low Q^2.

Rather strong variations, as large as 30% between $Q^2=5 \text{ GeV}^2$ and $Q^2=20 \text{ GeV}^2$ at x=0.8, have been calculated in the frame work of the x-rescaling model [26].

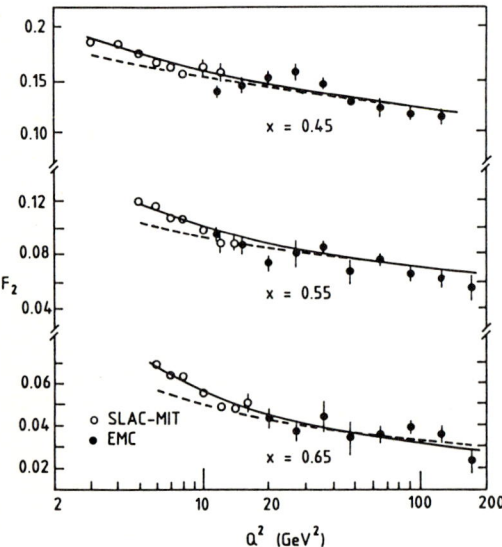

Fig. 7: Q^2 dependence of the proton structure function at large x measured by SLAC-MIT and EMC

Future experiments:

The existing data are not precise enough to look at changes below the one percent level, they only exclude Q^2 variations as big as a few percent. The "high luminosity run" of the NMC, however, will allow us to look at such effects on the few permille level.

II. 5   The behaviour of $F_2^{A_1}/F_2^{A_2}$ at x>1

This region is best suited to studying effects of nucleon-nucleon correlations or of multiquark clusters. As an example the predictions of a multiquark cluster calculation [33] are shown in Fig. 8. The height of the plateau in the range 1<x<2 is proportional to the ratio of probabilities of finding 6-quark clusters in nuclei A and B, the range 2<x<3 reflects the 9-quark cluster probabilities and so on. Figure 9 shows the preliminary cross section ratio for Fe and He obtained by a recent experiment at SLAC [34]. The data are not compatible with "conventional" calculations even with very high momentum components. Probably there is an indication of a step function similar to the behaviour in Fig. 8.

Future experiments:

This kinematic region is very difficult to explore because the counting rates are down by orders of magnitude compared to the low x region and the data are plagued by kinematical smearing corrections. More data can be expected from the 4 GeV nuclear physics facility at SLAC and in some years from CEBAF. Reasonable statistics can also be expected from the high luminosity run of the NMC.

II. 6. F_2^n/F_2^p for nuclei

One might ask the question whether nuclear effects are the same for neutron and proton. This must not necessarily be the case because neutrons and protons have,

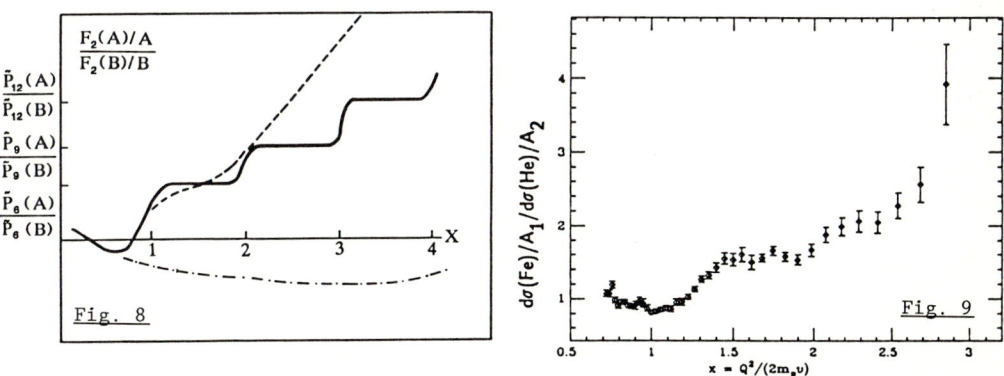

Fig. 8:   Theoretical predictions for the structure function ratio of two nuclei at x>1 (from ref. [33])

Fig. 9:   Ratio of cross sections per nucleon for iron and helium as measured by Day et al. [34]

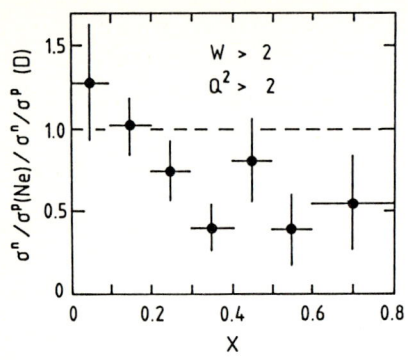

Fig. 10: The ratio σ^n/σ^p for neon and deuterium from antineutrino bubble chamber experiments.

for instance, different separation energies and Fermi momenta and might also occupy different areas of space in the nuclear volume.

In neutrino bubble chamber experiments one is able to distinguish whether the interaction took place on a proton or a neutron by looking at the summed charge of the hadronic final state. A comparison has been made with antineutrino data from an experiment performed at the 15 Ft bubble chamber at FNAL filled with neon [35] and from BEBC at CERN filled with deuterium [36]. In both cases the cross section ratio for neutrons and protons has been extracted. Figure 10 shows the ratio of these ratios for neon and deuterium.

Obviously at large x, σ^n/σ^p is much smaller in the neon data than in the deuterium data, while at low x it is larger. This might tell us that neutrons are more affected by nuclear effects than protons.

Future experiments:

The quoted result has to be taken with some scepticism because two experiments with completely different systematic errors are being compared. A dedicated neutrino experiment would be highly desirable. On the other hand the information might be already available in the WA 25/29 data [19], where the systematic errors are much better under control.

One promising way to look for such effects is the comparison of data from $^{40}Ar_{18}$ and $^{40}Ca_{20}$ in deep inelastic charged lepton scattering. A measurement of this kind is being envisaged by the NMC.

II. 7 The influence of the nuclear medium on hadron distributions

A detailed study of the hadronic final state, created in the deep inelastic process, will be very helpful for the understanding of nuclear effects. Quantities like multiplicities, energy and transverse momentum distributions will reflect modifications in the quark and gluon distributions. They will, however, also be influenced by multiple scattering and gluon bremsstrahlung of the struck quark through the strong force before the fragmentation and/or internuclear cascading of the produced hadrons (for a more detailed discussion see references [37, 38]).

More specific information can be obtained by the measurements of identified hadrons. High z hadrons ($z=E_h/\nu$ with E_h=energy of observed hadron, ν=photon energy) carry (with some dilution due to the fragmentation process) the information on which quark has been struck. Therefore changes in the K/π ratio with A at low x would reflect a nuclear dependence of the (strange) sea, the π^+/π^- ratio would be a measure for variations in the valence/sea ratio.

Future experiments:

Until now there has been hardly any experimental information available [36]. Since the hadronisation process is very complex and different mechanisms can contribute, very high statistics over a wide range of kinematic parameters are needed. The NMC will collect about 10^6 hadrons from different nuclei in its thin target run, but apart from a very crude calorimetry there is no hadron identification. Possibly this kind of study will be the domain of the E 665 muon experiment at FNAL, since this experiment has nearly full acceptance for the complete hadronic final state and very good particle identification by a series of Cherenkov, time-of-flight and shower counters. The experimental setup is shown in Fig. 11.

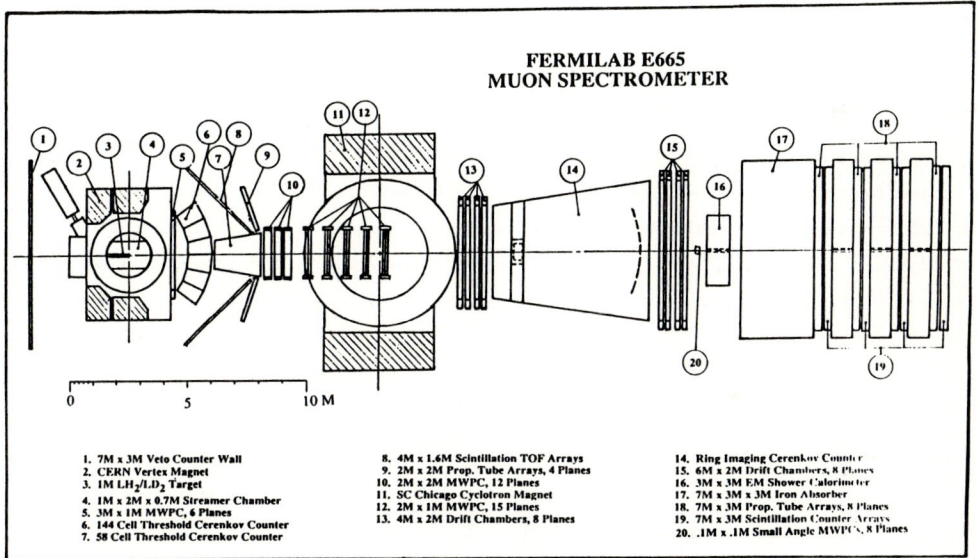

Fig. 11: The FERMILAB E 665 muon spectrometer

III The NMC experiment at CERN

Nearly all of the open experimental questions indicated in the previous paragraphs will be investigated during the following years by the New Muon Collaboration, NA 37, at CERN [16]. This collaboration inherited the Forward Muon Spectrometer previously used by the EMC [39]. The apparatus has been completely upgraded. Many wire chambers have been rewired, some detectors have been modified or replaced by new ones, some more added. A special low angle trigger has been added to improve the acceptance in the low x region and also a new beam momentum spectrometer has been installed which allows a better determination of the incident muon energy. Data taking started in 1986.

The setup is shown in Fig. 12. It allows us to cover the whole kinematic range $x>0.005$ and $1 \leq Q^2 \leq 200$ GeV^2. Special care is taken to minimize systematic errors. Data are always collected from pairs or triples of nuclei simultaneously. The targets are split into several alternating short sections and the whole setup is frequently exchanged by its complementary one with the different target materials having interchanged their positions along the beam. Due to this procedure overall normalization errors and point to point systematic errors will be kept below a level of 1%.

In 1986 about 10^7 raw deep inelastic triggers were collected on hydrogen and deuterium (for a detailed ivestigation of the Q^2 dependence of the neutron structure function and related quantities), 5×10^6 on helium and deuterium and 15×10^6 on lithium, carbon and calcium. More nuclei will be studied in 1987. The A dependence of R and of the gluon distribution, the Q^2 dependence of the EMC effect at $x>0.3$ and the behaviour of nuclear structure functions at $x>1$ will be studied in the "high luminosity run", starting in 1988, where a 600g/cm^2 target with some calorimetry will be used. Furthermore measurements on polarized holmium and the comparison of $_{40}$Ar^{18} and $_{40}$Ca^{20} are envisaged.

My hope is that in a few years the additional information will be sufficient to understand better how and why the nuclear environment influences quark and gluon distributions in nuclei and that the data will help to develop a fundamental understanding of nuclear forces in terms of quarks and gluons and their interactions.

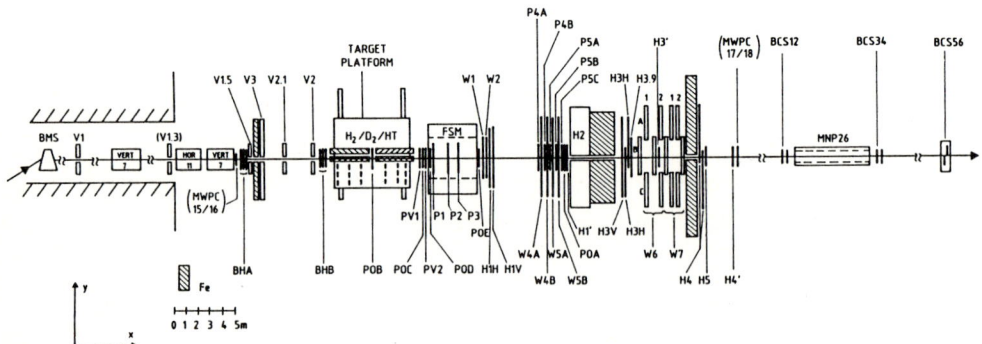

Fig. 12: The NMC muon spectrometer at CERN. V1-V3: Veto counters; BMS, BHA-B: beam counters: P0A-P0E, P1-P5: multiwire proportional chambers; W1-W7: drift chambers; H1-H4: trigger hodoscopes; H1'-H4': small angle trigger; H2: hadron calorimeter; MNP26, BCS12-BCS56: new beam spectrometer.

IV Acknowledgements

I would like to thank O. Nachtmann and H. Pirner and my colleagues from the NMC, especially S. Kullander, B. Povh, D.v. Harrach and Th. Walcher for many inspiring discussions.

V References

1) G. D'Agostini, these proceedings
2) G. Miller, these proceedings
3) K. Rith, Proceedings of the International Nuclear Physics Conference, Harrogate 1986, J.L. Durrel, J.M. Irvine and G.C. Morrison ed., Inst. of Physics Series 86, 395
4) R.G. Arnold et al., Phys. Rev. Lett. $\underline{52}$ (1984) 1431
5) For a rather complete list of references see for instance J. Franz et al., Z. Phys. $\underline{C10}$ (1981) 205
6) D.O. Caldwell et al., Phys. Rev. Lett. $\underline{42}$ (1979) 553
7) J.J. Aubert et al., Phys. Lett. $\underline{123B}$ (1983) 275
8) H. Abramowicz et al., Z. Phys. $\underline{C25}$ (1984) 29,
   V.V. Ammosov et al., PISMA v. ZHETF $\underline{39}$ (1984) 327,
   A.M. Cooper et al., Phys. Lett. $\underline{141B}$ (1984) 133,
   M.A. Parker et al., Nucl. Phys. $\underline{B232}$ (1984) 1
9) S. Stein et al., Phys. Rev. $\underline{D12}$ (1975) 1884
10) A.C. Benvenuti et al., CERN-EP/87-13 (Jan. 87)
11) N.N. Nikolaev and V.I. Zakharov, Phys. Lett. $\underline{55B}$ (1975) 397
12) A.H. Müller and J. Qiu, Nucl. Phys. $\underline{B268}$ (1986) 427
    J. Qiu, Columbia Univ. preprint CU-TP-361
13) L. Gribov et al., Phys. Rep. $\underline{100}$ (1984) 1
    C.H. Llewellyn-Smith, Nucl. Phys. $\underline{A434}$ (1985) 35c
    J.D. Bjorken, Springer Tracts in Modern Physics, Vol. $\underline{108}$ (1986) 17
14) for a review see G. Grammer, J. Sullivan in A. Donnachie, G. Shaw, Electromagnetic Interactions of Hadrons, Vol. 2 New York (1978)
15) E.L. Berger and F. Coester, Phys. Rev. $\underline{D32}$ (1985) 1071
16) NMC, Proposal CERN/SPSC/85-18, SPSC/P210
17) E 665, Proposal to FERMILAB (1982)
18) H. Abramowicz et al., Z. Phys. $\underline{C25}$ (1984) 29
19) J. Guy et al., CERN-EP/86-217 (Dec. 1986)
20) P.P. Bickerstaff et al., Phys. Rev. $\underline{D33}$ (1986) 3228
21) G.T. Garvey, private communication
22) E.L. Berger, Nucl. Phys. $\underline{B267}$ (1986) 231
23) G.T. Garvey, Nucl. Phys. $\underline{B279}$ (1987) 221

24) J.P. Leveille and T. Weiler, Nucl. Phys. B147 (1979) 147
    M. Glück and E. Reya, Phys. Lett. 83B (1979) 98
25) M.D. Sokoloff et al., Phys. Rev. Lett. 57 (1986) 3003
26) R.D. Smith, Phys. Lett. 182B (1986) 283
27) E.V. Shuryak, Nucl. Phys. A446 (1985) 259C
28) M. Glück et al., Z. Phys. C13 (1982) 119
29) D.v. Harrach and Y. Mizuno, private communication
30) R.G. Arnold et al., SLAC proposal E140 (1984)
31) M. Harada, Contribution to PANIC 1987
32) P.P. Bickerstaff and G. Miller, Phys. Rev. D34 (1986) 2890
33) J. Vary, Proceedings of the 7th. Int. Conf. on High Energy Physics Problems, Dubna 1984, 147
34) D.Day et al., Contribution h-23 to PANIC 1987
35) A.E. Asratyan et al., preprint ITEP-115 (1985)
36) D.Allasia et al., Phys. Lett. 107B (1981) 148
37) A.Arvidson et al., Nucl. Phys. B246 (1984) 381
38) K. Rith, Nuclear Physics B279 (1987) 195
39) O.C. Allkofer et al., Nucl. Instr. and Meth. 179 (1981) 445

The EMC Effect and Related Issues

Hong Jung and G.A. Miller

Department of Physics, FM-15, University of Washington,
Seattle, WA 98195, USA

The nucleon (plus pion) explanation of the EMC effect is investigated, and shown to be reasonable. The subjects of dynamical rescaling and the related Drell-Yan process are also discussed.

1. Introduction

Any conference report on the theory of the EMC effect should start with the EMC's nuclear deep-inelastic scattering (DIS) measurement[1] of the iron and deuterium structure functions per nucleon, Fig. 1. Nearly everyone expected a ratio of unity. The observation of significantly different values has led to many new ideas about Nuclear Physics.

Indeed there are N_c theories[2], where one may work in the limit of $N_c \to \infty$. These theories are quite nice—swollen nucleons, increased confinement volume, nuclei that conduct color, and six quark bags are all original and seem interesting and plausible.

Furthermore, the EMC observations stimulated much activity. This work is reviewed in this volume by d'Agostino. Plans for future experiments are discussed in Rith's review also presented at the workshop.

The different theories of nuclear DIS do reproduce the data as shown in Fig. 2. Curves labeled six quark bag, rescaling, pion and hybrid (a mixture of six quark bags and pionic effects) all provide a reasonable description of the data. More details about the theories can be found in Ref. 3, and in Ref. 2. In the past one could say that there are many nice theories and we need new (Drell-Yan) experiments to disentangle them.

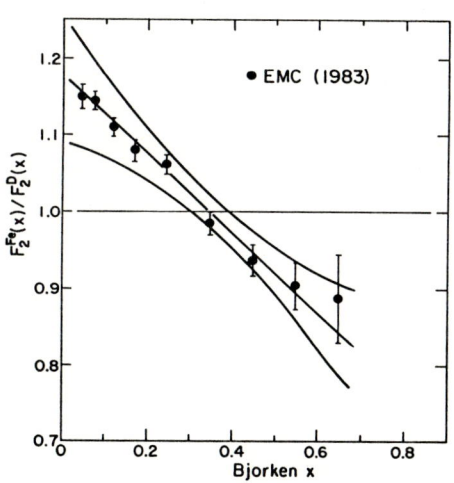

Fig. 1. EMC data for structure functions per nucleon

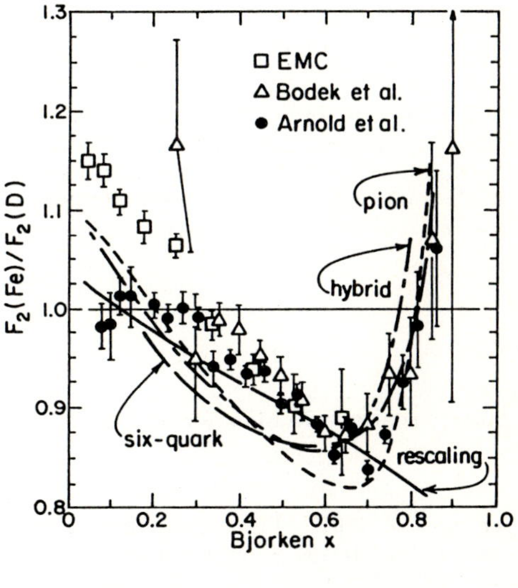

Fig. 2. Different theories of the EMC effect

However there is a growing realization[4] that the models called pions or "nuclear pionic excess" are really models in which the important and best understood effects are nuclear binding as prescribed by the standard nuclear shell model and its extensions. This brings us to the philosophy of the present report. We know that nucleons moving within the nucleus as described by the shell model are present. The corresponding contributions to nuclear DIS therefore provide an "experimental background". One should compute these nucleonic effects, and remove them. The remainder should be interesting new physics.

When one studies the nucleonic effect one sees that the pions which cause nuclear binding must also be included. The result that we wish to discuss (already mentioned in Rith's report) is that including nucleons and their concomitant pions leads to the correct qualitative x-dependence (in the region $0.2 \lesssim x \lesssim 0.7$) needed to reproduce the data. Several critical assumptions are made in arriving at this conclusion. Thus future experiments can change this result.

Now we may present the outline of the remainder. We follow the order:

1. Phenomenology of the nucleonic contributions.

2. Theory of the nucleonic contributions. There is a current disagreement between STRIKMAN & FRANKFURT[5] and the authors in Ref. 4. We find that the results of Ref. 4 are more correct.

3. Phenomenology of the pionic contribution.

4. Theory of the pionic contribution. The discrepancy between the work of Berger & Coester and Ericson & Thomas is resolved in favor of the latter authors.

5. Dynamical rescaling and the Cheshire Cat.

2. Phenomenology of the Nucleonic Contribution

The <u>nucleon</u> contribution to the nuclear structure function (per nucleon) $F_2^A(x)$ is given by

$$F_2^A(x) = \int_x^A dz f_{N/A}(z) F_2^N(x/z) \tag{1}$$

in its simplest form. The function F_2^N is the structure function of a nucleon, and $f_{N/A}(z)$ is the probability (normalized to unity) that a nucleon of mass m_N has a longitudinal momentum fraction $z = (k^0 + k^3)/m_N$. It is given by

$$f_{N/A}(z) = \int d^3k \, \rho^A(k) \, \delta(z - (k^0 + k^3)/m_N) \,, \tag{2}$$

where the probability that a nucleon in a nucleus A has momentum of magnitude k is $\rho^A(k)$. The function $f_{N/A}(z)$ is normalized to unity. Equations (1) and (2) have been used by Akulinichev et al., Llewellyn-Smith, Ericson & Thomas and others. In the shell model nucleons move in orbitals labeled by α, with eigenenergies $-\epsilon_\alpha$. The total energy of a shell model nucleon (α) is then $k^0 = m_N - \epsilon_\alpha$. For simplicity we use an average over orbitals

$$\bar{\epsilon} \equiv \sum_\alpha N_\alpha \, \epsilon_\alpha / \sum_\alpha N_\alpha \,.$$

Then

$$k^0 = m_N - \bar{\epsilon} \,. \tag{3}$$

Since $\bar{\epsilon}$ is about 30 MeV for an Fe nucleus one may write

$$k^0 = m_N \eta \,, \tag{4}$$

where $\eta \approx 0.97$. This small three percent effect has a strong implication for nuclear deep inelastic scattering as we shall show. (Our justification for using (1), (2) and (4) is presented in the next section.)

Let us try to understand the evaluation of (2). First we must ask what is the form of $\rho^a(k)$? We know it is a function peaked about $k = 0$. For spherically symmetric nuclei

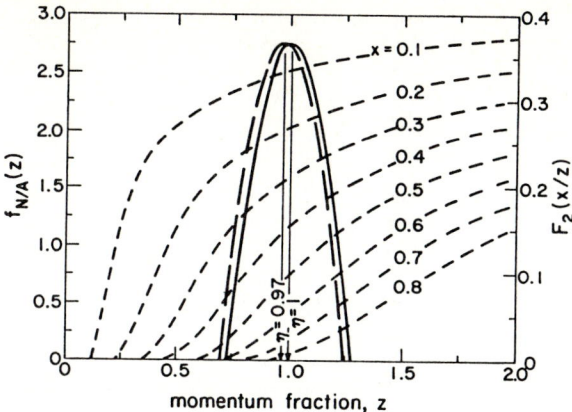

Fig. 3. The Fermi smearing function $f_N(z)$ for a free Fermi gas (solid line $\eta = 1$) and nucleons bound in nuclei (dashed line $\eta = .97$). In both cases $k_F/m_N = .273$ is used. Superimposed are some structure functions $F_2^N(x/z)$.

$$P^A(k) = \rho^A \left(k_\perp^2 + (k^3)^2\right)$$
$$\Rightarrow \rho^A(k_\perp^2 + (m_N z - k^0)^2) \; , \qquad (5)$$

where the second step incorporates the delta function of (2). Thus $f_{N/A}(z)$ obtains its maximum value when $z = k^0/m_N = \eta$. Furthermore the width of the function is determined by the Fermi momentum k_F. Since $k_F/m_N \approx 1/4$, $f_{N/A}(z)$ has a narrow width.

These features are illustrated in Fig. 3. The solid parabolic curve shows $f_N(z)$ if binding is ignored ($\eta = 1$). The better curve is the (long-dashed) parabola peaked at $\eta = 1$. Also plotted (short-dashed line) is $F_2^N(x/z)$. Note the falloff in magnitude as x increases.

Consider the region $0.2 \lesssim x \lesssim 0.6$. There $F_2^N(x/z)$ varies slowly in the region where $f_N(z) \neq 0$. One may then approximate the integral of (2) by evaluating $F_2^N(x/z)$ at $z = \eta$. Then

$$F_2^A(x) \approx F_2^N(x/\eta) \quad (0.2 \lesssim x \lesssim 0.6) \; . \qquad (6)$$

For larger values of x the situation is different. The spreading of $f_N(z)$ (Fermi motion) allows $F_2(x/z)$ to be evaluated at values smaller than x (for $1 < z < 1 + k_F/m_N$). Then

$$F_2^A(x) \gg F_2^N(x) \quad (x \gtrsim 0.6). \qquad (7)$$

Equation (6) represents the binding effect and (7) the Fermi motion effect.

Let us discuss the relevance of (6) with $\eta = 1 - <\epsilon>/m_N = 0.97$. Why should 3% matter? Consider $F_2^N(x/(1-<\epsilon>/m_N)/F_2^N(x)$. In a first-order Taylor series expansion

$$\frac{F_2^N(x/(1-<\epsilon>/m_N))}{F_2^N(x)} = 1 + \frac{<\epsilon>}{m_N} \frac{1}{F_2^N} \frac{dF_2^N}{dx} \; . \qquad (8)$$

The structure function of a nucleon falls rapidly with x. A typical representation of $F_2^N(x)$ for the relevant values of x is $F_2^N(x) \sim (1-x)^4$ so that

$$\frac{<\epsilon>}{m_N} \frac{1}{F_2^N} \frac{dF_2^N}{dx} = -4 \frac{<\epsilon>}{<m_N>} \qquad (9)$$

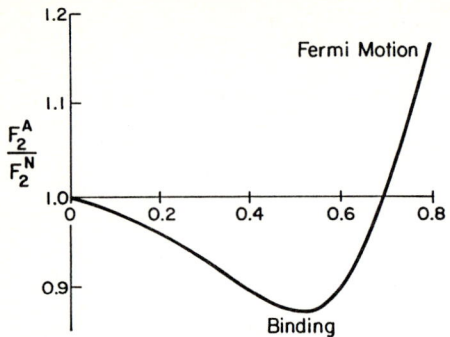

Fig. 4. Schematic influence of nuclear binding and Fermi motion on F_2^A/F_2^N

and a 3% effect is enhanced into a 12% effect. This is the typical size of the deviation of the nuclear structure function from unity.

The binding correction gives a depletion in the mix-x region. Fermi motion gives an increase. The *qualitative* behavior of $F_2^A(x)/F_2^N(x)$ is then given in Fig. 4. This is in good accord with the trends of the data.

3. <u>Theory of the Nucleonic Contribution</u>

FRANKFURT & STRIKMAN[5] (FS) have challenged the use of (1). They say that instead one should use[6].

$$F_2^{AFS}(x) = \int_x^A dy \left(\frac{m_N - <\epsilon> + P^3}{m_N - <\epsilon>} \right) f_{N/A}(y) F_2^N(x/y) . \tag{10}$$

This form arises from the requirements that $F_2^{AFS}(x)$ satisfy the baryon-number and the momentum-energy sum rules. (The nucleons carry all of the energy of the nucleus.) The use of (10) reduces the binding effect discussed above by about a factor of 2. The numerical results of FS have been confirmed in an explicit model calculation by G. CHANFRAY.[7]

The differences between (1) and (10) are subtle. One needs a more fundamental derivation to distinguish them. Thus we must consider the theory.

Start with definitions and kinematics. As shown in Fig. 5, deep inelastic scattering occurs when a photon of 4-momentum q^μ is absorbed by a target labeled by momentum P. The resulting final state is labeled by P_X. The electromagnetic current operator (J_μ) causing the transition is given by

$$J_\mu(\xi) = \bar\psi(\xi)\gamma_\mu Q \psi(\xi), \tag{11}$$

where $\psi(\xi)$ is a quark field operator; Q is the quark charge operator.

The cross section for inclusive DIS is proportional to a $L^{\mu\nu}W_{\mu\nu}$ where $L^{\mu\nu}$ incorporates the lepton physics and the essential quantity of interest $W_{\mu\nu}$ is given by[8]

$$W_{\mu\nu} = \frac{1}{4\pi} \int d^4\xi \, e^{iq\xi} <P|[J_\mu(\xi), J_\nu(0)]|P>_c . \tag{12}$$

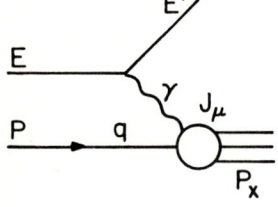

Fig. 5. Deep inelastic lepton-nucleus scattering

The second term $(-J_\nu(0)J_\mu(\xi))$ in the commutator appearing in (12) vanishes, but it is convenient to keep it. The subscript c means that only connected graphs are kept.

It is convenient to define a positive quantity $Q^2 = -q^2$. Another variable (lab frame value) of the lepton energy loss, $\nu = q^0$. For unpolarized measurements $W_{\mu\nu}$ depends on only two variables, ν and the lepton scattering angle. Instead one uses x_A and Q^2 where

$$x_A \equiv Q^2/2P \cdot q = -q^2/2m_A\nu . \tag{13}$$

The second equality follows from using $P^\mu = (m_A, \vec{O})$ in the lab frame.

The quark physics becomes apparent when one works at high Q^2 ($Q^2 \to \infty$) and in a limit in which ($\nu \to \infty$) such that x_A is finite. This is the famous Bjorken limit. Then the essential physics of $W_{\mu\nu}$ is contained in two independent structure functions $F_{1,2}(x_A, Q^2)$. The variation of these functions with Q^2 is caused mainly by effects of perturbative QCD. (See however K. Rith's talk at this meeting.) In the following the Q^2 dependence is ignored to simplify the notation.

One can choose the direction opposite to the photon's three momentum $(-\hat{q})$ as the 3-axis. Then in the Bjorken limit

$$q^\mu = (\nu, 0, 0, -\nu - m_A x_A). \tag{14}$$

It is useful to introduce light cone variables (defined by $A^\pm \equiv (A^0 \pm A^3)/\sqrt{2}$) so that

$$q^+ = -(q^0 + q^3)/\sqrt{2} = -m_A x_A/\sqrt{2}. \tag{15}$$

In the Bjorken limit $q^- = -\sqrt{2}\nu \Rightarrow -\infty$ so that integrations over space in the variable conjugate to q^- (ξ^+) are essentially delta functions at $\xi^+ = 0$. Thus q^+ is the only kinematic variable remaining.

The function $F_2^A(x_A)$ may be written in terms of quark (flavor a) $q_a(x_A)$ and quark $q_{\bar{a}}(x_A)$ distributions as

$$F_2^A(x_A) = \sum_a (q_a(x_A) + q_{\bar{a}}(x_A)) Q_a^2 . \tag{16}$$

The function $q_a(x_A)$ is the probability that a quark of flavor a momentum k^+ given by $x_A = k^+/P^+$. Jaffe[8] has obtained a convenient expression for $q_{a,\bar{a}}(x_A)$. One has

$$q_a(x_A) = \int \frac{d^4k}{(2\pi)^4} \delta\left(x_A - \frac{k^+}{P^+}\right) \frac{1}{2P^+} Tr\left(\gamma^+ \chi_a^A(k, P)\right) \tag{17}$$

where $\chi_a^A(k, P)$ is a matrix in the quark's Dirac indices given by[9]

$$\chi_a^A(k, P) = -\int d^4\xi \, e^{+ik\cdot\xi} < P|T(\psi_a(\xi)\bar{\psi}_a(0))|P>_c . \tag{18}$$

Note that $q_a(x_A)$ is a Lorentz scalar. This means that in Lorentz transformations the quantity $Tr \, \gamma^+ \chi_a^A(k, P)$ is the + component of a four-vector. JAFFE[8] has stressed the utility of (17) and (18) in deriving convolution model formulae and their limitations.[10]

At this stage we must discuss our philosophy regarding the light cone and the infinite momentum frame. The basic goal is to evaluate the quantity $<P|[J_\mu(\xi), J_\nu(0)]|P>_c$ in an accurate manner. According to FEYNMAN[11] this can be done if we take $P \Rightarrow \infty$. In this so-called infinite momentum frame time dilation effects remove the influence of interactions. Quarks may then be regarded as free particles during the time of the photon-target reaction.

However, working with $P \to \infty$ is difficult since one does not (in general) know the wave functions in that frame. There is an equivalent treatment which seems to avoid this problem. This is explained in the book by De Alfaro et al.[12] and already exploited in Ref. 8. The idea

263

is to simply use the target rest frame but replace J_μ by $J_\mu^{LC} = \bar{\psi}^{LC}\gamma_\mu Q \psi^{LC}$ where $\bar{\psi}^{LC}, \psi^{LC}$ are quantized at equal light cone times (i.e., at equal values of $(x^0 + x^3)/\sqrt{2}$). DeAlfaro et al. demonstrate that for the important matrix elements $<P=0|\left[J_\mu^{LC}(\xi), J_\nu^{LC}(0)\right]|P=0>_c$ and $\lim_{P\to\infty} <P||J_\mu(\xi), J_\nu(0)]|P>_c$ are equivalent by solving a few examples.[13]

This brings up another difficulty. The operator J_μ^{LC} are made of light cone quantized field operators, but the state $|P=0>$ is made of objects quantized at equal times. This problem is avoided by expanding $|P=0>$ into its nucleon and pionic components and then obtaining the necessary matrix elements of light cone quantized field operators from experimentally measured nucleon $q_a^N(x)$ and pion $q_a^\pi(x)$ quark distribution functions.

A necessary first step in deriving convolution model formulae such as (1) is to discuss the distribution function of a free nucleon. In that case we wish to employ (7) and (8) with $P = p$ (the nucleon's 4-momentum) and $A = N$. Applying the LSZ reduction procedure to $\chi_a^N(k,p)$ yields the result

$$\chi_a^N(k,p) = \frac{1}{2}\sum_s \bar{u}(p,s)G_a(k)\Gamma(k,p)G_a(k)u(p,s) \tag{19a}$$

$$\equiv \frac{1}{2}\sum_s \bar{u}(p,s)\hat{\chi}_a^N(k,p)u(p,s) \tag{19b}$$

where $G_a(k)$ is the quark Feynman propagator and $\Gamma(k,p)$ is the Fourier transform of the quark-nucleon four-point function. Nucleons in nuclei have spins that generally average to zero, so we have made an average over spin in writing (19). A Feynman graph representing $\chi_a^N(k,p)$ is given in Fig. 6. The quantity $\hat{\chi}_a^N(k,p)$ is an operator in the Dirac spaces of both nucleons and quarks.

We want to use the transformation properties of $Tr\gamma^+\hat{\chi}_a^N(k,p)$ to constrain its form. The d^4k integration of (17) sets k^+ equal to xp^+. Thus only two vector components γ_N^+ and p^+ are needed. We have

$$Tr\gamma^+\hat{\chi}_a^N(k,p) = C_a(k,p)\gamma_N^+ + D_a(k,p)(p^+ - m_N\gamma_N^+), \tag{20}$$

where γ_N^+ acts on the nucleon Dirac indices, and where C_a and D_a are unknown functions to be determined.

The following relations are needed to proceed:

$$2m_N P^+ = \frac{1}{2}\sum_s \bar{u}(p,s)p^+ u(p,s), \tag{21a}$$

$$2p^+ = \frac{1}{2}\sum_s \bar{u}(p,s)\gamma^{N+}u(p,s). \tag{21b}$$

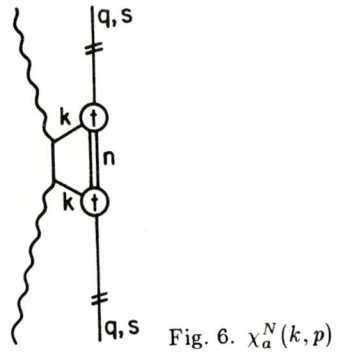

Fig. 6. $\chi_a^N(k,p)$

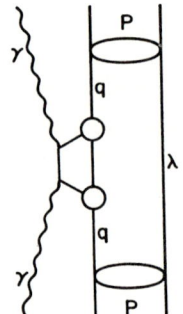

Fig. 7. $\chi_a^A(k,P)$

The use of (20) in (19b) then leads to the result that the coefficient of the $D_a(k,p)$ term vanishes. This means that $D_a(k,p)$ can not be measured in lepton-nucleon DIS. This is because D_a is the coefficient of an operator that vanishes for on-shell nucleons. To proceed, we simply set D_a to 0. Computing this coefficient can only be done with great difficulty.

The use of (20) in (19a) leads immediately to

$$Tr\gamma^+ \chi_a(k,p) = C_a(k,p) 2p^+, \qquad (22)$$

which is a useful result.

Since we have "understood" the nucleon quark distribution function, we may proceed to the nuclear structure function. An application of the Feynman rules to the diagrams of Fig. 7 leads to the result

$$Tr\gamma^+ \chi_a^A(k,P) = \int \frac{d^4p}{(2\pi)^4} Tr\gamma^+ \left[\hat{\chi}_a^N(k,p)\right]_{\alpha\beta} \left[\chi_N^A(p,P)\right]_{\beta\alpha}, \qquad (23)$$

where only the Dirac indices (α,ρ) of the nucleon are made explicit. The definition of $\chi_N^A(p,P)$ for nucleons N in a nucleus A is analogous to the one for quarks a in a nucleon:

$$\chi_N^A(p,P) = -\int d^4\xi\, e^{+ip\xi} <P|T\left(\Psi_N(\xi)\overline{\Psi}_N(0)\right)|P>_c. \qquad (24)$$

Now, however, the Ψ_N are Dirac field operators quantized at equal times.

We wish to apply (20) to (23). The use of (20) with $D_a = 0$ leads immediately to

$$Tr\gamma^+ \chi_a^A(k,P) = \int \frac{d^4p}{(2\pi)^4} \frac{Tr\gamma^+ \chi_a^N(k,p)}{2p^+} Tr_N \gamma_N^+ \chi_N^A(p,P), \qquad (25)$$

then $q_a^A(k,P)$ is given via (17) as

$$q_a^A(x_A) = \int d^4k\, \delta\left(x_A - \frac{k^+}{P^+}\right) \frac{d^4p}{(2\pi)^4}$$
$$\times \frac{Tr(\gamma^+ \chi_a^N(k,p))}{2p^-} \frac{Tr_N \gamma_N^+ \chi_N^A(p,P)}{2P^+}. \qquad (26)$$

One may bring out the connection to $q_a^N(x_A)$:

$$q_a^N(x_A) = \int d^4k\, \delta\left(x_A - \frac{k^+}{p^+}\right) \cdot \frac{Tr\gamma^+ \chi_a^N(k,p)}{2p^+} \qquad (27)$$

by inserting

$$1 = \int dy_A \delta(y_A - p^+/P^+) \qquad (28)$$

into the integral over d^4p appearing in (26). The result is

$$q_a^A(x_A) = \int \frac{dy_A}{y_A} \int d^4k\, \delta\left(\frac{x_A}{y_A} - k^+/p^+\right) \int \frac{d^4p}{(2\pi)^4} \frac{Tr\gamma^+ \chi_a^N(k,p)}{2p^+}$$
$$\times \frac{Tr_N}{2p^+} \gamma_N^+ \chi_N^A(p,P) \delta(y_A - p^+/P^+). \qquad (29)$$

The $\int d^4k$ integral appearing in (29) looks suspiciously like that of (27). However there is a difference; in (29) p^2 takes on all real values, whereas in (27) $p^2 = m_N^2$. We assume that $Tr\gamma^+ \chi(k,p)$ varies slowly as $p^2 = m_N^2$. We take this quantity out of the integral and get

$$q_a^A(x_A) = \int_{x_A}^A \frac{dy_A}{y_A} q_a^N(x_A/y_A) \tilde{f}_{N/A}(y_a), \qquad (30)$$

where a new quantity $\bar{f}_{N/A}(y_A)$ has been defined as an abbreviation to the quantities remaining in (29):

$$\bar{f}_{N/A}(y_A) = \int \frac{d^4p}{(2\pi)^4} \delta(y_A - p^+/P^+) \frac{\gamma_N^+ \chi_N^A(p,P)}{2P^+} \quad . \tag{31}$$

All that remains is to compare $\bar{f}_{N/A}(y_A)$ with $f_{N/A}(y_A)$ of (1).

Consider the quantity $\chi_N^A(p,P)$ given in (24). Use the well-known relation

$$\Psi_N(\xi) = e^{iH^0\xi^0} \Psi_N(0,\vec{\xi}) e^{-iH\xi^0} , \tag{32a}$$

and the non-relativistic approximation for the field expansion

$$\Psi_N(0,\vec{\xi}) = \int \frac{d^3p_1}{(2\pi)^{3/2}} b_{p_1} e^{i\vec{p}_1 \cdot \vec{\xi}} , \tag{32b}$$

to obtain

$$f_{N/A}(y_A) = \int d^4p \, \delta\left(y_A - \frac{p^+}{P^+}\right)$$

$$\times \frac{1}{2m_A} \int \frac{d^3p'}{(2\pi)^3} <P|b_{p'}^+ \delta(-p^0 + P^0 - H) b_{p'} |P> \quad . \tag{33}$$

The non-relativistic approximation $\gamma_N^+ = 1/\sqrt{2}$ for upper components and 0 for lower components has been used. Furthermore[14] an identity (2.4) of JAFFE[8] and $P^+ = \frac{m_A}{\sqrt{2}}$ in the target rest frame has been used.

Conservation of momentum may be invoked in (33) to find

$$\bar{f}_{N/A}(y_A) = \int d^4p \, \delta(y_A - p^+/P^+) S(p^0, \vec{p}) , \tag{34}$$

where $S(p^0, \vec{p})$ is the nucleon spectral function.

$$S(p^0, \vec{p}) = <P|b_p^+ \delta(-p^0 + P^0 - H) b_p |P> / <P|P> \quad . \tag{35}$$

The function $\bar{f}_{N/A}(y_A)$ is, apart from a trivial redefinition of variables ($m_A y_A = m_N y$) the same as that of Akulinichev et al. Thus (30), (34) and (35) embody the key result: the formula of Akulinichev et al. is correct.

A useful example is the shell model evaluation of the spectral function. Then

$$S(p^0, \vec{p}) = \sum_\alpha |\phi_\alpha(\vec{p})|^2 \delta(p^0 - (M - \epsilon_\alpha)) , \tag{36}$$

where the sum on α is over occupied orbitals and $\phi_\alpha(\vec{p})$ is the momentum space wave function for the orbital α.

It is useful to compare our formalism with that of Ref. 5. The essential difference is the factor $2p^+$ appearing in (22), (25-27) and (29). The result of Ref. 5 may be obtained by replacing p^+ by p^0 in those equations. In our formalism the treatment of the Lorentz properties of $\chi_a^N(k,p)$ leads to the definitive result that the factor is p^+ and not p^0.

4. Phenomenology of the Pionic Contribution

The need for components in addition to the nucleon can be seen immediately by considering the momentum sum rule. The total momentum carried by quarks in nucleons is given by $\sum_a \int dx_A \, x_A (q_a(x_A) + q_{\bar{a}}(x_A)) \equiv \lambda P^+$. For a nucleon, $\lambda \approx 1/2$. Using (1) and (6) one finds

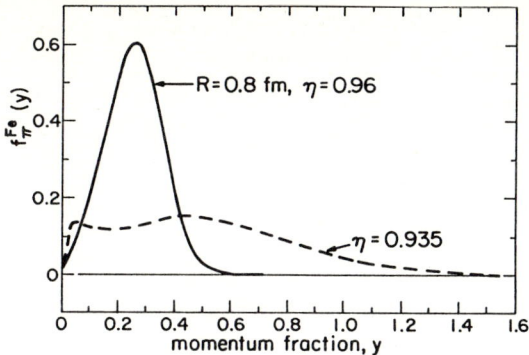

Fig. 8. The functions $f_\pi(y)$ of Ericson & Thomas (solid line) and Berger & Coester (dashed line)

$$\lambda^A = \eta\, A\, \lambda\ . \tag{37}$$

Since $\eta < 1$ some of the P^+ is missing.

This is not a surprise. The origin of nuclear binding is the exchange of pions between nucleons. It is natural to expect that pions carry some of the momentum, and that there should be a nuclear pionic contribution $(\delta_\pi q_a^A(x_A))$ given by a formula analogous to (1) or (30). That is, we need to include e.g.

$$\delta_\pi q_a^A(x_A) = \int \frac{dy_A}{y_A}\, q_a^\pi\left(\frac{x_A}{y_A}\right)\, f_{\pi/A}(y_A)\ . \tag{38}$$

See the paper of PANDHARIPANDE et al.[15] for a nice discussion of the need for nuclear pions.

Although pions must be included within our framework, there is a disagreement about the nature of $\pi^A(y_A)$. Consider for example, the results of Ericson & Thomas and Berger & Coester. As displayed in Fig. 8 there is a difference. The DIS results depend mainly on the integral of $\int f_{\pi/A}(y_A) dy_A$, and are not sensitive to the exact form of $f_{\pi/A}(y_A)$. However the Drell Yan predictions[3] differ significantly. Moreover, such a huge difference is unsettling. Understanding it better is necessary to have confidence in the entire approach.

5. Theory of the Pionic Contribution

We may understand $f_\pi^A(y_A)$ better by using the same formalism for the nucleon and the pionic contributions. We have

$$\delta_\pi q_a^A(x_A) = \int d^4k\, \delta(k^+ - x_A P^+) \frac{Tr}{2} \gamma^+ \chi_{a/A}^\pi(k, P)$$
$$- A(\text{pionic contribution to } q_a^N(x_A))\,, \tag{39}$$

where

$$\chi_{a/A}^\pi(k, P) = \int \frac{d^4p}{(2\pi)^4} \chi_{a/\pi}(k, p) \chi_{\pi/A}(p, P) \tag{40}$$

and

$$\chi_{\pi/A}(p, P) = \int d^4\xi\, d^{ip\xi} < P|T(\Phi_\pi(\xi), \Phi_\pi(0))|P>_c\,, \tag{41}$$

where Φ_π is a pion field operator.

The function $\chi_{\pi/A}(p, P)$ is closely related to the familiar pionic response function R_π. One has

$$R_\pi(p) = \frac{Im}{\pi} \int d^4\xi e^{ip\cdot\xi} < P|T(J_\pi(\xi), J_\pi(0))|P>_c\,, \tag{42}$$

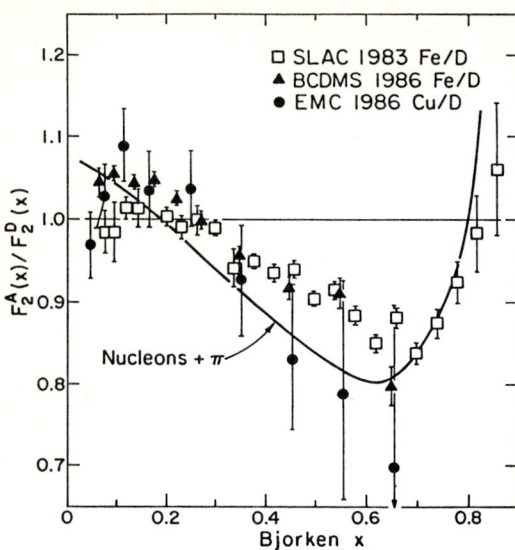

Fig. 9. Our calculation of the effects of nucleons and pions

where J_π is the source term in the pion field equation

$$(\Box + m_\pi^2)\Phi_\pi = J_\pi \ .$$

Equations (39) - (42) lead, after some algebra, directly to the result of Ericson and Thomas. The key point in the derivation is to realize that

$$\int d^4p \ e^{ip\xi} <P|\Phi(\xi)\Phi(0)|P> \propto \frac{R_\pi(p)}{(p^2 - m_\pi^2)^2} \ . \qquad (43)$$

Ericson and Thomas use R_π to obtain $f_{\pi/A}(y_A)$. Berger and Coester use $\rho(p) = <P|a^+(p)a(p)|P>$. Our result is that R_π should be used. We checked this numerically by using the Ericson Thomas $R(p)$ to obtain $\rho(p)$. Then we showed that the resulting $p(p)$ is just that of Berger and Coester. Details will appear in a forthcoming publication.

Our calculation of the conventional nucleon and pion dynamics is shown in Fig. 9. References to the data may be found in d'Agostino's talk. The essential features are

- rise at small x - pionic effect

- dip at medium x - nuclear binding

- rise at large x - Fermi motion.

With these qualitative features a reasonably good description of the data may be achieved.

6. Dynamical Rescaling and the Cheshire Cat

The dynamical rescaling of Close et al. and Jaffe arises from the simple observation that the change in F_2 obtained by varying the target from a nucleon to a nucleus is similar to the scaling violations seen as Q^2 increases. In both cases there is a depletion at mid x and an increase at small x.

QCD evolution is the same for all targets, A. However the evolution is governed by a differential equation. Some boundary condition involving the $q(x, Q^2 = \mu^2)$ must be used as input. μ^2 represents an infrared cutoff of gluonic effects and it may vary with $A: \mu^2 \Rightarrow \mu^2(A)$. If there is only one distance scale that matters one has

$$\mu^2(A) \propto 1/R^2(A), \tag{44}$$

where $R(A)$ is a measure of the confinement length scale in nucleus A. (It is not the radius of the nucleus.)

The use of (44) along with QCD evolution leads to the result

$$F_2(x, \xi^A Q^2) = F_2^A(x, Q^2), \tag{45}$$

where ξ^A varies around the region $\xi^A = 2$. As shown by CJRR the idea works well in describing the data.

In any case, the relationship (45) is an intriguing one that should result from any correct model. Indeed BICKERSTAFF and myself[3] used values of $<\epsilon>$ to derive ξ^A, thus adding validity to the conventional nucleon/pion picture. In the very same issue of Physics Letters B Close et al., using the same basic equation used the anomalous dimension d_n of QCD to obtain $<\epsilon>$ and k_F. Close et al. then suggested that there was a duality between nucleons and pions on the one hand and quarks and gluons on the other. This seemed amusing, but a bit too forced to me. However, after hearing the talk of Prof. Nielsen I am going to reconsider the idea of Close et al.'s duality as a possible example of the Cheshire Cat principle.

7. Summary

The shell model/π picture is consistent with present nuclear DIS data. This means that the nuclear quark presence is very difficult to detect. New data at low x, and high x obtained from forthcoming DIS and Drell-Yan measurements could change that unhappy statement.

8. References

1. J.J. Aubert et al., Phys. Lett. 123B, 275 (1983).
2. F.E. Close et al., Phys. Rev. D 31, 1004 (1985);
   E.L. Berger and F. Coester, Phys. Rev. D 32, 1071 (1985);
   R.L. Jaffe et al., Phys. Lett. 134B, 449 (1984);
   M. Ericson and A.W. Thomas, Phys. Lett. 132B, 112 (1983);
   C.H. Llewellyn Smith, Phys. Lett. 132B, 107 (1983);
   O. Nachtmann and H.J. Pirner, Z. Phys. C21, 277 (1984);
   Your Favorite.
3. R.P. Bickerstaff, M.C. Birse and G. Miller, Phys. Rev. D 33, 3228 (1986); Phys. Rev. Lett. 53, 2532 (1984).
4. S.V. Akulinichev et al., Phys. Rev. Lett. 55, 2239 (1985); G.V. Dunne and A.W. Thomas, Nucl. Phys. A455, 701 (1986); R.P. Bickerstaff and G.A. Miller, Phys. Lett. 168B, 409 (1986).
5. M.I. Strikman and L.L. Frankfurt, Phys. Lett. 183B, 254 (1987).
6. G. Chanfray has told me that (10) is a summary of the Strikman Frankfurt procedure. (Terms of the order of the average binding energy per nucleon (8 MeV) divided by the nucleon mass have been ignored in (10) for simplicity.)
7. G. Chanfray, private communication.
8. R.L. Jaffe, p. 537 in Relativistic Dynamics and Quark-Nuclear Physics, ed. by M.B. Johnson and A. Picklesimer, John Wiley & Sons, New York (1986).
9. Our definition of $\chi_a^A(k, P)$ has a slightly different form than Jaffe's, but the two are equivalent.
10. Jaffe (Ref. 8) asserts that the convolution model is not complete, and other terms are needed.
11. R.P. Feynman, Photon-Hadron Interactions, W.A. Benjamin Inc., Reading, MA (1972).
12. V. De Alfaro et al, Currents In Hadron Physics, North-Holland, Amsterdam (1972).
13. In the following notation we keep the symbol P, even though P is meant to be zero.
14. Jaffe's (Ref. 8) (2.4) is satisfied by the nuclear shell-model.
15. B.L. Friman, V.R. Pandharipande and R.B. Wiringa, Phys. Rev. Lett. 51, 763 (1983).
16. F. Close et al., Phys. Lett. 168B, 400 (1986).
17. J. Moss et al., Fermilab Expt 771.

Part IV

Strangeness

An Overview of Hypernuclear Physics

R. Bertini

Service de Physique Nucléaire-Moyenne Energie,
CEN Saclay, F-91191 Gif-sur-Yvette Cedex, France

1. INTRODUCTION

The quark models have obtained a large success in outlining many features of the baryon spectrum and it has been suggested that they would be more appropriate to describe the short range region of the nucleon-nucleon interaction. This suggests an alternative description of the nuclear matter : the nucleus being built up with bags of quarks that, because of confinement, at relatively large distances (> 1 fm) behave as baryons that interact by boson exchange like in the traditional picture. At shorter distances (< 1 fm) the bags overlap and fuse to form larger bags of six (or more quarks) where the interaction is carried by quark and gluon exchange.

The understanding of the relationship between this new approach and the traditional picture of nuclei, that is how one can go from the quark description, valid for very short interactive distances, towards the meson exchange representation, adopted for longer interaction distances, is related to the problem of confinement. What is the size of the confinement volume ? Is this the same when the interaction takes place between free baryons or when the baryons are embedded in the nuclear matter as constituents of the nuclei ?

A revival of interest for these questions has been caused by the discovery of the so-called E.M.C. effect /1/, that provides indications of a different behaviour of deep inelastic scattering on free nucleons and on nuclei. This discovery enhances the importance of intermediate energy nuclear physics, that can play here its fundamental role of bridge over the two fields of traditional nuclear physics and particle physics, provided that a careful selection is made of typical phenomena, strongly sensitive to the short range part of the interaction. In this sense the study of the strange particle-nucleus interaction is essential. First because it provides elements to extend the description of the nucleon-nucleon interaction to the more general and self-consistent frame of the baryon-baryon interaction. Secondly because the strange particle, just because of its strangeness, behaves in nuclear matter as a tagged probe, making easier the understanding of the phenomena.

An example of how the persisting of quark clusters deep inside nuclei can be tested with strange probes is shown in fig.1. The $^5_\Lambda$He ground state can be described /2/ in terms of the traditional picture with all the nucleons plus the Λ occupying the $S_{1/2}$ state. But in the quark picture, while the three quarks have the

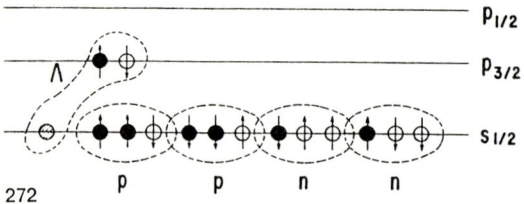

Fig.1. Quark structure of $^5_\Lambda$He hypernucleus

freedom to stay in the $S_{1/2}$ state, the u and d quarks do not because of the Pauli principle. This effect may appear not only as a shallow potential from the most conservative approach, but also as anomalous phenomena which can be understood in terms of somewhat deconfined characters of Λ and, by analogy, of Σ in Σ hyper nuclei.

In the perspective of future developments of this field, the present status of our knowledge of hypernuclear physics will be resumed in this lecture, without, because of lack of space, any pretention of completeness.

2. Hypernuclear systems

Nucleons are members of the baryon family, that includes nucleons and hyperons. An additional quantum number, strangeness S, characterizes the baryons : S=0 for nucleons, $-3 \leq S \leq -1$ for hyperons. This quantum number is conserved in strong and electromagnetic interactions but not in the weak interaction. Some properties of $S = -1$ hyperons are resumed in table 1. Mesons also can be characterized by strangeness S. In fact in addition to the mesons, responsible for the nucleon-nucleon interaction, and carrying strangeness S=0, there are strange mesons $(S = \pm 1)$. Note that unlike the baryons, mesons can have positive strangeness and that the maximum value of $|S|$ is 1.

Table 1: Properties of $S = -1$ hyperons

Hyp ($S = -1$)	Mass (MeV)	I	J^P	τ (s)	Mode	%
Λ	1115.6	0	$1/2^+$	2.6×10^{-10}	$p\pi^-$ $n\pi^0$	64 36
Σ^+	1189.4	1	$1/2^+$	$.8 \times 10^{-10}$	$p\pi^0$ $n\pi^+$	52 48
Σ^0	1192.5	1	$1/2^+$	5.8×10^{-20}	$\Lambda\pi$	100
Σ^-	1197.4	1	$1/2^+$	1.5×10^{-10}	$n\pi^-$	100

At this point it is a logical question to ask if, like the nuclear matter is built up with nucleons, other systems can be formed as aggregates of baryons, that is, nucleons and hyperons (Y).

In order to discuss this topic, let us consider first the case of the Λ hyperon. Its lifetime (table I), characteristic of a strange non conserving weak decay, is very short (2.6×10^{-10}s). However this lifetime is very long in the time scale of a strong interacting process ($\sim 10^{-21}$s) and therefore Λ can be considered as a stable particle in this respect. In nuclear matter, Λ mainly decays through the reaction $\Lambda N \to NN$, that is again a $\Delta S \neq 0$ weak process with a time scale of the order of 10^{-10}s or even more under the assumption that Pauli blocking hinders some decay channels. Therefore one could expect that the Λ hyperon could combine with nucleons provided that the ΛN interaction is ruled by some attractive potential. Aggregates of Λ and nucleons have effectively been observed for the first time in 1953 by Danysz and Pniewski in an emulsion experiment /3/ and have been named Λ hypernuclei. More recently also Σ hypernuclei have been discovered /4/.

In this lecture only aggregates of one hyperon Y (Λ or Σ) and nucleons will be considered.

Very simple mindedly the YN and Y nucleus interaction can be described, like the NN and N-nucleus interaction, in terms of a potential

$$V(r) = -V_0 \, f(r) + V_{LS} \, \vec{l}.\vec{\sigma} \, \left(\frac{h}{m_\pi c}\right)^2 \frac{1}{r} \frac{df(r)}{dr} + (V_1/A) \, \vec{t}.\vec{T}_{A-1} \qquad (1)$$

that is a function of the interacting distance r. In the eq.(1) the first term is the central part of the potential, the second one (the spin orbit term) depends on the relative spin ($\vec{\sigma}$) and angular ($\vec{l}$) momenta, the third one, also called Lane term, compensates the Coulomb repulsion of the charged particles inside the nucleus and gives the same spatial distribution for charged, ex. protons, and neutral, ex. neutrons, particles in the nucleus. In this last term $\vec{t}$ means the isospin of the baryon and $\vec{T}$ that of the remaining part of the nucleus. The symbol f(r) stands for an analytic function of the interacting distance r and nuclear radius R like in the Saxon-Woods potential for example. Of course, depending on the quantum numbers of the interacting particles, the spin-orbit or the Lane term can cancel.

In schematic description of the NN potential /5/ in terms of the lowest order pion exchanges the long-range one-pion exchange gives the NN interaction, its strong spin-spin and tensor character. The medium range attraction stems from the scalar (2π, I=0) and vector (ρ, I=1) parts of the exchange. The short range repulsion comes from the ω exchange. The spin orbit force is principally due to the vector exchange (ρ and ω).

The Λ has isospin zero, therefore only isoscalar (I=0) meson exchange is allowed in the ΛN interaction. This fact is illustrated in fig. 2 where the first diagram represents the longest range component : the two-pion isospin zero (σ) exchange, which gives a strong central character to the force. The second diagram represents the small isospin mixing, which would contribute to the long range part of the interaction. Thus the ΛN interaction is to be expected rather different from the NN interaction with shorter range, and weaker spin dependence. Furthermore the absence of ρ exchange will greatly modify the character of the ΛN spin orbit force.

The mass three system, the hypertriton $^3_\Lambda$H, is bound by 130 keV and has spin and parity $J^\pi = 1/2^+$. The mass four systems $^4_\Lambda$H and $^4_\Lambda$He have $J^\pi = 0^+$ and 1^+ for the ground state and the first excited state, respectively /6/. This indicates that the ΛN interaction is more attractive in the singlet spin state than in the triplet state by about 1 MeV. Just the opposite happens in the NN interaction.

The Σ hyperon has isospin I=1 and can couple with a nucleon to I=1/2 and 3/2. Therefore the whole complexity of the NN system is present in the ΣN interaction.

Data on YN interaction are quite scarce, however theoretical descriptions have been attempted in the general frame of a baryon-baryon interaction, that is considering both the nucleon-nucleon and the hyperon-nucleon interactions. This topic will not be discussed more intensively here. A detailed description, together with the relevant references, can be found in ref. /7/, where also the interaction of kaons with nucleons and nuclei is treated.

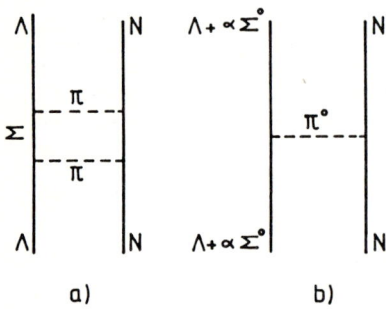

Fig.2. Lower order meson exchange diagrams of the ΛN interaction.

With the potential (1) the different quantum states of the hyperon-nucleus system can be described. This is done by extending to the hypernuclei the shell model, that has been so successful to explain the properties of ordinary nuclei.

As hypernuclei are produced transforming a nucleon of a nucleus in an hyperon the usual convention is to consider particle-hole states. Their angular momenta will be obtained coupling the angular momentum of the hole state, the nucleon hole in the nucleus, with the hyperon angular momentum. As an example a $(1p_{3/2}, 1p_{3/2})_{\Lambda n}$ state in $^{12}_{\Lambda}C$ would be a state obtained removing a $1p_{3/2}$ neutron in ^{12}C and replacing it with a Λ in the same quantum state $1p_{3/2}$. Its total angular momentum J can assume all the possible values that can be obtained coupling the two 3/2 angular momenta, that is 0, 1, 2, 3. The parity of the state is given by the rule $P = (-)^J$. Therefore one could get $J^P = 0^+$, 1^-, 2^+ or 3^-.

3. Production of hypernuclei

Hypernuclei can be produced by reactions involving one of these two elementary processes : associated production and strangeness exchange. In associated production a strongly interacting (S=0) probe hits a nucleon and creates an hyperon (S=-1) and a meson K^+ (S=+1). The simplest example of this process is the reaction

$$p + p \rightarrow p + K^+ + \Lambda . \quad (2)$$

Its cross section is not very large (~ 50 μb) but high intensity beams are available. Unfortunately the momentum transferred to the Y is quite high and thus the probability to form bound states quite low. However hypernuclei have been produced with a high energy ^{16}O beam /8/. The development of high energy heavy ion beams offers a very promising way to measure hypernucleus lifetimes by the recoil distance method.

The most efficient way to produce hypernuclei has been up to now the strangeness exchange process, which transfers the strangeness from kaon to nucleon in the reactions

$$K^- + N \rightarrow \Lambda + \pi \quad (3)$$

and

$$K^- + N \rightarrow \Sigma + \pi. \quad (4)$$

This success is related to the peculiarities of the reaction mechanism that leads to a simple interpretation of the hypernuclear states.

If we neglect two-step processes the differential cross section for the formation of a hypernuclear state at 0° is related /9-10/ to the elementary cross section of reactions (3) or (4) through the formula

$$\frac{d\sigma}{d\Omega}(0°) = N_{eff} \cdot \frac{d\sigma}{d\Omega}\bigg|_{K^-N \rightarrow Y\pi}(0°). \quad (5)$$

The effective number N_{eff} of nucleons in the target nucleus is proportional to the product of the distorted waves of the incoming kaon $\chi_K^{(+)}(\vec{r})$ and outgoing pion $\chi_\pi^{(-)}(\vec{r})$. This product can be expanded in partial waves in the usual way :

$$\chi_K^{(+)}(\vec{r}) \chi_\pi^{(-)}(\vec{r}) = \sum_{L=0}^{\infty} [4\pi (2L+1)]^{1/2} i^L \tilde{j}_L(qr) Y_L^0(\vec{r}), \quad (6)$$

where the radial function $\tilde{j}_L(qr)$ reduces to the usual Bessel function in the absence of any distortion (q is the momentum transfer). In Λ hypernucleus production (see fig. 3) the momentum transfer in the forward direction vanishes for kaon

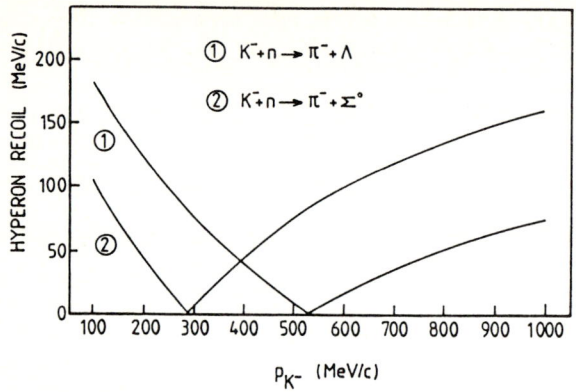

Fig.3 Kinematics of the kaon-neutron, unbound, reactions:
1) $K^- + n \to \pi^- + \Lambda$
2) $K^- + n \to \pi^- + \Sigma^\circ$

momenta of 530 MeV/c. At 900 MeV/c the momentum transfer is about 65 MeV/c, and the Bessel functions are very small inside the nuclear volume unless L=0. Distortion does not essentially change this argument.

Moreover if the pions are detected in the forward direction, spin-flip transitions are forbidden. Thus ΔL=0 transitions will predominate for small q values, ΔL=1 transitions, will become important at higher transfer momenta. A suitable choice of the kaon momentum will enhance one kind of transition selecting the corresponding hypernuclear state.

By analogy with the Mössbauer effect the ΔL=0 process in which the hyperon maintains the same spin and orbital wave function of the transformed nucleon has been called by Povh recoilless production /11/. When the hyperon jumps in a higher or lower hypernuclear state we have quasi-free production /12/. The reaction mechanism is illustrated in fig. 4.

Of course this simple interpretation does not exhaust all the physics contained in the hypernuclei spectra but constitutes a good starting point for the interpretation of the hypernuclei.

The behaviour of the elementary cross section /13/ $\frac{d\sigma}{d\Omega}\Big|_{K^-N \to Y\pi} (0°)$ of expression (5) as a function of the kaon momentum is shown in figs. 5 and 6 for Λ and Σ production respectively (dashed lines). The bumps correspond to excitation of strange baryonic resonances in the I=0 (Λ) or I=1 (Σ) channels. As the nucleons are constantly moving inside the nucleus because of the Fermi motion, the elementary cross section has to be averaged over the momentum of the hit nucleon. The result of such an averaging is to smear out the resonant behaviour of the elementary cross section of the free process $K^- N \to Y\pi$. However such a calculation is not easy and the output depends on some theoretical assumptions, and, as we'll see discussing Σ hypernuclei, discrepancies still exist between the calculations and the experimental data. Results of such Fermi motion averages are also shown in figs. 5 and 6 (solid lines).

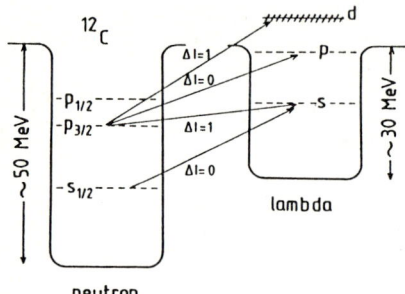

Fig.4. Simplified scheme of the reaction mechanism leading to recoilless production (ΔL = 0) and quasi-free production (ΔL = 1).

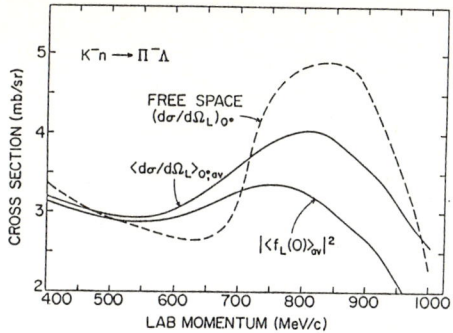

Fig.5. Fermi-averaged forward $K^- n \to \pi^- \Lambda$ lab. cross sections as a function of K^- momentum (solid lines). The free space cross section of Gopal et al. /13/ is shown as a dashed line.

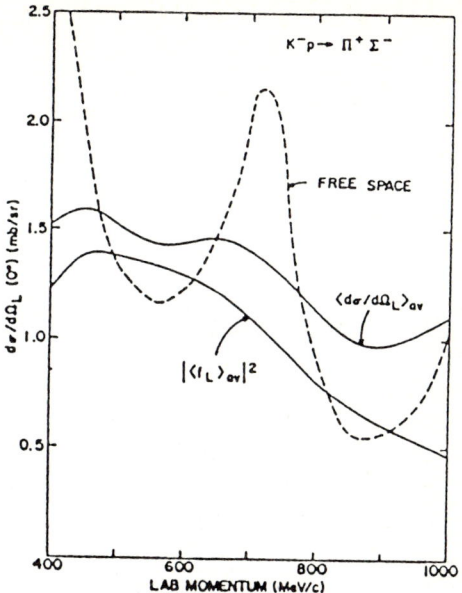

Fig. 6. Fermi averaged $K^- p \to \pi^+ \Sigma^-$ lab. cross sections at 0° as a function of momentum (solid lines), compared to the free space cross section (dashed lines) taken from ref./13/.

In addition to the reaction 2 another associated production process has been recently used to produce hypernuclei. Data /14/ on $^{14}C_\Lambda$ have been obtained via the reaction $\pi^+ \, ^{12}C \to K^+ \, ^{12}C_\Lambda$. In order to illustrate this process we show in figs.7 and 8 respectively the kinematics and the cross section of the elementary reaction $\pi^+ n \to K^+ Y$. Comparing these data with those of the strangeness exchange reaction it appears that cross sections are lower and therefore one could expect this process to be less efficient to produce hypernuclei. However pion beams are much more intense ($\simeq 10^8$ pions/10^{12} protons) than kaon beams ($\sim 2 \times 10^4 K^-/10^{12}$ protons at 700 MeV/c). As a matter of fact a full set of hypernuclei, from ^{9}Be to ^{89}Y, has been recently studied at Brookhaven /15/.

An important characteristic of this reaction is that the transfer momentum q to the hyperon is higher (fig. 7) and is well matched to states of high spin. Since the spin flip is small near 0°, one populates states of natural parity. This high spin selectivity picks out preferentially a series of states in which the various Λ shell model orbitals couple to the valence holes. This series strikingly dominates /16-17/ over other particle-hole combinations and gives relatively simple spectra.

Under this aspect this process appears complementary to the strangeness exchange reaction, that leads to low spin states in forward production.

As a larger set of hypernuclear systems, both Λ and Σ hypernuclei, has been studied with the K^-, π reaction, we will discuss hereafter only data obtained by this method.

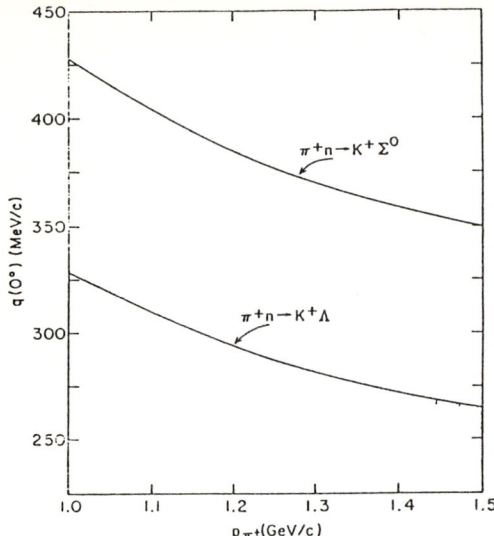 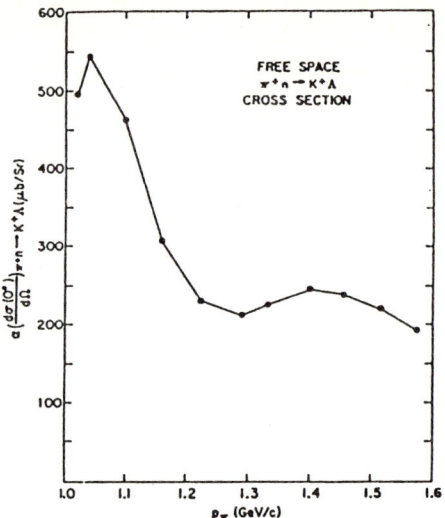

Fig. 7. The lab. momentum transfer q at $\theta_L = 0°$ as a function of incident lab. momentum for processes Λ or Σ involving production.

Fig. 8. Momentum dependence of the free space lab. differential cross section at $\theta_L = 0°$ for the $\pi^+ n \to K^+ \Lambda$ reaction.

4. The Λ hypernuclei

With this reaction many hypernuclear systems have been studied from $^6_\Lambda$Li to $^{209}_\Lambda$Bi (refs./18-21/). In fig. 9 are given the transformation energy spectra and the binding energy B_Λ for the (K^-, π^-) reaction on ^{12}C and ^{16}O. The transformation energy $M_{HY} - M_A$ is defined as the mass difference between the excited hypernuclear system and the target nucleus. From the measured kaon and pion momenta and the relative angle of their trajectories, the total energy of the kaon E_K and of the pion E_π and the recoil energy of the hypernucleus T_{HY} can be calculated. The transformation energy is given by the relation :

$$E_K - E_\pi - T_{HY} = M_{HY} - M_A = M_C + M_Y - B_Y - (M_C + M_N - B_N) = M_Y - M_N - (B_Y - B_N), \quad (7)$$

where B_Λ and B_n are respectively the binding energy of the Λ in the hypernucleus and of the neutron in the target nucleus.

From the transformation energy one can deduce the potential, eq.(1), that, in the case of Λ hypernuclei, and because the Λ has an isospin 0N. contains only two terms, the central part and the spin-orbit term. The central part was obtained first by Dalitz and Gal /12/ analysing a first generation experiment on Λ hypernuclei. The data /22/ are shown in fig. 10 as a function of $\omega = E_K - E_\pi$. As the energy resolution is very poor many states are mixed up. It is therefore a reasonable assumption to assert that baryons behave inside the nucleus like Fermi gas particles. It is also assumed that the (K^-, π) reaction occurs on a single nucleon, the rest of the nucleus being a "spectator". This reaction mechanism is named quasi-free scattering. If the momentum of the struck nucleon is k, then the relation between the energy transfer ω and the momentum transfer q for quasi-free scattering is

$$\omega = M_\Lambda - U_\Lambda + (\vec{k} + \vec{q})^2 / 2 M_\Lambda - [M_N - U_N + k^2 / 2 M_N], \quad (8)$$

where U_N and U_Λ are the single-particle well depths (> 0) for the nucleon and Λ, assumed to be momentum independent. If we now take a Fermi gas model ($|k| < k_F$)

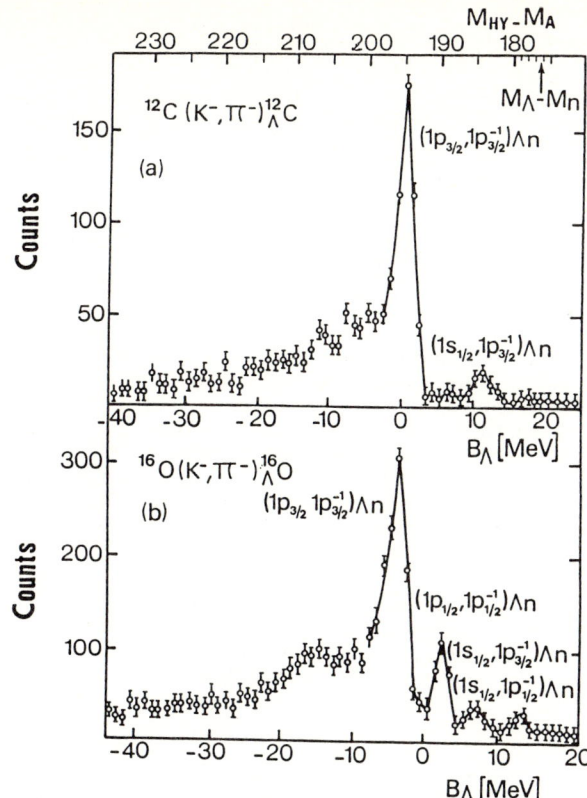

Fig. 9. Spectra obtained from the (K^-,π^-) reaction : a) on ^{12}C and b) on ^{16}O at a kaon momentum of 715 MeV/c, plotted as a function of the transformation energy $M_{HY}-M_A$. In addition the Λ binding energy B_Λ is given for each spectrum /18/.

for the nucleus, choose the z-axis along q, and average over k_x and k_y, we obtain

$$\omega \approx (M_\Lambda - M_N) + (U_N - U_\Lambda) - \left(\frac{1}{M_N} - \frac{1}{M_\Lambda}\right)\frac{k_F^2}{4} + \frac{q^2}{2M_\Lambda} + \frac{qk_z}{M_\Lambda}. \quad (9)$$

Since the k_z distribution dN/dk_z in a Fermi gas is parabolic, i.e.

$$\frac{dN}{dk_z} = \frac{3}{4k_F}(1 - k_z^2/k_F^2), \quad |k_z| \leq k_F \quad (10)$$

the resulting shape $dN/d\omega$ of the quasi-free spectrum is also approximately parabolic in ω :

$$\frac{dN}{d\omega} = \frac{dk_z}{d\omega}\frac{dN}{dk_z} \approx \frac{3M_\Lambda}{4\bar{q}k_F}\left[1 - \frac{M_\Lambda^2}{\bar{q}^2 k_F^2}(\omega-\bar{\omega})^2\right] \quad (11)$$

if $|\omega-\bar{\omega}| \leq \bar{q}k_F/M_\Lambda$. In eq.(11), $\bar{q}=q$ ($k_z=0$), and $\bar{\omega}$ corresponds to the peak of the distribution :

$$\bar{\omega} = M_\Lambda - M_N + U_N - U_\Lambda - \left(\frac{1}{M_N} - \frac{1}{M_\Lambda}\right)\frac{k_F^2}{4} + \bar{q}^2/2M_\Lambda. \quad (12)$$

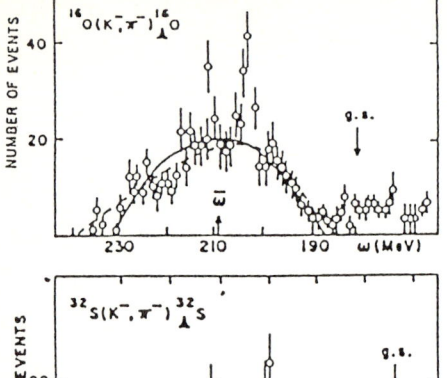
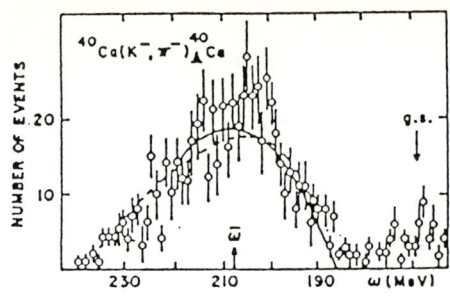

Fig. 10 - Hypernuclear excitation spectra for (K^-,π^-) reactions on ^{16}O, ^{32}S and ^{40}Ca. The solid line gives the quasi-free shape for a constant momentum transfer $\bar{q}$ and a linear dependence of the energy transfer ω on k_z. The dashed line gives the shape with the full k_z dependence for ω and q. The location of (normalized) shapes has been fitted (by eye) to the observed broad bump. $\bar{\omega}$ divides the area of the spectra into two equal parts /12/

In fig. 10 we show also the quasi-free spectrum of eq.(11) for ^{16}O, ^{32}S and ^{40}Ca targets for 900 MeV/c incident kaon momentum. This simple approximation gives a reasonable account of the shape of the experimental "background" (K^-,π) cross section, at least not too far from $\omega \approx \bar{\omega}$. To center the peak properly, a choice $U_N - U_\Lambda \approx 30$ MeV is indicated.

Note that in this treatment the Λ momentum k_z can assume all the possible values $\leq k_F$ (eq.(10)) because of its strangeness. In the analogous calculation performed to get the potential well depth in ordinary nuclei, some values of k_z of the nucleon are hindered by the Pauli principle. The relative potential well depth for a neutron in ^{12}C (50 MeV) and a Λ in $^{12}_{\Lambda}C$ (30 MeV) are also shown in fig. 4.

The better energy resolution of the data shown in fig. 9 allows a more precise interpretation of the excitation energy spectra and the determination of the spin orbit part of the potential as is discussed hereafter.

The ground states of ^{12}C and ^{16}O have $J^\pi = 0^+$. Then $\Delta\ell=0$ and $\Delta\ell=1$ transitions will lead respectively to $J^P = 0^+$ and $J^P = 1^-$ states (remember that spin flip is not allowed in a colinear geometry, see chapter 3). From the angular distributions it can be deduced that the more intense peaks have $J^P = 0^+$ and the smallest ones $J^P = 1^-$. Then it can be assumed that in the $^{12}_{\Lambda}C$ spectrum the $J^P = 0^+$ state belongs to the $(1p_{3/2}, 1p_{3/2}^{-1})_{\Lambda n}$ configuration, with both the Λ particle and the neutron hole in the $1p_{3/2}$ shell, whereas the state at $B_\Lambda = 11$ MeV $J^\pi = 1^-$) belongs to the $(1s_{1/2}, 1p_{3/2}^{-1})_{\Lambda n}$ configuration. Let us now compare the $^{12}_{\Lambda}C$ spectrum with that of $^{16}_{\Lambda}O$ in the $M_{HY} - M_A$ scale. The transformation energies for the levels corresponding to the $(1p_{3/2}, 1p_{3/2}^{-1})_{\Lambda n}$ configurations in $^{12}_{\Lambda}C$ coincide well with those of the two transitions in $^{16}_{\Lambda}O$ at $B_\Lambda = -3.5$ MeV and $B_\Lambda = 7$ MeV. Moreover the relative intensities are equal in both cases. They are therefore assigned to the $(1p_{3/2}, 1p_{3/2}^{-1})_{\Lambda n}$ and $(1s_{1/2}, 1p_{3/2}^{-1})_{\Lambda n}$ configurations respectively. In $^{16}_{\Lambda}O$ the two additional peaks at $B_\Lambda = 2.5$ MeV $(J^P = 0^+)$ and at $B_\Lambda = 13$ MeV $(J^P = 1^-)$ are then related to the $(1p_{1/2}, 1p_{1/2}^{-1})_{\Lambda n}$ and $(1s^{1/2}, 1p_{1/2}^{-1})_{\Lambda n}$ configurations respectively.

The splitting between the hypernuclear states with the $1p_{3/2}^{-1}$ and $1p_{1/2}^{-1}$ nuclear core is 6 MeV for both the $J^P = 0^+$ and $J^P = 1^-$ states. This discloses directly that the spin-orbit interaction of the Λ particle in the p shell nuclei is small. In fact the energy difference between the $(1p_{3/2}, 1p_{-1}^{3/2})_{\Lambda n}$ and the $(1p_{3/2}, 1p_{-1}^{1/2})_{\Lambda n}$ configurations and between the $(1s_{1/2}, 1p_{3/2}^{-1})_{\Lambda n}$ and the $(1s_{1/2}, 1p_{1/2}^{-1})_{\Lambda n}$ states is 6 MeV in both cases. The energy splitting between the $1p_{1/2}$ and $1p_{3/2}$ hole states in ^{15}O is 6.1 MeV. Thus an upper limit of .3 MeV for the spin- orbit splitting in the Λ-nucleus interaction can be inferred /18/. An analogous result is obtained for the hypernuclei in the s-d shell by comparison of the $^{32}_\Lambda S$ and $^{40}_\Lambda Ca$ transformation energy spectra /19/.

The neutrons of the inner shells (1s for ^{12}C and ^{16}O) do not give relevant contribution to the narrow structure of the Λ hypernuclei spectra. The width of the Λn particle hole state is dominated /23/ by the neutron hole width, that, as shown by p,2p and e,e'p experiments, is large for inner shells. Thus the Λn^{-1} particle-hole strength is spread over a large region of the hypernuclear spectrum. Moreover, since both the K^- and π^- mesons are strongly absorbed, the reactions take place mainly at the nuclear surface.

Since the pioneer work of Hüfner et al./10,24/ relevant contributions to the understanding of the Λ hypernuclei have come from many shell model /25-27/ and cluster model /28/ calculations. In particular Bouyssy /27/ has performed a shell model evaluation of a large set of hypernuclei and has obtained a nice agreement with the CERN results. The parameters of the Λ-nucleus potential used in his calculation are presented in table 2.

Table 2:

Comparison of the central and spin orbit potential depth for a neutron and a Λ particle in p-shell nuclei /27/.

	n	Λ
V^c	53	32 ± 2
V^{so}	17-18	4 ± 2

Other experimental results have been obtained at Brookhaven /29/. The $^{13}_\Lambda C$ data have been analysed /25/ in a shell model, where the Λ is weakly coupled to ^{12}C nuclear core. The results on the spin orbit part of the Λ-nucleus potential are in agreement with those of table 2.

The data on Λ hypernuclei so far shown have been obtained at kaon momenta (700 to 900 MeV/c) (refs./18-21/) where the value of the transfer momentum q is large enough to cause sizeable amount of quasi-free transitions. At lower q these contributions should almost vanish and only substitutional states will be excited. In the figure 11 are shown Λ hypernuclear spectra fed by the (K^-,π^-) reaction on a ^{12}C and a ^{13}C target (90 % enriched) at a kaon momentum of p_K = 550 MeV/c. The measurement has been performed near 0° with transfer momenta q_Λ ranging from 20 to 40 MeV/c.(ref./30/).

The spectra are almost free of background. The $K^- \rightarrow \mu^- \pi^0 \nu$ and $K^- \rightarrow e^- \pi^0 \nu$ three-body decays are suppressed owing to the addition of anticounters. On the low mass side (< 180 MeV) few events still remain coming from the $K^- \rightarrow \pi^0 \pi^-$ decays.

The ^{12}C spectrum shows the well-known narrow $(\nu\ p_{3/2}, \lambda\ p_{3/2})_{0^+}$ resonance /18/ at the transition energy $M_{HY}-M_A$ = 194.5.

As expected the quasi-free $\Delta l=1$ transition to the $(\nu\ p_{3/2}^{-1}, \lambda\ s_{1/2})_{1^-}$ ground state of ^{12}C, located at $M_{HY}-M_A$ = 183.5 MeV, is suppressed.

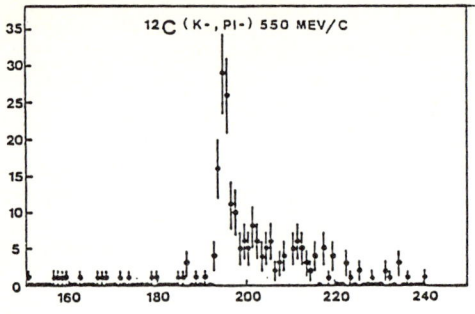 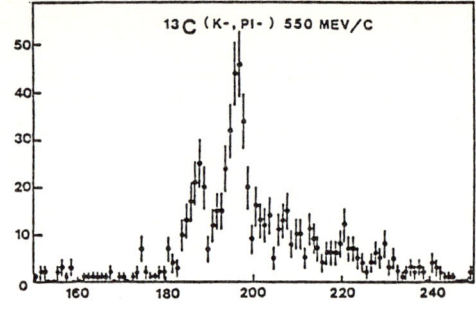

Fig. 11. Hypernuclear spectra of $^{12}C_\Lambda$ and $^{13}C_\Lambda$ measured /30/ at 550 MeV/c ($20 < q_\Lambda < 40$ MeV/c).

The broad distribution at higher masses should not be due to quasi-free transitions and contains only ($\nu\ s_{1/2},\ \lambda\ s_{1/2})_{0^+}$ strength. Its width should correspond to the neutron hole width in the 1s shell /23/.

The $^{13}_\Lambda C$ spectrum shows a strong resonance at the same transformation energy ($M_{HY}-M_A = 195$ MeV) as the ($\nu\ p^{-1}_{3/2},\ \lambda\ p_{3/2})_{0^+}$ state in $^{12}_\Lambda C$. Therefore it can be related to a substitutional process where the Λ replaces a $p_{3/2}$ neutron of the ^{12}C ground state core leading to the configuration ($\nu\ p_{3/2},\ \lambda\ p_{3/2})_{0^+} \otimes \nu\ p_{1/2}$).

Two weak narrow peaks are observed at 174.5 MeV and 180.5 MeV. An unresolved complex of several resonances is present between 184 MeV and 188 MeV. Almost all of these states must be of substitutional nature. We don't observe the $^{13}_\Lambda C$ ground state, that must be fed by a quasi-free transition. Otherwise our data are very similar to those obtained at $p_K = 800$ MeV/c in Brookhaven /29/.

The large structure observed around 184-188 MeV may be explained by the splitting between the 1/2⁻ and 5/2⁻ members of the ($\nu\ p^{-1}_{3/2} \otimes \lambda\ p_{3/2},\ p_{1/2}$) multiplet.

5. Σ hypernuclei

It was quite a surprise to find out that light Σ hypernuclear states show up as well-defined peaks in the (K^-,π^-) reaction /4/. In fact if in the nuclear matter the Λ decay can proceed through a weak interacting process ($\Delta S \neq 0$)

$$\Lambda + N \rightarrow \Lambda + N \qquad (13)$$

the Σ hyperon can decay through a strong interacting process ($\Delta S=0$)

$$\Sigma + N \rightarrow \Lambda + N \ . \qquad (14)$$

Thus the Λ escape width is small /23/ whereas the Σ escape width is expected to be large

However the peaks observed in the (K^-,π^-) reaction /4/ on ^{9}Be and ^{12}C show widths comparable to those of the corresponding Λ hypernuclear states.

The character of the ($\Sigma\ N^{-1}$) particle-hole states is somewhat more complicated than the Λ case since the Σ has isotopic spin 1. Thus the resulting configurations can have I=1/2 or 3/2 and be a mixture of Σ^-, Σ^0, Σ^+ particles with neutron and proton holes.

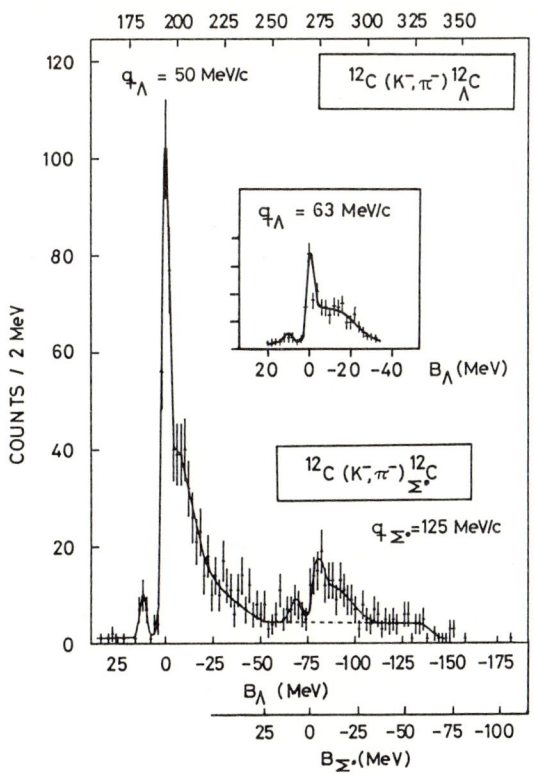

Fig. 12 - Comparison of the Σ hypernuclear spectrum obtained in the (K^-,π^-) reaction on ^{12}C at 715 MeV/c q_Σ = 125 MeV/c with the Λ spectrum measured in ref./22/ at 900 MeV/c (q_Λ= 63 MeV/c).

The spectrum shown in fig. 12 has been measured at a kaon momentum of 715 MeV/c. There the ratio between the cross sections of the elementary process

$$K^- + n \to \Sigma^0 + \pi^- \qquad (15)$$

and

$$K^- + p \to \Sigma^+ + \pi^- \qquad (16)$$

is about 10. It can then reasonably be assumed that only reaction (15) contributes to the formulation of the hypernuclear states at this energy. A support to hold to this interpretation is provided by the data obtained at 400 and 450 MeV/c figs.13a and 13b) (ref./31/). Here too, the shape of the hypernuclear spectrum reflects the energy dependence of the elementary cross section. Reaction (16) dominates at 400 MeV/c and therefore the peak showing up in fig. 13a can be interpreted as Σ^+ hypernuclear state. At 450 MeV/c the cross sections for reactions (15) and (16) are about the same.[a] As shown in fig. 13b contributions from Σ_0 hypernuclear states appear on the high mass side. In the reaction (K^-,π^+) only Σ^- hypernuclei can be produced and therefore the peak of fig. 13c must correspond to a pure I=3/2 state. In the Σ^+ hypernuclear spectrum (fig. 14a) a peak appears at the transition energy $M_{HY}-M_A$ = 270 ± .5 MeV, in the Σ^- hypernuclear spectrum (fig. 13c) a similar peak shows up at $M_{HY}-M_A$ = 279 ± .5 MeV.

Two important results stem from this comparison :

- first the difference between the position of the two peaks (about 9 MeV) is almost equal to the mass difference between the Σ^- and Σ^+ hyperons (8 MeV) ;
- second in both cases the binding energy $B_{\Sigma^\pm}$ is equal to about 4 MeV.

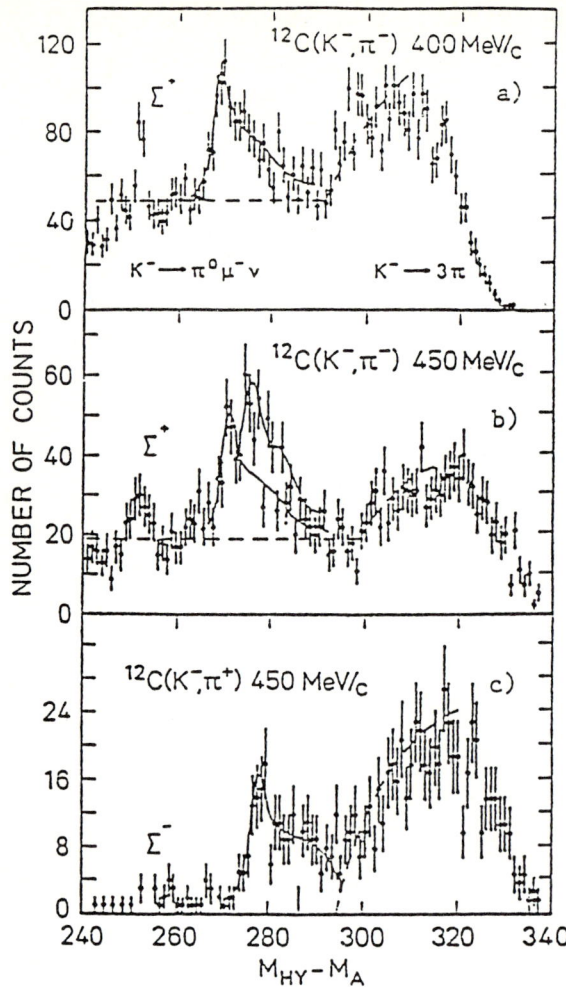

Fig. 13 - Σ hypernuclear spectra of ^{12}C obtained in the reaction (K^-, π^-)
a) at 400 MeV/c
b) at 450 MeV/c
c) and in the reaction (K^-, π^+) at 450 MeV/c.

In ordinary nuclei Coulomb effects are compensated by the symmetry energy related to the isospin-dependent potential $(V_1/A)\vec{t}\cdot\vec{T}_{A-1}$ where t is the nucleon isospin and $\vec{T}_{A-1}$ that of the rest of the nucleus.

Therefore if the peaks shown in fig. 13 are of substitutional nature $(p_{3/2}^{-1}, p_{3/2})$ the coupling between the isospin of the hole state and the Σ hyperon should be taken into account in a self-consistent calculation.

Now assume charge independence, and therefore the nuclear binding energies to be $B^{\Sigma^-} = B^{\Sigma^+}$ and the Coulomb energies $B_C^+ = - B_C^- = B_C$ and let us call m_3 and m_2 the mass of the Σ^- and Σ^+ hypernuclei respectively, then we derive from the standard mass formula :

$$\Delta_{3/2} = m_3 - m_2 = 7.97 - 2 B_C , \qquad (17)$$

which has to be compared to the experimental value $\Delta \approx 9$ MeV. From this, we get $B_C \approx .5$ MeV. The B_C of Σ is difficult to derive theoretically because the Σ lies in the continuum. For a proton of the p-shell in ^{12}C $B_C^+ = 2.71$. From that a value of $B_C^\Sigma \gtrsim 2.7$ MeV can be estimated. The accuracy of our mass measurement, which has been calibrated on the free Σ production from the protons in the LH$_2$

Cerenkov detector is ± .5 MeV. However the discrepancy between the experimental value (B_C = .5 MeV) and the theoretical one (B_C > 2.7) is larger than the measurement uncertainty and shows the existence of a symmetry term related to an isospin dependent strong interacting force that compensates the effect of the Coulomb interaction /30/. Therefore the simple assumption of complete charge independence is doubtful.

In fact, since it is expected that the charge exchange reaction plays a role, the two hypernuclear states with Σ^0 and Σ^+ should mix.

However, if we call m_1 the mass of the Σ^0 hypernucleus and we apply in the same way the mass formula, we get

$$\Delta_{21}\, m_2 - m_1 = -4.57 + B_C . \qquad (18)$$

Taking the value B_C = .5, just found, we obtain Δ_{21} = - 4.07. This number has to be compared to the experimental result Δ_{21} = - 5. obtained by unfolding the spectra (fig. 13b) with two peaks (Σ^+ and Σ^0 states). Therefore this splitting and the fact that the population of the states follows the free production cross section at 400 and 450 MeV/c rather well, indicates that the mixing is likely to be small. For a production of states with good isospin /32/ the mixing had to be rather strong. The isospin classification therefore seems to be unlikely, but cannot be totally excluded on the basis of these data.

The result for $B_{\Sigma^\pm}$ is similar to what has previously been found at 715 MeV/c. There the Σ_0 binding energy B_{Σ_0} was about 4 MeV less than the B_Λ.

The Fermi gas model successfully reproduces the quasi-free part of the transformation energy spectra in Λ hypernuclei /12/. If this analysis is applied for ^{12}C, the observed difference in the binding energy between Λ and Σ^0 would imply a Σ^0 central potential ($V_C^{\Sigma_0}$) of about 21 MeV (ref./33/). This value, compared to V_C^Λ = 30 MeV for $^{12}_\Lambda C$, is consistent with a Σ^0-nucleus interaction globally attractive although less so than the Λ-nucleus and N-nucleus interactions.

The analysis /34/ of Σ^- atomic level shifts, widths and yields has been performed with an optical potential, that has real central depth $V_C^\Sigma \simeq$ 26 MeV. A different fit /35/ of the Σ atomic data gives $V_C^\Sigma \simeq$ 22 MeV. Recent theoretical calculations /36-38/ give values in the same range.

The width of the Σ single particle states in nuclei, due to the $\Sigma N \to \Lambda N$ conversion, is estimated from the optical model expression

$$\Gamma = v\, \sigma_c\, \rho . \qquad (19)$$

Here σ_c the conversion cross section at velocity v in nuclear matter and ρ the two-body ΣN density.

A direct application of this formula would lead to widths too large for a direct experimental observation /32/

Different explanations have been suggested /39-44/ to solve this discrepancy between theoretical predictions and the observed widths, without finding, so far, a satisfactory solution.

6. The spin orbit interaction in Σ hypernuclei

As already mentioned, because of isospin, the interpretation of the excitation energy spectra of Σ hypernuclei is more involved than in the case of Λ-hypernuclei. Therefore we have focussed our attention on the (K^-,π^+) reaction that feeds only T=3/2 states.

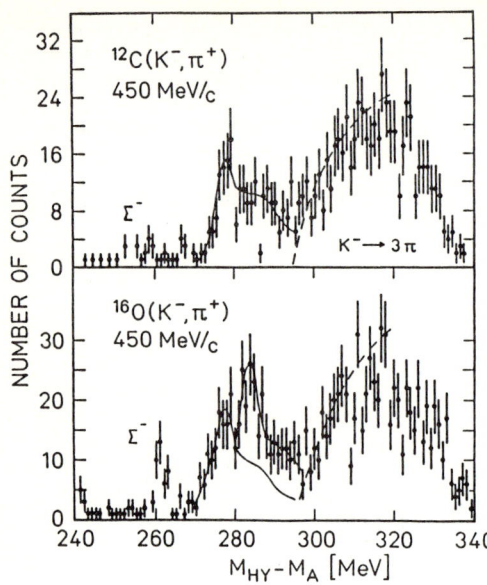

Fig. 14 - Σ hypernuclear spectra from ^{12}C (a) and ^{16}O (b) measured at 450 MeV/c K momentum. The spectra are plotted as a function of the difference of the hypernuclear mass M_{HY} and the target mass M_A in MeV/c². The continuum on the right-hand side of the dashed line is due to the $K^- \to 3\pi$ decay. At 261 MeV/c² the residue of the $H(K^-,\pi^+)\Sigma^-$ reaction from the LH_2 Cerenkov detector can be seen. The solid lines indicate the line shapes used for fitting the spectra.

Fig. 14b shows a spectrum from the (K^-,π^+) reaction on ^{16}O leading to Σ^- production /45/. This spectrum is to be compared to the one obtained on ^{12}C (ref./31/), shown in fig. 14a. The Σ hypernuclear states are in the nuclear continuum and can decay by strong interactions or by escape of a Σ. Therefore, this shape is theoretically unknown. We have adopted a method to analyse the spectra which is suggested by the Λ hypernuclei /18/. The peak in the $^{12}_\Sigma Be$ spectrum observed at 279 ± 1 MeV is ascribed to the substitutional state with a configuration $(p_{3/2}, p^{-1}_{3/2})_{\Sigma^-,p}$. In the $^{16}_\Sigma C$ spectrum two transitions are observed, one at 277.5 ± 1 MeV and the other at 284 ± 1 MeV. They presumably belong to the $(p_{3/2},p^{-1}_{3/2})_{\Sigma^-,p}$ and $(p_{1/2},p^{-1}_{1/2})_{\Sigma^-,p}$ configurations. In order to deduce their relative intensities, the line shape of the structure in $^{12}_\Sigma Be$ was assumed to be composed of a narrow line at 279 MeV due to the $(p_{3/2},p^{-1}_{3/2})_{\Sigma^-,p}$ configuration and a broad shoulder around 285 MeV due to quasi-free transitions, i.e. transitions to higher shells. The curve through the lower mass peak in $^{16}_\Sigma C$ shows an estimate of this shape. In particular, the quasi-free contribution has to be regarded as approximative only. On top of this, a normal resonance shape at 284 MeV reproduces the measured spectrum. From this fit an intensity ratio of 1.5 ± 0.3 for the lower peak to the upper peak is estimated.

A comparison between the mass spectra of neighbouring Σ hypernuclei is particularly instructive. For moderate residual ΣN interaction, i.e. smaller than the shell spacing for the Σ particle and smaller than the nuclear spin-orbit splitting, the transitions to the states in neighbouring nuclei with the same configuration appear at the same $M_{HY}-M_A$.

In order to compare the transitions to the $^{12}_\Sigma Be$ and $^{16}_\Sigma C$, we have to correct the mass values for the trivial Coulomb energy shift. By replacing a proton by the negative Σ in $^{16}_\Sigma C$, the Σ^- will be more bound by 1.5 MeV as compared to the $^{12}_\Sigma Be$. Taking this shift into account, the 279 MeV state of $^{12}_\Sigma Be$ coincides with the 277.5 MeV state of the $^{16}_\Sigma C$ within the experimental errors. Within the assumptions mentioned above, it can be concluded that the 277.5 MeV peak corresponds to the $(p_{3/2}, p^{-1}_{3/2})_{\Sigma^-,p}$ and the peak at 284 MeV to the $(p_{1/2}, p^{-1}_{1/2})_{\Sigma^-,p}$ configuration. The intensity ratio of 1.5 : 1 instead of the value derived from the statistical weights of 2 : 1 may be explained by a mixing of the two configurations by residual interaction.

It is important to note that the $(p_{1/2}, p_{1/2}^{-1})_{\Sigma^-,p}$ state appears at a mass which is about 6 MeV higher than that for the $(p_{1/2}, p_{1/2}^{-1})_{\Sigma^-,p}$ state. This is just the reverse of the situation in Λ hypernuclei.

The spin-orbit interaction for the Λ particle in the nucleus is much smaller than the nucleon one /18/. Therefore the splitting between the $p_{3/2}$ and $p_{1/2}$ substitutional state is determined by the spin-orbit splitting for the p-shell nucleon in the ^{16}O nucleus. As the $p_{3/2}$ state is bound more strongly than the $p_{1/2}$, the $(p_{1/2}, p_{1/2}^{-1})_{\Lambda,n}$ configuration appears at a lower mass than the $(p_{3/2}, p_{3/2}^{-1})_{\Lambda n}$ one. Following the same argument for the Σ hypernuclei, one would conclude that the spin-orbit interaction for the Σ particle is so large as to reverse the order of the $(p_{3/2}, p_{3/2}^{-1})_{\Sigma^-,p}$ and the $(p_{1/2}, p_{1/2}^{-1})_{\Sigma^-,p}$ states and is thus larger than that of the nucleon one. In fact, by neglecting the residual interaction fully, the spin-orbit interaction for the Σ particle would be just 12 MeV or twice that for the nucleon.

The above considerations lose their validity if the expectation value for the residual interaction matrix element is larger than the nuclear spin-orbit splitting. In this case also for a zero spin-orbit interaction for the Σ particle in the nucleus one could fit the observed energy spectra. It should be pointed out, however, that in the p shell the nuclear matrix elements for the residual interaction have a typical value of 2 MeV (ref./46/). It seems to be very unlikely that the residual interaction for the ΣN system would be stronger than that for the NN system. The ΣN interaction /47/ is, in all spin and isospin states, much weaker than the NN one, with the exception of the spin 1 and isospin 1/2 state for which it is comparable in strength to the NN interaction.

Therefore, it seems reasonable to conclude that the results on the $p_{3/2}$ and the $p_{1/2}$ orbit splitting in $^{16}_{\Sigma}$C presented here favour the solutions for the spin-orbit interaction for the Σ particle being stronger than that for the nucleon. One should, however, be aware of the fact that this result is obtained by the assumption that the residual interaction is not anomalously strong, i.e. not stronger than the nuclear spin-orbit splitting in the p-shell nuclei.

Another result on the Σ hypernuclear spin-orbit interaction, derived from a stopped kaon experiment has been published /48/. The value for the $p_{3/2}-p_{1/2}$ splitting, also derived under the assumption of negligible residual interaction, is 5 MeV. This is in clear disagreement with the value of 12 MeV derived here. However, the structures observed do not seem to have been confirmed in more recent runs /49/.

On the theoretical side, only the additive quark model /50/ predicts a spin-orbit splitting of energy levels in Σ hypernuclei larger than in ordinary nuclei. Therefore it seems to give the best description of the experimental data. However, because of the many simplifications of the model and of the poor statistics of the data, this possible "agreement" has to be considered with caution.

More recent calculation /51/, also with quark models, find splittings that, in contradiction with this estimation, are of the same order of those predicted by the boson exchange models /52/, that is smaller for Σ hypernuclei than ordinary nuclei. Shell model calculations /38/, made with the assumption of a negligible ΣN residual interaction, fit the data /45,48/ with a spin orbit potential that is again larger than for ordinary nuclei. As a matter of fact the situation is rather confused and only new data with better statistical significance could provide new clues.

7. Non-leptonic weak decays of hyperons

Free hyperons mainly decay (see table 1) via a strangeness non-conserving weak process to the πN channel. In this sense they have been extensively discussed in the literature as a typical example of non-leptonic weak interactions. In nuclear matter an other decay channel is open for the Λ decay through the non-mesonic

ΛN → NN channel. The study of this decay mode can be essentially performed measuring the lifetimes and branching ratios of the Λ hypernuclear states. This topic has been extensively discussed in ref. /53/. Let us just present here the main theoretical ideas and experimental facts.

The mesonic decays of the Λ hyperon shown in table 1 may be described in terms of the exchange of a W boson between the s and u, d quarks.

The hamiltonian that describes this process leads to transitions with both isospin changes $\Delta I = 1/2$ and $\Delta I = 3/2$ and corresponding transition amplitudes $A_{1/2}$ and $A_{3/2}$. However, both kaon and hyperon decays are in agreement with the so called "$\Delta I = 1/2$ value", that shows that the $A_{1/2}$ transition amplitudes are enhanced by a factor fifteen to thirty. For example, the branching ratios of table 1 can be satisfactorily reproduced taking account of for the $A_{1/2}$ amplitude only. The mechanism leading to the enhancement of the $A_{1/2}$ transition amplitude is not understood. It is therefore, of the greatest interest in order to produce an hamiltonian with the correct isospin character to see if the $\Delta I = 1/2$ rule also applies to the non-mesonic weak decays of the Λ.

In nuclear matter there are two non-mesonic channels for the Λ decay : the proton (Λp → np) and neutron (Λn → nn) stimulated processes. In LS coupling the initial state is assumed to be $L = 0$, $J = 0,1$ and $I = 1/2$. Let the elementary rates for the neutron stimulated process be R_{nJ} and for the proton stimulated process R_{pj}. The $\Delta I = 1/2$ rules enter as conditions on these elementary rates namely : $R_{n0} = 2 R_{p0}$ and $R_{n1} < 2 R_{p1}$.

Recently the decay modes of the ground state of ^{12}C have been studied at Brookhaven /53,54/.

This state has been fed by the (k^-,π^-) reaction and corresponds to the peak at $B_\Lambda = 11$ MeV of figure 9 a). The authors reported on the measurement of the total width $\Gamma = 1/\tau$, the partial widths $\Gamma_{\pi°}$ and γ_{π^-} for the mesonic channels and Γ_n and Γ_p for the neutron and proton stimulated processes. Their results are shown in table 3, where Γ_Λ is the free decay width.

Table 3: Decay widths of ^{12}C

$\Gamma \;\;= (1.25 \pm 0.18) \; \Gamma_\Lambda$

$\Gamma_n = (0.65 ^{+\;0.2}_{-\;0.3}) \; \Gamma_\Lambda$

$\Gamma_p = (0.49 ^{+\;0.3}_{-\;0.2}) \; \Gamma_\Lambda$

$\Gamma_{\pi} = (0.05 ^{+\;0.06}_{-\;0.03}) \; \Gamma_\Lambda$

$\Gamma_{\pi^+} = (0.06 ^{+\;0.08}_{-\;0.05}) \; \Gamma_\Lambda$

An inspection of these data shows that the total decay width is almost equal to free Λ decay width, even though the decay mode is very different : the mesonic decay is negligible and the neutron and proton stimulated processes are almost equal important. Error bars are quite important, so that the $\Delta I = 1/2$ rule can be tested only insofar as the relation $\Gamma_n < 2 r_p$ (that derives from the selection rule on the R_{NJ}) is satisfied. These data, however, represent a serious test for any theory that must be able to reproduce the measured widths.

The theoretical approaches can be grouped in three categories. The first /55/ adds strong interaction corrections to the elementary quark-quark weak interaction diagram of W exchange. The second /56/ assumes that the interaction takes place via a single meson exchange (π or ρ) from a weak vertex. This approach is closely related to the scheme of NN parity non-conserving interaction. The third one /57/ develops a composite of these two approaches and treats the ΛN interaction through the one-pion exchange for separations greater than 0.8 fm and as a six-quark state with the quark-quark interaction given in ref./55/, for shorter distances.

All the authors find values of Γ in the right range (1 to 3 × Γ_Λ) but no one estimates the neutron stimulated fraction. Therefore more complete calculations are needed to check the theoretical approach. For completeness it should be added that lifetimes of heavy hypernuclei have also recently been measured /58/ by observation of the delayed fission induced by antiproton annihilation in ^{238}U and ^{209}Bi.

8. Conclusion

The study of hypernuclei has provided surprising results about the behaviour of hyperons inside nuclear matter. Narrow Σ hypernuclear states, that is states where a Σ hyperon (I=1) is embedded in a nucleus, have been observed, whereas strong interaction inside nuclear matter was expected to prevent even the formation of such states. No satisfactory theory exists at the moment to explain why such narrow states can be produced. Related to this topic is the question of the lifetime of these states and of the branching ratios of the different decay modes. For Σ hypernuclei measurements of these quantities do not exist at the moment.

Spin-orbit splitting is mainly due to a short-range interacting process. The new results on the Λ-nucleus and Σ-nucleus spin-orbit potential discussed here have given rise to new theoretical attempts to give a complete description of the phenomena in the frame of a general baryon-nucleus theory.

The experiments performed so far on Δ production in nuclei in order to study the resonance propagation inside nuclear matter, and possibly the formation of narrow Δ-nucleus states, did not give satisfactory results, mainly because of the width of the Δ-resonance. The study of the hyperon resonances, like Λ (1520), inside nuclei can provide useful clues for the understanding of this problem. Here the resonance width is expected to be about ten times narrower than for the Δ-production and therefore differences between free production and propagation inside nuclear matter more easily investigated.

All the examples discussed in this talk can be classified as first generation experiments. They, usually, suffer from poor statistics, due to an inadequate beam quality. However they can be considered as very useful clues to design future experiments devoted to the investigation of the short range region of the baryon-baryon interaction. These experiments will require sophisticated and expensive detectors, high intensity kaon beam with a very good pion rejection, pion beams. In that sense the facility, proposed at this workshop /59/, appears to be very well suited. An important aspect to be considered, in my view, is that, with such a facility, many experiments requiring different kinds of beams will be possible at the same time.

Modern experiments become more and more costly, require big efforts of the large teams, develop new techniques that become quickly obsolete. Therefore they must be performed in a reasonably short time.

At present there is a shortage of beam time in Europe. Since the shutdown of kaon beams at CERN, very few experiments on strange particle physics are in progress in Europe.

This field, in which European physicists have been so active in the last decades, and that appears so promising for the future of intermediate energy physics should not be deserted.

References

1. J.J. Aubert et al.: Phys. Lett 123B, 277 (1983) ;
   A. Bodek et al.: Phys. Rev. Lett. 50, 1431 (1983) ;
   A. Bodek et al.: Phys. Rev. Lett. 51, 534 (1983)
2. K. Blenler et al.: Z. Nature Forsch, 38a, 705 (1983)
3. M. Danysz and J. Pniewski : Phil. Mag. 44, 348 (1953)
4. R. Bertini et al.: Phys. Lett. 90B, 375 (1980)
5. M. Lacombe et al.: Phys. Rev. C21, 861 (1980)
6. R.L. Jaffe : Phys. Rev. Lett. 38, 195 (1977) ; Errata 38, 717 (1977).
   M. Bedjdian et al.: Phys. Lett. 83B, 252 (1979)
7. C.B. Dover and G.E. Walker : Phys. Rep. 89, 1 (1982)
8. K.J. Nield et al.: Phys. Rev. C13, 1263 (1976)
9. R.J. Esch : Can. J. Phys. 51, 1524 (1973)
10. J. Hüfner et al. : Nucl. Phys. A234, 429 (1974)
11. B. Povh : Z. Phys. A279, 159 (1976)
12. R.H. Dalitz and A. Gal.: Phys. Lett. 64B, 154 (1976)
13. G.P. Gopal et al.: Nucl. Phys. B119, 362 (1977)
14. C. Milner et al., Phys. Rev. Lett. 54, 1237 (1985)
15. P.H. Pile et al.: In R.E. Chrien : Studies of Hypernuclei by associated production, inv. talk at XIth Int. Conf. on Part. and Nuclei (PANIC), Kyoto (1987) preprint BNL 39723
16. C.B. Dover et al.: Phys. Rev. C22, 2073 (1980)
17. H. Bando and T. Motoba : Prog. Th. Phys. 76, 1321 (1986)
18. W. Bruckner et al.: Phys. Lett. 79B, 157 (1978)
19. R. Bertini et al.: Phys. Lett. 83B, 306 (1979)
20. R. Bertini et al.: Nucl. Phys. A360, 315 (1981)
21. R. Bertini et al. : Nucl. Phys. A368, 365 (1981)
22. W. Bruckner : Phys. Lett. 62B, 481 (1976)
23. N. Auerbach et al.: Phys. Lett. 68B, 225 (1977) ;
    N. Auerbach and N. Van Giai : Phys. Lett. 90B, 354 (1980)
24. J. Hüfner et al.: Phys. Lett. 49B, 409 (1974)
25. C.B. Dover et al.: Phys. Lett. 89B, 26 (1979) ;
    E.H. Auerbach et al.: Phys. Rev. Lett. 47, 1110 (1981)
26. R.H. Dalitz and A. Gal.: Ann. Phys. 116, 167 (1978)
27. A. Bouyssy : Nucl. Phys. A290, 324 (1977) ;
    A. Bouyssy : Phys. Lett 84B, 41 (1979) ;
    A. Bouyssy : Phys. Lett. 91B, 15 (1980)
28. T. Motoba et al.: Progr. Theor. Phys. Sup. 81, 42 (1985) and refs. therein
29. M. May et al.: Phys. Rev. Lett. 47, 1106 (1981)
30. R. Bertini, Heidelberg-Saclay collaboration : in Proc. of Int. Conf. on kaon physics, Heidelberg (Germany) June 20-24, 1972, MPI-H-1982-V.20, p.362
31. R. Bertini et al. : Phys. Lett. 136B, 29 (1984).
32. A. Gal : Nukleonika 25, 44 (1982)
33. R. Bertini : In Meson nuclear physics, Houston, III, 703 (E.V. Hungerford, American Institute of Physics, New York, 1974)
34. C.J. Batty : Phys. Lett. 87B, 374 (1979)
35. S. Wyceck et al.: Nucl. Phys. A324, 288 (1979).
36. A. Gal et al.: Ann. of Phys. 137, 341 (1981)
37. J.A. Johnston and A.W. Thomas : Nucl. Phys. A392, 409 (1983)
38. R. Hausmann and W. Weise : Z. Phys. A324, 355 (1986)
39. C.J. Batty et al.: Nucl. Phys. A402, 349 (1983)
40. L.S. Kisslinger : Phys. Rev. Lett. 44, 968 (1980)
41. W. Stepien-Rudzka and S. Wyceck : Nucl. Phys. A362, 349 (1981)
42. C.B. Dover and A. Gal : Comm. Nucl. Part. Phys. 12, 155 (1984)
43. R. Brockmann and E. Oset : Phys. Lett. 118B, 33 (1982)
44. J. Dabrowski : Nucl. Phys. A434, 373c (1985)
45. R. Bertini et al. : Phys. Lett 158B, 19 (1985)
46. A. Bohr and B. Mottelson : Nuclear Structure, Vol.I, sect.2.5, (Benjamin, New York 1979)
47. H.G. Dosch et al.: Phys. Lett. 21, 236 (1966)

48. T. Yamazaki et al. : Phys. Rev. Lett. 54, 102 (1985) ; and Nucl. Phys. A450, 1c (1986)
49. B. Povh, private communication
50. H.J. Pirner : Phys. Lett. 85B, 190 (1979)
51. Y. He, F. Wang, L.W. Wong : Nucl. Phys. A451, 653 (1986) ;
    O. Morimatsu et al.: Nucl. Phys. A420, 573 (1984)
52. R. Brockmann and W. Weise : Nucl. Phys. A355, 365 (1981) ;
    A. Bouyssy : Nucl. Phys. A381, 445 (1982)
53. P.D. Barnes : Nucl. Phys. A450, 43c (1986)
54. R. Grace et al.: Phys. Rev. Lett. 55, 1055 (1985)
55. R. Gilman and M. Wise : Phys. Rev. D20, 2392 (1979) ;
    R. Gilman and M. Wise : Phys. Lett. 83B, 83 (1979) and refs. therein
56. B.H.J. Mc Kellar and B.F. Gibson : Phys. Rev. C30, 322 (1984)
57. C.Y. Cheung et al. : Phys. Rev. C27, 335 (1985)
58. J.P. Bocquet et al.: Phys. Lett. B182, 146 (1986).
    M. Epherre-Rey Campagnolle, in Proc. of 1986 INS Int. Symp. On Hypernuclear Physics, eds H. Bando, O. Hashimoto and K. Ogawa (Toshicha, Tokyo 1986) p.207
59. F. Bradamante : The European hadron facility contribution, to this workshop.

---

a) Note that, as a function of the energy, the forward differential cross section of Σ production on nuclei behaves like the cross section of the elementary process (reactions (15, 16)) and not as the Fermi averaged cross section discussed in chapter 3, fig.6. For the moment that is a riddle.

Perspectives in Strange Particle Physics

P.D. Barnes

Physics Department, Carnegie Mellon University,
Pittsburgh, PA 15213, USA

The strangeness degree freedom has been studied extensively in nuclear and particle physics and has provided new insight into both strong and weak interactions. I will concentrate on nuclear related topics. The prospects for new information on the strong hyperon-nucleon interaction from new hypernuclear spectroscopy measurements is discussed in section I. The weak hyperon-nucleon interaction, observed in hypernuclear decay, is dealt with in Section II, while Section III deals with strange dibaryon physics.

I. Hypernuclear Formation and the Strong Hyperon-Nucleon Interaction

Investigation of the production and spectroscopy of Λ hypernuclei has progressed steadily over the past twenty years at BNL, CERN, and KEK. By now the techniques for studying their properties are varied and sophisticated. As a result, hypernuclear spectroscopy has become a very useful tool for studying the strong hyperon-nucleon interaction. We only discuss here some new techniques which are becoming available for the production and tagging of hypernuclei.

The conventional method for converting a neutron into a Λ, in order to create a neutron hole Λ particle state, utilizes the reaction

$$K^- + n \rightarrow \Lambda + \pi^- . \qquad (1)$$

This has been widely used for nuclear targets resulting in an overall mass resolution of 3 - 5 MeV. Recently there has been a developing interest [1,2] in the pion induced reaction

$$\pi^+ + n \rightarrow \Lambda + K^+ . \qquad (2)$$

Spectra obtained at BNL [3] for these two reactions on a ^{12}C target are compared in Fig. 1. In reaction (2), the momentum is typically 330 MeV/c with a cross section of about 10 μb/sr. This small value can be compensated for by the high pion fluxes available. Recent (π, K) measurements [4] at BNL on targets masses from ^9Be up through ^{89}Y give

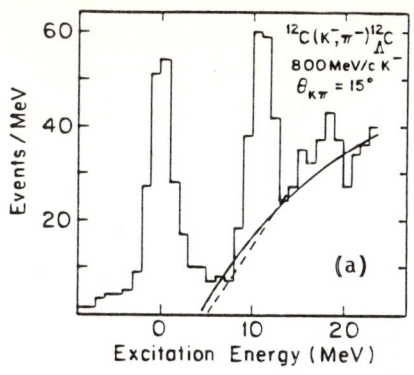
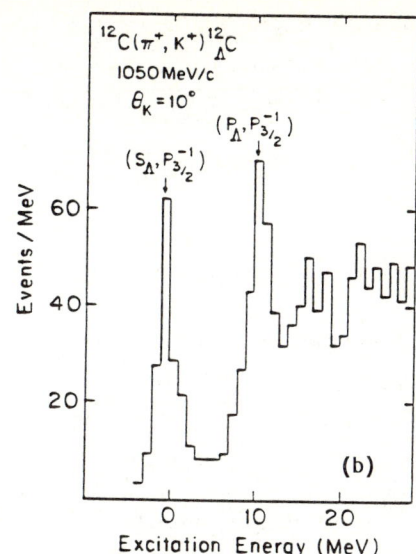

Fig. 1. Comparison of the production of $^{12}_\Lambda C$ with a) the (K,π) reaction and b) the (π, K) reaction

evidence for states involving orbitals from l = 0 up to possibly l= 4. Thus although the production rates are comparable in the two reactions, the (π,K) reaction has the higher momentum transfer as required for formation of higher spin states. Hashimoto has reported developments in progress for a new spectrometer at KEK for (π, K) studies and which may ultimately be used for the production and study of polarized hypernuclei[5].

With the development of CEBAF, the electromagnetic production of hypernuclei has received renewed interest. The cross section for photo production of lamda hyperons is shown in Fig. 2. For Kaons observed at forward angles, the momentum transfer can be 300 MeV/c as shown in Fig. 3. Mecking [6] has proposed a scheme for CEBAF shown in Fig. 4. for electroproduction of tagged hypernuclei. With a 100 nA electron beam and a 10 mg/cm^2 target, the two spectrometer system may be expected to tag about 200 hypernuclei/day for each state with a mass resolution 200 KeV.

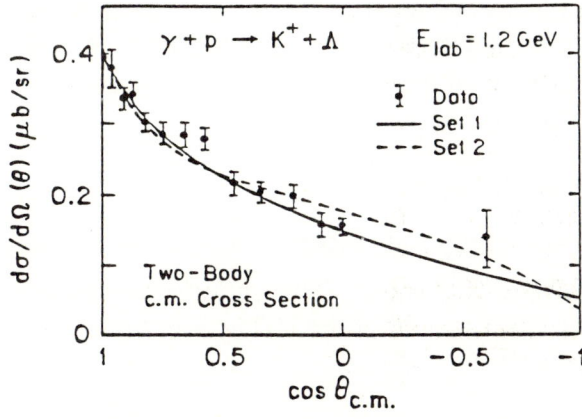

Fig. 2. Differential cross section for Λ photoproduction

293

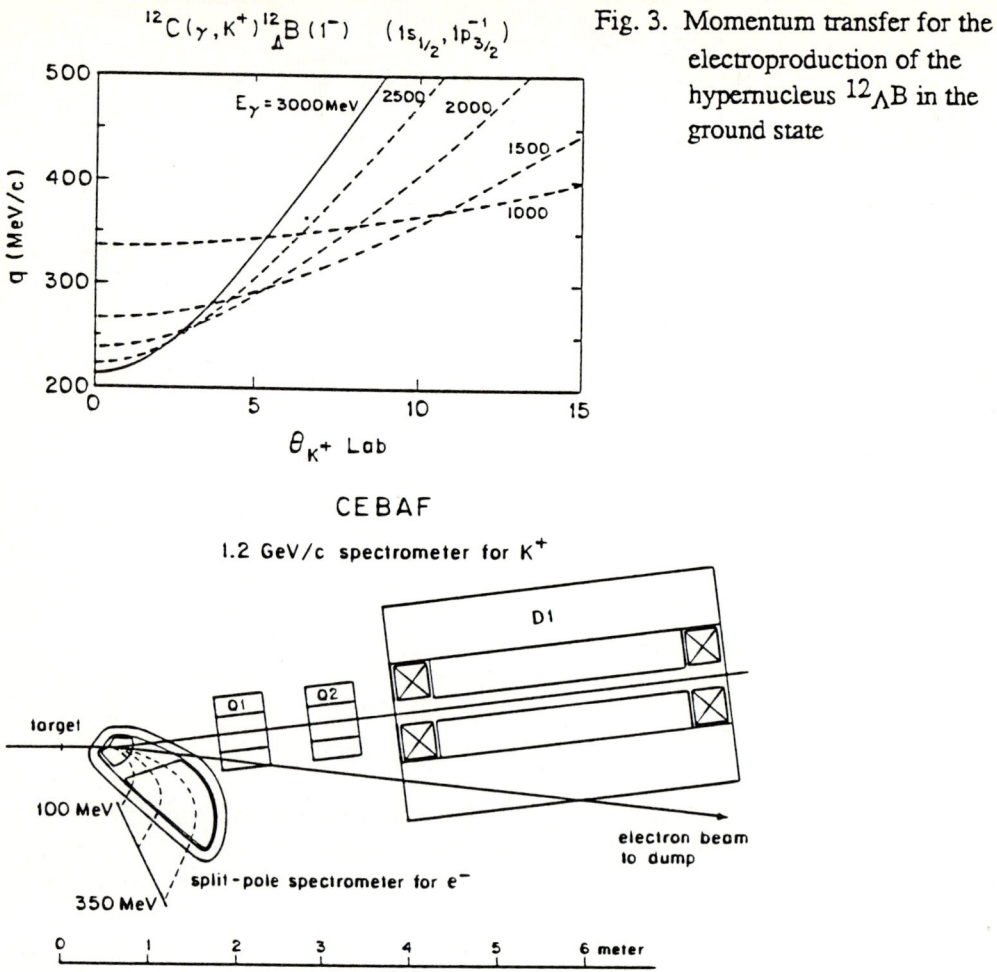

Fig. 3. Momentum transfer for the electroproduction of the hypernucleus $^{12}_\Lambda B$ in the ground state

Fig. 4. Possible magnetic spectrometer configuration for electroproduction of hypernuclei at CEBAF

Thus there are several methods by which the scope and quality of hypernuclear spectroscopy data can improve in the coming years. I will turn now to the ultimate fate of the Λ in the hypernuclear ground state, that is, a strangeness changing decay through the weak interaction.

II. Hyperon - Nucleon Weak Interactions.

The mesonic decay of free hyperons occurs with isospin changes of 1/2 and 3/2 and lifetimes, τ, of 100 to 200 $\times 10^{-12}$ sec. The underlying physics for these nonleptonic weak decays is illustrated at the left in Fig. 5 in which

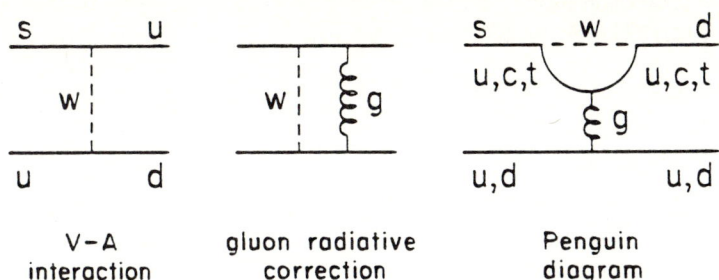

Fig. 5. The lowest order W boson exchange diagram for the weak quark-quark interaction plus strong interaction corrections in the form of soft gluon exchange and penguin diagrams

an s quark is converted to a u quark through the exchange of the charged W boson. The V-A Hamiltonian for this weak interaction is expected to be of the form

$$H_{V-A} = (G_F/\sqrt{2}) \sin\theta_c \cos\theta_c \, Q_{V-A} ,$$

where

$$Q_{V-A} = u\gamma_\mu(1-\gamma_5)s \; d\gamma^\mu(1-\gamma_5)u$$

and θ_c is the Cabbibo angle. This Hamiltonian describes transitions with isospin change, ΔI, of 1/2 and 3/2 with about equal strength. It is well known that both free hyperon decay and kaon decay branching ratios suggest that, in the effective weak Hamiltonian, the $\Delta I = 1/2$ component is enhanced by about a factor of twenty. For example an analysis of hyperon decay [7] shows that the largest $\Delta I = 3/2$ amplitude, $A_{3/2}$, which is consistent with the data, is

$$A_{3/2} \approx \text{Semi-leptonic hyperon decay amplitude} \qquad (3)$$

and

$$A_{1/2} \approx (15 \text{ to } 30) \times A_{3/2} . \qquad (4)$$

This enhancement of the $\Delta I = 1/2$ amplitude is referred to as the "$\Delta I = 1/2$ Rule". It is an empirical rule which is not understood in a fundamental way.

Several authors have added strong interaction corrections to the simple W boson exchange diagram, including both soft gluon radiative corrections and penguin diagrams (see Fig. 5), in order to understand the role of the strong field in these weak interactions. The penguin diagram is emphasized because of its $\Delta I = 1/2$ character. For example, Gilman and Wise [8] developed an effective quark-quark weak Hamiltonian:

$$H_{qq} = (G_F/\sqrt{2}) \sin\theta_c \cos\theta_c [C1\, Q1 + C2\, Q2 + ...], \quad (5)$$

where

$$Q_1 = u\gamma_\mu(1-\gamma_5)s\ d\gamma^\mu(1-\gamma_5)u, \quad (6)$$

$$Q_2 = d\gamma_\mu(1-\gamma_5)s\ u\gamma^\mu(1-\gamma_5)u, \quad (7)$$

and C1 and C2 are Wilson coefficients. Gilman and Wise find that $C1 = +1.51$ and $C2 = -0.856$. This gives an enhancement of ≈ 4, well below the experimental value.

The Λ hyperon in free space, Fig. 6 a, has the two mesonic decay modes

$$\Lambda \rightarrow p^+ + \pi^- \quad (8)$$

and

$$\Lambda \rightarrow n + \pi^0 \quad (9)$$

in a ratio consistent with the $\Delta I = 1/2$ rule. When bound in the nuclear medium it has the opportunity to also decay by the proton and neutron stimulated nonmesonic processes

$$\Lambda + p \rightarrow n + p \quad (10)$$

and

$$\Lambda + n \rightarrow n + n. \quad (11)$$

as shown in Fig. 6 b. It is of particular interest to explore whether the latter decay modes, which have no final state pions, also display the same very strong isospin dependence which characterizes the mesonic decay.

It would be very difficult to explore these weak interaction modes (10 - 11) in a scattering experiment at a hyperon beam. A more realistic approach is to investigate the decay modes of the Λ when bound in the 1S orbit of a light nucleus. Since the system is otherwise stable, the Λ will ultimately decay by a weak interaction into either the mesonic or nonmesonic channels. These partial widths which have not previously been systematically measured, are now the focus of a series of measurements [9, 10] at the BNL-AGS (Experiments E-759 and E-788).

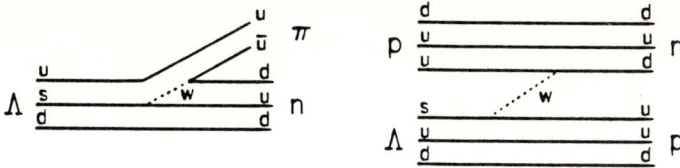

Fig. 6. Quark flow diagrams for mesonic and nonmesonic hyperon weak decay

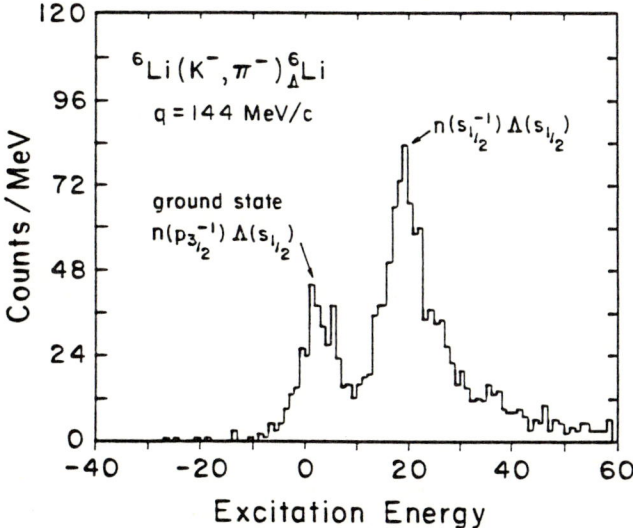

Fig. 7. Hypernuclear excitation spectrum obtained for $^6_\Lambda$Li

The measurements have focused on targets of Lithium and Carbon. Fig. 7 shows an excitation spectrum of $^6_\Lambda$Li in which the ground state is cleanly identified. As it turns out this state is proton unbound by ≈ 1 MeV and quickly suffers a strong decay into $^5_\Lambda$He. A cut on the $^6_\Lambda$Li ground state allows one to define a sample of hypernuclei whose mean lifetime and decay modes can be measured. This technique has now been applied to both Li and Carbon targets and the general trends in the data can now be discussed.

A Λ bound in a nucleus can decay through the same four processes listed above (eqn. 8 - 11) in proportions given by the four partial widths : Γ_{π^-}, Γ_{π^0}, Γ_p, and Γ_n. In mesonic decays the energy release is 38 MeV - ($B_\Lambda - B_N$) where B_Λ and B_N are the binding energy of the Λ and recoil nucleon, respectively. This reduction in phase space combined with the Pauli blocking of the recoil nucleon tends to suppress these mesonic decay modes. In $^5_\Lambda$He it represents about half of the decay rate[10] while in $^{12}_\Lambda$C it is almost fully suppressed [9].

In nonmesonic decay, $\Gamma_{nm} = \Gamma_n + \Gamma_p$, the energy release is large, 176 MeV - $(B_\Lambda - B_N)$, and serves as the experimental signature for a strangeness changing weak decay. These processes (10 - 11) are characterized by the emission of very energetic neutrons and protons which can be detected by Time of Flight counters and range spectrometers as shown in Fig. 8 for experiment E- 788.

Thus a measurement of the proton, neutron, and π^- yields from the hypernucleus event sample, when combined with a direct measurement of the hypernuclear mean life time, τ :

$$\tau = 1 / [\Gamma_{\pi^-} + \Gamma_{\pi 0} + \Gamma_n + \Gamma_p] ,$$

fully determines the four nonleptonic decay widths. The ratio $\Gamma_{\pi^-} / \Gamma_{\pi 0}$, is strongly influenced by the isospin of the hypernuclear states involved[11]. Of special interest here is the total nonmesonic rate, Γ_{nm}, and the ratio: Γ_n / Γ_p.

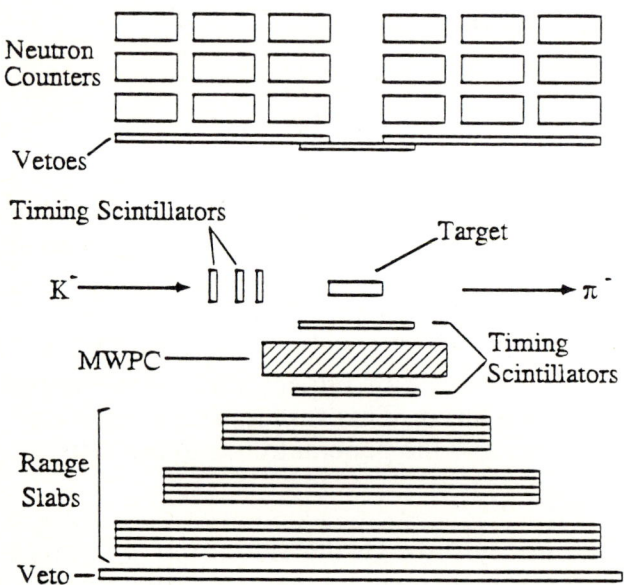

Fig. 8. Target region of Experiment E-788 showing the production and decay time detectors, the charged particle range spectrometer, and the neutron time of flight detector system

Table I gives a comparison for these various rates in units of the free lambda decay rate, Γ_Λ, as measured [9,10] for the ground states of $^{12}_\Lambda C$, and $^5_\Lambda He$. The $^5_\Lambda He$ results are preliminary [10].

Table I Comparison of $^{12}_\Lambda C$ and $^5_\Lambda He$ Results

Target	$\Gamma_{nm}/\Gamma_\Lambda$	Γ_n/Γ_p	$\Gamma_{\pi^-}/\Gamma_{nm}$
$^{12}_\Lambda C$	$1.14 \pm .20$	$1.36\ (+1.09,-0.84)$	0.045 ± 0.04
$^5_\Lambda He$	$0.44\ (+.15,-.31)$	$1.30\ (+.65,-1.60)$	$0.97\ (+2.26,-0.24)$

For $^5_\Lambda He$ the various widths are

$$\Gamma_{total} = 1.03 \pm 0.08\ \Gamma_\Lambda$$
$$\Gamma_{\pi^-} = 0.43 \pm 0.10\ \Gamma_\Lambda$$
$$\Gamma_p = 0.19 \pm 0.07\ \Gamma_\Lambda$$
$$\Gamma_n = 0.25 + 0.11, -0.30\ \Gamma_\Lambda$$

From these one can calculate:

$$\Gamma_{\pi^0} = 0.16 + 0.34, -0.21\ \Gamma_\Lambda.$$

In order to interpret these results it is useful to compare them to some of the many calculations that have appeared in the literature on this subject recently[12]. There are two treatments of the problem to be considered: a) meson exchange descriptions in which there is weak coupling of the meson to the lamda-nucleon vertex in a formalism which enforces the $\Delta I = 1/2$ rule and b) six quark calculations in which two quarks interact through modified versions of the Gilman and Wise Hamiltonian [7] in which both $\Delta I = 1/2$ and $3/2$ transitions are allowed.

The results of Takeuchi, Takaki, and Bando [13] are typical of π and ρ exchange calculations in which the uncertainty comes from knowledge of the weak ρ coupling constant and the relative phase between the two types of meson exchange. Their predictions for $^5_\Lambda He$ are listed in Table II with two choices of the relative phase.

Table II Calculated Weak Decay Widths

Nucleus	$\Gamma_{nm}/\Gamma_\Lambda$	Γ_n/Γ_p
Bando, Takaki and Takeuchi		
$^5_\Lambda$He		
π only	0.144	0.058
ρ only	0.097	0.000
$\pi + \rho$	0.450	0.017
$\pi - \rho$	0.033	2.32
Dubach et al.		
Nuclear Matter	1.23	0.345
$^{12}_\Lambda$C	1.2	
$^5_\Lambda$He	0.5	
Heddle, Kisslinger, and Cheung		
$^{12}_\Lambda$C		
Cabbibo	3.87	
Corrected	2.25	
$\Delta I = 1/2$	1.5	small
Experiment		
$^5_\Lambda$He	0.44	1.30
	(+0.15, -0.31)	(+0.65, -1.60)
$^{12}_\Lambda$C	1.14	1.36
	± 0.20	(+1.09, -0.84)

Although the phase choice: $\pi + \rho$, gives a total nonmesonic rate, $\Gamma_{nm}/\Gamma_\Lambda$ consistent with the experimental value, it predicts that the yield from the neutron stimulated process should be close to zero.

The most complete version of the meson exchange calculations has been generated by Dubach et al. [14] who use a SU(6)$_{weak}$ representation to predict the magnitude and phase of many of the meson exchange weak coupling constants and therefore are able to include the π, η, K pseudoscaler mesons and the ρ, ω, K^* vector mesons in a calculation that enforces the $\Delta I = 1/2$ rule. Their predictions for nuclear matter, $^{12}_\Lambda$C, and $^5_\Lambda$He are also listed in Table II. Note that inclusion of the heavy mesons raised the value of Γ_n / Γ_p.

Heddle, Kisslinger, and Cheung [15] have calculated Γ_{nm} in the Hybrid Quark-Hadron model in which they use the one pion exchange $\Delta I = 1/2$ treatment for baryon baryon separations greater than one fermi and at shorter distances, the Gilman and Wise effective interaction (equation 5) between quarks. They explore several choices of the Wilson coefficients, C1 and C2, and find that as the coefficients are adjusted to enforce the $\Delta I = 1/2$ rule in the interior region, the prediction for Γ_{nm} comes into agreement with experiment (see Table II). In the latter case the dominate contribution comes from the one π exchange and predicts only small contributions from neutron stimulated decay, Γ_n.

Thus the nonmesonic rate has been measured in two nuclei with 20-30 % errors. The theoretical calculations tend to reproduce these results in the context of the $\Delta I = 1/2$ rule. The measured nonmesonic rate increases in passing from Helium to Carbon in proportion to the mass, A. It appears[16] that per ΛN pair:

$$\Gamma_{nm} / (\Lambda N \text{ pair}) = (\Gamma_n + \Gamma_p)/(A-1) \approx 1/10 \; \Gamma_\Lambda.$$

Although the measured value of Γ_n is not as small as predicted, the statistical and systematic errors are large. Because all the theoretical calculations suggest that Γ_n is significantly less than Γ_p it is important to design an experiment that can provide a good measure of Γ_n. This could be an important test of the isospin dependence of the effective weak Hamiltonian.

In conclusion, we are faced with a large enhancement in the effective weak Hamiltonian whose origins are not understood. The experimental evidence comes from:
 a. comparison of $\Delta I = 1/2$ to $3/2$ transitions in kaon and hyperon decay.
 b. Comparison of nonleptonic to semileptonic transitions for both Kaon and hyperon decay.
 c. analysis of nonmesonic decays in hypernuclear weak processes.
Attempts to explain this enhancement of more than an order of magnitude have centered around the addition of gluonic corrections to H_{V-A}, especially the penguin diagram of Fig. 2 because of its $\Delta I=1/2$ character. This approach is constrained by the fact that this penguin diagram will also increase the direct CP violation in $\Delta S = 1$ nonleptonic K^0 decay processes, as characterized by the parameter ε'. Experiment has already shown this parameter to be small[17]:

$$\epsilon' / \epsilon = + 0.0017 + 0.007 \text{ (Stat)} + 0.004 \text{ (syst.)}.$$

It appears that corrections at long range, due to the strong interaction, are needed. These are of course difficult to calculate in the quark sector. Perhaps the problem is more appropriately treated in the meson exchange framework, but with $\Delta I = 3/2$ treated explicitly. Lattice gauge calculations are also in progress [18], by groups both in Europe and the U. S. Alternatively one could turn to "New Physics" as for example the the Super String Models. It has been suggested [19] that, by the introduction of new penguin diagrams using the new quarks and bosons of the Super String Models, one could decouple consideration of the ϵ' limit from the $\Delta I = 1/2$ enhancement.

Finally one can look for new experiments which might give a hint as to the origin of this phenomena. A precision measurement of Γ_n has been mentioned above. In addition there are the possibilities of studying nonmesonic decay in polarized hypernuclei as proposed by Fukuda et al. [5].

III. Strange Dibaryon States

Over the past ten years there have been many suggestions and calculations regarding the possible existence of different types of multiquark states. I would like to focus on six quark states with nonzero strangeness and discuss some recent and planned measurements.

A. Six Quark States with S = -1

There have been a number of calculations of the hyperfine splitting of the S = -1 six quark system, see for example the spectrum generated by the Nijmegen group [20]. Several experimental groups [21,22] have focussed on the lowest p wave states near 2.1 GeV. and have searched for a resonance in the ΛN system. It has been known for twenty years that there is structure in the excitation spectrum at 2130 MeV near the opening of the Σn threshold. The searches have concentrated on ascertaining whether there is additional resonance structure in this mass region.

Both Pigot[21] at CERN and Piekarz[22] at BNL have studied the process

$$K^- + d \rightarrow X + \pi^-$$
$$\downarrow$$
$$\Lambda p$$

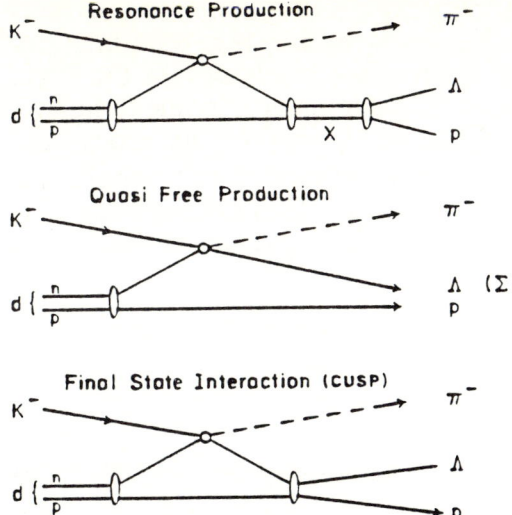

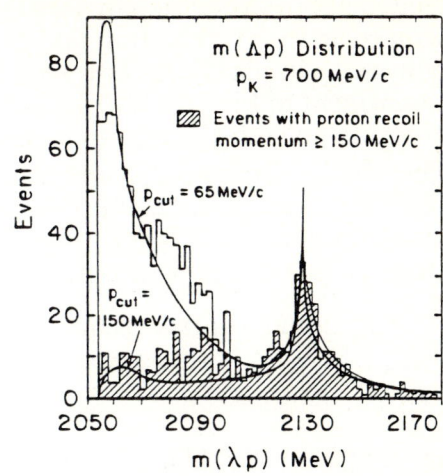

Fig. 9. Various reaction channels which contribute to the $K^- + d \to X + \pi^-$ process

Fig. 10. Cusp structure observed [22] in the 2130 MeV mass region due to the opening of the ΣN threshold

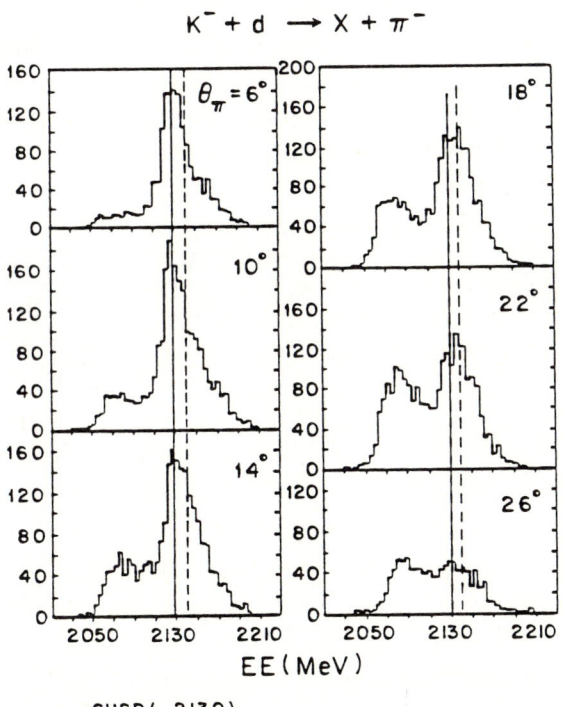

Fig. 11. Variation in shape observed [22] in the cusp structure as the angle of the outgoing pion is changed

One can think of this as resonance production (Fig.9 top) in competition with quasifree production and final state interactions (Fig. 9 middle and bottom). In particular one sees the cusp structure as shown for the BNL experiment in Fig. 10 By placing cuts on the proton recoil momentum, this group is able to discriminate against some of the competing processes. They have studied the shape of the structure as a function of π^- angle and find that the shape changes as shown in Fig. 11. This corresponds to a momentum transfer range up to 400 MeV/c. Further analysis and interpretation of this data is in progress.

Additional studies in this mass region are in progress at Saturne [23] using SPES IV for the associated production reaction

$$p + p \rightarrow K^+ + X$$

at 2.3 GeV . Evidence for structure in the mass region near 2080 MeV has been reported by Frascaria [23].

B. Six Quark State with S = - 2 : The H Particle

Jaffe [24] pointed out in 1977 that the hyperfine splitting in the S = -2 six quark system pushes one state well below the $\Lambda\Lambda$ mass. He estimated that this dihyperon, the H particle , would be stable against strong decay with a binding energy, $B_H = M_{\Lambda\Lambda} - M_H = 80$ MeV. This is the ten year anniversary of those calculations. In the interval there have been many additional estimates including corrections and different models which push the state both up and down. Several of these estimates are shown in the mass spectrum of Fig. 12. Different lattice gauge calculations [25] have given both higher and lower masses. It is of great interest to prove the existence (or nonexistence) of this system because it would demonstrate the formation of a radically different form of matter,would set the mass scale for six quark systems, and thereby would give a measure of the SU(3) flavor breaking effects.

In addition to the mass, M_H, the life time , τ, of the H particle has also been calculated. Depending on the binding energy, the decay can proceed by strong interactions ($B_\Lambda \leq 0$), by $\Delta S = 1$ weak processes for $M_{\Lambda\Lambda} > M_H > M_{\Lambda N}$, and by $\Delta S = 2$ processes for $M_{\Lambda N} > M_H > M_{NN}$. The physics of these decays is presumably very similar to the hyperon-nucleon nonmesonic decays discussed above. Donoghue , Golowich, and Holstein [26] have calculated this lifetime as a function of mass as shown in Fig. 13. They find that for $B_H > 0$, the lifetime is substantially longer than that of the hyperons due to the isospin structure of the H particle wave function, with $\tau = 10^{-8}$ sec to 10 days.

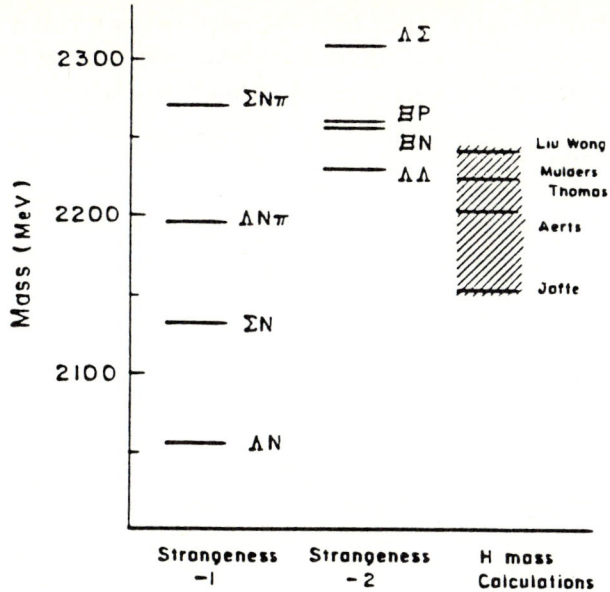

Fig. 12. Comparison of the S = - 2 mass spectrum with selected estimates of the the H particle mass

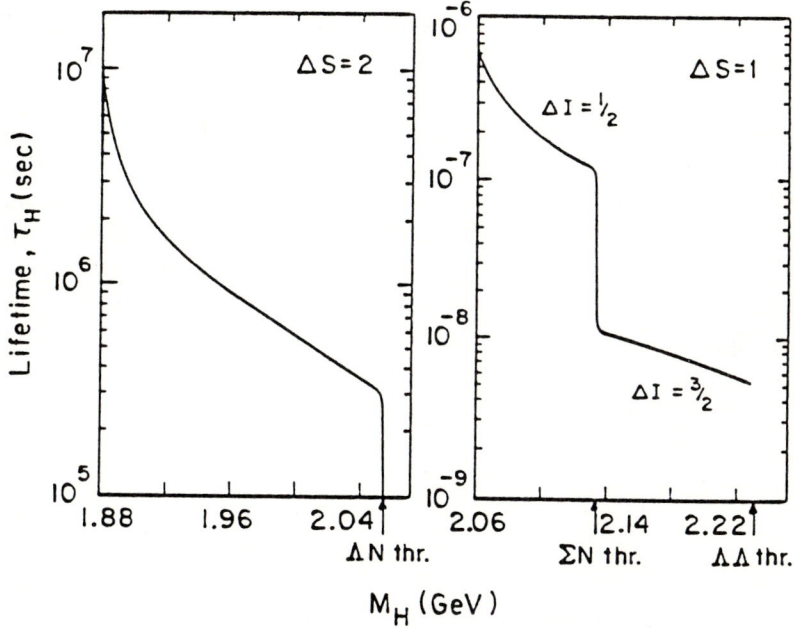

Fig. 13. Calculated lifetime of the H particle as a function of assumed mass as obtained by Donoghue, Golowich, and Holstein [26]

Efforts at BNL [27] to find this object in a counter experiment, have given negative results. One bubble chamber event has been reported and interpreted as evidence for this object [28]. One the other hand a few events indicating the possible existence of double lamda hypernuclei [29] have been interpreted as evidence against it.

A major new program is now developing at BNL to search for this neutral object using the two elementary processes [30]

$$K^- + p \rightarrow K^+ + \Xi^-$$

and

$$\Xi^- + (pn) \rightarrow H^0 + n \ .$$

These two steps can be achieved as a two target process (Hydrogen + Deuterium) as shown in Fig. 14 left and proposed in AGS experiment E-813 or as a two step one target process (Helium - 3) as shown in Fig. 14 right and proposed [31] in AGS experiment E-836.

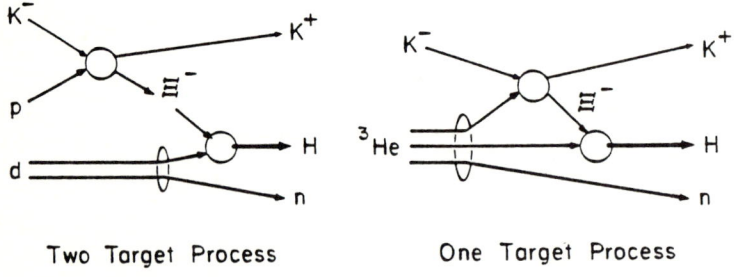

Fig. 14. Comparison of two-step processes for H particle formation using a composite proton/deuterium target (left) and a Helium-3 target (right)

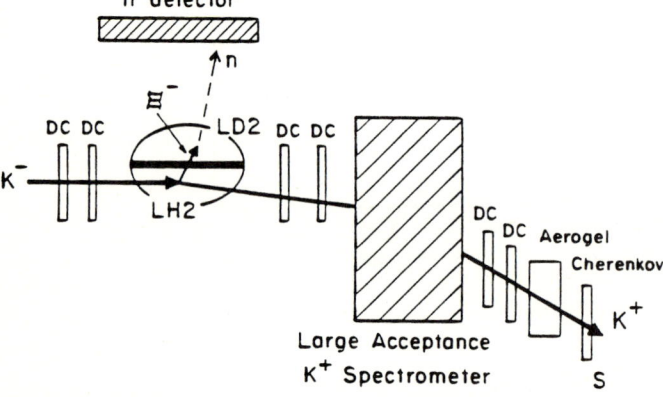

Fig. 15. Schematic layout of the E-813 experiment showing the K^+ tagging spectrometer, the composite target, and the neutron TOF detectors

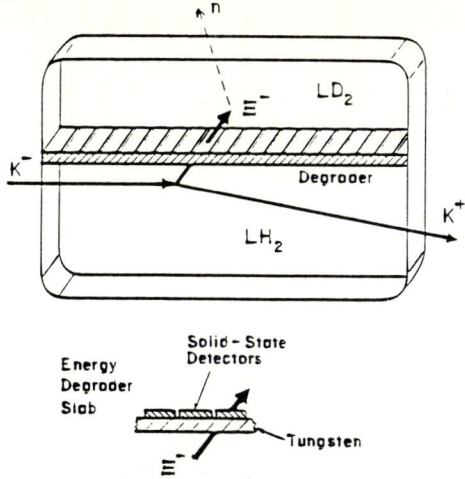

Fig. 16. Schematic drawing of the composite liquid hydrogen / deuterium target showing the tungsten degrader and the segmented solid state detector

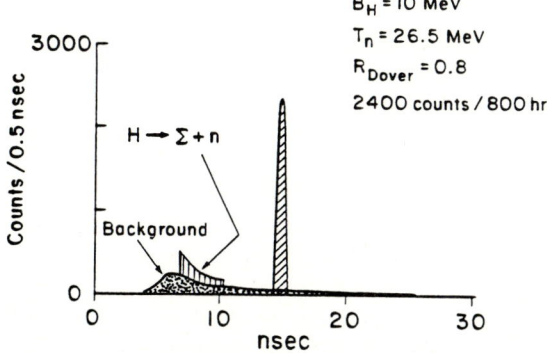

Fig. 17. A simulated neutron time-of-flight spectrum in which the H particle appears as a sharp peak

We discuss the two target technique first as shown in Figs. 16 and 17. This is essentially a tagged Ξ^- production experiment in which the recoiling Ξ^- passes out of the hydrogen, through a tungsten degrader and solid state detector, into the deuterium region where it comes to rest. Here it forms a Ξ^- - deuterium atom which ultimately decays through a strong interaction into the H + neutron channel. The segmented solid state detector behind the tungsten degrader provides a positive check on the position and dE / dx of the low energy recoiling Ξ^-. A simulated Time-of-Flight spectrum for the neutrons is shown in Fig. 17 in which the signature of the H particle is a sharp peak. The major sources of background in the triple coincidence spectrum come from interactions in the degrader and from the $\Lambda\Lambda N$ final states. A peak can also be generated by the decay of the H particle but is smeared out by its recoil momentum.

The second approach, shown in Fig. 14, utilizes a ^{3}He target in the reaction

$$K^- + {}^3He \rightarrow K^+ + n + H .$$

This reaction has the advantage of greater sensitivity at low masses since the momentum transfer can be accommodated more easily by the ^{3}He wave function. If both the outgoing K^+ and the neutron are detected then the missing mass due to the H can be calculated. Alternatively, if the neutron is only a spectator as suggested by the diagram in Fig. 14 then it is sufficient to perform a one-arm spectrometer measurement. The H particle will show up as a peak in the K^+ momentum spectrum with about a 30 MeV/c width.

Aerts and Dover [32] have pointed out that the cross section for this process could increase for lower binding energies. This technique has been exploited in AGS experiment 836. Although it potentially has good sensitivity at low masses, this method suffers from 1) the competition due to high cross section background reactions due to quasi free Ξ^- production, and 2) the possible mis-identification of the kaons as pions in either the incident or out going channel. Since these processes have very large cross sections and can generate structure in the background, this is a serious consideration in an exotic particle search. In order to deal with this we are using a combination of Cerenkov and Time-of-Flight detectors, as well as momentum constraints and vertex reconstruction techniques, to suppress the competing reactions. The beam line and detectors required for both these projects are under design and construction.

---

1   C. Milner et al., Phys. Rev. Lett. 54 (1985) 1237.

2   O. Hashimoto, Proceedings of 1986 INS International Symposium on Hypernuclear Physics, H. Bando, O Hashimoto, and K Ogawa, Editors, Institute for Nuclear Study, Univ. of Tokyo, August, 1986, p. 196; H. Bando and T. Motoba, Prog. of Th. Phy. 76, (1986) 1321.

3   R. E. Chrien et al., Phys. Lett. 89B (1979) 31 and reference 1.

4   P. Pile, et al., Abstract submitted to the BAPS, April 1987; R. E. Chrien, Abstract submitted to the PANIC conference, Kyoto, 1987.

5   H. Ejiri et al. Osaka Univ. Lab. of Nuclear Studies preprint OULNS 86-6, 1986; T. Fukuda et al., Proceedings of 1986 INS International Symposium on Hypernuclear Physics, H. Bando, O Hashimoto, and K Ogawa, Editors, Institute for Nuclear Studies, Univ. of Tokyo, August, 1986, p. 170.

6   B. Mecking, CEBAF report # R-86-013, 1987.

7   E. D. Commins and P. H. Bucksbaum, Weak Interaction of Leptons and Quarks, (Cambridge Univ. Press, 1983), p. 227.

8   F. J. Gilman and M. B. Wise, Phys. Rev., D20 (1979) 2392.

9   R. Grace, Weak Decay of P Shell Lambda Hypernuclei, (PhD Dissertation, Carnegie Mellon University, 1986).

10   J. J. Szymanski, The Weak Decay of Lambda Hypernuclei, (PhD Dissertation, Carnegie Mellon University, 1987); P. D. Barnes and J. J. Szymanski, Proceedings of 1986 INS International Symposium on Hypernuclear Physics, H. Bando, O Hashimoto, and K Ogawa, Editors, Institute for Nuclear Studies, Univ. of Tokyo, August, 1986, p. 136.

11   T. Motoba, et al., preprint, OUAM 87-6-3, 1987, T. Motoba, et al., Proc. 1986 INS Int. Symp. on Hypernuclear Physics (INS, Tokyo, 1986) eds. H. Bando, O. Hashimoto, and K. Ogawa, p. 160.

12   For a recent review of calculations see B.H.J. McKellar, Proceedings of 1986 INS International Symposium on Hypernuclear Physics, H. Bando, O Hashimoto, and K Ogawa, Editors, Institute for Nuclear Studies, Univ. of Tokyo, August, 1986, p. 146.

13   K. Takeuchi, H. Takaki, and H. Bando, Prog. Theor. Phys. 73(1985) 841.

14   J. V. Dubach, Weak and Electromagnetic Interactions in Nuclei, (Springer-Verlag, 1986), p.576, and references therein.

15   D.P. Heddle and L.S. Kisslinger, Phys. Rev., C33 (1986) 608; D.P. Heddle, Lamda Nonmesonic Decay in the Hybrid Quark-Hadron Model, (PhD dissertation, Carnegie Mellon University, 1985); C.Y. Cheung, D.P. Heddle, and L.S. Kisslinger, Phys. Rev., C27(1983) 335.

16   But also see results reported for heavy mass nuclei, M. E. Ray Campagnolle, Proceedings of 1986 INS International Symposium on Hypernuclear Physics, H. Bando, O Hashimoto, and K Ogawa, Editors, Institute for Nuclear Studies, Univ. of Tokyo, August, 1986, p. 207.

17   R. H. Bernstein et al, Phys. Rev. Lett. 54 (1985) 1631.

18   J.F. Donoghue, Univ. of Massachusetts preprint #263. 1986, and Proceeding of the XXII International Conference on High Energy Physics, Berkeley, CA., 1986.

19   E. Ma, Phy. Rev. Lett., 57 (1986) 287.

20   P. J. Mulders, A. T. M. Aerts, J. J. deSwart, Phys. Rev. D21 (1980) 2653.

21   C. Pigot et al., Nucl. Phys. B249 (1985) 172.

22   Brandeis, BNL, MIT, Osaka, Houston, Texas, Vassar, AGS Experiment 820, H. Piekarz, Spokesman; H. Piekarz, Proceedings of 1986 INS International Symposium on Hypernuclear Physics, H. Bando, O Hashimoto, and K Ogawa, Editors, Institute for Nuclear Studies, Univ. of Tokyo, August, 1986, p. 291.

23   R. Frascaria, Saturne Report, Nov. 1986; and R. Frascaria, and R. Siebert, preprint, 1987.

24   R. L. Jaffe, Phys. Rev. Lett. 38 (1977) 195.

25   P.B. MacKenzie and H.B. Thacker, Phys. Rev. Lett. 55 (1985) 2539.

26   E. Golowich, Conf. on Intersections between Particle and Nuclear Physics, AIP Conf. Proc. #150, 1986, p. 952; J. F. Donoghue, E. Golowich and B.R. Holstein Phys. Lett. B (1986) and Univ. Mass. preprint # UNHEP-253, 1986; also see M.I. Krivoruchenko and M.G. Shchepkin, Soviet Journal of Nuclear Physics 36 (1982) 769.

27   A. S. Carroll et al., Phys. Rev. Lett. 41 (1978) 777.

28   B.A. Shanbazian and A.O. Kechechyan, Joint Inst. for Nuc. Res. report. (1984).

29   M. Danysz et al., Nucl. Phys. 49 (1963) 121; D. J. Prowse, Phys. Rev. Lett. 17 (1966).782

30   P. D. Barnes, Proc. of the Second Lampf II Workshop, Los Alamos, 1982, edited by H.A. Thiessen et al., Los Alamos Nat. Lab. report #LA-9752-C Vol. I, p. 315.

31   CMU, BNL, Erlangen-Nurnberg, Freiburg, Kyoto, Kyoto-Sangyo, New Mexico, Pittsburgh, CEN Saclay, Vassar, AGS Experiments 813 and 836, G. Franklin and P. D. Barnes, Spokesmen.

32   A.T.M. Aerts and C. Dover, Phy. Rev. Lett. 49 (1982) 1752.

Hyperon-Hyperon Interaction and the H Particle

J. Carbonell, B. Silvestre-Brac, and C. Gignoux

Institut des Sciences Nucléaires, 53, av. des Martyrs,
F-38026 Grenoble Cedex, France

1. Introduction

Since its appearance about twenty years ago, the quark model has been at the origin of two undeniable progresses. On one hand, to provide a strong interaction theory (QCD) by assuming a local SU(3) gauge symmetry of its color degree of freedom. On the other hand, to put some order into the ever thickening jungle of the review of particle properties. Viewed through this model each new entry in the booklet appears no more as an intruder but as a requirement of the theory (e.g. the Ω^-) or as a natural extension of it (e.g. the J/ψ). However, this theory allows, besides the traditional meson ($q\bar{q}$) and baryon (qqq) structures, the existence of an infinite number of objects: those with a quark content of the form

$q^m \bar{q}^n$ with m-n = multiple of 3

There is no general principle to forbid them and its reality should only depend on its internal dynamics. But, due to the enormous complexity of the QCD calculations behind its perturbative domain, this dynamics can only be treated in an approximate way, either by performing lattice calculations, with their finite size problems and some ambiguities concerning the inclusion of fermions, or through some simplified models assumed to account for the most relevant facts of the whole theory.

In 1977, JAFFE [1] used the bag model to predict a 6 quark bound state with the flavor content (uuddss): the so-called H. It would have the quantum numbers $J^\pi = 0^+$, I = 0, Y = 0 (S = -2), flavor singlet and 2150 Mev of mass, 80 MeV below the $\Lambda\Lambda$ threshold, the lowest accessible threshold to the strong interaction decay. Since, a great number of results has been published on this subject. Interesting by itself as a possible new particle, it is also a test for the different models which give on this point scandalously varying results.

On the bag models side, the situation is confused. Some results confirm the JAFFE's prediction [2] whereas some others which take into account the center of mass motion [3] or a better parametrisation of the model [4] give contradictory results. Concerning the SU(3) Skyrme model, all results agree in predicting a bound state, but its binding energy varies from 20 MeV to 1300 MeV [5].

Concerning the lattice calculations, the only published result up to now to our knowledge predicts an unbounded H with a mass in the range (300, 1400) MeV above the

$\Lambda\Lambda$ threshold [6]. Concerning the non relativistic potential models (NRPM) there is also only one published result from OKA et al. [7]. They found an unbounded state which can appear as a sharp resonance in the $\Lambda\Lambda$ scattering, 26.3 MeV above the threshold.

Experimentally, the H has not been seen. The negative result of the CARROLL et al. research [8] leads to the conclusion that in the energy range (2.0,2.48 GeV) the production cross section of the H in the reaction

$$p + p \longrightarrow H + K^+ + K^+$$

is less than 30 nb.

The few detected events on the $_{\Lambda\Lambda}^{6}He$ nuclei [9] have been considered as an indirect evidence of an H binding energy less than 10 MeV. There exist several experimental projects which should give us more information on this question in the near future. A review is given in the Barnes contribution to these lectures.

In the following we present our conclusion on this subject together with some results concerning the nucleon-hyperon elastic scattering in the frame of NRPM.

2. The model

In this model, quarks are considered as non-relativistic particles interacting via the two-body potential

$$v_{ij}(r) = -\frac{3}{16}\vec{\lambda}_i\vec{\lambda}_j\left\{-\frac{k}{r}+\frac{r}{a^2}+\frac{\hbar^2 k}{m_i m_j c^2}\frac{e^{-r/r_0}}{r_0^2 r}\vec{\sigma}_i\cdot\vec{\sigma}_j\right\}, \tag{1}$$

where $\vec{\lambda}$ = SU(3) Gell-Mann matrices
$\vec{\sigma}$ = SU(2) Pauli matrices

The different parameters have been fitted by BHADURI [10] and give a good overall spectroscopy of mesons and baryons.

$$\begin{aligned} m_u &= m_d = 337 \text{ MeV} \\ m_s &= 600 \text{ MeV} \\ k &= 102.67 \text{ MeV fm} \\ a &= 0.0326 \text{ (MeV}^{-1}\text{ fm})^{1/2} \\ r_0 &= (1/2.2)\text{ fm} \end{aligned} \tag{1'}$$

Each quark is characterised by a color, flavor, spin and space quantum number and we look for stationary states of the intrinsic hamiltonian

$$H = \sum_{i=1,6}\frac{p_i^2}{2m_i} + \sum_{i\neq j} v_{ij} - T_{cm} \tag{2}$$

carrying the quantum numbers of the H.

The effect of the T_{cm} term is only to factorize the 6 q wave function in an intrinsic part and a plane wave associated with the center of mass free motion. We can thus forget it and work only with the intrinsic wave function.

In order to solve this 6-body problem we first re-write the hamiltonian (2) in the following way :

$$H = H_1 + H_2 + K + V_{12}, \tag{3}$$

where

H_1, H_2 are the intrinsic 3 q cluster hamiltonian
K is the intercluster kinetic energy
V_{12} is the intercluster potential energy.

In order to solve the Schrödinger equation we use the Resonating Group Method techniques [11] by choosing a trial wave function of the form

$$|\psi\rangle = A \sum_{\alpha\beta} |B_\alpha(123)\rangle |B_\beta(456)\rangle |\chi(123/456)\rangle \; , \qquad (4)$$

where

A is the 6 q antisymmetrizer

$|B_\alpha\rangle$ is an eigenstate of a 3 q cluster hamiltonian H_i, i = 1,2

α meaning the set of quantum numbers carried by a baryon
(i.e. color (singlet), flavor (octet), spin (1/2), space)
so we have $H_i |B_\alpha\rangle = \mathcal{E}_{i\alpha} |B_\alpha\rangle$

$|\chi(123/456)\rangle$ is the variational unknown.

The baryon states are expanded in an harmonic oscillator (HO) basis

$$|B_\alpha\rangle = \sum_{c,f,s,n_x,\ell_x,n_y,\ell_y} d_\alpha \, |Color,c\rangle |Flavor,f\rangle |spin,s\rangle [\varphi_{n_x,\ell_x}(\vec{x}) \varphi_{n_y,\ell_y}(\vec{y})] \qquad (5)$$

up to a number of quanta N = $(2n_x + 1_x) + (2n_y + 1_y)$. See figure below.

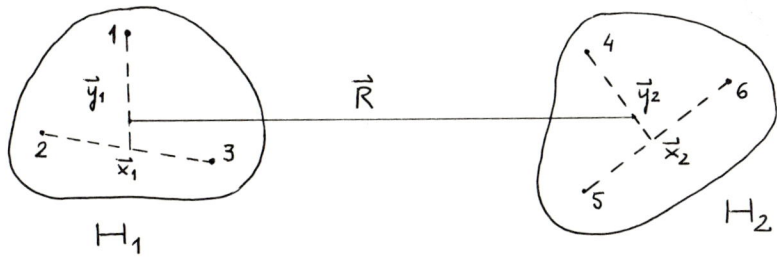

In our calculation we take N = 0,2 leading to the different possibilities

N = 0 $|B_\alpha\rangle = |B_0\rangle$ gaussian in $\vec{x}$ and $\vec{y}$ coordinates

N = 2 $|B_\alpha\rangle = |B\rangle, |B^*\rangle, |B^{**}\rangle$ "dressed" ground and 2 first radially excited states

The trial wave function (4), when inserted in the Schrödinger equation and suitably projected, leads to a Hill-Wheeler equation. Its non-local kernels have also been expanded on an HO basis up to a number of quanta 2n. After some calculations we are left with a Lippmann-Schwinger like equation providing the transition amplitudes T for a given intercluster kinetic energy E

$$[\mathbb{1} - \overline{V}(E) G_0(E)] T = T^\circ \qquad (6)$$

where

$\overline{V}$ is an effective potential built with the exchange kernels

G_0 is the usual free propagator

A more detailed explanation of this technique is given in [12]. Equation (6) allows us to solve on the same footing both the scattering and the bound state problem. In the scattering case the phase shifts are obtained from the T matrix by standard methods. In the bound state problem the inhomogeneous term T^0 is absent and the energy is obtained by vanishing the determinant of $[1-\overline{V G_0}]$. We emphasize that this procedure allows us to treat correctly the center of mass motion.

3. Results

Let us first begin by showing some results on nucleon-hyperon (S = -1) elastic scattering in the SU(3) flavor symmetry limit. In this limit, the low lying baryon octet $8 \equiv (N, \Lambda, \Sigma, \Xi)$ is degenerate in mass. We translate this degeneracy in our calculations by putting equal quark masses, m_q, in the hamiltonian (1),

$$m_u = m_d = m_s = m_q = \frac{1}{3} \frac{2(m_N + m_\Xi) + m_\Lambda + 3 m_\Sigma}{8} = 384 \text{ MeV}.$$

The decomposition in irreducible representations (IRs) of the two-octet product is written as follows:

$$8 \times 8 = (1 + 8_s + 27) + (8_a + 10 + \overline{10}),$$

where the first 3 IR are symmetric under the baryon exchange and the 3 last IR are antisymmetric. We remark however that one cannot always associate one SU(3) IR to each scattering channel. Some of them belong to only one IR (e.g. NN(I=1) and $\Sigma N(I = 3/2)$ are purely 27) whereas some others are a mixing (e.g. Λ which is a $1, 8_s$ and 27 mixing).

The phase shifts obtained for the different S = -1 channels are plotted in Figure 1. These results are qualitatively the same and in a sense almost trivial. They all show the same typical feature of a repulsive hard core potential which is essentially due to the Pauli repulsion. The corresponding hard core radii are given below.

ΛN		$r_c(^1S_0) = 0.51$ fm;	$r_c(^3S_1) = 0.38$ fm
ΣN	J = 3/2	$r_c(^1S_0) = 0.49$ fm;	$r_c(^3S_1) = 0.59$ fm
ΣN	I = 1/2	$r_c(^1S_0) = 0.60$ fm;	$r_c(^3S_1) = 0.38$ fm

They provide a microscopic justification of the phenomenological values used in the meson exchange calculations [13].

Still in the SU(3) flavor symmetry limit, the H particle belongs to the singlet. Its flavor wave function is

$$H = \frac{1}{\sqrt{8}} \{ p\Xi^- + \Xi^- p - n\Xi^0 - \Xi^0 n - \Sigma^+\Sigma^- - \Sigma^-\Sigma^+ + \Sigma^0\Sigma^0 + \Lambda\Lambda \}$$

When we put this special combination in (4) we obtain the results showed in Fig. 2. There, it appears that $\delta(0) - \delta(\infty) = \pi$, which according to the

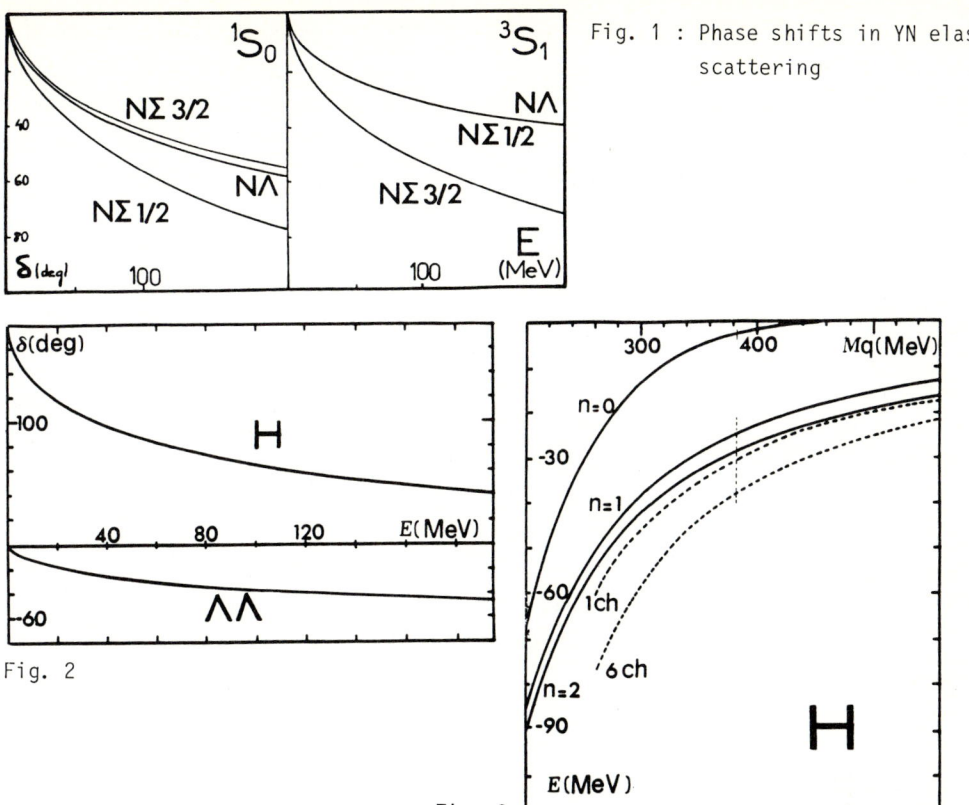

Fig. 1 : Phase shifts in YN elastic scattering

Fig. 2

Fig. 3

Levinson theorem is a signature of a bound state. In the same figure we have plotted the $\Lambda\Lambda$ phase shifts displaying once again the same typical repulsive hard core behaviour with r_c = 0.44 fm. All the calculations showed in Fig. 1 and 2 were done with N = 0 (i.e. with gaussian baryon wave functions) and n = 4 in the HO kernel expansion.

In Fig. 3 we have plotted the total energy of the H particle relative to the $\Lambda\Lambda$ threshold as a function of the quark mass m_q. The solid curves were all obtained with N = 0. The different results correspond to increasing values of n. The curves with n = 2, 3, 4 are superimposed, showing that the convergence of the kernel expansion in the HO basis is already reached for n = 2. Keeping this value of n we have "dressed" the baryon wave function up to N = 2. The dotted lines show the results thus obtained by including only the BB channel or by the inclusion of the six channels BB, BB*, BB**, B*B*, B*B**, B**B**. This treatment provides a gain of about 10 MeV in the H binding energy.

At this stage, i.e. in the SU(3) flavor symmetry limit, the H appears as a 6 q bound state (37 MeV of binding energy for the m_q = 384 MeV value).

However, SU(3) is not an exact symmetry and the octet degeneracy works only within 20%. It is then an essential point to break this symmetry in any realis-

tic calculation. In our formalism the effect of this breaking has to be considered at two stages :
1) in the 6 q hamiltonian (2) by taking into account the different quark masses. Following Bhaduri we take the values given in the preceding section.
2) in the flavor space, the H being no longer a singlet but a configuration mixing.

The last point was studied by ROSNER [14] in the case of a purely chromomagnetic interaction. He showed that the effect of configuration mixing is negligible (about 1/1000) in the mass range we have considered. The first point is treated only partially in our calculations. The masses in the kinetic terms of the hamiltonian are left identical and equal to a reference mass m_q. The only change we have performed is in the chromomagnetic term V_{cm}. We keep in it the reference mass and we introduce the flavor-dependent parameters $x_{ij} = \frac{m_q^2}{m_i m_j}$ in such a way that

$$V_{cm} = x_{ij} V_{cm}^{SU(3)}$$

The results on the H total energy thus obtained are showed in Fig. 4 as a function of m_q. All these results were obtained with n = 2 and the different curves correspond to:
(1) one channel calculation with gaussian baryon $|B_0>$
(2) one   "   "   " "dressed" " $|B>$
(3) six   "   "   " including "dressed" ground state baryons and two first radially excited states.

We remark that the m_q dependence of the energy is much more weak than in the SU(3) limit and that the effect of including several channels is still of providing additional binding. Nevertheless the consequence of breaking this symmetry is dramatic. The H mass is now above the $\Lambda\Lambda$ threshold and consequently no longer a 6 q bound state. For the value of m_q = 384 MeV the H mass is 68 MeV above the threshold and this result is stable enough in the range $300 < m_q < 500$ MeV to confirm our conclusion.

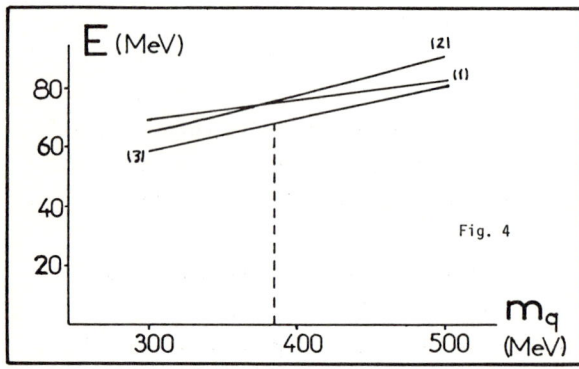

Fig. 4

References
1. R.L. Jaffe, Phys. Rev. Lett. 38 (1977) 195
2. P.J. Mulders, A.T. Aerts, J.J. de Swart, Phys. Rev. D 21 (1980) 2653
   P.J. Mulders, A.W. Thomas, J. Phys. G 9 (1983) 1159
3. K.F. Liu, C.W. Wong, Phys. Lett. 113B (1982) 1
4. A.T.M. Aerts, J. Rafelski, Phys. Lett. 148 B (1984) 337
5. A.P. Balachandran et al, Phys. Rev. Lett. 52 (1984) 887
   S.A. Yost, Ch. R. Nappi, Phys. Rev. D 32,3 (1985) 816
   R.L. Jaffe, C.L. Korpa, Nucl. Phys. B 258 (1987) 468
6. P.B. Mackenzie, H.B. Thacker, Phys. Rev. Lett. 55 (1985) 2539
7. M. Oka, K. Shimizu, K. Yazaki, Phys. Lett. 130 B (1983) 365
8. A.S. Carroll et al. Phys. Rev. Lett. 41 (1978) 777
9. M. Danysz et al, Nucl. Phys. 49 (1963) 121
   D.J. Prowse, Phys. Rev. Lett. 17 (1966) 782
   J.B. Franklin, Nucl. Phys. A 450 (1986) 117c and references therein
10. R.K. Bhaduri, L.E. Cohler, Y. Nogami, Nuovo Cimento 65A (1981) 376
11. Y C. Tang, M. Lemere, D.R. Thompson, Phys. Rep. 47 (1978) 167
12. B. Silvestre-Brac, J. de Physique 46 (1985) 1987
    C. Gignoux, B. Silvestre-Brac, rapport interne ISN 86.42
    B. Silvestre-Brac, C. Gignoux, Y. Ayant, submitted to J. of Physics A
13. M.M. Nagels, T.A. Rijken, J.J. de Swart, Ann. of Phys. 79 (1973) 338; Phys. Rev. D15 (1977) 2547; Phys. Rev. D20 (1979) 1633
14. J. Rosner, Phys. Rev. D 33 (1986) 2043

Part V

Relativistic Heavy Ions

A Review of Quark-Gluon Plasma and High Energy Heavy Ion Collisions

L. McLerran

Fermi National Laboratory, P.O. Box 500, Batavia, IL 60510, USA

The quark-gluon plasma, and how it might be produced in ultra-relativistic nuclear collisions is reviewed. I briefly introduce the quark-gluon plasma, and what we might learn from studying it. I discuss various possible experimental probes of the quark-gluon plasma as it might be seen in nuclear collisions. I discuss hydrodynamical probes, strangeness and charm production as well as electromagnetic probes of the plasma.

1 Introduction

High energy nuclear interactions may allow for tests of novel features of QCD. These features reflect non-perturbative phenomenon such as confinement and chiral symmetry breaking. In this lecture, I shall give an overview of current theoretical understanding of these non-perturbative phenomena. I will then turn to a description of various proposed experimental probes of these phenomena. Due to lack of space, the large number of papers which have contributed to this field cannot all be cited. I therefore refer to reviews or conference proceedings for recent discussions of the work discussed below.

To study matter at densities of the order of and larger than those typical of QCD, we must study either the collisions of ultra-relativistic nuclei, or very high multiplicity fluctuations in hadron-hadron collisions. We shall see that simple arguments suggest that densities far in excess of those typical of ordinary nuclei may be achieved under such extreme conditions. I will later briefly discuss a few suggested experimental probes of high density matter as it might be produced in such collisions.

In this section I shall discuss the properties of hadronic matter at high energy density. The word high implies a scale for the measurement of the energy density. Such a scale may be provided by a variety of estimates, all of which agree on the order of magnitude of a typical density scale for hadronic matter. The first is the energy density of nuclear matter. With m the proton mass, R_A the nuclear radius, and A the nuclear baryon number, the density of nuclear matter is

$$\rho_A \sim \frac{Am}{\frac{4}{3}\pi R_A^3} \sim 0.14 \ GeV/Fm^3. \tag{1}$$

We can also use (1) to estimate the energy density inside a proton. If we use a proton radius of $0.8 Fm$, (1) gives

$$\rho_p \sim 0.5 \; GeV/Fm^3 \; . \tag{2}$$

There is a good deal of uncertainty in this estimate of ρ_p. We might have instead used the MIT bag radius, or a proton hard core radius, corresponding to an order of magnitude uncertainty in (2). Finally, another estimate comes from dimensional grounds using the value of the QCD Λ parameter, suitably defined as Λ_{ms} or Λ_{mom}, as the dimensional scale factor. Using the Λ parameter, we find

$$\rho_{QCD} \sim \Lambda^4 \sim 0.2 \; GeV/Fm^3 \; . \tag{3}$$

Again there is an order of magnitude uncertainty both due to the lack of precise experimental knowledge of Λ, and differences induced by using alternative sensible definitions of Λ.

In all of the above energy density estimates, the typical scale was in the range of several hundreds of MeV/Fm^3 to several GeV/Fm^3. At energy densities low compared to this scale, we presumably have a low density gas of the ordinary constituents of hadronic matter, that is, mesons and nucleons. At densities very high compared to this scale, we expect an asymptotically free gas of quarks and gluons. At intermediate energy densities, we expect that the properties of matter will interpolate between these dramatically different phases of matter. There may or may not be true phase changes at some intermediate densities.

To understand how such a transition might come about, consider the example of QCD in the limit of a large number of colors, N_C. Recall that extensive quantities such as the energy density, ϵ, or entropy density, σ, measure the number of degrees of freedom of a system. The dimensionless quantities ϵ/T^4 or σ/T^3 should be of the order of the number of degrees of freedom. For hadronic matter, the number of degrees of freedom relevant at low density is the number of low mass hadrons. Since matter is confined at low density, the number of such degrees of freedom is $N_{dof} \sim 1$ in terms of the number of colors. At high energy density, the relevant number of degrees of freedom are those of unconfined quarks and gluons. The gluons dominate and give $N_{dof} \sim N_C^2$. Therefore in the large N limit, the number of degrees of freedom changes by an infinite amount.

Assuming that the transition occurs at finite temperature in the large N_C limit, as is verified by Monte Carlo simulation, this result can be interpreted in two ways. From the vantage point of a high density world of gluons, the asymptotic energy density is finite, but at low energy density at some finite temperature the energy density goes to zero. The energy density itself is therefore an order parameter for a phase transition, and there is a limiting lowest temperature. Viewed from the low density hadronic world, there is some limiting temperature where the energy density and entropy density become infinite. Here there is a Hagedorn limiting temperature.

For $N_C = 3$, the above statements are only approximate. The number of degrees of freedom of low mass mesons is

$$N_{dof} \sim N_F^2 \sim 4 , \tag{4}$$

where we have taken the number of low mass quarks to be $N_F \sim 2$ for the up and down quarks. The number of degrees of freedom of a quark-gluon plasma is on the other hand

$$N_{dof} \sim 40 . \tag{5}$$

The number of degrees of freedom might change in a narrow temperature range, or there might be a true phase transition where the degrees of freedom change by an order of magnitude, if our speculations concerning the large N_C limit are applicable.

Results of a Monte Carlo simulation of the energy density are shown in Fig. 1.[1] These results are typical of the qualitative results arising from lattice Monte Carlo simulation. The precise values of the energy density are difficult to estimate as is the scale for the temperature. The figure does make clear the essential point, on which all Monte Carlo simulations agree, that the number of degrees of freedom of hadronic matter changes by an order of magnitude in a narrowly defined range of temperature. There is apparently a first order phase transition for SU(3) Yang-Mills theory in the absence of fermions, and a rapid transition which may or may not be a first order transition for SU(3) Yang-Mills theory with two or three flavors of massless quarks.

For Yang-Mills theory in the absence of dynamical quarks, there is a local order parameter which probes the confinement or deconfinement of a system. This order parameter measures the exponential of the free energy difference between the thermal system with and without the presence of a single static test quark inserted as a probe,

$$<L> = e^{-\beta F_q} . \tag{6}$$

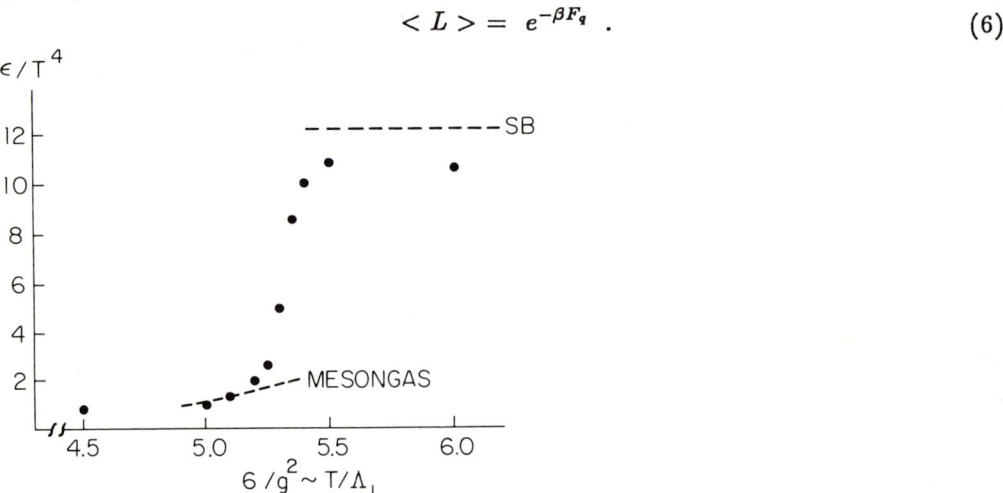

Fig. 1. Energy density scaled by T^4 as a function of T

As originally proposed by Polyakov and Susskind, and developed in Monte Carlo studies, the Polyakov loop is a Wilson loop at the position of the quark which evolves only in time and is closed by virtue of the thermal boundary conditions which make the system have a finite extent in Euclidian time. The two phases of the theory are the confined and unconfined phases where

$$e^{-\beta F_q} \sim \text{ finite if confined, or 0 if deconfined .} \qquad (7)$$

This quantity is an order parameter for a confinement-deconfinement in theories without fermions or in the large N_C limit in theories with fermions (in the fundamental representation of the gauge group). If there are fermions in the fundamental representation, in the 'confined phase' dynamical fermions may form a bound state with a heavy test quark, so the free energy is finite in what would be the confined phase. Since it is already finite in the deconfined phase, the free energy of a static test quark does not provide an order parameter.

Although $<L>$ is not an order parameter, Monte Carlo simulations with dynamical fermions show that $<L>$ changes very rapidly in a narrow range of temperatures. For SU(3) lattice gauge theory without dynamical quarks, when $<L>$ is a true order parameter, there is a noticeable discontinuous change. It is not entirely clear whether there is a discontinuous change corresponding to a true phase change for the theory with fermions.

In the limit of large dynamical quark mass the quarks are no longer important at any finite temperature and decouple. In this limit the confinement-deconfinement phase transitions is a well defined concept with an order parameter which measures a phase change. At zero quark masses there is another phase transition which may be carefully defined, that is, the chiral symmetry restoration phase transition. Chiral symmetry is a continuous global symmetry of the QCD lagrangian in the limit of zero quark mass. Its realization would require that all non-zero mass baryons have partners of degenerate mass and opposite parity. Since this is far from true for the spectrum of baryons observed in nature, chiral symmetry must be broken. Breaking the continuous global symmetry generates a massless Goldstone boson, which we identify with the light mass pion. As a consequence of the breaking of chiral symmetry, the quarks acquire dynamical masses, which may be seen by computing $<\overline{\Psi}\Psi>$. For the chiral symmetric phase, $<\overline{\Psi}\Psi> = 0$, and is non-zero in the broken phase.

For not unreasonable values of the quark masses, there appears to be a rapid change in $<\overline{\Psi}\Psi>$ at about the same place where the order parameter $<L>$ changes rapidly. We conclude therefore that chiral symmetry is approximately restored at the same temperature where quarks stop being approximately confined. The word approximately is important here since absolute confinement or absolute chiral symmetry is impossible for finite mass dynamical quarks.

We can now conjecture on the phase diagram in the temperature mass plane. It is important to realize that we may physically vary the temperature, but not the masses of quarks. Theoretically in a Monte Carlo simulation, these masses may be changed, but they cannot be changed in nature. It is also important to realize that the mass-temperature diagram represents an over simplification to the case of equal mass quarks. With different mass quarks, the diagram has more variables and is more complicated.

To plot this diagram, we first discuss the limiting case $m = \infty$. Here there should be a first order confinement-deconfinement phase transition along the T axis. Since a discontinuous change will not be removed by a large but finite quark mass, this first order phase change must be a line of transitions in the $m - T$. Along the $m = 0$ axis there is a chiral symmetry restoration transition. By the arguments of Pisarski and Wilczek, this transition is first order, and therefore must generate a line of transitions which extends into the $m - T$ plane.

Of course, we do not know what happens with these two lines of transitions, whether they join or never meet, or pass through one another etc. There may be no true phase transition at the values of masses which are physically relevant, or there may be one or two which are the continuation of the chiral transition from zero mass and the confinement-deconfinement transition from infinite mass. The weight of the evidence from Monte Carlo numerical simulation suggests a very large transition in the properties of matter in a very narrow temperature range, and not much more than that can be said at present. There are a variety of conflicting claims as to whether or not there is a true first order transition at physically relevant masses.[1]

There have been serious attempts to obtain reliable quantitative measures of the properties of matter from Monte Carlo simulation. The only truly reliable numbers have been extracted for the unphysical case of $N_F = 0$, that is, no dynamical fermions. It has been shown that the critical temperature of the confinement-deconfinement transition is

$$T_C = 220 \pm 50 \; MeV \qquad (8)$$

by fitting the potential computed in these theories and comparing it with the potential which fits charmonium. This corresponds to an energy density of $1-2 \; GeV/Fm^3$ required to make a quark-gluon plasma. These results now appear to be valid for the continuum limit, and seem to be fairly good.

The numerical situation for QCD with $N_F = 2-3$ is not nearly so good. The qualitative results have been summarized above, but it is premature to draw any firm conclusions about numbers.

2 How to Make a Plasma

The collisions of ultra-relativistic nuclei and fluctuations in $\bar{p}p$ collisions provide the possibility of producing a quark-gluon plasma in a controlled experimental environment. Such a collision is shown in Fig. 2 where two nuclei of transverse radius R collide in the center of mass frame. The longitudinal size of the nuclei is Lorentz contracted.

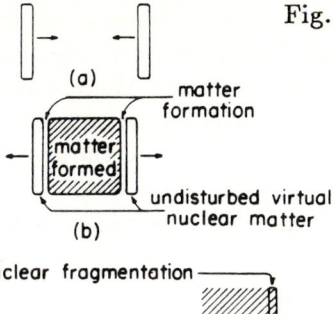
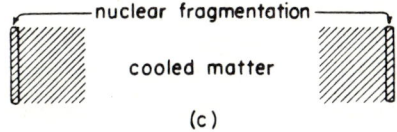

Fig. 2 AA collision in the center of mass frame

There is a scale implicit in the Lorentz contraction. Once the nuclei have a large enough Lorentz gamma factor so that they would be contracted to a size less than some typical hadronic length scale, possibly a fermi, the Lorentz contraction of virtual quanta with energy corresponding to this length scale stops. Below the beam energy appropriate for this gamma factor, the nuclei Lorentz contract. This energy is

$$E_{CM}^0 = m\gamma = \frac{mR}{l_0} = 7 - 70 \; GeV \qquad (9)$$

for uranium nuclei and the hadronic distance scale $l_0 \sim .1 - 1 \; Fm$. Here and in the rest of this paper, we shall quote the center of mass energy in GeV per nucleon in each nucleus.

We expect qualitative differences in the scattering above E_{CM}^0. Another equivalent estimate of E_{CM}^0 is given by estimating the energy at which the fragmentation regions of the two nuclei separate. At energies greater than E_{CM}^0 there will be a central region between the two colliding nuclei, which will have small net baryon number density. This separation is shown in the dual-parton model computations of O-Pb collisions[2], shown in Fig. 3.

An important fact to remember about the matter formed in the collision of two ultra-relativistic nuclei is that it is born expanding in the longitudinal direction. This is because particles are formed with a more or less uniform density in rapidity. Since

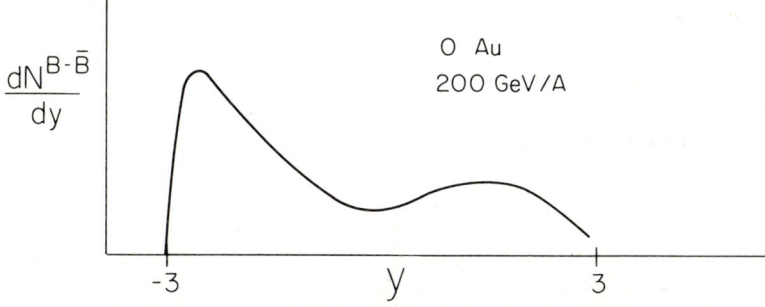

Fig. 3 A dual parton model computation of the rapidity distribution of baryons minus anti-baryons in O-Pb collisions.

these particles follow a trajectory which has its origin approximately at $x = t = 0$, and there is a large dispersion in particle velocities, there will be a large longitudinal velocity gradient built into the initial matter distribution. There should be no transverse expansion in the initial condition since we expect a random orientation in the transverse momentum of produced particles. It can be shown that if the distribution of produced particles is uniform in rapidity, the expansion is initially a 1+1 dimensional similarity expansion, and the density of particles decreases like $1/t$.

The initial energy density may be estimated on dimensional grounds. The initial energy density should be proportional to the initial rapidity density per unit transverse area. The energy per particle should be of the order of the typical transverse momentum per particle. The longitudinal distance scale and p_T are correlated at early time by the uncertainty principle, since initially the matter appears in a quantum mechanical state, $p_T \sim 1/l_o$. We therefore have

$$\epsilon_i \sim \frac{dN}{dy} \frac{1}{\pi R^2} p_T^2 |_{t=t_i} \quad . \tag{10}$$

The initial time t_i will be chosen as the earliest time we believe that the matter may be described as approximately expanding as a perfect fluid.

If the matter expands approximately as a perfect fluid, then ϵ_i may be bounded by parameters which are experimentally measured at late times after the matter decouples, that is, after the pions present in the late state of evolution of the matter have stopped scattering from one another, and are experimentally observed. We first use that the rapidity density in perfect fluid hydrodynamic expansion is proportional to the entropy and because entropy is conserved, one can prove that dN/dy is also conserved, at least in the central region. Since the system cools as it expands, p_T is a monotonically decreasing function of time. (Some of the transverse momentum is recovered by transverse flow, but p_T nevertheless monotonically decreases.) We find therefore that

$$\epsilon_i > p_t^2 \frac{1}{\pi R^2} \frac{dN}{dy} \quad . \tag{11}$$

In this equation, all quantities are experimentally observable.

It should be strongly emphasized that the above estimate only applies to the central region for collisions of equal A nuclei of large A at very high energy. Therefore, the above formula does not apply for the asymmetric A, low energy collisions at CERN. Estimates of the energy density in the fragmentation region for asymmetric nuclei have not yet been attempted.

The initial energy density might be much larger than this estimate for a variety of reasons. In fluctuations in $\bar{p}p$ collisions, the multiplicity may be much larger. In nuclear collisions, the initial p_T may be much larger than is typical of the final state. This initial p_T may be determined by kinetic theory arguments, and might be in the range of $0.4 - 2\ GeV$, corresponding to uncertainty in the energy density of at least an order of magnitude. The initial transverse momentum, and correspondingly, the initial time, may even depend upon the nuclear baryon number A. I think the best estimates of the achievable energy densities in central collisions of large nuclei is $2 - 200 \ GeV/Fm^3$. This corresponds to an initial temperature in the range of $T_i \sim 200 - 700 \ MeV$.

To achieve very high energy densities, however requires very high energy densities. If the initial formation time is $t_o = C/T$ where C is of order one, to acquire a temperature of 500 Mev would require that the nulcei be Lorentz contracted to a size of 0.4 Fm, which for uranium requires a center of mass energy of 40 GeV/A. Also, it is quite likely that the maximum possible temperatures are only achieved for very large A.

To make a convincing case that there is sufficient time for the formation and evolution of a quark-gluon plasma as an approximate perfect fluid, the expansion rate of the system should be compared to a typical particle collision time. When the collision time is much less than the expansion time, the system should expand approximately adiabatically as a perfect fluid. Since entropy is conserved, the initial and final times for expansion in d dimensions are related by

$$\left(\frac{t_f}{t_i}\right)^d = \frac{N_{dof}^i}{N_{dof}^f} \frac{T_i^3}{T_f^3} \sim 10 - 10^4 \ , \tag{12}$$

where N_{dof} is the number of particle degrees of freedom. At early time, the expansion is one dimensional, and later times becomes three dimensional. We estimate therefore that $t_f/t_i \sim 10 - 10^3$. Detailed hydrodynamic computations show that the final decoupling time is probably somewhere in the range of $t_f \sim 20 - 50 \ Fm/c$.

Large nuclei are clearly the more favored system for producing and studying a quark-gluon plasma. This follows simply from the facts that the average energy density achieved is larger, and that the system is physically larger in transverse extent. We require $\lambda_{scat} << R_{nuc}$ in order for a perfect fluid hydrodynamic treatment to be sensible. Estimates of λ_{scat} give $0.1 - 1 \ Fm$.

Experimental data exist which throw some light on the size of systems necessary for fluid dynamic effects to become important. At Bevalac energies, the flow of hadronic matter was studied in nuclear collisions. In collisions of nuclei of small impact parameter, single particle collisions occur at large transverse momentum. The nuclei do not collectively flow in a given transverse direction unless there are subsequent rescatterings among the constituents of the nuclei. If these subsequent rescatterings do not occur, the transverse momentum of each particle is randomly oriented. To get collective flow, one needs rescattering, and this should be enhanced in collisions at small impact parameter, and collisions of large A nuclei.

In Fig. 4, the flow angle is plotted for various measures of the impact parameter (large impact parameters at the top and small at the bottom of the figure) for various nuclei (small on the left and large on the right). Little evidence of flow is shown

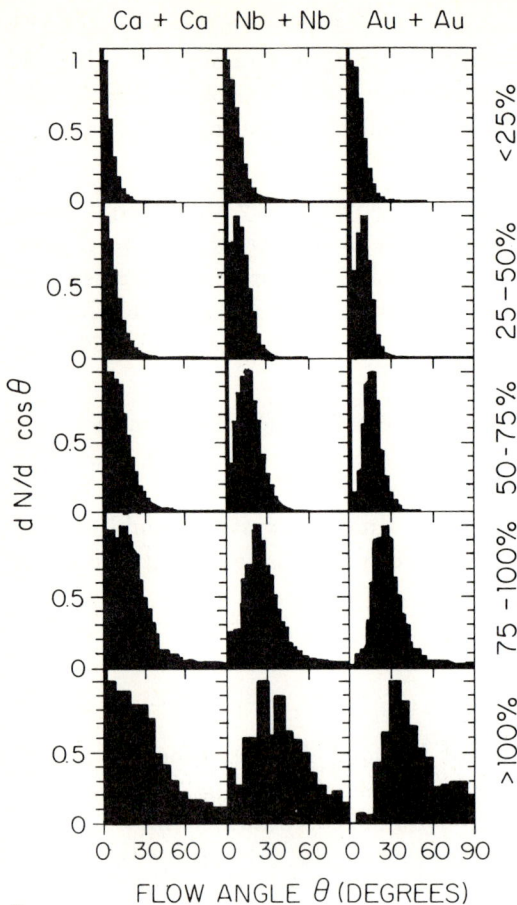

Fig. 4 Flow distributions as measured by Gustafsson et. al.

Table 1 Models of Nuclear Collisions

Model	Thermalization?	p_t Enhancement?
DPM, Hi-Jet	no	no
Lund, Rope model	no	some
Nuclear cascade	some	yes
QGP	yes	yes

for nuclei as large as calcium, and collective effects begin to become important for nuclei of the size of niobium.

The current experiments at CERN may allow for some determination of the energy densities which might be achievable in high energy heavy ion collisions at asymptotically high energy and for very large A. To sort this out from the data, one must have models to compare the data with. The data which are now available are primarily for E_t and dN/dy distributions. In principle the correlation between these variables can determine whether there is thermalization. For thermal models the p_t

is enhanced due to rescattering. This is a small effect for pions, but is a larger effect for nucleons, as will be discussed in later talks. In Table 1, I give a list of various models which attempt to describe nuclear collisions and the distinguishing features which may allow their resolution.

3 Probes of the Quark-Gluon Plasma

In Table 2, various experimental probes of the quark-gluon plasma are presented.

Table 2 Probes of the Quark-Gluon Plasma

Probe	Physics
Photons and dileptons	Plasma expansion, impact parameter meter resonance melting
p_t distributions	Equation of state, evidence of fluid flow
Strangeness and charm	Dynamics of expansion
Pion correlations	Size and lifetime of plasma
Jets	Scattering cross section of quarks or gluons with plasma and hadronic matter

In this presentation, I do not have space to discuss in detail the various possible experimental probes of the quark gluon plasma as they may appear in heavy ion collisions. I give a few paragraphs of discussion of the above proposed probes in the following pages. For an up to date review of various proposed signals, the interested reader is referred to the review article by K. Kajantie and myself.[3] The bottom line on all of these probes is that they all will involve correlations between several variables. For example, just the requirement of head-on, small impact parameter collisions requires a cut either on total multiplicity or nuclear fragmentation. Because of this often complicated analysis of correlated variables, it is difficult to argue that any one of the probes will yield an unambiguous signal for a plasma. Nevertheless, in several cases such as photon and dilepton probes or J/Psi production, with a little luck it may be possible to construct a convincing case that a plasma has been formed, and to measure some of its properties.

The correlation between p_T and dN/dy reflects properties of the equation of state of matter. This is easily seen from the example of a spherically expanding gas. We assume that at some initial time, there is a spherically symmetric drop of hadronic matter of uniform density matter at rest. We then allow the system to hydrodynamically expand. We assume we know the volume of the initial system, V_o. We measure the total energy of all particles and the total multiplicity of particles in the final state. Since the system is slowly expanding at late times, the entropy of particles in the final state is known, assuming the particles were produced thermally from a weakly interacting gas. Since energy and entropy are conserved in the expansion of a perfect fluid, the energy and entropy of the final state is that of the initial state.

We can therefore experimentally measure the correlation between say p_T, which is proportional to E/S, and the energy density. We can compare this to a theoretically predicted correlation determined by knowing the equation of state.

Quark-antiquark annihilation produces dilepton pairs in the plasma. If we sum over all possible quark-gluon interactions in the initial and final state, then the overall rate for production of dileptons and photons per unit time and volume is proportional to

$$\frac{dN}{dt d^3x d^4q} \sim Im \int d^4x \; <J^\mu(x)J^\nu(0)> \; e^{iqx} \tag{13}$$

This assumes emission from a plasma at a fixed temperature T. The brackets $<>$ denote a thermal expectation value. The current $J^\mu(x)$ has a real, Minkowski time argument.

There are of course a variety of non-thermal sources for dileptons and photons. There are backgrounds for photons from π° decays, which in the low q region obscure the signal. There may also be backgrounds for the dileptons arising from decays of charmed particles. For large q, hard scattering processes from the initially unthermalized beams of quarks and gluons presumably dominate. As the momentum is softened, the contributions arise from an ever more thermalized system which eventually may come from a plasma, provided backgrounds from soft hadronic decays do not become too large. In this intermediate range of q, there are several thermal regions which contribute. At the higher q values, there is presumably a contribution from a quark-gluon plasma, at lower q a mixed phase of plasma and hadronic gas, and at the lowest q values larger than that for which background becomes important, there is a contribution from a hadronic gas.

To compute these distributions of photons and dileptons, a knowledge of the space-time history of the evolution of the quark-gluon plasma is required. Detailed estimates of the space-time evolution of matter produced in head-on collisions of nuclei at large A have now been carried out, and the dilepton distributions have been computed in detail.[3] There has as yet been no attempt to treat non-zero impact parameter collisions. Techniques have also been developed to study the fragmentation region. No attempt has been made to treat the pre-equilibrium region, although the cascade computation of Boal may be useful for this.[3] A treatment of the late stages in the evolution of the matter are best treated by cascade simulation of pion interactions, and again could easily be used to compute dilepton and photon distributions.

Strangeness has been widely suggested as a possible signal for the production of a quark-gluon plasma.[4] The argument for large strangeness in its most naive form follows from the observation that there are equal numbers of up, down and strange quarks in the plasma. One might naively expect that there would be roughly equal numbers of kaons and pions produced, and that the ratio of strange to non-strange baryons would be proportional to their statistical weight, $N_S/N_{NS} \sim 2/3$.

For the case of mesons, the above argument may be easily seen to be false. In the expansion of the quark-gluon plasma, and later the hadron gas, entropy is conserved, and the pions are a result of this entropy. A better measure of the strangeness of a plasma is therefore the K/S ratio, where S is the entropy. This may be computed and shown to be smaller in a plasma than in a hadron gas for all temperatures larger than 100 MeV. The K/π ratio is therefore not a direct signal for a plasma. Further, the K/π ratio may be computed in a variety of hydrodynamic scenarios. The result is typically $K/\pi \sim 0.3$. This number is a little larger than is typical of $\bar{p}p$ interactions. As has been suggested by Rafelski and Muller, perhaps only if a plasma is formed will the dynamics allow for such a large K/π ratio, and therefore is a signal of interesting dynamics, or perhaps even the production of a plasma.

Strange baryons and anti-baryons may also provide a signal. Direct computations of the ratio of the ratios of strange to non-strange baryons in a plasma to that in a hadronic gas shows however that a hadronic gas is (if at all) only a little less strange than a plasma. These estimates are done for net baryon number zero plasma, and an enhancement may exist for the plasma in the baryon number rich region. At RHIC and SPS energies, the baryon number density is effectively small at all rapidities, and this should be a good approximation. Again, although this ratio of ratios indicates a lack of a signal for equilibrium quark-gluon plasmas, the ratio of non-strange to strange baryons is large, 0.3-2, in either scenario for $100 MeV < T < 300 MeV$. This number is far larger than is typical of $\bar{p}p$ interactions, and again by the arguments of Rafelski and Muller, perhaps the only way to achieve this dynamically is by production of the plasma. This ratio is therefore interesting for dynamical reasons.

I conclude therefore that a large strangeness signal is not a direct signal for production of a quark-gluon plasma. It is almost certainly a signal for interesting dynamics, and it may be true that the only reasonable dynamical scenarios where large strangeness may be produced involve the formation of a quark-gluon plasma.

The Hanberry-Brown-Twiss effect arises from the interference of the matter waves of identical particles as they are measured in coincidence experiments. The measurement of identical particles closely correlated in momentum therefore allows the possibility of measuring properties of the space-time evolution of matter produced in heavy ion collisions. One can in principle measure the size and shape of the matter at the temperature when decoupling occurs, and perhaps verify the existence of an inside-outside cascade description.

The rescattering of jets after production in a quark-gluon plasma in principle provides a probe of the plasma and hadronic matter as the jet propagates through the evolving system.[3] The jets will scatter from the constituents of the plasma as well as the constituents of hadronic matter which forms later. The degree of scattering is a measure of the quark-matter or gluon-matter cross section. This scattering can dramatically change quantities such as the jet acoplanarity, and can produce phenomenon such as single jets.

References

[1] For a review see e.g. L. D. McLerran, *Reviews of Modern Physics* **58**, 102 (1986).

[2] A. Capella, C. Pajares, and A. V. Ramallo, *Nuc. Phys.* **B241**, 75 (1984); A. Capella, J. A. Casado, C. Pajares, A. V. Ramallo, and J. Tran Tanh Van, Orsay Preprint US-FT-27 (1986).

[3] K. Kajantie and L. McLerran, to be published in Annual Reviews.

[4] Proceedings of Quark Matter 86, Asilomar Ca. May, 1986, to be published in Nuclear Physics A.

New Processes and Old Spin Physics

M. Jacob

Theoretical Physics Division, CERN,
CH-1211 Geneva 23, Switzerland

Polarization may offer a very useful tool, even in cases where the dynamics which creates it, or responds to it, is not yet well understood. This is particularly the case in some of the processes studied in detail at this Workshop. The purpose of this contribution is to emphasize this point further, showing how some simple analyses, borrowed from techniques which bloomed in the late Fifties and early Sixties, can be fruitfully applied to some topical questions of the Eighties.

We here illustrate this point with two examples. They are:
(i) the determination of the spin of the W particle, which was measured to be 1 by UA1;
(ii) the measurement of the $\bar{\Lambda}$ longitudinal polarization in nucleus-nucleus collision at SPS energies (200 GeV per nucleon). This could provide a direct and unambiguous test of the formation of a quark-gluon plasma in some of the collisions.

We conclude with some general remarks about spin physics.

The techniques used are rather simple. They had a period of glory in the early Sixties, when the study of the properties of the many newly discovered resonances called for an analysis of polarization phenomena in order to determine their spins and parities. These techniques are, however, no longer well known to many active practitioners of particle physics in the Eighties. Yet they remain useful.

1. The Spin of the W

The discovery of the W and Z particles was a great achievement at the $p\bar{p}$ collider[1]. The production cross-section peaks at zero transverse momentum and, to a good approximation, we can assume that the W(Z) is produced along the $p(\bar{p})$ beam direction, and use this direction as the polarization axis. In the standard model, a W^- (say), created by the annihilation of a $\bar{u}$ antiquark and a d quark, when produced in the direction of the antiquark, assumed to have the larger momentum, will have helicity +1. It would have helicity -1 if the d quark had the largest momentum in absolute value.

We now consider the electron antineutrino decay in the W rest frame. The electron emerges at an angle θ with respect to the W direction in the lab. This angle is easily measured, since it is simply related to the transverse momentum p_T of the electron

$$\sin\theta = \frac{2p_T}{M_W} \tag{1}$$

One can thus directly determine the mean value of the cosine of this angle. As shown in the following it is given by

$$\langle\cos\rangle = \frac{\langle\lambda\rangle \langle\mu\rangle}{J(J+1)} . \tag{2}$$

Here $\langle\lambda\rangle$ and $\langle\mu\rangle$ are respectively the mean value of the global helicities in the final and initial states and J is the spin of the W.

The global helicity in the final state is the difference between the electron helicity and the antineutrino helicity. The same definition applies to the quark-antiquark system in the initial state. In the standard model (V-A coupling) these two quantities should both be equal to -1. One thus expects that $\langle\cos\theta\rangle$ should be equal to 0.5. It would obviously be zero for spin zero.

A first experimental determination, in 1983, gave $\langle\cos\theta\rangle$ = 0.5±0.1[2]. On the basis of that, it could be announced that the W spin is one and that parity is maximally violated in W production and decay[3]. If parity were not maximally violated, $\langle\lambda\rangle$ and /or $\langle\mu\rangle$ should be less than 1 in absolute value. Even with spin 1, $\langle\cos\theta\rangle$ would then be less than 0.5. If the spin were not 1, the value would furthermore be less than 0.5. Relation (2) thus kills three birds with one stone. One notices, however, that one can keep the same value changing the sign of both $\langle\lambda\rangle$ and $\langle\mu\rangle$. One cannot distinguish between V_-A and V_+A couplings without a separate polarization experiment. This is well known, and should be remedied soon when the polarization of the τ is measured with an increased sample of W. The analysis of the τ decay pattern in $\pi\,\nu_\tau$, or in $\ell\bar{\nu}\nu_\tau$, indeed allows a determination of the helicity of the τ [4].

At present, precision has increased and results of UA1 and UA2 now give

$\langle\cos\theta\rangle$ = 0.43 ± 0.07.

The derivation of (2) can be found in an old paper written in connection with the disintegration of strange particles[5] and it is most easily presented in the framework of the helicity formalism[6].

The angular part of the two-body ($\ell\bar{\nu}$) decay amplitude of the W is written as [5,6]

$$A_{\mu\lambda}(Q\phi) = \sqrt{\frac{2J+1}{4\pi}}\, a_\lambda\, D^{J*}_{\mu\lambda}(\phi,\theta,-\phi). \tag{3}$$

Here λ (and μ) are the global helicities in the final (and initial) states and ϕ and θ are the polar azimuthal angles of the electron direction in the W rest frame. There are as many different decay amplitudes a_λ as there are final helicity states, 4 most generally and only 1 in the standard model.

From (3) one readily obtains the angular distribution:

$$I(\theta,\phi) = \sum_{\mu\mu'} \rho_{\mu\mu'} \frac{2J+1}{4\pi} \sum_\lambda (a_\lambda)^2\, D^{J*}_{\mu\lambda}(\phi,\theta,-\phi)\, D^{J}_{\mu'\lambda}(\phi,\theta,-\phi), \tag{4}$$

where $\rho_{\mu\mu'}$ is the density matrix which describes the polarization of the W. If the initial $p(\bar{p})$ beams are not polarized,

$$\rho_{\mu\mu'} = \delta_{\mu\mu'} P_\mu, \qquad \sum_\mu P_\mu = 1. \tag{5}$$

Any beam polarization, if present, would in any case be averaged out integrating over ϕ.

The amplitude as written in (3) has been normalized to unity, with

$$\sum_\lambda |a_\lambda|^2 = 1. \tag{6}$$

Integrating over ϕ, and expressing the product of the two Wigner functions in terms of the Clebsch-Gordan series [5,6], one has

$$I(\theta) = \frac{2J+1}{2} \sum_\mu (-1)^\mu P_\mu \sum_\lambda (-1)^\lambda |a_\lambda|^2$$
$$\sum_{\ell=0}^{2J} c(JJ\ell/\mu,-\mu) \; c(JJ\ell/\lambda,-\lambda) \; P_\ell(\cos\theta). \qquad (7)$$

The mean value of $\cos\theta$ (2) is thus the projection of the $\ell = 1$ term in (7). To complete the derivation one needs some well-known relations among Clebsch-Gordan coefficients, namely,

$$c(JJ\ell)/m,-m) = (-1)^{J-m} \sqrt{\frac{2\ell+1}{2J+1}} \; c(J\ell J/m,0), \qquad (8)$$

$$c(J1J/m,0) = \frac{m}{\sqrt{J(J+1)}} \; . \qquad (9)$$

Using (8) and (9) in (7), one readily obtains

$$\langle \cos\theta \rangle = \sum_\mu \mu P_\mu \sum_\lambda \lambda (a_\lambda)^2 \frac{1}{J(J+1)} \; ,$$

which, in view of the normalization chosen in (5) and (6), is identical to (2).

Relation (2) is quite natural. In order to get an asymmetry, one has to violate parity, hence to have a non-zero value of $\langle\lambda\rangle$ and $\langle\mu\rangle$. The larger these values, the larger the asymmetry. If, however, the spin is too high, there is a large centrifugal effect which tends to wash out the asymmetry. The presence of the denominator is thus also natural.

We know that $\langle\lambda\rangle$ is bound by 1 in absolute value (electron-antineutrino state). We think (in practice know) that $\langle\mu\rangle$ is also bound by 1 (quark-antiquark state). However, to be quite general, we could say that $\langle\mu\rangle$ is bound by J, with $\langle\cos\theta\rangle \leq \langle\lambda\rangle/J+1$. This does not change our conclusions since one measures a value 0.5.

Having dealt with the W, it is tempting to do the same with the Z. The coupling to e^+e^- is, however, not favourable. The vector coupling is indeed proportional to $(-1+4\sin^2\theta_W)$ which, with a value of $\sin^2\theta_W$ of the order of 0.22, is almost zero as compared to the axial coupling. The coupling of the Z to e^+e^- is therefore almost purely axial and, as a result, the global helicity is small (-0.135 as opposed to -1 in the W $e\bar{\nu}$ case). The angular distribution in Z decay is thus expected to be almost symmetrical. In $p\bar{p}$ annihilation, its average polarization is also such that the angular distribution is expected to show little anisotropy. This is indeed the case as shown by Fig. 1. Figure 1 puts together the angular distribution in W $e\bar{\nu}$ decay, which follows the $(1+\cos\theta)^2$ distribution of the standard model, and the angular distribution in Z e^+e^- decay, where the lack of asymmetry can be related to the value of $\sin^2\theta_W$.

In the former case one sees the very strong peaking which, as is now well known, gives more electrons on the proton side and more positrons on the antiproton side as a result of the polarization and the decay asymmetry of the W. In the latter case, one sees that, with present statistics, it can merely be said that the results are compatible with the standard model[7] and the known value of $\sin^2\theta_W$. As previously mentioned, the angular distribution shows very little peaking even if the Z is expected to be longitudinally polarized, its coupling to quark-antiquark pairs mixing vector and axial terms. However, the global helicity is opposite in the $u\bar{u}$ (0.27) and $d\bar{d}$ (-0.96) cases and, starting with a mixture of $u\bar{u}$ and $d\bar{d}$ processes as expected in $p\bar{p}$ annihilation, angular effects average to something rather smooth. One thus does not expect prominent features in the angular distribution of the Z, even if one could observe it in its jet decay mode, thus overcoming the result of its almost purely axial coupling to electrons.

Yet, even though one cannot obtain a simple relation as in the W case, the striking difference between the two distributions and their respective structure can be considered as a good self-consistency test of the standard model.

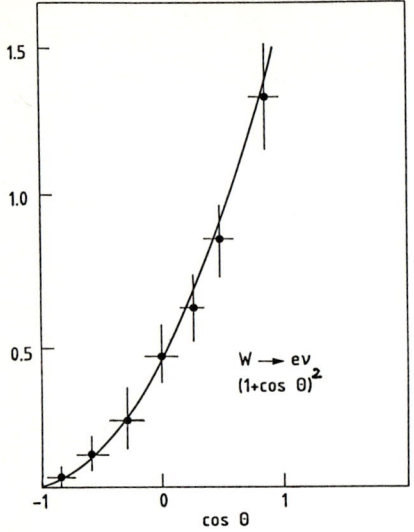
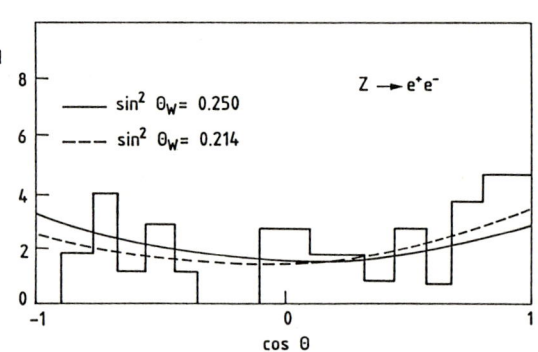

Fig.1. Asymmetry in the angular distribution in W decay and in Z leptonic decay

The dominant jet decay mode of the W should give a forward-backward asymmetry identical to that observed for the electron, or a forward-backward peaking if one does not distinguish the charge (or nature) of the two jets. This $(1+\cos^2\theta)$ dependence is, however, rather weak as compared with the very strong forward-backward peaking of the QCD background.

Rutherford scattering in QCD, with a typical $(\sin^4 \theta/2)^{-1}$ angular dependence, gives a distribution which is flat, up to non-scaling effects in the variable $\chi = (1+\cos\theta)/(1-\cos\theta)$ which absorbs the $\sin^4 \theta/2)^{-1}$ in the relevant Jacobian. This is very well satisfied by the jet data. However, the analysis of the angular dependence has so far emphasized the large jet-jet mass region 150 GeV $< M_{jj} <$ 350 GeV [8].

With improved calorimetry, their analysis will be extended to lower M_{jj} values and the presence of the W and Z should show up as a deviation from the rather flat χ dependence. There is already some evidence from UA2 that the W, Z peak shows up in the jet-jet mass distribution.

2. Longitudinal $\bar{\Lambda}$ Polarization in Nucleus-Nucleus Collisions

Parity violation in Λ decay leads to an up-down asymmetry with respect to a plane normal to the polarization of the hyperon. Here again the question was the subject of much discussion in the late Fifties[9].

We consider the decay of a parent particle of spin½ into a daughter particle of spin½ and a pion. The angular part of the decay amplitude is written as (3)

$$A_{\mu\pm\frac{1}{2}}(\theta,\phi) = \frac{1}{\sqrt{2\pi}} a_{\pm\frac{1}{2}} D^{\frac{1}{2}*}_{\mu\pm\frac{1}{2}}(\phi,\theta,-\phi) \tag{10}$$

and the angular distribution reads (4)

$$I(\theta,\phi) = \frac{1}{2\pi} \sum_{\mu\mu'} \rho_{\mu\mu'} \sum_\lambda |a_\lambda|^2 D^{\frac{1}{2}*}_{\mu\lambda}(\phi,\theta,-\phi) D^{\frac{1}{2}}_{\mu'\lambda}(\phi,\theta,-\phi). \tag{11}$$

Integrating over ϕ one has

$$I(\theta) = \sum_\mu p_\mu \sum_\lambda |a_\lambda|^2 (d^{\frac{1}{2}}_{\mu\lambda}(\theta))^2 . \tag{12}$$

Or, explicitly,

$$I(\theta) = p_{\frac{1}{2}}|a_{\frac{1}{2}}|^2 \cos^2\tfrac{\theta}{2} + p_{-\frac{1}{2}}|a_{\frac{1}{2}}|^2 \sin^2\tfrac{\theta}{2} + p_{\frac{1}{2}}|a_{-\frac{1}{2}}|^2 \sin^2\tfrac{\theta}{2} +$$
$$+ p_{-\frac{1}{2}}|a_{-\frac{1}{2}}|^2 \cos^2\tfrac{\theta}{2} . \tag{13}$$

The two final helicity states are orthogonal and one thus obtains in the same stroke the longitudinal polarization of the daughter baryon. One has simply to include a factor -1 in front of the terms involving $|a_{-\frac{1}{2}}|^2$, namely,

$$P_L(\theta) = p_{\frac{1}{2}}|a_{\frac{1}{2}}|^2 \cos^2\tfrac{\theta}{2} + p_{-\frac{1}{2}}|a_{\frac{1}{2}}|^2 \sin^2\tfrac{\theta}{2} - p_{\frac{1}{2}}|a_{-\frac{1}{2}}|^2 \sin^2\tfrac{\theta}{2}$$
$$- p_{-\frac{1}{2}}|a_{-\frac{1}{2}}|^2 \cos^2\tfrac{\theta}{2} . \tag{14}$$

Regrouping terms, one writes

$$I(\theta) = \tfrac{1}{2} (1 + \alpha p \cos\theta), \tag{15}$$

$$P_L(\theta) = \alpha + p \cos\theta, \tag{16}$$

where the initial polarization is $p = p_{\frac{1}{2}} - p_{-\frac{1}{2}}$ and $\alpha = |a_{\frac{1}{2}}|^2 - |a_{-\frac{1}{2}}|^2$ is the decay asymmetry parameter. The use of the helicity formalism[5,6] allows for an easy derivation of these relations.

The up-down asymmetry is equal to

$$\frac{\text{up-down}}{\text{up+down}} = \tfrac{1}{2} \alpha p \tag{17}$$

or

$$\langle\cos\theta\rangle = \tfrac{1}{3} \alpha p . \tag{18}$$

It is related to the fact that the parent particle is polarized ($p \neq 0$) and that parity is violated in the decay ($\alpha \neq 0$).

We now consider a cascade decay of the type

$$\Xi \to \Lambda\pi, \Lambda \to N\pi .$$

Parity violation in Ξ decay will induce a non-vanishing longitudinal polarization of the Λ $\langle p_\ell \rangle = \alpha_\Xi$. Any transverse polarization of the Ξ in its production process would be averaged away in its contribution to p_Λ (16), taking Ξ produced to the right and to the left of the incident beam. The acceptance of the experiment can be assumed not to favour one of the sides, even if it is not fully symmetric in azimuth. The eventual decay of the Λ will thus show an average up-down asymmetry

$$\frac{\text{up-down}}{\text{up+down}} = \tfrac{1}{2} \alpha_\Lambda \alpha_\Xi . \tag{19}$$

The asymmetry parameters are known to be $\alpha_\Lambda = 0.64$ and $\alpha_\Xi = -0.44$ respectively. One is therefore considering a 15% effect.

Using PC invariance, the asymmetry parameters change sign going from particle to antiparticle. Relation (19) therefore also applies as such to the $\bar\Xi \to \bar\Lambda\pi$, $\bar\Lambda \to \bar N\pi$ cascade. Only such a cascade process could lead to an up-down asymmetry in $\bar\Lambda(\Lambda)$ decay. It therefore tests the relative abundance of $\bar\Xi(\Xi)$ daughters among the $\bar\Lambda(\Lambda)$ sample.

This can provide a test for the formation of a quark-gluon plasma in nucleus-nucleus collisions, which has the advantage of providing a straight yes or no answer[10].

The formation of a quark-gluon plasma is indeed expected to result in a strong abundance of $s\bar{s}$ pairs, due to the thermalization of a gluon-quark system with a chemical potential of u and d quarks imposed by the squashing nuclei[11]. In the following hadronization the production of $\bar{\Xi}(\Xi)$, involving two s quarks, should not be greatly suppressed as compared to that of $\bar{\Lambda}(\Lambda)$ involving only one s quark, as is the case in typical hadronic processes.

If γ is the ratio between the number of $\bar{\Xi}$ produced to that of the number of antihyperons, $\bar{\Sigma}$ and $\bar{\Lambda}$, the mean up-down asymmetry in $\bar{\Lambda}$ decay should be[10]

$$\frac{\text{up-down}}{\text{up+down}} = \frac{1}{2} \frac{\gamma}{\frac{1}{2}+\gamma} \alpha_\Lambda \alpha_\Xi \ . \tag{20}$$

One could expect γ to be as high as 0.5 in events with large transverse energy where a quark-gluon plasma could have been formed[11]. The measurement should be done with $\bar{\Lambda}$ and with Λ. One expects the asymmetry to be more pronounced in the former case, since, in the latter, many of the Λ's should be mere fragments of nucleons into ΛK systems.

A sizeable asymmetry would indicate a relatively large abundance of $\bar{\Xi}$ with respect to $\bar{\Lambda}$. This would be an unambiguous test of the formation of a quark-gluon plasma.

One can translate the test into an effective lifetime test, measuring the effective lifetime of $\bar{\Lambda}$ and Λ through the time required (after Lorentz correction) for the formation of a V. If a good fraction of the $\bar{\Lambda}$'s originate from $\bar{\Xi}$ decays, the effective lifetime should be longer, since part of the $\bar{\Lambda}$'s are only gradually produced through $\bar{\Xi}$ decay. Here again we have a yes or no answer. Different effective lifetimes would be a signature of the formation of a quark-gluon plasma.

3. Spin Physics, a Few Remarks

We see with these two examples how old spin techniques can be used to answer topical questions. We now take a more general attitude.

In QCD, spin selection rules are particularly simple. If one neglects quark masses:
  (i) gluon radiation does not change quark helicity;
  (ii) quarks and antiquarks annihilate with opposite helicities.

It may seem that spin physics could thus give striking results, in particular in processes at short distances where perturbative QCD applies. However, there is a long way between phenomena at the quark level and those directly measured in reactions involving hadrons[12]. One can also rephrase this, saying that those experiments which should give easily understandable effects, i.e., Drell-Yan production in longitudinally polarized $\bar{p}p$ collisions, are very hard to carry out, whereas those polarized experiments which are more practical give in general large effects, for which one is at a loss to find a simple, convincing explanation. This is particularly the case of the beautiful series of Argonne experiments with polarized beam and target[13]. Many polarization effects are surprisingly large as compared to naive expectations based on the two quark rules. It is, however, well known that many polarization effects involve an interference between two amplitudes and therefore emphasize the role of the smaller one. They are likely to respond to small effects which may be missed in the leading terms.

We refer the reader to recent review articles[12,13] for a detailed discussion of spin physics. We merely mention here one example which involves a simple quark rule while being outside the naive realm of perturbative QCD. This is the question of Λ and Σ polarization in inclusive proton-proton collisions at high energy (Fermilab)[14]. They both increase with transverse momentum to reach a value of the order of 30% at $p_T \simeq 2$ GeV/c. The transverse polarizations measured at fixed angle as a function of the incident energy are practically mirror images of each other.

It is very tempting to interpret it as resulting from the polarization of the s quark which is "picked up" by a ud diquark system in order to make the final Λ or Σ particle. There is no reason for this s quark not to undergo an $\vec{L}\cdot\vec{S}$ coupling which would polarize it even if one cannot calculate it with present techniques. The SU(6) wave functions then simply impose polarization of opposite signs for the Λ and the Σ. A proper and natural ratio of triplet to singlet ud systems then gives the opposite value observed, taking also into account the fact that some Λ's originate from radiative Σ decays.

This is one example pointing to diquark structures in hadronic collisions whereby a baryon behaves as a diquark ($\bar{3}$) quark (3) system. There are several other examples[15].

Despite the complications of hadron physics, this is a case where polarization effects give an interesting new view of the dynamics. As discussed at this Workshop, LEAR physics should offer several other interesting cases.

References

1. W discovery:
   G. Arnison et al.: Phys. Lett. 122B, 103 (1983);
   M. Banner et al.: Phys. Lett. 122B, 476 (1983).
   Z discovery:
   G. Arnison et al.: Phys. Lett. 126B, 391 (1983);
   P. Bagnaia et al.: Phys. Lett. 129B, 130 (1983).
2. C. Rubbia: Proceedings IX Topical Conference on Particle Physics, Honolulu (1983);
   See also C. Rubbia: in Old and New Problems in Fundamental Physics, Meeting in honour of G.C. Wick (1984), Scuola Normale Superiore Quaderni, Pisa (1986).
3. C. Rubbia: Nobel Prize Address, Stockholm (1984).
4. M. Davier: in LEP Summer Study, Les Houches and CERN (1978), CERN Report 79-01 (1979).
5. M. Jacob: Nuovo Cimento 9, 826 (1958).
6. M. Jacob and G.C. Wick: Ann. Phys. 7, 404 (1959).
7. A. Parker: Invited Talk, Rencontres de Physique de la Vallée d'Aoste, La Thuile (March 1987).
8. W. Scott: Invited Talk, Rencontres de Physique de la Vallée d'Aoste, La Thuile (March 1987).
9. T.D. Lee and C.N. Yang: Phys. Rev. 106, 1645 (1958);
   T.D. Lee, J. Steinberger, P. Kabir and C.N. Yang: Phys. Rev. 105, 1367 (1957).
10. M. Jacob and J. Rafelski: CERN Preprint TH. 4649 (1987), Phys. Lett. B, to be published.
11. J. Rafelski: Physics Reports 88, 331 (1982);
    Quark Matter Formation and Heavy Ion Collisions, Proceedings of the Bielefeld Workshop (1982), ed. by M. Jacob and H. Satz, WSPC.
12. M. Jacob: Round Table Discussion. Specialized Conference on High Energy Physics with Polarized Beams and Targets, Lausanne (1980), ed. by C. Joseph and J. Soffer, Birkhaüser 1981;
    N.S. Craigie, K. Hidaka, M. Jacob and F. Renard: Physics Reports 99, N° 2 (1983).
13. A. Yokosawa: Physics Reports 64, N° 2 (1980);
    C. Bourrely, E. Leader and J. Soffer: Physics Reports 59, N° 2 (1980).
14. L.G. Pondrom: Physics Reports 122, N° 2 (1985).
15. P.V. Landshoff: Proceedings of the EPS High Energy Conference, Lisbon (1981).

Hot Strange Matter in Relativistic Nuclear Collisions

J. Rafelski

CERN, CH-1211 Geneva 23, Switzerland and
University of Cape Town, Rondebosch 7700, Cape, RSA

Excitation and flow of the first heavy flavour-strangeness in energetic nuclear collisions is described. Strange particles emerging from the collision are witnesses of the microscopic processes of creation of hot strange matter, especially so in the new phase-quark gluon plasma.

1. INTRODUCTION

There is evidence from recent relativistic nuclear collision experiments[1] that the reaction proceeds through a stage in which a hot and dense region of nuclear matter has been formed. The key issue, if the quark-gluon plasma state has been excited, will not be easy to ascertain, although this reaction mechanism is viewed as a rather compelling intermediate stage given the current QCD-based hadronic phenomenology. As it now appears, following on preliminary reports by the CERN experimental groups[2], there is substantial probability that the 200 GeV/A projectile Oxygen is essentially stopped by a heavy target (e.g. Pb). As function of the rapidity, the preliminary result of the NA35 collaboration shows a clear peak of negative particle multiplicity dN^-/dy around rapidity $y \lesssim 2.5$, which broadens as it moves down to $y \lesssim 2$ for 60 GeV/A collisions.

This and other evidence presented[2] suggests strongly that in central (impact parameter nearly zero) collisions, oxygen projectiles collide with the tube of matter covered by its surface, containing perhaps $\bar{A}_t = 50$ nucleons: This number is found by comparing the volume of the target tube with the nucleonic volume:

$$\bar{A}_t = \bar{V}_t/V_N = 2\pi R_t R_p^2 \frac{4\pi}{3} (1.2 \text{ fm})^3 = 1.5 A_t^{1/3} A_p^{2/3} \quad . \tag{1.1}$$

Here A_t is the nuclear content of the tube. A_t (A_p) are target (projectile) atomic numbers and $R_t(R_p) = 1.2 A_t^{1/3}(A_p^{1/3})$ are target (projectile) radia. The energy per nucleon in the CM frame consisting of the projectile and the tube then is

$$\sqrt{s}/A = m_N \sqrt{2\gamma_p} \frac{\sqrt{1.5(A_t/A_p)^{1/3}}}{1+1.5(A_t/A_p)^{1/3}} \quad , \tag{1.2}$$

where γ_p is the Lorentz factor of the projectile and the projectile rapidity is $y_p = \cosh^{-1} \gamma_p =$ (6 for $\gamma_p = 200$ and $= 4.8$ for $\gamma_p = 60$). We note that there is an optimum value of A_t/A_p leading to the highest fireball energy per participating nucleon. We find this to be at $1.5(A_t/A_p)^{1/3} = 1$. Oxygen collisions at 200~GeV on A ~ 200 targets that the energy per nucleon in the central fireball reaches 8 GeV: should one have $A_p = 40$ available, then this value would rise to 8.5 GeV, aside from greatly increased maximum number of participating nucleons from ca 70 in the case of $A_p = 16$ to 140 participating baryons in the case of $A_p = 40$. As a final kinematic detail, we note that in the laboratory frame the rapidity of the central fireball is given by

$$y_f = \cosh^{-1} \gamma_f, \tag{1.3a}$$

$$\gamma_f = \frac{A_p \gamma_p + \bar{A}_t}{\sqrt{A_p^2 + 2\gamma_p A_p \bar{A}_t + \bar{A}_t^2}} \xrightarrow{A_p = A_t} \sqrt{\frac{1+\gamma_p}{2}}, \tag{1.3b}$$

and hence for relativistic projectiles

$$y_f = \frac{1}{2} y_p - \frac{1}{2} \ln(\bar{A}_t/A_p) \tag{1.3c}$$

and we find $y_f \gtrsim 2.4$ in the particular circumstances of 200 GeV/A oxygen beam. For 60 GeV/A beam we find, similarly, $y_N \gtrsim 1.8$. We see that for the existence of central fireball, a substantial reduction of the particle rapidity from 6 (res.4.8) to 2.5 (resp.1.9) is required, just what is already established experimentally. Note further that in a symmetric collision consisting of e.g. N-N scattering the rapidity of the fireball would be greater than observed, with the ln-term in eq.(1.3c) vanishing.

Whatever the microscopic mechanism leading to the stopping of the projectiles in a relatively 'thin' nuclear target may be, its presence opens up the opportunity to study the properties of hot, dense, nuclear matter created in such interactions. But it is difficult to estimate the energy density ε reached at the moment of greatest compression. Noting that we have 8-9 GeV energy available per nucleon and that nuclear density may possibly be as compressed as five times the usual value $\rho_0 \sim 1/6 \; 1/fm^3$ we arrive at $\varepsilon \sim 8 \; GeV/fm^3$. Such high energy density may suffice indeed for the formation of the Quark Gluon Plasma (QGP) phase which under these conditions could be as hot as 250-300 MeV, within a volume of ± 60 fm^3 for oxygen projectiles, and exceeding 100 fm^3 for sulphur projectiles scheduled for the September/October 1987 SPS run.

With these rough reaction mechanism estimates at hand we can attempt to understand how strangeness flavour is excited in nuclear collisions. We must consider the two different scenarios, namely with and without quark gluon plasma formation. As we will see in detail in the next section, the gluonic processes in the plasma phase are highly efficient in making strange particle pairs, while the conventional associate production of strange hadrons is hindered greatly, in view of higher threshholds, lower statistical equilibrium particle density, smaller available phase space (c.f. Section 3). The nett result is that should QGP be formed, then it will contain a greatly enhanced strange particle content facilitating the formation of multiply strange hadrons. In Section 4 we consider how this will influence the abundance of otherwise rare strange hadrons, such as $\bar{\Xi}$, and will explain how this can be used to identify the process of QGP formation and some of its properties.

2. STRANGENESS PRODUCTION IN QUARK-GLUON PLASMA

In lowest order in perturbative QCD, $(s\bar{s})$-quark pairs can be created in collisions of two gluons and by annihilation of light quark-antiquark pairs. The gluonic production rate dominates strangeness production and leads to equilibration times comparable to the expected plasma lifetime[3]. Given the averaged QCD reaction cross sections it is easy to calculate the rate of events per unit time summed over all final and initial states. Neglecting final state interactions (the influence of Pauli blocking is negligible) the strangeness production rate is obtained by averaging the QCD reaction cross section with the spectra of particles in the plasma: $<\sigma_{s\bar{s}} v_r>$, with v_r being the relative velocity of colliding particles. We shall assume here that each perturbative QGP quantum (light quark, gluon) will rescatter many times during the lifetime of the plasma. Hence, we approximate the momentum distribution functions by the statistical Bose, or respectively, Fermi distribution functions, where the statistical parameters, e.g. temperature T and chemical potential μ are a priori functions of the location in the fireball ('local equilibrium').

The invariant strangeness formation rate per unit time and volume for the dominant gluonic processes is found to be[3]

$$A_g \approx \frac{7}{6\pi^2} \alpha_s^2 MT^3 e^{-2M/T} \left(1 + \frac{51}{14}\frac{T}{M} + \ldots\right) \quad . \tag{2.1}$$

Once strange quark-antiquark pairs have been created abundantly the $s\bar{s}$-annihilation reaction will limit the strange quark population. Under the condition of statistical independence of the creation and annihilation process, this loss term is proportional to the square of the density ρ_s of strange and antistrange quarks. With ρ_s^∞ being the saturation density at large times, the differential equation for the evolution of ρ_s as a function of time is:

$$\frac{d\rho_s}{dt} = A[1 - (\rho_s(t)/\rho_s^\infty)^2] - \rho_s \frac{1}{V}\frac{dV}{dt} \quad . \tag{2.2}$$

The last term, the volume dilution term arises, should the volume occupied by the quark gluon plasma change in time (expansion). A and ρ_s^∞ are also time dependent quantities, through the parametric time- dependence of the temperature T(t). (Both $A_g \approx A$ and ρ_s^∞ do not depend on the chemical potential of light quarks.) In Fig.2.1 the evolution of strange particle density, calculated[4] assuming m_s = 170 MeV, α_s = 0.6 is shown. The simple functional relation $V \sim t$, $T \sim t^{-1/3}$, is assumed to parameterize the time dependence of T and V. Case I corresponds to a blob of plasma of size 3 fm and initial T_0 = 250 MeV, while case II corresponds to the initial plasma temperature being very modest, T_0 = 180 MeV, while the initial plasma radius being 4 fm. The energy content is almost equal in both examples. They hence represent two different conceivable scenarios of a collision of heavy nuclei with case I being favoured by current experimental results. The arrow marked 90% shows the time at which the phase space saturation reaches that level. The total strangeness is quite rapidly built up in the collision, and we can expect 80-120 strange quark pairs to be produced. The slow fall of the strangeness density is a consequence of the volume expansion, rather than strangeness annihilation. There is virtually no net strangeness annihilation $\bar{s}s \to gg$ as the plasma expands, because in the expansion process the temperature and strangeness density both drop rapidly, decoupling effectively strangeness abundance from statistical equilibrium. It is in order to illustrate this point that we have chosen the slow expansion, $V \sim t$. In fast expansion, $V \sim t^3$, the freezing out of strangeness is even more sudden. Points (a), (b), (c) and (d) indicate when the critical temperature is reached at which the chemical potential assumes the values 0, 200, 400, 600 MeV, respectively, in a particular dynamical model[5]. The phase transition to the hadronic gas may occur at these points, depending on baryon content of the plasma.

As we see from Fig.2.1 the strangeness density at the transition point is expected in the interval $0.15/fm^3 < \rho_s > 0.3/fm^3$. This value indicates that

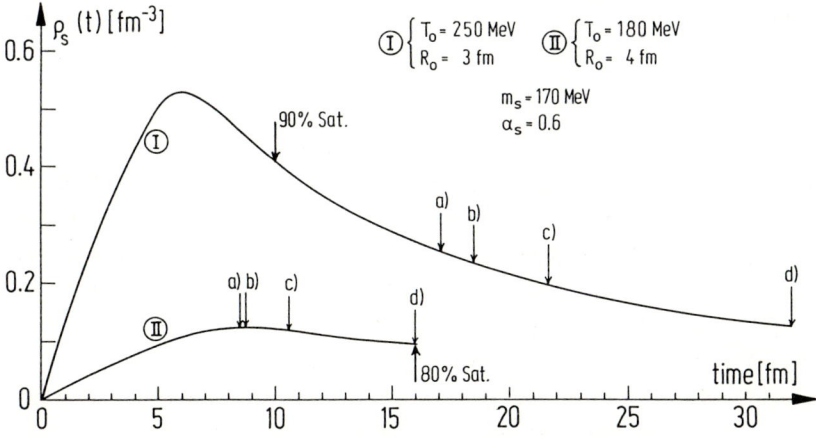

Figure 2.1 : Evolution of strangeness density for two typical temporal evolution scenarios of the plasma.

clustering of two strange quarks in one hadronic volume $V_N \sim 5$ fm^3 will be frequent, and further that the total abundance of strangeness is a rough measure of the plasma volume at the phase transition. The high particle density of strange quarks in plasma assures that when the quark matter hadronizes some of the numerous s and $\bar{s}$ quarks may form strangeness clusters such as Ξ, Ω, and particularly important, their antiparticles $\bar{\Xi}$, $\bar{\Omega}$, instead of being bound in kaons only. Consequently, this work emphasises our expectations about these multistrange hadrons.

3. STRANGE HADRON FORMATION OUTSIDE OF QGP (STATISTICAL MODEL)

We now study the dynamics of strangeness production in the hadronic gas phase as if no QGP were formed. The purpose of this section is to estimate the strange particle abundances expected if no quark gluon plasma has been formed in the hot fireball of the nulear collision. When a thermally equilibrated fireball has been established, the conventional reaction for strangeness production, e.g. $N + N \rightarrow N + Y + K$ where $Y = \Lambda$ or Σ, plays almost no role, because: (1) its threshold at ~ 670 MeV energy in the center-of-mass frame is higher than in reactions between a pion and a baryon, (2) the pions are the most abundant particles except perhaps at very high values of μ_b, (3) the phasespace of two-particle final states is more favourable than that of three-body final states. Hence, the strangeness producing reactions involving a pion are dominant.

The common reaction feature of all processes is the $q\bar{q} \rightarrow s\bar{s}$ reaction, where several quarks are spectators, and a $q\bar{q}$ pair is annihilated and an $s\bar{s}$ pair is formed. The experimental value of the OZI- rule forbidden cross section for the reaction $\pi + N \rightarrow Y + K$ (N ≈ nucleon), is only about 0.1 mb in the energy region of interest. Hence, the strangeness phase space saturation in HG should be a relatively slow process since the $s\bar{s}$ forming collision time of a pion with a nucleon, at $\rho \sim 3\rho = \frac{1}{2} \frac{1}{\text{fm}^3}$ is

$$\tau \sim (\sigma v \rho)^{-1} \sim 10^{-21} \text{ s}. \tag{3.1}$$

It is also clear that while the <u>direct</u> pair production of multiple strangeness carrying (anti-) baryons is strongly suppressed in nuclear collisions, these particles may be built up by exothermic strange quark <u>exchange</u> reactions. Experimentally, one finds for the strangeness exchange reaction $\bar{K} + N \rightarrow Y + \pi$, a cross section roughly ten times larger than the production cross section. This means that strangeness is much faster redistributed among the strange particle family than produced. Fast redistribution of strange s-quarks among different hadrons means that the <u>relative</u> chemical equilibrium of s-quarks is reached during the conceivable lifetime of the fireball, unless the fireball lives 10^{-21}sec, the <u>absolute</u> equilibrium. i.e. saturation of the phase space is not achieved.

Similar to eq.(2.2) we find a (now quite complex) family of evolution equations for the densities of the various particles in question. In actual calculations we used[6,5] parameterisation of measured cross sections to calculate average cross sections entering the (nonlinear) master equations which describes the evolution of the chemical composition of the hot hadronic matter phase. The initial condition for the strange particle densities at a given 'initial' time t_0 were all set to zero, thus neglecting the contributions of strangeness production reactions in the first moments of the collision before thermal equilibrium is reached. Further below in Section 4 we will use the same master equations in order to describe the break up of QGP and the subsequent evolution particle: the hadronic reaction then starts with the strangeness abundance that has been built up in the quark-gluon plasma, as further modified in the phase transition back to hadronic matter. For the required antinucleon density at $t=t_0 \sim 10^{-24}$ sec we use the chemical equilibrium value, being aware of the fact that the assumption of baryochemical equilibrium for antinucleons is largely overestimating the possible abundance. This leads to an overestimate of the abundance of strange antibaryons generated by hadronic interactions in the absence of quark gluon plasma. As the abundance of strange antibaryons we find is extremely small anyway, our approach is a 'conservative' estimate of the background of strange antibaryons expected, should no plasma be formed.

The time evolution of the total strangeness abundance ρ_s in hadronic gas is displayed in Fig.3.1 (due to exact strangeness conservation, ρ_s is equal to the total antistrangeness density $\rho_{\bar{s}}$) for T = 160 MeV and two values of the baryon chemical potential, μ_b = 0 and μ_b = 450 MeV. When looking at both curves, we see that after a typical break-up time of the hadronic fireball of about 5×10^{-23}s, the strangeness abundance predicted for the hadronic phase is about a factor of three smaller than its equilibrium value. Individual particles with higher strangeness content such as $\bar{\Omega}$ (see below) will be still more distant from equilibrium distribution: the density of anti-omegas as well as omegas is in HG a

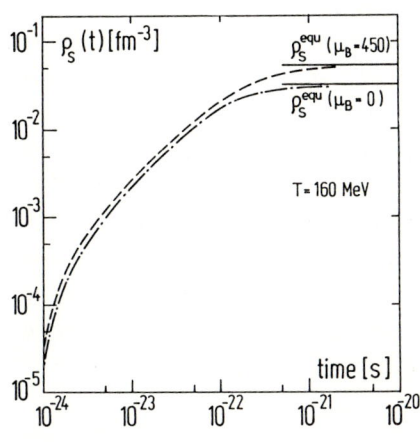

Figure 3.1 : Time evolution of strangeness density in hot hadronic matter at fixed temperature T = 160 MeV and baryochemical potentials μ_b = 0 and μ_b = 450 MeV.

factor of about 10^3 lower than their equilibrium values for a breakup time between 10^{-23} and 10^{-22} seconds. This is true for baryonless as well as for baryon rich hadronic gas. Anticascades ($\bar{\Xi}$) are suppressed by a factor of about 10^2 with respect to their equilibrium values. We emphasise that in more realistic calculations we expect to find an even lower density of strange antibaryons created outside of QGP.

For this reason, i.e. due to their effective absence in HQ, strange antibaryons appear as the most promising signal for the temporary existence of a deconfined quark-gluon phase during the course of the nuclear collision given the very large abundance of strangeness expected to be created in such experiments.

4. STRANGE ANTIBARYON FORMATION FROM QUARK-GLUON PLASMA

During the process of QGP hadronisation substantial fragmentation of gluons in particular is required: note that if quarks and antiquarks would just simply recombine[7] into mesons, mainly pions, there would be only half as many pions afterwards as there were quarks and antiquarks before: Each pion carries about 4 units of entropy[8], while each quark and antiquark carries 4.2 units of entropy. Hence in the recombination half of the entropy is lost, while one also ignores another factor of two in entropy when gluons are not accounted for. Thus in a recombination model[7] nearly 3/4 of entropy disappears. But we know that the total entropy should <u>increase</u> during the phase transition in view of the second law of thermodynamics. In order to appreciate what went wrong, we recall that the usual chemical reactions involve nonrelativistic particles, whose entropy per particle increases with temperature and also possibly on account of internal excitations. The entropy can, therefore, be made to increase while the number of particles decreases due to recombination. Such an effect is not possible for relativistic particles. The large amount of entropy residing in the quark-gluon plasma can only be disposed of by generating, ultimately, a sufficiently large number of pions[8].

There exists a simple mechanism permitting explicit implementation of the fragmentation process: recall that a quark-antiquark pair will only in one case out of nine be found in a colour singlet state that can form a single meson. In the other eight cases the quark-antiquark pair, forming a colour octet state, <u>must first radiate a gluon</u> in order to be able to recombine into a meson. Similarly, gluons will not, in general, simply disappear into the vacuum during the hadronization process in view of their colour octet nature, but will e.g. fragment into a quark-antiquark pair. We find that in order to conserve entropy during the hadronization process, every gluon and about one third of the quarks must fragment before coalescing into mesons. The fragmentation processes must be correctly weighted with regard to their flavour content. Although it is not entirely clear whether the concepts developed[9] for quark jet fragmentation are valid in the environment of the hadronizing quark-gluon plasma, we shall here adopt these concepts as a first guideline. The relative probability of glue fragmentation into a quark pair of mass m_i is controlled by the parameter f_i. For light quarks we can take $m_u=m_d=0$ leading to $f_u = f_d \sim 0.425$, $f_s \sim 0.15$ for $m_s \sim 170$ MeV. The number of quark-antiquark pairs of each flavour is now obtained by multiplying the number of fragmenting gluons with the probability f_i for producing the considered flavour.

For the description of the strangeness evolution and strange particle excitation during the spatial expansion of the fireball in its hadronization stage some further detailed model assumptions are required: there are two mechanisms that lead to a change in volume occupied by the fragmenting QGP in the mixed phase. The volume may change due to a hydrodynamical expansion at nearly constant entropy. Alternatively the volumes can change due to exchange of particles, i.e. entropy, with the hadronic gas phase. Simultaneously, the volume of the hadronic phase continuously increases, due to the accumulation of hadronic particles as well as due to (hydrodynamical) volume expansion. Note also that a large volume jump occurs between QGP and hadronic phases as suggested by the high entropy density of the plasma as compared to the hadronic gas phase. Consequently, the master equations describing the mixed phase are quite complicated.

We are mainly interested in the fate of strange quarks. The number of strange quarks in the plasma phase changes because of two reasons: on the one hand, strangeness-producing reactions make it approach the equilibrium abundance (c.f. Section 2). Furthermore, the number of strange quarks in the plasma phase changes due to their conversion into hadrons. In the hadronic phase the number of strange particles will further change through standard hadronic reactions and this part of the reaction can be treated as described in Section 3. We now summarise the results[5] of calculations involving, in particular, strange and nonstrange baryons and antibaryons in addition to pions and kaons. Note that these calculations are based on the following physical picture of the evolution of the fireball: initially a quark-gluon plasma is formed, which expands and cools until it reaches the phase transition point towards the hadronic gas phase. Temperature and baryochemical potential are then kept constant and the plasma is gradually converted into hadronic gas (mixed phase). During this process the volume of the mixed phase grows according to the requirement of entropy conservation. The calculation is stopped when the quark-gluon plasma is fully converted into hadrons. When plasma and hadronic phase coexist, complex master equations for the mixed phase are integrated in the hadronization process; mesons and baryons are assumed to form according to the fragmentation-recombination scenario - gluon fragmentation ensures approximate entropy conservation by providing the additional partons required to fill the larger hadronic gas volume with mesons and baryons.

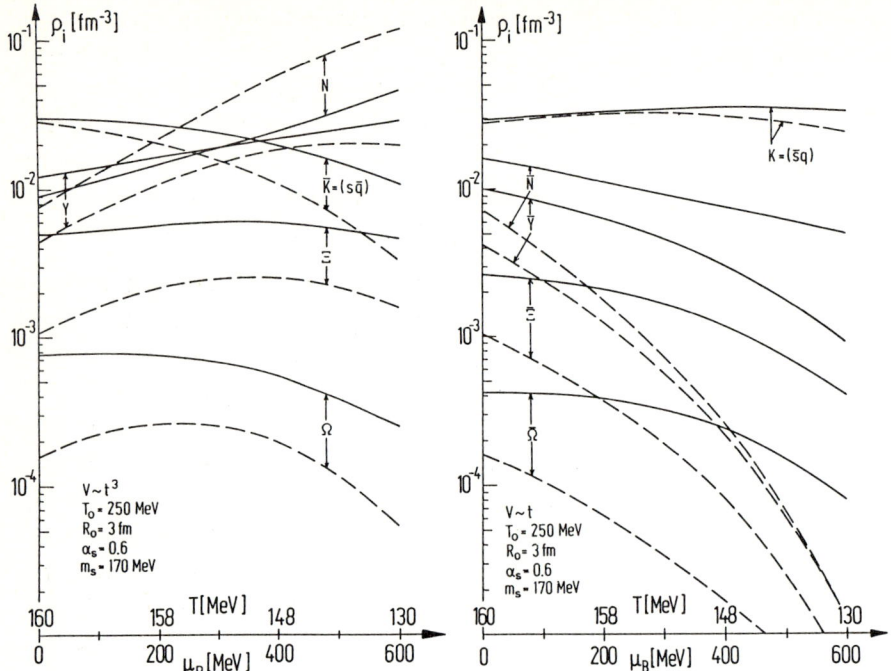

Figure 4.1 : Strange particle abundances assuming fast volume expansion ($V \sim t^3$) and recombination with gluon fragmentation: (a) s- and (b) $\bar{s}$-quark carrying hadrons at the end of coexistence of quark gluon plasma and hadronic gas phase, as a function of critical values of T and μ_b. Dashed: equilibrium abundances in the hadronic gas phase at the same values of T and μ_b.

The resulting particle abundances are presented in Fig.4.1. Parts (a) and (b) refer to abundances of hadrons containing s and $\bar{s}$ quarks, respectively. The abundances are presented as found at the time when the plasma phase has been completely converted to the hadronic gas phase and the particle density is already quite small. The curves are given as function of the values of the thermodynamical variables T and μ_b at the point where the critical curve is crossed. In the calculation a constant T/μ_b ratio was assumed. The dashed curves are the equilibrium hadronic gas abundances which, as we have discussed in the case of strange antibaryons, are gross overestimates of actual expectations for the background. As we see from the presented results, there is a very significant overabundance as compared to the hadronic gas phase space of strange antibaryons found upon hadronisation of the QGP phase. For reasons of experimental practicability it is further important that the densities of strange antibaryons $\bar{Y}$, $\bar{\Xi}$ are in the range 10^{-3} to $10^{-2}/fm^3$ at the time of hadronic break-up. Given the large fireball volume of more than a thousand fm^3 at this time, a significant number of these particles are expected to be produced in each single reaction event leading to formation of quark gluon plasma.

The synopsis of the results presented here confirm the particular suitability of the global abundance of strange antibaryons for diagnosis and study of the quark gluon plasma state. However, one can also consider the direct effect of the large s, $\bar{s}$ density in the early plasma before hadronisation, and in particular the possibility of strange particle radiation from such a hot plasma state[10]. Two channels of particle production may be considered. In the first premordial particle emission mechanism a fast quark (or coloured diquark) from the plasma impinges on the boundary between the plasma perturbative vacuum and the true vacuum. In the associated color string breaking process[9], at least one quark-antiquark pair is formed. We will refer to this process as a 'microjet' (not to be confused with the minijet processes). Second, in the recombination approach, several constituents of the plasma, clustered into colorless objects, penetrate the surface and hadronize. This latter process is similar to the model of the phase transformation of the plasma to the hadronic gas at the later stages of the plasma life which we have just considered. However, this recombination radiation is, in detail, different in that we may not allow gluon fragmentation. The equilibrium quark abundances in plasma contain the effect of continuous gluon and quark fragmentations and recombinations. Only at the point of phase transition is equilibrium lost and additional particle and entropy generating microscopic processes such as gluon fragmentation become necessary. We expect that the microjet picture will dominate the spectra at medium to high $E_\perp \sim (2-4)$m.

In Fig.4.2 we compare for the $\bar{\Xi}/(\bar{Y}/2)$ ratio the results[10] of the microjet and recombination mechanisms to global abundance ratios i.e. (including low $p_\perp$), c.f.Fig.4.1. The surprise is that in both early emission processes (microjet, recombination) populating medium to high $E_\perp$ portions of the spectra we are led to expect more anticascades than antilambdas (since $\bar{\Sigma}^0 \to \bar{\Lambda} + \gamma$ we included $\bar{\Sigma}^0$ into the 'lambda' abundance: the visible hyperon decays are just half of the total abundance).

But there is some difference which is difficult to differentiate between the two early emission processes, unless the value of μ_b is known to some accuracy. Although experimental observation of the ratio might appear difficult, longitudinal polarisation method[11] can be applied in visual experiments in e.g. CERN experiment NA35, while effective decay length of a indicates it as a descendant of a Ξ. This method may be applied in the CERN experiment WA85. Electronic particle track reconstruction will be attempted in CERN experiment NA36 for multiply strange, cascading decay particles.

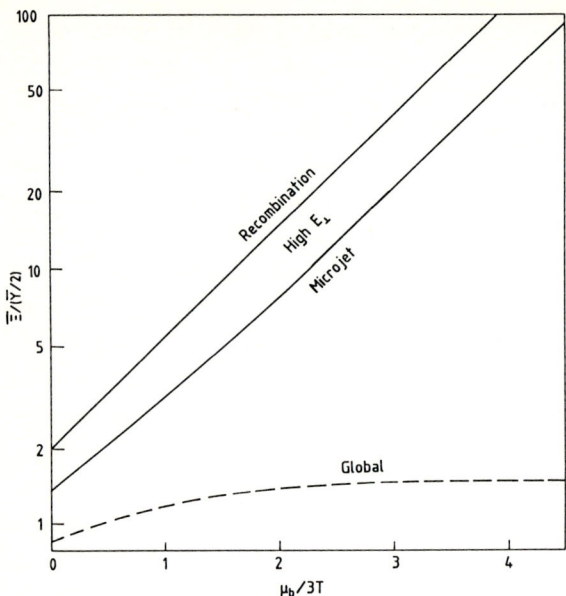

Figure 4.2 : $\bar{\Xi}/(\bar{Y}/2)$ particle ratio as a function of $\mu_b/3T$ for the microjet, recombination models of particle spectra at medium to high $E_\perp$. Dashed: global ratio.

The source of all these surprising results about strange hadrons from QGP can be traced back to the fact that strange quark-pair production in the plasma phase proceeds at a sufficiently fast rate to permit statistical equilibrium abundance to be established in times shorter than 10 fm/c. This is due to the abundant presence of gluonic excitations, allowing for quark-pair production in gluon-gluon collisions, c.f. Section 2. In a sense, therefore, abundant strange antibaryon production is indicative of an environment in which gluon collision processes are an essential element of the reaction picture.

In conclusion, we have shown on the basis of a dynamical theoretical model of strange particle production in nuclear collisions that normally rare strange antibaryon particles with multiple strangeness content provide a very promising experimental signal in the search of the quark-gluon plasma. Abundant strangeness production is indicative of the presence of gluon excitations, a characteristic property of the deconfined QCD phase.

REFERENCES

1. A. Bamberger et al. (NA35 collaboration) Phys. Lett. B184 (1987) 271. The different CERN experiments, presently using the oxygen beam provided by SPS are presented in the CERN report 'Experiments at CERN in 1986'.

2. R. Stock, for NA35 collaboration, lecture at DPG-meeting, Groningen, 23 March, 1987. H. Gutbrot, for WA80 dto; H. Specht for NA34, dto. Y. Sirois, for NA34 at the Rencontres des Moriond, Les Arcs, 19 March 1987; D. Vranic, for NA35 dto, Garpman, for WA80 dto.

3. J. Rafelski and B. Müller, Phys. Rev. Lett. 48 (1982) 1066 and 56 (1986) 2334(E).

4. P. Koch, B. Müller and J. Rafelski, Z.Phys. A324 (1986) 453.

5. P. Koch, B. Müller and J. Rafelski, Phys.Rep. 142 (1986) 167.

6. P. Koch and J. Rafelski, Nucl. Phys. A444 (1985) 678.

7. T. Biro and J. Zimanyi, Nucl. Phys. A395 (1983) 525.

8. N.K. Glendenning and J. Rafelski, Phys. Rev. C31 (1985) 823

9. A. Casher, H. Neuberger, and S. Nussinov, Phys. Rev. D20 (1979) 179. N.K. Glendenning and T. Matsui, Phys. Rev. D28 (1983) 2890.

10. J. Rafelski and M. Danos, "Possible Signature for/and Early Hadronization Mechanisms of Quark Gluon Plasma", CERN-TH 4686/87, Phys. Lett. B, in press.

11. M. Jacob and J. Rafelski, "Longitudinal $\overline{\Lambda}$-Polarization, Ξ Abundance and Quark Gluon Plasma Formation", CERN-TH4649/87, Phys. Lett. B, in press, and M. Jacob, these proceedings.

First Results from the CERN Light Ion Program

G.W. London

DPhPE, Saclay, F-91191 Gif-sur-Yvette, France

1. INTRODUCTION

Detailed studies of ultra-relativistic nucleus-nucleus interactions are only beginning. The aims of this on-going program are discussed in reference [1] and can be summarized as studies of:

- parton interactions in space-time in large (i.e. nuclear) volumes at extreme conditions of energy density and baryon density

- nuclear matter in and out of equilibrium

- the thermodynamics of dense hadronic matter

- the complicated QCD phase diagram which exhibits unconfined partons as a phase of hadronic matter.

The first results from the 17-day CERN 1986 oxygen run at 200 GeV/nucleon are beginning to become available. Since the theoretical framework for the analysis of these data is not compelling, we need experimental guides and models to learn how to proceed. In this paper, the first (April 1987) results from the CERN light ion program are presented and compared to what one expects from p-nucleus interactions.

2. EXPERIMENTAL POSSIBILITIES

2.a. Beam composition

In Fig.1, is a measurement of a typical beam composition (measured in the HELIOS beam) where the abscissa represents the total energy of the beam = 200 GeV/nucleon x A of the beam particle. The oxygen component, at 3200 GeV, represents (94±1.3)%, the largest background being He^4, $\simeq 1.6$%.

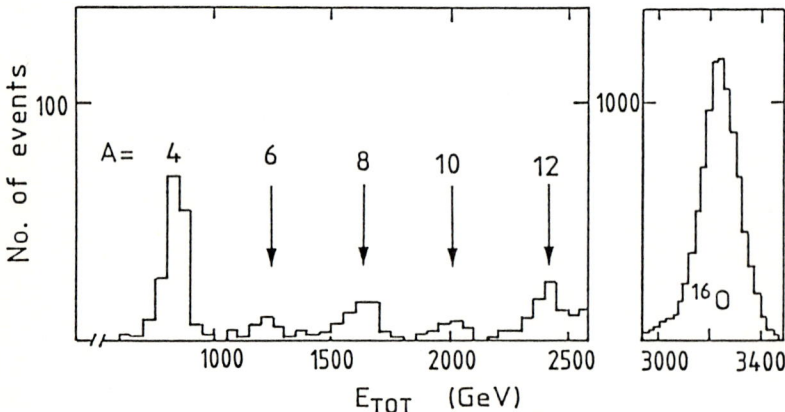

Fig.1. Beam composition

2.b. Target considerations (multiple interactions)

Among many experimental considerations, it is important to avoid multiple interactions in the target, in particular in order to measure accurately the transverse energy ($ET \equiv \Sigma E_i \cos\theta_i$, i=1,cells) of the reaction (see section 3.b for the importance of this parameter). The HELIOS experiment has measured the interaction rate as a function of ET (in the pseudo-rapidity region of $-0.1 \leq \eta \leq 2.9$) for two target thicknesses, 0.1 mm W and 1.0 mm W, respectively about 0.1% and 1% of an interaction length. The ratio, R=10*rate(0.1mm)/rate(1mm), is shown as a function of ET in Fig.2. Up to ET≃100 GeV, the ratio is consistent with 1 but, for example at ET≃175 GeV, the ratio is close to 2! Thus at this ET, 50% of the interactions on a 1% target are multiple and one cannot know what interaction produced the signal one is measuring as a function of ET, e.g. di-muons.

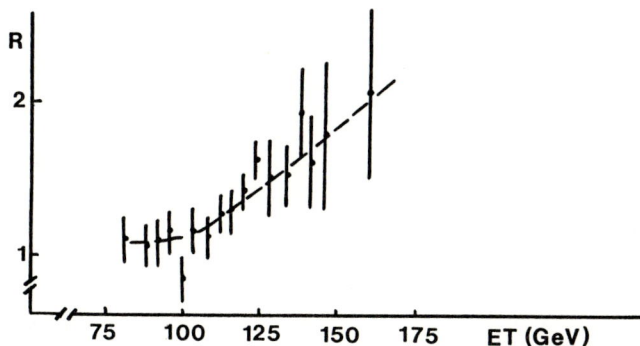

Fig.2. Multiple interactions in target vs thickness

2.c. Four large experimental setups (in 1986)

We describe briefly the four major experiments which took data in 1986.

The HELIOS (or NA34) experiment [2] consists of 4π fine-grained calorimetry which measures $d(E,ET)/d\eta$, silicon detectors which measure the charged multiplicity in the central region and over most of the forward and backward regions, a magnetic spectrometer which identifies charged particles and photons in the backward direction, and a di-muon spectrometer which covers all masses in the forward direction.

The NA35 experiment [3] consists of a streamer chamber which measures charged particles, Λ^0 and K^0 in the backward hemisphere, calorimetry which measures $d(E,ET)/d\eta$ in the central and forward regions, and a very forward calorimeter which measures the energy of the projectile fragments.

The NA38 experiment [4] consists of an electromagnetic calorimeter which measures $dET^{em}/d\eta$ in the central region and a di-muon spectrometer which covers the forward hemisphere for all masses. A notable feature of this experiment is its high data-rate capabilities.

The WA80 experiment [5] consists of the Plastic Ball detector which identifies positive particles and fragments in the very backward direction, time-of-flight counters which give particle identification in the backward region, a lead glass array which measure gammas in backward region, scintillator detectors which measure the charged multiplicity in the central and forward regions, calorimeters which measure $d(E,ET)/d\eta$ in the central and forward regions, and a very forward calorimeter which measures the energy of the projectile fragments.

In addition, a number of emulsion exposures were made in 1986, one in conjunction with the HELIOS spectrometer.

A summary of the various detectors is given in Table 1:

TABLE 1: Comparison of the Capabilities of the Different Detectors

	target λ	ET,E flow	Eveto	μμ	chgd mult	γ	identified particles
NA34	.01-.001	-0.1<η<6	±1.4°	3<y<6	0.9<η<5	1<y<2	0.7<y<2 π,K 0.2<y<1.5 p,p̄
NA35	.01	2.25<η<5.9	±0.3°		0<η<3	2.3<y<3.5	0<y<3 π⁻,Λ,K_s
NA38	.01 x10	1.8<η<4 e.m.		3<y<4			
WA80	.005	2.4<η<5	±0.3°		η<5	1.7<y<2.4	y<1.3 π⁺,p,d 1.3<y<2.4 chgd
emulsion					4π		

Therefore, when the data are finally analysed, we can expect the following measurements from the combination of the different experiments:

TABLE 2. Physics capabilities of the combined experiments

$dE/d\eta$	$-0.1 \leq \eta$
$dE_T/d\eta$	$-0.1 \leq \eta \leq 5$
$d\sigma/dE_T$	$-0.1 \leq \eta \leq 5$
$d\sigma(\pm)/d\eta$	$0 \leq \eta \leq 5$
$d\sigma(\pi\pm)/dydp_T$	$0.7 \leq y \leq 2$
$d\sigma(K\pm)/dydp_T$	$0.7 \leq y \leq 2$
$d\sigma(p)/dydp_T$	$-1.5 \leq y \leq 1.5$
$d\sigma(\bar{p})/dydp_T$	$0.2 \leq y \leq 1.5$
$d\sigma(\Lambda)/dydp_T$	$0 \leq y \leq 3$
$d\sigma(K_S)/dydp_T$	$0 \leq y \leq 3$
$d\sigma(\mu\bar{\mu})/dydp_T dM$	$3 \leq y \leq 6$
$d\sigma(\gamma)/dydp_T$	$1 \leq y \leq 3.5$

In addition, correlations among the above quantities are possible and no doubt necessary for an understanding of these complex processes.

3. REFERENCE MODEL

To understand whether the results (or those to come) are unusual, one must compare them to reasonable extrapolations of p-A physics. This is best done via p-A data or, lacking that, via a physics generator which reproduces p-p and p-A physics and which keeps its parameters fixed after having chosen an extrapolation method to B-A interactions. We have chosen the IRIS generator. The generator and its results are described in detail in the Saclay report, DPhPE 86-06 and in the Asilomar proceedings [6]. A very brief description is to be found in the Appendix.

IRIS is a model incorporating the Dual Parton Model for non-diffractive small PT processes, with higher order terms, and extended to diffractive and hard processes. The hadronization is made via the LUND program.

4. PRELIMINARY RESULTS

4.a. Charged multiplicity vs pseudo-rapidity

The energy density is directly proportional to the multiplicity density, $dN/d\eta$, which is an important parameter in producing large ET (see section 4.c.).

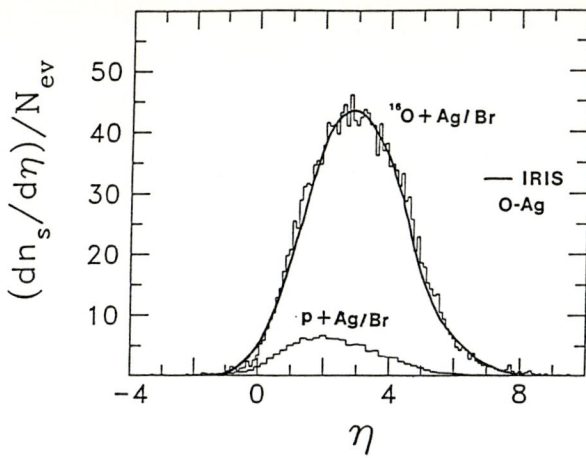

Fig.3. KLM Collaboration: charged multiplicity vs pseudo-rapidity

The multiplicities produced in nucleus-nucleus collisions are much larger than those in proton-nucleus collisions at the same energy. At 200 GeV/nucleon, the Krakow-Louisiana-Minnesota (KLM) collaboration has measured in emulsions [7] that the O-Ag/Br multiplicities are about 7 times that for p-Ag/Br.

In Fig.3, the normalized pseudorapidity distributions for relativistic singly charged secondaries, n_s, produced at 200 GeV/nucleon in (p, ^{16}O)+Ag/Br are presented. The events are required to have 15 heavily ionizing tracks at the vertex, thus selecting violent collisions. One observes a large increase in the pseudorapidity density over most of the pseudorapidity range. As in the following, the IRIS predictions are superimposed on the data, in this case relatively normalized since the selection criterion is difficult to reproduce absolutely.

4.b. Charged multiplicity vs forward energy

The energy measured in the very forward direction (±0.3°) is basically the energy of the non-interacting projectile fragments; i.e. if n nucleons of the 16 projectile nucleons interact, then (16-n) x 200 GeV goes in the very forward direction. The small number of produced particles makes a small correction on this. Thus if there is a lot of forward-going energy, few nucleons have interacted and a small multiplicity is expected. Conversely, if there is a small amount of forward-going energy, many nucleons have interacted and a large multiplicity is expected. This is what we see in Fig.4 [8].

We note that for targets with decreasing A, the data approach the y-axis less. This is probably due to the geometry wherein a small target (carbon is even smaller than oxygen!) intercepts less of the projectile nucleons. The IRIS predictions are for O-Pb and O-Al.

4.c. Transverse energy

Another parameter which measures the violence of the collision is the transverse energy which is related to the number of interacting beam <u>and</u> target nucleons. Very large transverse energy indicates not only that all the beam nucleons have interacted but also that an unusually large number of target nucleons have interacted. These events determine the tail of the ET distribution. This parameter has been measured in different regions of rapidity, for example, in the backward hemisphere (Fig.5a) [9] and in the central region (Fig.5b) [10].

In addition to comparisons with IRIS, in Fig.5b, a comparison has been made to p-Pb data measured in the same apparatus, convoluted 16 times to simulate a central collision. One sees that the data are well reproduced; to a certain extent, ET scale errors are minimized.

4.d. ET/particle in backward region

It is important to determine whether high ET events are due primarily to an increase in multiplicity, as indicated in section 4.a., or to an increase also in ET/particle. The latter has been measured in the backward region, near the peak of the ET distribution (vs η). In Fig.6 [9], the charged multiplicity is plotted vs ET (charged + neutral).

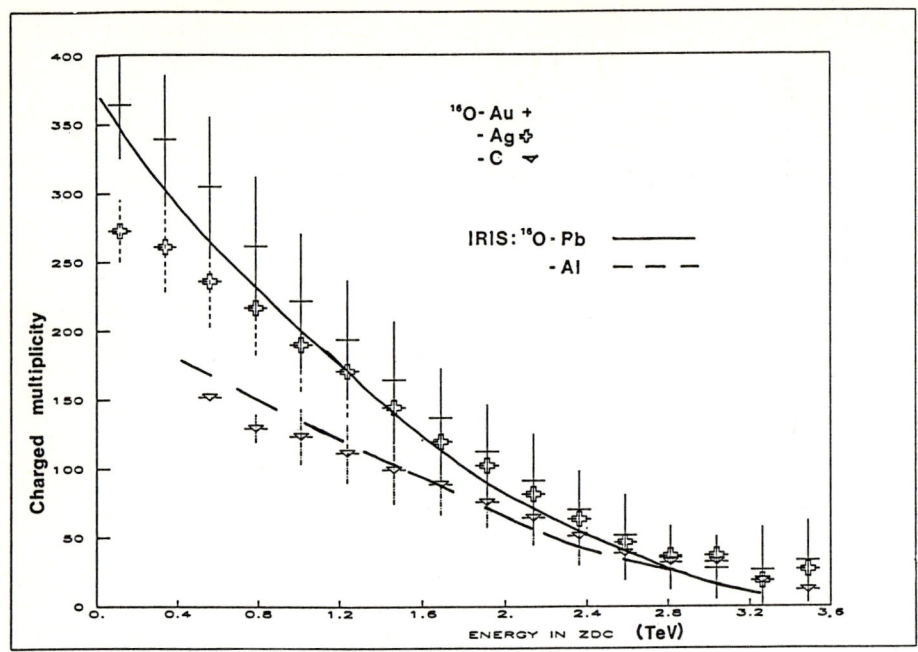

Fig.4. WA80: charged multiplicity vs energy in very forward direction

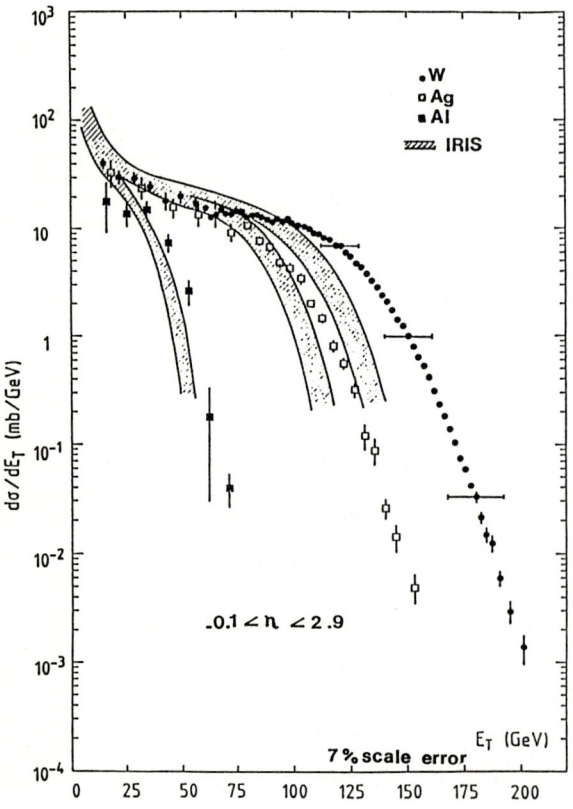

Fig.5a. (caption see opposite page)

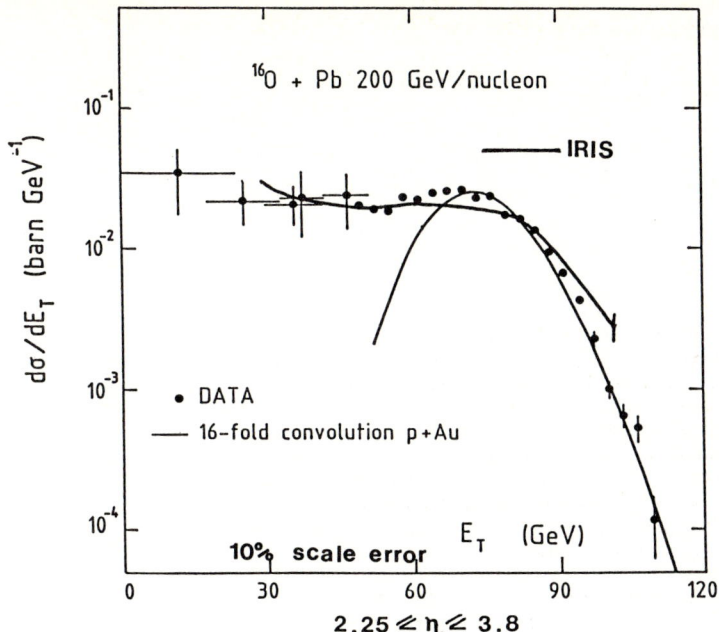

Fig.5. ET distributions: a) HELIOS (backward hemisphere) and b) NA35 (central region)

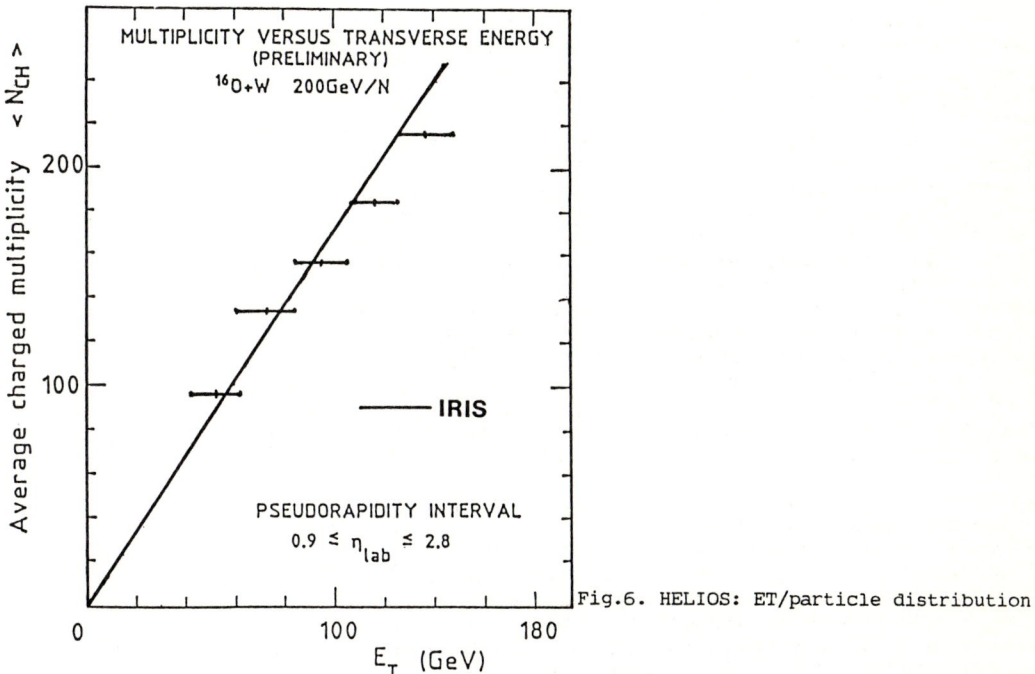

Fig.6. HELIOS: ET/particle distribution

We see that there is a linear relationship. The charged contribution to ET is 59% as given by the IRIS physics generator, No striking dependence on ET is observed. giving an <ET>/charged particle = 350 MeV/c.

4.e. PT spectra for charged particles and photons

The transverse momentum spectrum should reflect the temperature, if there is thermalization, at which the particles become non-interacting [11], parametrized as exp(-PT/T). For hadrons, and γ from neutral pion decay, this temperature is that after hadronization and reflects the latter stages of the interaction which are somewhat independent of the early stages. This implies for example that T(hadrons) should be approximately independent of ET. On the other hand, for direct or virtual photons, T is that essentially at the parton production level, thus reflecting strongly the early stages of the interaction, giving a strong dependence on ET.

We are only in the very beginning phase of this type of analysis. In particular, in Fig.7 [9] we observe the transverse momentum spectra of a) positive, b) negative particles and c) photons.

The negative particle distribution is dominated by π^- It is not yet well measured above 1 GeV/c and in particular a non-exponential behavior cannot be ruled out. The positive particle distribution includes π^+ and protons. The flatter distribution is probably due to the latter, which in IRIS have a exp(-4.5 PT) distribution. In Fig.7c, one compares the pW and OW photon PT distributions. Since the pW distribution is not well measured above 1-1.2 GeV/c, we cannot yet claim that there is a difference between pW and OW → photons. In Fig.8 [9] the average transverse momentum of charged particles as a function of ET is plotted. There is little dependence on ET, as expected.

4.f. Di-muon spectrum

A promising signal for the quark-gluon plasma comes from the detection of di-muons. In particular, a signature [12] would be the disappearance or reduction of the Ψ production rate relative to the continuum rate at about the same mass. This arises from the inability of charmed quarks in the quark-gluon plasma to form bound states, and the large mass of the Ψ prevents it from being formed later. In Fig.9 [13], we see the di-muon spectrum for events with a) low ET^{em} and b) high ET^{em} where the Ψ peak is quite apparent. Detailed comparisons have not yet been made to see if it is suppressed, in particular in comparison to comparable pA data.

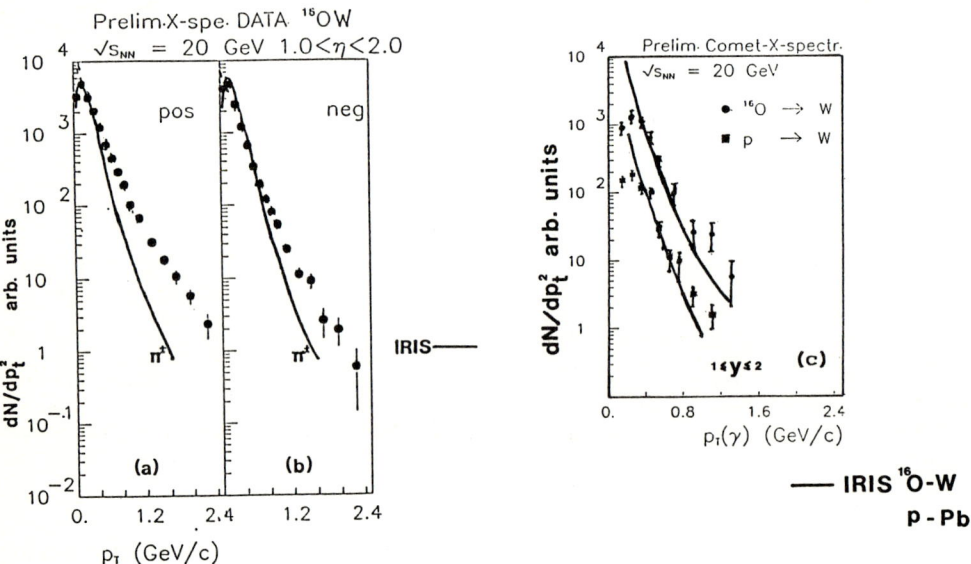

Fig.7. HELIOS: PT distributions a) positive, b) negative particles and c) photons

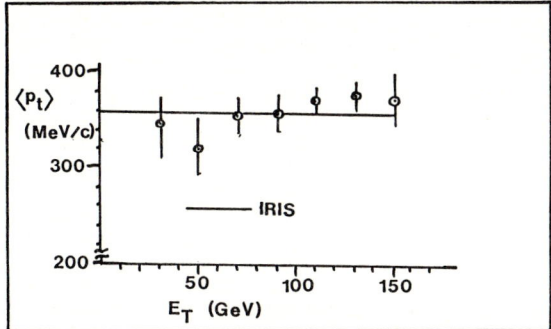

Fig.8. HELIOS: <PT> for charged particles vs ET

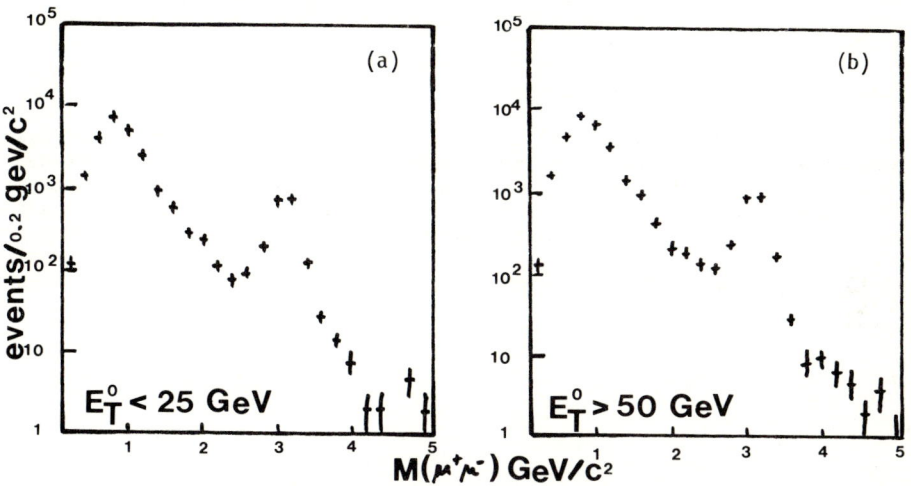

Fig.9. NA38: dimuon spectrum for a) low ET^{em} and b) high ET^{em}

5. TENTATIVE CONCLUSIONS: ARE THE PRELIMINARY RESULTS UNUSUAL?

We are just at the beginning of the analysis of the 1986 oxygen data; the measurements of Table 2 are not yet in, nor are the correlations among them. It is therefore premature to draw definitive conclusions.

The predictions of IRIS have been given in Figs.3-7. One observes that the data are reproduced quite satisfactorily within the expected 10 or 15%. The first preliminary results seem to be mere extrapolations of p-nucleus physics with additional geometrical effects. The collective effects necessary for thermalization have not yet been observed.

In 1987, similar data will be taken with a sulfur beam. There will be two new large experiments which will in particular extend the type of measurements to charged hyperons and anti-hyperons.

In my opinion, it is not surprising that new physics (i.e. collective effects) has not yet been uncovered. It has always been known, but sometimes forgotten, that oxygen (and even sulfur) nuclei are rather small and that 200 GeV/nucleon is a rather low energy. Clearly, we must extend our analysis into the ET tails for example to see something new. This is an exciting and hopefully rewarding task.

6. APPENDIX: IRIS event generator [6]

IRIS is a Monte Carlo generator of nucleon-nucleus and nucleus-nucleus interactions at high energy (≥ 50 GeV per nucleon for the projectile on a fixed target). The physics basis of the generator is the exchange of color.

The main goal of IRIS is the prediction of what to expect in nucleus-nucleus collisions as a result of a simple extrapolation from nucleon-nucleus collisions, i.e. the situation in which there is <u>no new physics</u>, for the various signals suggested for detecting the quark-gluon plasma: (1)$<p_t>$ of the produced particles; (2)rapidity density fluctuations; (3)lepton pairs; (4)strangeness production, as well as the E_T distribution, considered to measure the centrality (i.e. number of interacting nucleons) of the collision. The predictions and the descriptions of the data have to be done with a single set of parameters.

In IRIS, a nucleus-nucleus collision is an incoherent superposition of elementary collisions at the parton level. Individual collisions are described by the exchange of color and emission of gluons.

The starting point is the Dual Parton Model as developed by A. Capella and J. Tran Thanh Van. The hadronization of colorless strings is done with the Lund program. To this description of non-diffractive inelastic processes, we have added the diffractive process <u>and</u> hard processes. Hadronization follows the LUND program.

There are obvious parameters such as the atomic number, the charge and the momentum of the interacting particles. The other parameters are: A)the total and inelastic p-p cross sections used to compute the number of participating nucleons and the p-nucleus and nucleus-nucleus cross sections. B)the QCD parameter Λ; C)the non-perturbative cutoff, Q_0, of Odorico's program; D)the coefficients of the structure functions for low p_t physics; E)the quark and gluon masses; F)the intrinsic p_t of quarks and gluons, important for the E_T distributions; G)the cutoffs for hard scattering; H)the coefficients of the structure functions for hard scattering; I) the parameters of the hadronization are those obtained from the study of e+e- $\rightarrow$ hadrons at PETRA and are considered to be fixed, except for the diquark production which has been roughly adjusted in order to reproduce the production rate of antiprotons in the central region at the ISR. It is important to realize that the parameters are quite constrained and unchanged once set by e+e- and p-p collisions.

IRIS has not settled down yet and there are probably $\simeq 10\%$ prediction errors.

7. REFERENCES

1. L.McLerran, these proceedings.
2. NA34, SPSC 84-43
3. NA35, SPSC 85-36
4. NA38, SPSC 85-42
5. WA80, SPSC 85-35
6. J.-P.Pansart, Nucl. Phys. A461 (1987) 521c and references therein.
7. KLM Collaboration, preprint (May 1987)
8. WA80, private communication
9. HELIOS, private communication
10. NA35, Phys.Lett. B184 (1987) 271
11. G.W.London, Perspectives on Heavy Ion Physics at CERN, these proceedings,2.c.iii
12. T.Matsui and H.Satz, Phys. Lett. 178 (1986) 416
13. NA38, private communication

Perspectives on Heavy Ion Physics at CERN in the 1990s (or What Can We Gain from a Lead Beam?)

G.W. London

DPhPE, Saclay, F-91191 Gif-sur-Yvette, France

1. INTRODUCTION

Systematic studies of ultra-relativistic nucleus-nucleus interactions are just beginning. Apart from a few cosmic ray events, which have been quite instructive, and some studies of very light nucleus-nucleus (d-d and α-α) collisions at the ISR, this is virgin territory.

The aim of this program is the study of the space-time development of hadronic interactions in nuclear volumes under extreme conditions of energy density and baryon density. Bigger nuclei and higher energies enable us to explore these two variables over a very wide range selectively, exploring non-equilibrium and equilibrium aspects of nuclear matter. While we have studied the strong interaction of elementary particles over a large energy domain, we have not yet studied in parallel strongly interacting macroscopic systems at high energy and baryon densities. The thermodynamics of very dense matter is largely an unexplored field.

From the point of view of an experimentalist, the question is what new physics can be observed when we achieve high densities over large (i.e. nuclear) volumes. On the other hand, theorists have been predicting that there is a complicated phase diagram of QCD in which, at high temperatures and/or at high net baryon densities, partons become unconfined ("deconfinement") and massless ("restoration of chiral symmetry")[1]. Perhaps these two phase transitions coincide.

These two points of view are far apart in practical terms. The thermalized QCD world with its chemical potential (Fig.1) is quite separate from the measurements made in experiments such as the rapidity density of charged particles, Fig.2a, or the (proton-antiproton) rapidity density, Fig.2b. Nonetheless, it is necessary to find some (model-dependent) way to connect $dN^{ch}/d\eta$ to the energy density and, eventually, to the temperature, and to connect $d(p-\bar{p})/dy$ to a measure of the chemical potential. Lattice gauge calculations give a deconfinement temperature $\simeq$ 200 MeV (at zero net baryon density) with an error of, say 20% and thus an energy density ($\propto T^4$) within a factor of 2. The critical net baryon density (at zero temperature) may vary between 4 and 10 times the nucleon density in ordinary nuclear matter.

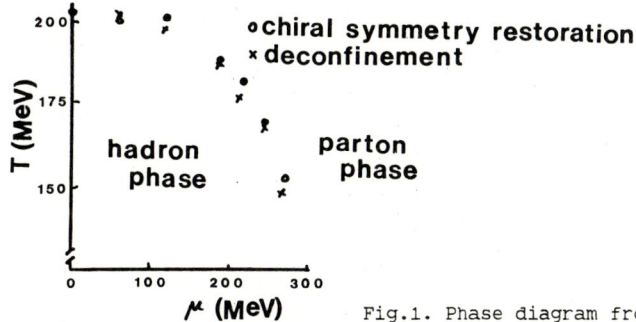

Fig.1. Phase diagram from H.Satz, Nature 324 (1986) 116

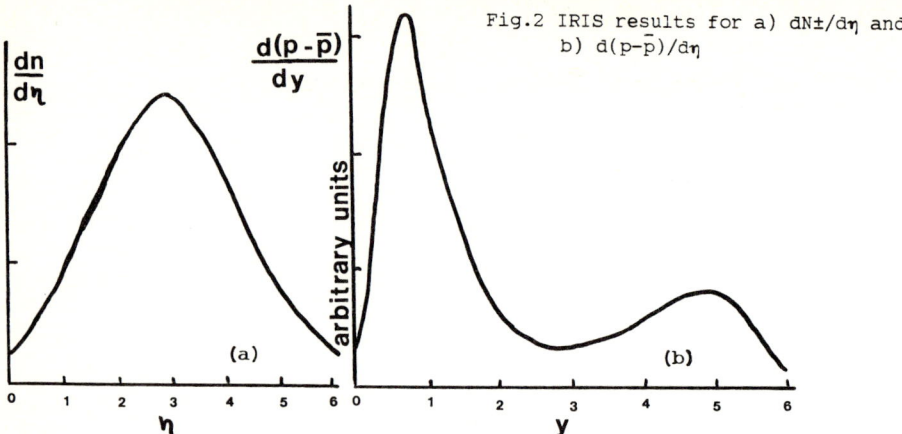

Fig.2 IRIS results for a) $dN\pm/d\eta$ and b) $d(p-\bar{p})/d\eta$

All this argues for a systematic study of ultra-relativistic nucleus-nucleus interactions. In the first place, the present program needs to be extended beyond 1987. In the longer run, continuing from the Oxygen and Sulfur studies at CERN energies, we need to extend the range up to very heavy projectiles, such as lead. This leads to the subject of this report: "perspectives in the 1990's at CERN for heavy projectiles".

Using the hydrodynamic model, we shall give the arguments which demonstrate the interest of nuclear beams with large atomic mass even at a small sacrifice of beam momentum/nucleon. In particular the A-dependences of the energy density and net baryon density as well as possible initial temperature and plasma and mixed phase lifetimes will be given.

It should be obvious in a qualitative sense that large nuclei will better approximate the hydrodynamic model which demands that the scattering length of the constituents be small relative to the transverse size of the system, facilitating thermalization.

2. SOME INTERESTING A-DEPENDENCES IN NUCLEUS-NUCLEUS INTERACTIONS

2.a. Simulation of Pb-Pb events and comparison to O-Pb and Ca-Pb

Minimum bias events were generated using the IRIS generator [2] for O-Pb and Ca-Pb interactions at 200 GeV/nucleon and Pb-Pb interactions at 160 GeV/nucleon (see next section for momentum choice). To give some feeling for the experimental difficulties and opportunities, we present in the following table some of the characteristics of the events.

Note that for "central collisions", $dN/dy \propto A^1$. and that $d(B-\bar{B})/dy \propto A**0.9$, where A is that of the projectile.

2.b. Beam momentum/nucleon, longitudinal beam size and minimal formation time

Theoretical treatments have generally considered the case of extreme energies where the γ_{cm} is very high. One then worries about quantum fluctuations, wee partons and minimal formation times not given by the geometry [3]. At CERN energies, we are not quite in this regime. In this case, the beam momentum/nucleon is Z/A 400 GeV/c. This gives 200 GeV/nucleon for beam projectiles up to Ca; for lead beams, we have 160 GeV/nucleon. The γ factors are 10 and 9, respectively, in the nucleon-nucleon system, with longitudinal dimensions of $R/\gamma = 0.28$ (oxygen) and 0.72 fm (lead). These dimensions give the minimal formation time for a plasma formed in A-A collisions at these energies.

TABLE 1: Comparison of O-Pb, Ca-Pb and Pb-Pb generated events

	O-Pb	Ca-Pb	Pb-Pb
MINIMUM BIAS INTERACTIONS			
<beam participants>	8	16	53
<target participants>	20	29	53 (symmetrical)
<ET>	70 GeV	130	345
<N charged>	130	245	665
N charged for 1‰σ	430	920	2950
dN(charged)/dy at y*=0	60	115	340
<N baryons>	30	45	105
d(B-B̄)/dy at y*=0	1.5	3	12
CENTRAL INTERACTIONS ≡ beam participants > 90% beam nucleons			
% of events	20	13	2
dN(charged)/dy at y*=0	145	370	1610
d(B-B̄)/dy at y*=0	10	20	90

2.c. A-dependencies for the interactions of identical nuclei

Within the context of a hydrodynamical model, whose hypotheses we shall list, we wish to derive some A-dependencies for the interactions of identical nuclei.

2.c.i. Hypotheses

We list the hypotheses which we have used:

- identical nuclei

- thermalization

- cylindrical symmetry (sometimes known as Lorentz boost invariance), applicable to a plateau in rapidity, for example in the central region

- neglect of transverse expansion

- conservation of entropy

At the end of this section, we shall make some comments on these hypotheses.

2.c.ii. Temperature-time scenario

Let us recall the standard temperature-time scenario (see Fig.3):

- The interaction occurs at proper time $\tau \equiv 0$ and the partons are formed at τ_p.

- The partons interact (one must use kinetic theory as a description) and thermalize at τ_i with temperature T_i.

- If this temperature is superior to the confinement temperature, T_c, the partons remain unconfined. The temperature falls mainly via the longitudinal expansion which can be described by hydrodynamics. This is the pure plasma phase.

- The partons begin to be confined at τ_c (and T_c) via a first order transition and enter the mixed phase. During this phase, the longitudinal expansion occurs, as well as some transverse expansion, at least for oxygen. See section 2.c.ix.

- At the end of this mixed phase, at τ_h, the partons have all hadronized and we enter the interacting hadron phase.

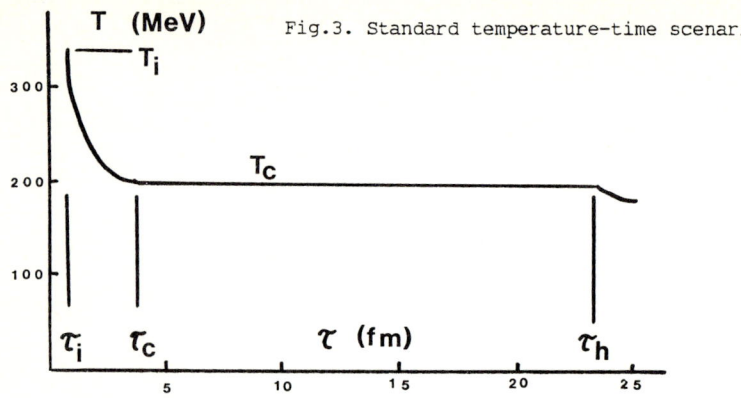

Fig.3. Standard temperature-time scenario

- At some later time and temperature, τ_f and T_f, the hadrons are free-streaming and are ready to enter our detectors. In this idealized picture, the transverse momentum distribution of the hadrons should reflect T_f (while that of real and virtual photons should reflect their formation temperature), see section 3.e [2a], and interferometry should measure the space-time volume at τ_f.

In the next sections, we shall make an estimate of the A-dependence of the initial net baryon density and energy density, of τ_i, T_i, τ_c and τ_h, and thus of the plasma phase lifetime, $\tau_c - \tau_i$, and the mixed phase lifetime, $\tau_h - \tau_c$.

2.c.iii. A-dependence of initial net baryon density

The two independent variables of the phase diagram (Fig.1) are temperature and net baryon density. For the latter we make an estimate relative to the ordinary nucleon density in nuclear matter: $n_o \simeq 1/7$ nucleons/fm^3. A simple calculation gives for the central region

$$d(B-\bar{B})/dV = d(B-\bar{B})/dy * dy/dV.$$

From cylindrical symmetry, we have $dV=\pi R^2 dz$ and, since $z=\tau_i \sinh y$, $dz/dy=\tau_i \cosh y \simeq \tau_i$ at $y \simeq 0$. Therefore

$$d(B-\bar{B})/dV \simeq d(B-\bar{B})/dy * (1/\pi R^2 \tau_i)$$

$$\propto A**0.9/(A**2/3 * A**-1/6)$$

[see sections 2.c.iv for A-dependence of τ_i and 2.a for A-dependence of $B-\bar{B}$]

or, $d(B-\bar{B})/dV \propto A**0.4$. (1)

This gives a factor of 2.8 for Pb-Pb relative to O-Pb. For Pb-Pb, the net density relative to the nucleon density in ordinary matter is

$$d(B-\bar{B})/dV*(1/n_o) = 7 * 90/\pi \tau_i (7.1)^2 = 4/\tau_i \quad \text{(see Table 1)}.$$

Thus for theoretical values of $\tau_i \leq 1$ fm, the baryon density for central Pb-Pb interactions in the central region is <u>at least 4 times</u> the ordinary nucleon density.

2.c.iv. A-dependence of thermalization time and initial temperature [4]

For the following, we have taken the gluon degrees of freedom = 16 (8 color states x 2 spin states), the quark degrees of freedom = 24 (3 color states x 2 spin states x 2 flavors x 2 for quarks and anti-quarks) and the hadron degrees of freedom = 3 for pions (surely an underestimate, but unimportant for the A-dependence). Thus the energy density for a massless parton gas is (see Appendix)

$$\varepsilon = \frac{3\zeta(4)T^4}{\pi^2(\hbar c)^3} [16 + 24 \times 7/8]$$ where $\hbar = h/2\pi$. (4),

For massless bosons, the entropy density is related to the nucleon density: $s = 4n\zeta(4)/\zeta(3) = 3.6n$. In particular, for a pion gas (i.e. the final state)

$$n(\pi) = s^f/3.6 \quad \text{or}$$

$$dN/d\eta = (s^f/3.6) \, dV/d\eta, \quad \text{where } n = dN/dV.$$

As above, we have ($dV/d\eta \simeq dV/dy$ in the central region)

$$dN/d\eta = \pi R^2 s^f \tau_f / 3.6.$$

From entropy conservation we have $s^f \tau_f = s^i \tau_i$ and therefore

$$dN/d\eta = \pi R^2 s^i \tau_i / 3.6 \quad \text{or}$$

$$dN/d\eta \propto R^2 T_i^3 \tau_i, \quad \text{since } s \propto T^3.$$

For central collisions, $dN/d\eta \propto A^1$, see section 2a. Therefore

$$T_i^3 \tau_i \propto A^{**}1/3.$$

T_i and τ_i are sensitive to quantum fluctuations, which can be ignored if $T_i \tau_i \gg \hbar$. In particular, the product is dimensionless and therefore independent of A. Thus $T_i^2 \propto A^{**}1/3$ or

$$T_i \propto A^{**}1/6, \tag{2a}$$

$$\tau_i \propto A^{**}-1/6. \tag{2b}$$

The A-dependence of the energy density ($\propto T^4$) is $A^{**}2/3$ and not $A^{**}1/3$, which is from an earlier analysis [3]. This is due to the A-dependence of τ and $\langle p_t \rangle$ ($\propto A^{**}1/6$) [5], which Bjorken took as A-independent. Note that the parton model [6] and the flux tube model [7] give the same A-dependencies.

In specific hydrodynamic models [7-9], one gets $T_i \equiv T_0 A^{**}1/6 \simeq (140-200) A^{**}1/6$ MeV and $\tau_i \equiv \tau_0 A^{**}-1/6 \simeq (0.2-1.4) A^{**}-1/6$ fermi, which gives for oxygen and lead beams, taking the more conservative values:

TABLE 2: *Model-dependent values for initial energy density, T and proper time*

	oxygen	lead
initial energy density (MeV/fm^3)	3.7	21
initial temperature (MeV)	220	340
thermalization time (fm)	0.9	0.6

One must remember that the thermalization times can be limited by kinematics as indicated in section 2b, in particular for the lead case.

2.c.v. A-dependence of confinement time: lifetime of plasma phase

If entropy is conserved during the expansion and cooling from the time of initial thermalization to the confinement time, and if we neglect the transverse expansion, justified in 2.c.ix., we have

$$\tau_i T_i^3 = \tau_c T_c^3.$$

Since T_c is fixed, we have

$$\tau_c = \tau_0 (T_0/T_c)^3 A^{**1/3}. \quad (3)$$

The conservative oxygen and lead confinement times are 1.2 and 3.8 fm respectively.

The plasma lifetime, $\tau_{plasma} \equiv \tau_c - \tau_i$, is

$$\tau_{plasma} = \tau_0 [(T_0/T_c)^3 A^{**1/3} - A^{**-1/6}]. \quad (4)$$

Comparing two different nuclei for their plasma lifetimes, we have

$$R \equiv \frac{\tau_{plasma}(A)}{\tau_{plasma}(B)} \geq (A/B)^{**1/3} \quad \text{(neglecting the } A^{**-1/6} \text{ term)}$$

$R_{min} = 2.3$ for A=208 and B=16.

2.c.vi. A-dependence of hadronization time: lifetime of mixed phase

Likewise, if entropy is conserved during the mixed phase and the transverse expansion is neglected, we have $\tau_h s^h = \tau_c s^c$. The only difference between the entropy densities, at fixed temperature T_c, is the number of degrees of freedom. Therefore

$$\tau_h = \tau_c \, 37/3 \quad \text{or}$$

$$\tau_h \propto A^{**1/3} \quad (5)$$

and the mixed phase lifetime, $\tau_{mixed} \equiv \tau_h - \tau_c$,

$$\tau_{mixed} \propto A^{**1/3}. \quad (6)$$

We remind the reader that the factor 37/3 is overestimated by maybe a factor 2 since the meson degrees of freedom are greater than 3.

2.c.vii. A-dependence of di-lepton signal

The production of dileptons is very sensitive to the initial conditions. This is because of the exponential Maxwell-Boltzmann factor $\exp[-E/T(t)]$ which must be integrated over the time evolution of the temperature. For example, in reference [10], an explicit form is given for this integral:

$$d\sigma/[d(P_T c)^2 \, dy \, d(M^2 c^4)] = 2\pi \sigma_{tot} \alpha^2 F(Pc/Mc^2) \Sigma e_q^2 / (hc)^4 \int V(\tau) \exp[-E/T(\tau)] \, d(c\tau),$$

where
- σ, M, P_T, y, E and P refer to the dilepton
- σ_{tot} = total nucleus-nucleus cross section
- $F(x) = [1 + \arctan(x)(1+x^2)/(2x)]/2$
- $T(\tau)$ = time evolution of the temperature
- $V(\tau)$ = time evolution of the volume = $\pi R^2 \tau$
- α = fine structure constant
- Σe_q^2 = sum over charges of different flavors.

Within the context of a given hydrodynamical model [9], this type of integral was performed. As we can see in Fig.4, the production of dileptons, e.g. around 2 GeV/c^2, is particularly sensitive. At this mass, the plasma phase production dominates by 4 to 20 depending on initial conditions.

2.c.viii. A-dependence of mean free path

The mean free path, $\lambda = 1/\sigma n$, is inversely proportional to the parton cross section and the number density of the partons, given by

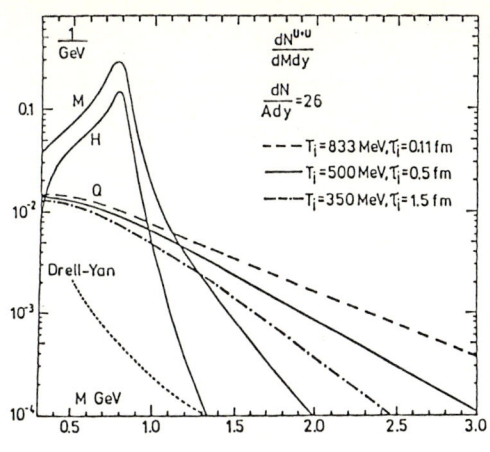

Fig.4. Sensitivity of di-lepton spectrum to the initial conditions

$$n = \frac{\zeta(3)T^3}{\pi^2(\hbar c)^3} [16 + 24 \times 3/4] \quad \text{where } 3/4 = c(3)/\zeta(3).$$

This is to be compared to the transverse dimension, $R \propto A^{**1/3}$. With $T \propto A^{**1/6}$, we have therefore

$$\lambda/R \propto A^{**(-5/6)}. \tag{7}$$

This gives a reduction of nearly an order of magnitude in the parton mean free path relative to the size of the system in going from oxygen to lead beams.

2.c.ix. Comments on hypotheses: work to be done

Even though we have restricted ourselves in general to A-dependences, there are a number of points which have to be worked on:

- Is there thermalization? Some work has begun on this difficult problem in the context of kinetic theory [11].

- What is the effect of a lack of plateau in the central region? Perhaps the cylindrical symmetry should be replaced by spherical symmetry as another extreme case. It appears [12] that for times ≤ radius/speed of sound the cooling is slower in the spherical case.

- How does the transverse expansion influence the mixed phase? The setting in of the transverse expansion occurs at a time = the radius of the system divided by the speed of sound, which for an ideal gas = $1/\sqrt{3}$, or at $\Delta\tau$=4.9 (11.3) fm for oxygen (lead). Thus the increase is 2.3 for lead relative to oxygen. Fig.5 shows the result of a hydrodynamic calculation [12] for the expansion as a function of time.

- What is the effect of entropy production, especially in the mixed phase?

- What is the influence of the limitation on the formation time given by the beam momentum?

- What signals other than di-leptons are very affected by increase in the plasma and mixed phases?

- What signals are affected by the large net baryon density?

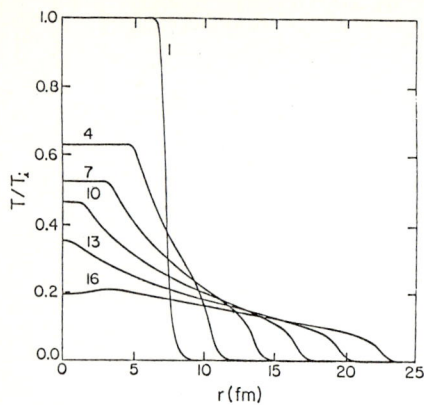

Fig.5. Temperature distribution as a function of r for cylindrical transverse expansion coupled to longitudinal expansion. Each curve is labeled by the time in fm that has elapsed since the collision with $\tau_i = 1$ fm

2.c.x. Conclusion

In Table 3, we collect all the A-dependencies discussed in the previous section:

TABLE 3: Summary of A-dependences for central collisions in the central region

	α for $A^{**}\alpha$
initial net baryon density	0.4
energy density	2/3
initial temperature	1/6
plasma lifetime	>1/3
mixed phase lifetime	1/3
thermalization time	-1/6
confinement time	1/3
hadronization time	1/3
relative mean free path	-5/6
onset of transverse expansion	1/3

In the comparison of oxygen and lead beams, we have shown in the context of a model that, for central collisions in the central region,

- the initial net baryon density increases by 2.8
- the initial energy density increases by 5.5
- the initial temperature increases by 1.5
- the plasma lifetime increases by at least a factor of 2.4
- the mixed phase lifetime increases by 2.3
- the di-lepton signal is very sensitive to the initial conditions
- the relative mean free path is decreased by nearly an order of magnitude

3. COMMUNITY INTEREST. HOW? WHEN? HOW MUCH?

The community interest for lead beams is quite large as expressed in the participation of the Heavy Ion Discussion Group which has produced a document for the CERN SPSC [13] with the following conclusions:

- Lead beams provide a large increase over oxygen beams (as indicated in this article)

- The lead injector is technically feasible with little technological development
- The cost for the injector is about 30 MSF
- Experiments can probably be extended
- The injector and experiments could be ready by 1991

4. APPENDIX: A very short reminder about thermodynamics

Let us calculate the number density of uninteracting massless bosons at temperature T. The average number with energy E is $1/[\exp(\beta E)-1]$ where $\beta=1/kT$, k being the Boltzmann constant. The number of states in volume V is $V\, d^3p/h^3$. This gives a density

$$dn = d(N/V) = d^3p/(hc)^3 \cdot 1/[\exp(\beta E)-1].$$

Integrating, we get

$$n = 4\pi/(hc)^3 \int p^2\, dp/[\exp(\beta E)-1] \quad \text{or}$$

$$n = 4\pi/(hc)^3\, T^3 \int_0^\infty x^2\, dx/[\exp(x)-1],$$

where the integral is the Riemann zeta function $2!\,\zeta(3)$. Therefore

$$n = T^3\, \zeta(3)/[\pi^2(\hbar c)^3] \text{ where } \hbar = h/2\pi.$$

In the following table, we have listed similar useful relations, derived in the same manner.

TABLE 4: Some useful thermodynamics relations

per degree of freedom	number density, n	energy density, ε	entropy density, s
uninteracting massless bosons	$\dfrac{\zeta(3)T^3}{\pi^2(\hbar c)^3}$	$\dfrac{3\zeta(4)T^4}{\pi^2(\hbar c)^3}$	$\dfrac{4\zeta(4)T^3}{\pi^2(\hbar c)^3}$
uninteracting massless fermions	$\dfrac{c(3)T^3}{\pi^2(\hbar c)^3}$	$\dfrac{3c(4)T^4}{\pi^2(\hbar c)^3}$	$\dfrac{4c(4)T^3}{\pi^2(\hbar c)^3}$

As can be seen, the zeta function, $\zeta(n)$, for bosons is replaced by a modified function for fermions, $c(n)$. Some values are $\zeta(3)\simeq 1.2$, $c(3)\simeq 0.9$, $\zeta(4)=\pi^4/90\simeq 1.08$ and $c(4)=7\pi^4/720\simeq 0.947$.

5. REFERENCES

1. Asilomar Quark Matter Conference, Nucl.Phys. A461 (1987) and references therein.
2. a) G.W.London, "First Results from the CERN Light Ion Program, these proceedings, and b) J.-P.Pansart, Nucl. Phys. A461 (1987) 521c and references therein.
3. J.D.Bjorken, Phys.Rev. D27 (1983) 140
4. J.-P.Blaizot, Saclay preprint PhT/86-141 (Zakopane 1986)
5. J.-P.Blaizot and A.Mueller, Saclay preprint PhT/87-22
6. A.Kerman, T.Matsui, B.Svetitisky, Phys.Rev.Lett. 56 (1986) 219
7. T.Matsui, Nucl.Phys. A461 (1987) 27c
8. R.Hwa, K.Kajantie, Phys.Rev.Lett 56 (1986) 696
9. P.V.Ruuskanen, Jyvaskyla JYFL 15/86 (Zakopane 1986)
10. J.Badier, G.W.London, M.Winter, Saclay preprint DPhPE 82-11 (1982); note that equations A.8 and A.9a must be multiplied by 2 and equation A.9b by 2π.
11. for example, G.Baym, Phys. Lett. 138B (1984) 18
12. G.Baym et al., Nucl.Phys. A407 (1983) 541
13. G.W.London, N.A.McCubbin and R.Stock (in preparation)

Part VI

Axions

The Emission of Isoenergetic Electron Positron Pairs from Very Heavy Ion-Atom Collisions*

K.E. Stiebing

Institut für Kernphysik der Johann Wolfgang Goethe-Universität,
D-6000 Frankfurt/Main, Fed. Rep. of Germany

In experiments at the heavy-ion accelerator UNILAC of the Gesellschaft für Schwerionenforschung (GSI) in Darmstadt the very strong dependence of dynamic atomic positron production processes on the total nuclear charge Z_{UA}, contrasting with positron emission by nuclear pair conversion, has been demonstrated [1,2] ($Z_{UA} = Z_{projectile} + Z_{target}$). In collisions of heaviest atoms like U+U, Th+Th, U+Th etc. the atomic processes of positron creation clearly dominate the total positron production providing a useful tool to test the behaviour of the quantum-electrodynamical vacuum under the influence of the strongest electric fields accessible in experiments [3]. The continuous positron energy spectra measured are in very good agreement both in shape and in magnitude with theoretical coupled-channel calculations [4,5,6]. The most interesting theoretical prediction, however, the emission of positrons by the spontaneous filling of vacant atomic levels embedded as a resonance in the negative energy continuum, when a critical value $Z_{UA} \approx 173$ is exceeded, has not been observed yet. This process is predicted to manifest itself by the emission of monoenergetic positrons, with an energy, reflecting just the resonance energy in the "Dirac Sea" [7], under the favorable condition that nuclear sticking at a time scale of $\tau_s \approx 10^{-19}$s prolonges the presence of the strong field.

First clear positron line structures, distinctly superimposed onto the continuous energy spectra of dynamic atomic and nuclear positrons, observed at GSI [8,9] supported the above explanation. However, systematic studies in overcritical as well as in undercritical [10,11,12,13] systems, investigating the expected extremely strong Z_{UA}-dependence of spontaneous positron emission revealed an independence of the line energies on the combined nuclear charge. A collection of positron-energy spectra for various collision systems is displayed in fig. 1. The measurements have been performed by the EPOS-collaboration at GSI, using a straight solenoidal transport system [8], to well separate the positron detection system from the target area. By means of the high resolution of the kinematic heavy-ion coincidence obtained in these measurements, a reduction of the background from the continuous positron processes was achieved by choosing slightly inelastic events ($\Delta Q \leq 20$ MeV) for the selection of the spectra of fig.1. The obvious near degeneracy in energy of these lines, all essentially positioned around $E_{e^+} \approx 340 \pm 30$ keV, is demonstrated in fig. 2, where the line energies are plotted as a function of the combined nuclear charge. Also shown are results obtained by the ORANGE-collaboration at GSI, working with an orange-type β-spectrometer [2]. The incompatibility with the spontaneous positron model, predicting a systematic decrease of the line energies with Z_{UA} and an absence for subcritical systems is obvious. Moreover, it appears that more than one line per collision system may have been identified.

In order to test obvious nuclear origins, in all experimental studies the γ-ray spectra have been measured simultaneously. No associated structures have been found to explain the lines by e.g. nuclear pair conversion. In the solenoidal system EPOS, one side of the spectrometer was built up to also simultaneously measure the high-energy electron spectrum ($E_{e^-} > 1$MeV). Again no signature for related structures from the conversion of radiatively forbidden E0-transitions could be identified in the e^--spectra. Additional information has been obtained from the line widths of the positron structures. For all systems a width of $\Delta E = 70$ keV independent of the associated heavy-ion scattering angles has been measured [11]. Under the assumption, that the natural linewidths

---

*work supported by Bundesministerium für Forschung und Technologie

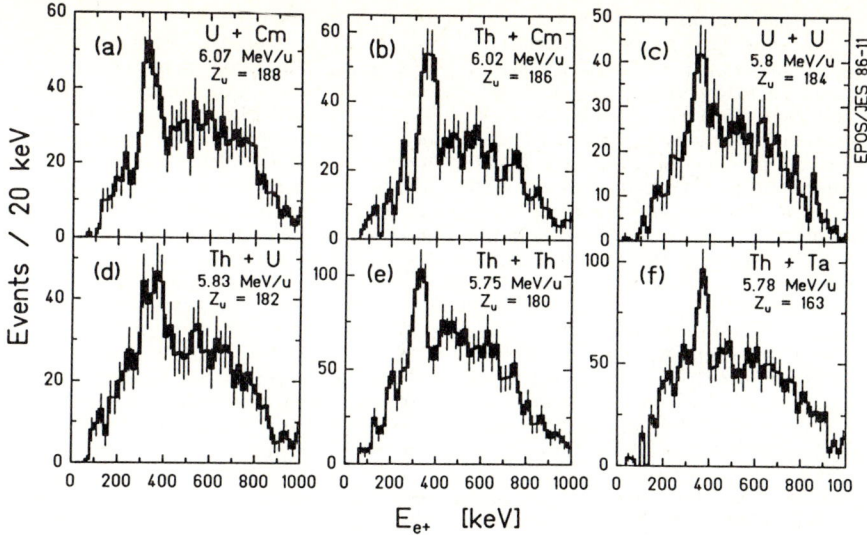

Fig.1: Positron-energy spectra for various collision systems and bombarding energies. The conditions of selection are given in the text.

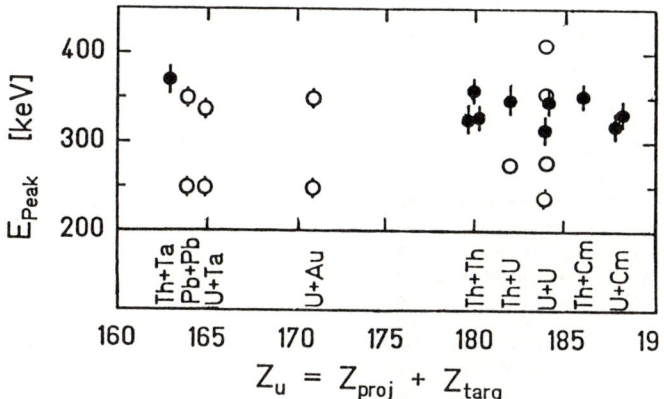

Fig.2: A summary of peak energies as a function of Z_{UA} for all single-positron stuctures closed circles: lab. energies measured by the EPOS-group ($E_{lab}= E_{inr.}$ for isotropic emission) open circles : data from the ORANGE-group in CM-units.

of the observed structures are negligible compared to the observed widths, the latter are essentially direct measures of the emitter velocity in the laboratory frame of observation (see also below). The constant velocities derived from the measured widths are inconsistent with the kinematic velocities for elastic scattering of the heavy ions. Therefore, emission from the separating nuclei, having been excited during the collision, has to be ruled out at a high degree of confidence.

Considering all results described so far leads to the conclusion that one or possibly more sources have been observed which are common to all systems investigated. The most direct picture of such a source emitting positrons isoenergetically for differently strong central potentials Z_{UA} is the decay of a neutral object into an e^+,e^- pair, occurring well separated from the nuclei to avoid final-state interactions. However, besides conceptional difficulties arising from the small line width ($\Delta E_{e^+} \approx 70 keV$), demanding extremely low kinetic energies of the object with respect to the heavy-ion rest frame, up to now no viable production mechanism has been found for such a source.

The clear experimental signature of such a hypothetical decay must be the simultaneous emission of an electron and positron of identical kinetic energy with an angular correlation of $\Theta_{e^+,e^-}=180°$ in the rest frame of the source. The EPOS apparatus therefore has been modified to allow the coincident detection of electrons and positrons in the energy range between 150keV and 800keV [14,15]. In fig. 3 the experimental set up is displayed. Special care has been taken to effectively suppress the intense background of δ-electrons created by the heavy-ion collision. For the positron detection it is a technique already used in the experiments described above, suppressing electrons within the energy range of interest by blocking them at a propeller-shaped baffle taking advantage of the different senses of rotation for electrons and positrons around the guiding center of their trajectories in the magnetic field. An unambiguous identification of positrons by their annihilation radiation is then performed in a NaJ assembly, the azimutal eight-fold separation of which can be used to also select conditions for the angular correlations of the coincident photons. On the electron side the well defined characteristic radial extension of the electron's and positron's spiralling path in the homogeneous magnetic field is utilized for a sharp cut off at low energies by means of two planar Si(Li)-detectors mounted upstream and downstream from the solenoid axis respectively (see illustration in fig.3). Their relative up-and-down dislocation enhances the efficiency for electron detection by a factor of ~4 compared to the positron efficiency, again taking advantage of the different senses of rotation. A blade baffle close to the target here additionally blocks the electrons and positrons of very low energies and those having a too large emission angle with respect to the solenoid axis. The use of one hemisphere for positron detection and the other for electron detection results in an enhanced coincidence efficiency of totally $\approx 4\%$ for pairs emitted "back to back" ($\Theta_{e^+,e^-}=180°$) compared to one of totally $\approx 2\%$ for spatially uncorrelated pairs. Another important feature of the system, already used also for the previous experiments, is its identical acceptance for leptons emitted backward and forward with respect to the heavy-ion CM-motion, which is a consequence of the rotational symmetry of the apparatus around an axis perpendicular to the beam direction. Therefore,the kinematic energy shifts observed in the laboratory frame when leptons of intrinsic energy $E_{lept.}$ are emit-

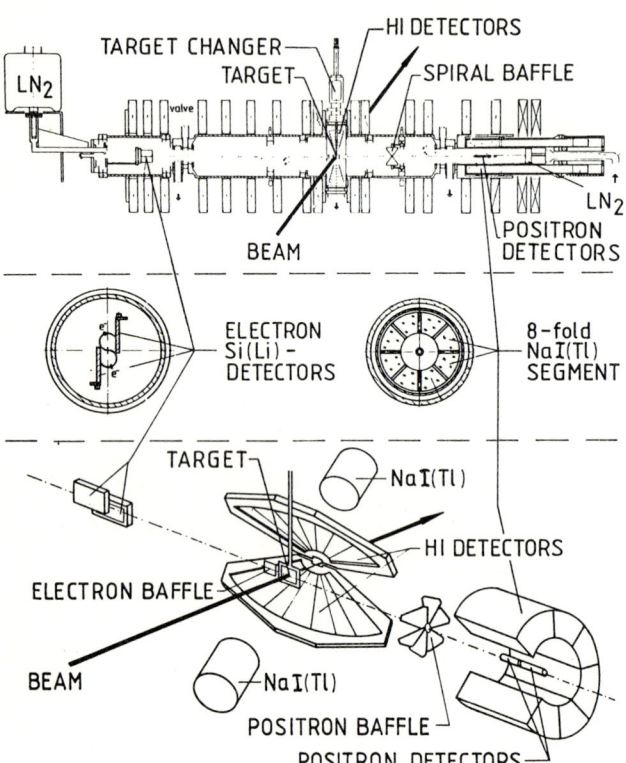

Fig.3: Schematic view (top) and perspective drawing (bottom) of the EPOS-spectrometer together with cross sections of the electron and positron detection systems (middle).

ted from a moving source add up to the observation of a practically unshifted line at $E_{lab}=E_{lept}$ to first order in β, if an integration over all emission angles, accepted by the spectrometer is performed. This line then is broadened according to the source velocity and the range of acceptance angles of the detection system. For the case of electrons and positrons measured in coincidence, an exact "back-to-back" correlation of isoenergetic electrons and positrons leads to a mutual cancellation of the kinematic line shifts in the sum-energy spectra $E_\Sigma = E_{e^+} + E_{e^-}$. In this case the resulting widths of the sum lines are given by the detector resolution and the intrinsic width. This cancellation is exactly correct for emission at rest in the heavy ion CM-system but also holds approximately for source velocities not much larger than that of the heavy-ion CM-system [16]. The difference-energy spectra $E_\Delta = E_{e^+} - E_{e^-}$ then exhibit a width being a convolution of the individual line widths and thus reflect the source velocity. The **ratio** of the sum-difference-energy widths, on the other hand, to a far extent is independent of the velocity but is a measure for the opening angle of the emission. In order to study the instrumental properties in detail, the E0-pair decay of a ^{90}Sr-source has been measured (E_γ = 1.74 MeV). Excellent agreement has been obtained with the predictions from Monte-Carlo simulations containing all important properties of the apparatus [16,17].

In a first series of experiments with the coincidence set up in summer 1985 a correlated emission of isoenergetic electron-positron pairs has been found in U + Th collisions at 5.83 MeV/u [14]. As shown in fig. 4 a sum-energy line at 760 keV with a width of 40 keV, being only 25% broader than the electron-detector resolution and distinctly smaller than expected from a convolution of the individual electron- and positron-line widths, could be identified. The sum spectrum shown in fig.4 is generated by projecting the events in the wedge-shaped cut, which takes into account the increase of the kinematic energy shift for the individual events with increasing lepton energies. In the second row of fig.4 it is demonstrated, that no structures could be found when neighbouring cuts were applied. The dashed lines represent the Monte-Carlo calculations normalized to the total coincidence yield. In another, shorter run during this experiment, investigating the collision system Th+Th, indications for correlated e^+,e^--emission of a sum-energy of $E_\Sigma \approx 600$ keV have been found [16].

More detailed studies of the U+Th collision system have been carried out in 1986. For these experiments the resolution of the electron detectors has been improved by LN_2-cooling to a value of ≈ 10 keV being now compatible with that of the positron detector. The mutual cancellation of the kinematic energy shifts in the sum-energy lines, already observed in the previous measurements, should therefore show up much more clearly now. In addition to this, the total coincidence efficiency was increased by a factor of ≈ 3 by enlarging the azimutal acceptance of the heavy-ion detectors. In these new experiments the correlated emission of e^+,e^--pairs has been reproduced from the body

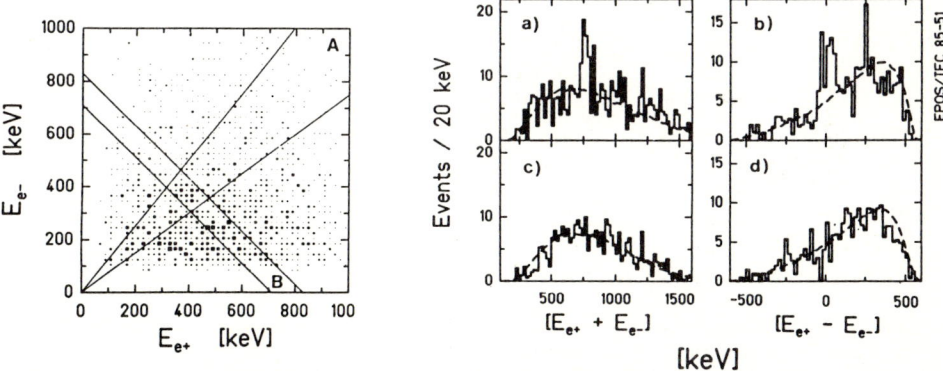

Fig.4: Sum- and difference-energy spectra of the electron-positron coincidences from 5.83 MeV/u U + Th collisions (right panel). The projection windows for the spectra denoted with a) and b) are indicated in the coincidence distribution (left panel; A, B). The spectra c) and d) are projections of cuts adjacent to A and B respectively.

of a much higher total coincidence statistic. The existence of more than one correlated structure, already indicated in the previous measurements, was also confirmed, including the 760 keV sum-energy line. The most distinct sum-energy lines, have been identified, however, at energies of $E_\Sigma=610$ keV and $E_\Sigma=810$ keV. The sum- and difference-energy spectra of these two lines are shown in fig.5. In these spectra all data from two independent measurements, separated in time by several months, are added. A selection has been made on the beam energy as given in the figure captions. Besides this energy dependence an almost complete separation of the two correlations could be achieved by a further selection according to the time-of-flight difference of the positrons relative to the start signal in one of the heavy ion detectors. For the positrons associated with the 610 keV correlation a window is applied, which is delayed by ≈2 ns compared to the events of dynamic and nuclear positrons, whereas the positrons belonging to the 810 keV correlation are prompt. The most probable explanation for this behaviour is a difference in the emission characteristics of both correlations, implying that the positrons associated with the 610 keV sum-energy line are emitted preferentially with large angles relative to the solenoid axis. The small sum-energy widths of the two lines of $\Delta E_\Sigma=25$ keV and $\Delta E_\Sigma=40$ keV for the lower and higher energy respectively are associated with broad difference-energy lines. This clearly identifies the mutual cancellation of kinematic shifts and demonstrates that the emitting source indeed is moving in the laboratory frame of observation. The ratios between corresponding sum- and difference-energy widths are compatible with emission correlations of $\vartheta_{e^+,e^-} \geq 165°$ for the 610 keV line and $\vartheta_{e^+,e^-} \geq 145°$ for the 810 keV line. This result again rules out processes emitting the pair with small opening angles like nuclear pair conversion, if not exotic angular correlations are involved. It favours on the other hand the assumption of common neutral sources. Besides the difficulty to understand the existence of a group of lines from the scope of an elementary object, up to now no independent information about the existence of such a new source has been found, in spite of substantial experimental effort undertaken at many places [18,19,20].

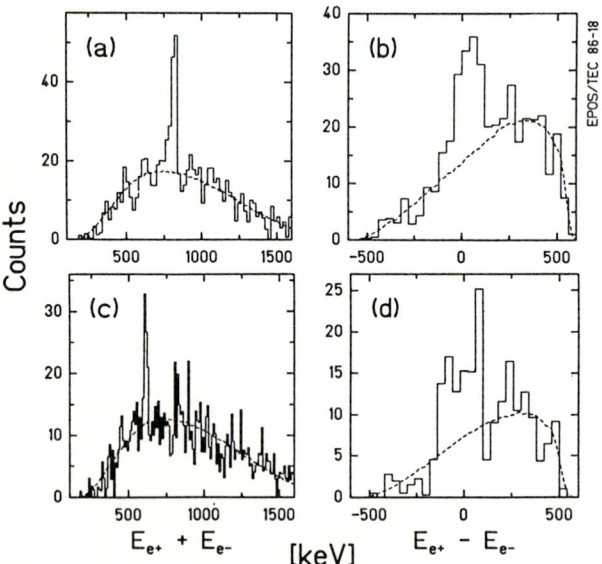

Fig.5: Sum- and difference-energy spectra of electron-positron coincidences for the U + Th collision system, projected from two-dimensional windows similar to those shown in fig.4. The spectra for the 800-keV correlation (top) contain the data from a beam-energy interval between 5.87 and 5.9 MeV/u. For the spectra of the 600-keV correlation (bottom) all data in a beam-energy range between 5.86 and 5.9 MeV/u have been added. Additionally there are windows set on the time-of-flight difference between positrons and associated heavy ions as described in the text.

In the next series of experiments it is planned to investigate the time-of-flight difference for the positrons, which is presently ascribed to the different emission characteristics of both correlations. Also the sharp energy dependence suggested by the data has to be rigorously established. Finally an extension of the measurements to other also lighter systems is needed. A further promising approach is the double orange set up already installed at GSI by the ORANGE-collaboration to also investigate coincident electron-positron emission. Another effort to solve this puzzle is presently undertaken at GSI by the TORI-collaboration [21]. The S-shaped solenoidal spectrometer, used by this group separates electrons and positrons by their opposite drifts in an inhomogeneous magnetic field. It therefore principally allows to also measure correlations where the electrons are emitted colinear to the positrons and thus provides another rigorous test of the apparent anticolinearity found in the EPOS-experiments.

Most of the results reported here have been worked out by the EPOS-collaboration, whose present members are: H. Backe, K. Bethge, H. Bokemeyer, T. Cowan, H. Folger, J.S. Greenberg, K. Sakaguchi, P. Salabura, D. Schwalm, J. Schweppe, K.E. St.

References:

1. H.Backe, L.Handschug, F.Hessberger, E.Kankeleit, L.Richter, F.Weik, R.Willwater, H.Bokemeyer, P.Vincent, Y.Nakayama, J.S.Greenberg; Phys.Rev.Lett.**40**(1978)1443
2. C.Kozhuharov, P.Kienle, E.Berdermann, H.Bokemeyer, J.S.Greenberg, P.Vincent, H.Backe, L.Handschug, E. Kankeleit; Phys.Rev.Lett.**42**(1979)376
3. Quantum Electrodynamics of Strong Fields,ed. W.Greiner,Plenum Press, New York(1983)
4. J.Reinhardt, B.Müller, W.Greiner; Phys.Rev.**A24**(1981)103
5. W.Pieper, W.Greiner, Z.Physik **218**(1969)327
6. J.Rafelski, B.Müller, W.Greiner; Z. Physik **257**(1972)62,183
7. J.Reinhardt, U.Müller, B.Müller, W.Greiner; Z.Physik **A303**(1981)173
8. H.Bokemeyer, K.Bethge, H.Folger, J.S.Greenberg, H.Grein, A.Gruppe, S.Ito, R.Schule D.Schwalm, J.Schweppe, N.Trautmann, P.Vincent, M.Waldschmidt, in ref. 3 p.273
9. P.Kienle; in ref.3 p.293
10. J.Schweppe, A.Gruppe, K.Bethge, H.Bokemeyer, T.Cowan, H.Folger, J.S.Greenberg, H. Grein, S.Ito, R.Schule, D.Schwalm, K.E.Stiebing, N.Trautmann, P.Vincent, M.Waldschmidt; Phys.Rev.Lett.**51**(1983)2261
11. T.Cowan, H.Backe, M.Begemann, K.Bethge, H.Bokemeyer, H.Folger, J.S.Greenberg H.Grein, A.Gruppe, Y.Kido, M.Klüver, D.Schwalm, J.Schweppe, K.E.Stiebing, N.Trautmann, P.Vincent;Phys.Rev.Lett.**54**(1985)1761
12. M.Clemente, E.Berdermann, P.Kienle, H.Tsertos, W.Wagner, C.Kozhuharov, F.Bosch, W.König; Phys.Lett. **B137**(1984)41
13. H.Tsertos, E.Berdermann, F.Bosch, M.Clemente, P.Kienle, W.König, C.Kozhuharov, W.Wagner; Phys.Lett.**B162**(1985)273
14. T.Cowan, H.Backe, K.Bethge, H.Bokemeyer, H.Folger, J.S.Greenberg, K.Sakaguchi, D.Schwalm, J.Schweppe, K.E.Stiebing, P.Vincent; Phys.Rev.Lett.**56**(1986)444
15. H.Bokemeyer; GSI-Preprint 87-1, and contribution to: Int. Adv. Courses on Physics of Strong Fields, Maratea (1986), edit. W.Greiner
16. T.Cowan and J.Greenberg; Yale Prep. 3074-927, and in the proceed. cit. in ref. 15
17. P.Salabura; priv. comm.
18. K.A.Erb, I.Y.Lee, W.T.Milner, Phys.Lett.**B180**(1986)
19. R.Peckhaus, Th.W.Elze, Th.Happ, Th.Dresel; submitted to Phys.Rev.
20. K.Meyer, W.Bauer, J.Briggmann, H.D.Carstanjen, W.Decker, V.Heinemann, J.Major, H.E.Schaefer, A.Seeger, H.Stoll, P.Wesolowski, E.Widmann, F.Bosch, W.König; to be publ.
21. E.Kankeleit, B.Blanck, G.Klotz, M.Kollatz, M.Krämer, R.Krieg, U.Meyer, H.Oeschler, P.Senger; Nucl.Instr.and Meth.**A234**(1985)81

Introduction to Axions

Kyungsik Kang

Division de Physique Théorique, Laboratoire Associé au C.N.R.S.,
Institut de Physique Nucléaire, F-91406 Orsay Cedex, France and
Laboratoire de Physique Théorique des Particules Elémentaires,
Université Pierre et Marie Curie, F-75252 Paris, France and
Permanent address: Department of Physics, Brown University,
Providence, RI 02912, USA

1. Introduction

There are a number of pseudoscalar particles in the literature[1] that originate from the Nambu-Goldstone phenomenon[2]. They include the *axion*[3], the *majoron*[4] and the *familon*[5]. Of these, the axion has been studied most intensively and also has received renewed interest recently[6] because of the report on the narrow states[7] with mass arount 1.8 MeV produced in heavy ion collisions at GSI in Darmstadt and decay into e^+e^-. Historically, axions arose when the strong CP-problem[8] was resolved by the Peccei-Quinn (PQ) mechanism[9] with an additional chiral $U(1)$ symmetry in an enlarged electroweak gauge model with two doublets of Higgs scalars. As observed by Weinberg and Wilczek, the PQ symmetry would be spontaneously broken at the electroweak scale and a Goldstone boson, the axion, emerged to reflect the existence of the PQ symmetry. Since the color current has an anomaly[10] in QCD arising from the gluon interactions that break the chiral PQ symmetry, the axion aquires a small mass thus becoming a pseudo-Goldstone boson. The color anomaly which is the cause of the strong CP-problem breaks the chiral PQ symmetry, as well as the QCD $U(1)_A$ symmetry. Thus while the old $U(1)_A$ problem[11] is lifted, there will be mixings of the axions with π^0, η, etc. due to the color anomaly. In addition, the axion couplings of light quarks are affected by the existence of the color anomaly or equivalently the QCD instantons. Because of the lack of experimental support, the standard axion resulting from the breaking of the PQ symmetry at the electroweak scale is less promising than the so called invisible axion[12] which is associated with the PQ symmetry broken at a significantly higher scale than the electroweak one. The mass and coupling strength of the invisible axion are much smaller than those of the standard axion. Though interacting weakly, the axions could be potentially important in cosmology and astrophysics since a light and weakly coupled particle could provide an important mechanism for the dark matter of the universe and for the stellar emission of energy.

Another Nambu-Goldstone-type particle searched for in recent experiments[13] of the neutrinoless double beta decay is the majoron, which is associated with the global lepton number (more generally B-L number) symmetry that is broken spontaneously. The familon is due to the spontaneous breaking of a global family symmetry.

This lecture is organized as following: after a brief discussion on the Nambu-Goldstone phenomena with pions as an example, I discuss the strong CP-problem and its resolution through the PQ mechanism in the standard model. Various predictions of the standard axion are compared to experiments to motivate the invention of an invisible axion. I then present the invisible axion model of Dine, Fischler and Srednicki and summarize the mass and coupling constants of the invisible axion from the consideration of astrophysical and cosmological constraints. In passing, I touch briefly upon the role of the invisible axion in the grand unification model building[14]. Finally I present an electroweak gauge model[15] that contains an invisible axion and majoron which is motivated by the recent solutions to the solar neutrino problem[16] in $^{37}Cl \to {}^{37}Ar$ by the resonance amplification method[17] of the oscillation in the matter.

2. The Nambu-Goldstone Theorem

Spontaneous breaking of a continuous symmetry induces a massless excitation, the Goldstone boson[2]. Consider the general Lagrangian density made of some multiplet of real scalar fields ϕ

$$\mathcal{L} = \frac{1}{2}\partial_\mu \phi \partial^\mu \phi - V(\phi). \tag{1}$$

Suppose that (1) is invariant under some global gauge group G having the generators t_i ($i = 1, 2, \ldots, n$), i.e., $\delta\mathcal{L} = 0$ under

$$\delta\phi_k = i\theta_j(t_j)_{kl}\phi_l. \tag{2}$$

Since the kinetic term is invariant under (2), the potential term is invariant by itself:

$$\delta V = \frac{\delta V}{\delta \phi_k} i\theta_j(t_j)_{kl}\phi_l = 0, \tag{3}$$

which gives after differentiating with respect to ϕ_m

$$\frac{\delta^2 V}{\delta\phi_m \delta\phi_k}(t_j)_{kl}\phi_l + \frac{\delta V}{\delta\phi_k}(t_j)_{km} = 0. \tag{4}$$

Assuming the ϕ-field to have a vacuum expectation value (VEV) $<\phi_k> = \lambda_k$

$$\left(\frac{\delta^2 V}{\delta\phi_m \delta\phi_k}\right)_{\phi_k = \lambda_k}(t_j)_{kl}\lambda_l = 0, \tag{5}$$

where the second derivative of the potential is the mass-squared matrix M^2_{mk} for the ϕ-field, so that we may write

$$M^2 \cdot t_j \cdot \lambda = 0. \tag{6}$$

This is trivially satisfied for the generators t_j ($j = 1, 2, \ldots, n'$) in the unbroken subgroup $G' \subset G$, i.e., $(t_j)_{kl}\lambda_l = 0$ for $j = 1, 2, \ldots, n'$. However, for the $n - n'$ generators which do not leave the vacuum invariant, (6) requires that $(t_j \cdot \lambda)$ is an eigenvector of M^2 with zero eigenvalues, i.e., there are $n - n'$ massless excitations, the Goldstone bosons.

Another example is the chiral symmetry breaking of the σ-model[18], in which the pion emerges as a Goldstone boson. Consider an isospin doublet ψ of massless quarks, u and d. The kinetic energy term is invariant under the chiral $SU(2)_L \times SU(2)_R$:

$$\delta\psi_{L,R} = i\vec{\theta}_{L,R} \cdot \left(\frac{\vec{\tau}}{2}\right)\psi_{L,R}. \tag{7}$$

In terms of the infinitesimal parameters $\vec{\theta} = (\vec{\theta}_L + \vec{\theta}_R)/2$ representing the isospin and chiral rotations, (7) can be rewritten as

$$\delta\psi = i(\vec{\theta} - \gamma_5\vec{\theta}_5) \cdot \left(\frac{\vec{\tau}}{2}\right)\psi. \tag{8}$$

The presence of a mass term $\bar{\psi}\psi$, however, will break the chiral symmetry, leaving only the isospin invariance. The σ-model has a Lagrangian with chiral symmetry which is made to spontaneously break to accommodate a nucleon. In the process, the pion triplet is created corresponding to $\vec{\theta}_5$ as the Goldstone bosons, the original chiral symmetry parameters $\vec{\theta}_5$ thus becoming dynamical. The Lagrangian in the σ-model contains a 2×2 matrix of scalar fields Σ which transform under $SU(2)_L \times SU(2)_R$

$$\delta\Sigma = i\vec{\theta}_L \cdot \frac{\vec{\tau}}{2}\Sigma - i\Sigma\vec{\theta}_R \cdot \frac{\vec{\tau}}{2} \tag{9}$$

and has the following structure:

$$\mathcal{L} = i\bar{\psi}\partial_\mu\gamma^\mu\psi - g\bar{\psi}_L\Sigma\psi_R - g\bar{\psi}_R\Sigma^\dagger\psi_L + \mathcal{L}(\Sigma). \tag{10}$$

The chiral invariance of the Yukawa term is evident from (7) and (9) or more explicitly from their integrated forms,

$$\psi_L \to L\psi_L, \qquad \psi_R \to R\psi_R, \qquad \Sigma \to L\Sigma R^\dagger, \tag{11}$$

where L and R are 2×2 unitary unimodular matrices

$$L = e^{i\vec{\alpha}\cdot\vec{\tau}/2}, \qquad R = e^{i\vec{\beta}\cdot\vec{\tau}/2} \tag{12}$$

with arbitrary real vectors $\vec{\alpha}$ and $\vec{\beta}$. With the Σ field that is constrained to depend only on four fields,

$$\Sigma = \sigma + i\vec{\tau}\cdot\vec{\pi}, \tag{13}$$

$\mathcal{L}(\Sigma)$ can be identified as

$$\mathcal{L}(\Sigma) = \frac{1}{2}\partial_\mu \Sigma^\dagger \partial^\mu \Sigma - V(\Sigma^\dagger \Sigma). \tag{14}$$

Note that the potential is just a function of $\Sigma^\dagger \Sigma$, which is proportional to the identity times the only invariant quantity of non-derivative type,

$$\text{Det}\Sigma = \frac{1}{2}\text{Tr}(\Sigma^\dagger \Sigma) = \sigma^2 + \vec{\pi}^2 \equiv F_\pi^2. \tag{15}$$

To break the chiral symmetry spontaneously the potential is chosen to have a minimum where (15) is satisfied, so that

$$V(\Sigma^\dagger \Sigma) = \frac{\lambda}{4}\left[(\Sigma^\dagger \Sigma) - F_\pi^2\right]^2 \tag{16}$$

and a vacuum expectation value is given to Σ such that $<\sigma>= F_\pi$ and $<\vec{\pi}>= 0$. Here λ is a dimensionless constant. By perturbing around the minimum by shifting $\sigma' = \sigma - F_\pi$, one then obtains from (10) a nucleon mass $m_N = gF_\pi$ which is interacting with the scalar field σ' with a mass $\sqrt{2\lambda}F_\pi$ and massless triplet pseudoscalar fields $\vec{\pi}$, the Goldstone bosons.

Obviously we can extend the σ-model to the chiral $SU(3)_L \times SU(3)_R$ symmetry[19] for quark triplets u, d and s. The corresponding scalar field Σ will transform

$$\Sigma \to \Sigma' = L\Sigma R^\dagger, \tag{17}$$

where L and R are the unitary unimodular matrices of $SU(3)_L$ and $SU(3)_R$ transformations. Unlike in $SU(2)$ however, the fundamental representation of $SU(3)$ is complex and the Σ-field cannot be constrained as in (13). We may at the best require the reality of $\text{Det}\Sigma$. In analogy with the σ-model, one may expect that the chiral symmetry is spontaneously broken below the scale $<\bar{\psi}\psi>= \Lambda \sim F_\pi$. Here in fact there are two possibilities, i.e., either $SU(3)_L \times SU(3)_R \to SU(3)$ or $SU(2) \times U(1)$. In the first case, only the axial-$SU(3)$ generators are broken spontaneously so that there will be an $SU(3)$ octet of massless pseudoscalars, the

Goldstone bosons. In the second case, the strangeness-changing generators of ordinary $SU(3)$ do not leave the vacuum invariant, thus causing a strange scalar Goldstone boson with zero mass. In view of the existence of the light pseudoscalar octet as a candidate for the octet Goldstone bosons and no light scalar meson with strangeness, the first case of the breaking patterns is favored. We thus express Σ as the unitary unimodular matrix made of the eight Goldstone bosons in the exponent

$$\Sigma = e^{2i\pi/f}, \tag{18}$$

where $\pi = \sum_{\alpha=1}^{8}(1/2)\lambda_\alpha \pi_\alpha$ and f is a constant with dimensions of mass. The linear transformation (17) of Σ is then reflected by the non-linear transformation of π_α. To construct the Lagrangian, we note that there are no nonderivative-type terms other than a constant since $\Sigma^\dagger \Sigma = \mathbf{1}$, i.e., the nonderivative terms can have no dependence at all on the Goldstone fields. There is only one term with two derivatives,

$$\frac{f^2}{4}\mathrm{Tr}(\partial_\mu \Sigma^\dagger \partial^\mu \Sigma), \tag{19}$$

which gives the conventionally normalized kinetic energy term for the Goldstone pions. In the real world, the pseudoscalar octet mesons are massive, which may be due to the chiral $SU(3) \times SU(3)$ breaking. The chiral symmetry will be broken in QCD by the presence of the quark mass term.

$$\bar{\psi}_L M \psi_R + \mathrm{h.c.} \tag{20}$$

with nonzero quark masses m_u, m_d and m_s. A constant M would not be transformed like

$$M \to L M R^\dagger, \tag{21}$$

which could restore the invarianceof the mass term (20) under $SU(3) \times SU(3)$. But one can use (21) to built an effective theory including symmetry breaking below the chiral symmetry breaking scale Λ of QCD, and calculate perturbatively[20] the effect of M if M is small compared to the QCD scale Λ, which is certainly the case for m_u and m_d. For example, if one wants to construct an invariant function of Σ and M which is linear in M, the choice is unique,

$$v^3 \mathrm{Tr}(\Sigma^\dagger M) + \mathrm{h.c.}, \tag{22}$$

which gives a mass term to the pseudoscalar mesons, thus making them the pseudo-Goldstone bosons. From (22) in the limit of exact isospin invariance, Georgi[19] derived the Gell-Mann–Okubo mass formula for the mass squared of the pseudoscalar octet mesons. Actually, the free massless quark triplets described by the

Lagrangian (19) possess the $U(3) \times U(3)$ symmetry so that one can expect naively an $SU(3)$ singlet Goldstone boson associated with the spontaneously broken chiral $U(1)_A$ symmetry. Similar analysis used above with (22) would then predict the mass of the ninth meson to be very close to the eighth member of the $SU(3)$ octet in contradiction to experiments which show η' to be much heavier than η. This is nothing but the old $U(1)$ problem[11]. The resolution of this problem is due to the existence of the color anomaly term[8] in QCD which in fact breaks the chiral $U(1)_A$ symmetry. This is good news. On the other hand, as we will trace in the next section, this QCD anomaly term gives rise to a new difficulty, i.e., the strong CP problem, which is the origin of the axion.

To sum up this section, we see from the examples worked out above that when the exact global symmetry is broken spontaneously, there will be massless excitations, as the only allowed interaction terms are of the derivative type. Thus their coupling to the matter will generate a spin-dependent long-range potential of r^{-3} type. If an approximate global symmetry is broken spontaneously, one can built nonderivative-type interactions with the symmetry breaking term in the Lagrangian, thus giving rise to masses of the Goldstone bosons, which we call the pseudo-Goldstone bosons. Obviously such a scheme will be a perturbative one and will make sence only when the symmetry breaking term is small compared to the symmetry breaking scale. As mentioned in the previous section, the majorons and familons are examples of the Goldstone boson and the axions are pseudo-Goldstone bosons due to the QCD instanton term that does not leave the PQ symmetry intact. It is this QCD anomaly term coupling to the axion that gives the axion a small mass, as we will see in the next section. The role of the global symmetry in connection with Goldstone or pseudo-Goldstone bosons is to be emphasized once more, for if it were the local gauge symmetry that was broken spontaneously we could have had an exception to the Nambu-Goldstone theorem. Namely, the massless gauge bosons could absorb the field degrees of the massless Goldstone bosons and become massive, which is the well-known Higgs phenomenon[21]. Global symmetries in particle physics manifest themselves rather differently from the gauge symmetry, usually to supplement the local gauge invariance so as to avoid unwanted processes. The Goldstone-type particles that arise as a result of spontaneous breakdown of such global symmetries can nevertheless play an important role in astrophysical and cosmological problems.

3. The Strong CP Problem and Axion

In non-Abelian gauge theories the vacuum has a nontrivial topological structure. This is because the gauge fields do not vanish fast enough at infinity so that the surface term cannot be ignored. The simplest example is the classical solution of the pure $SU(2)$ Yang-Mills field equation in the four-dimensional Euclidean space obtained by Belavin et al.[22] This solution suggested that a new gauge-invariant term $(g^2/16\pi)\mathrm{Tr}F^{\mu\nu}\tilde{F}_{\mu\nu}$ made of the field tensor and its dual tensor should be added to the Lagrangian of the general Yang-Mills fields, even though it appears to be a total derivative, i.e.,

$$\mathrm{Tr}F^{\mu\nu}\tilde{F}_{\mu\nu} = 2\epsilon_{\mu\nu\lambda\sigma}\partial^\mu \mathrm{Tr}\left(A^\nu \partial^\lambda A^\sigma + \frac{2g}{3i}A^\nu A^\lambda A^\sigma\right) \equiv \partial^\mu K_\mu, \qquad (23)$$

where $A^\mu = \sum_\alpha (\lambda_a/2)A^\mu_\alpha$. The reason is because the surface term cannot be ignored but in general

$$\nu[A] = \frac{g^2}{16\pi^2}\int d^4x \partial^\mu K_\mu = \frac{g^2}{16\pi^2}\oint_{S^3} K_\mu d^3\sigma^\mu \qquad (24)$$

is nonzero integer. Furthermore the gauge fields with different ν are gauge-inequivalent, therefore the solutions of the field equations must specify the topological number ν, which is called the instanton number or the Pontryagin index. The solution obtained by Belavin et al. in $SU(2)$ corresponds to $\nu = 1$ and the solution corresponding to arbitrary instanton number[23] in a general non-abelian theory such as QCD can also be constructed in pure gauge $A_\mu = (i/g)U\partial_\mu U^{-1}$ with the boundary condition $F_{\mu\nu} \to 0$ as $x_\mu \to \infty$.

Existence of the different classes of the classical solutions for non-Abelian gauge fields corresponding to different instanton number suggests also that we distinguish the vacuum configuration by the class of the instanton numbers, i.e., the homotopy class $|n\rangle$ where $n = 0, \pm1, \pm2, \cdots$. In terms of the gauge transformation U that takes the vacuum $|n\rangle$ of the homotopy class n to another one of the homotopy class $n+1$, i.e.,

$$U|n\rangle = |n+1\rangle, \qquad (25)$$

one can construct an arbitrary gauge transformation that takes the vacuum $|n\rangle$ to another one $|n+N\rangle$ to be $(U)^N$. Since the true vacuum must be invariant under all such gauge transformations, it must be a linear superposition of $|n\rangle$-vacua. Furthermore, U being unitary, its eigenvalue is just a phase factor, which we will denote as $\exp(-i\theta)$. Then the so-called θ-vacuum defined by[8]

$$|\theta\rangle = \sum_n e^{in\theta}|n\rangle \qquad (26)$$

can be shown to satisfy the invariance under all such gauge transformations among the different homotopy classes. With this, it is easy to see how the instanton solutions modify the effective Lagrangian of QCD. Consider the vacuum to vacuum amplitude, the S-matrix, in Euclidean space,

$$\langle \theta'|e^{-Ht}|\theta\rangle = \sum_n e^{in(\theta'-\theta)} \sum_\nu e^{-i\nu\theta} \langle n+\nu|e^{-Ht}|n\rangle, \tag{27}$$

where the S-matrix between the $|n\rangle$-vacua is

$$\langle n+\nu|e^{-Ht}|n\rangle = \int [dA_\mu]_\nu e^{-\int d^4x \mathcal{L}}. \tag{28}$$

Using the completeness relation for the summation over n, we then get

$$\langle \theta'|e^{-Ht}|\theta\rangle = \delta(\theta'-\theta) \sum_\nu \int [dA_\mu]_\nu e^{-[\int d^4x \mathcal{L}+i\theta\nu]} \tag{29}$$

so that the effective Lagrangian in Minkowski space becomes

$$\mathcal{L}_{\text{eff}}[A] = \mathcal{L}[A] + \frac{\theta g^2}{16\pi^2} \text{Tr} F_{\mu\nu}\tilde{F}^{\mu\nu}, \tag{30}$$

where (24) is used for ν. This is the origin of the QCD anomaly term. Although it is a total divergence, its surface contribution cannot be neglected due to the existence of the instantons, i.e., there are nonperturbative effects due to instantons. In fact this anomaly term is odd under P and T and therefore violates the CP invariance in QCD if θ is nonzero, i.e., in strong interactions, the θ-parameter is an observable quantity.

If, however, the quarks are massless, then one can rotate away the θ-term by a chiral rotation of the quark fields. This is because the gauge-invariant current corresponding to $U(1)_A$

$$j_5^\mu = \sum_i \bar{q}_i \gamma^\mu \gamma_5 q_i \tag{31}$$

is not conserved but has the same anomaly, i.e.,

$$\partial_\mu j_5^\mu = 2N_f \left(\frac{g^2}{16\pi^2}\right) \text{Tr} F_{\mu\nu}\tilde{F}^{\mu\nu}, \tag{32}$$

where N_f is the number of the quarks. Thus the corresponding charge

$$Q_5 = \int d^3x\, j_5^0$$

cannot be used as a generator of the chiral transformation. This resolves therefore the old $U(1)_A$ problem but we now have to face the observability of θ through the

CP violation effects in strong interactions. In order to see how the chiral rotation gets rid of the θ term for the massless quarks, define the chiral current that is conserved,

$$\tilde{j}_5^\mu = j_5^\mu - 2N_f \left(\frac{g^2}{16\pi^2}\right) K^\mu, \tag{33}$$

where K^μ is defined in (23). The corresponding generator

$$\tilde{Q}_5 = Q_5 - 2N_f \nu \tag{34}$$

with ν given by (24) is a constant of motion though not gauge invariant. In particular it transforms under the homotopy class one gauge transformation as

$$U\tilde{Q}_5 U^\dagger = \tilde{Q}_5 - 2N_f \tag{35}$$

and it follows that

$$U\{e^{i\tilde{Q}_5\alpha}|\theta\rangle\} = e^{-i(\theta - 2N_f\alpha)}\{e^{i\tilde{Q}_5\alpha}|\theta\rangle\}. \tag{36}$$

In other words,

$$e^{i\tilde{Q}_5\alpha}|\theta\rangle = |\theta - 2N_f\alpha\rangle, \tag{37}$$

which shows that θ can be chirally rotated away as long as there is a massless quark. However, in the real world even the u-quark does not seem to be massless.

If the quarks are massive and have the mass matrix M as given in (20), which is in general nondiagonal and not free of the γ_5 terms, the mass matrix M can in general be diagonalized by a biunitary transformation. The eigenvalues can be made real by chiral transformations on each flavor $q_j \to \exp(i\alpha_j \tilde{Q}_5) q_j$ such that $\phi_j = -2\alpha_j$ or $\arg \text{Det} M = -\sum_j (2\alpha_j)$. On the other hand these chiral transformations modify the θ-parameter effectively by

$$\bar{\theta} = \theta - \sum_j (2\alpha_j) = \theta + \arg \text{Det} M. \tag{38}$$

Then it is this $\bar{\theta}$ that is physically observable. Note that there is a CP violation contribution from the mass matrix even if the θ parameter from QCD is put to zero because the chiral rotation to make the mass matrix real will generate a θ out of the θ-vacuum through the QCD anomaly. Since the anomaly term is singlet in flavor, the effects should be seen from the static and CP-noninvariant properties of hadrons. One such measurement is the electric dipole moment d_n of the neutron and the current experimental upper bound[24] on d_n is

$$|d_n| \leq 5 \times 10^{-25} \quad \text{ecm}. \tag{39}$$

A number of theoretical calculations[25] have been performed on d_n thereby replacing the QCD anomaly term by the effective CP-violating interaction term arising from the quark mass matrix and with different approximations having to do with the intermediate states and hadronic matrix elements. They all give $|d_n|/\bar{\theta}$ to be less than 10^{-15} ecm so that we may say

$$\bar{\theta} \leq 10^{-9}. \tag{40}$$

Then the strong CP problem is now to understand such a small $\bar{\theta}$ naturally.

The most attractive resolution of the strong CP-problem is the Peccei-Quinn (PQ) mechanism[9]. Basically it extends the chiral solution for the case of massless quarks to the case of massive quarks by imposing an additional global chiral $U(1)_A$ to the full Lagrangian so that the $U(1)_A$ current has the QCD anomaly. The standard minimal electroweak theory which uses one Higgs doublet has a global $U(1)$ symmetry but it is just the B-L symmetry and not color anomalous. Therefore one has to introduce another Higgs doublet[27] and make them give the masses to the up- and down-quarks seperately. However, since this additional symmetry, which we will denote henceforth $U(1)_{PQ}$, is also broken spontaneously at the scale where $SU(2) \times U(1)$ gauge group is broken, there will be a pseudo-Goldstone boson, the axion. In other words, the θ-parameter will become a dynamical degree closely related to the axion field and the small value of $\bar{\theta}$ is due to the axion that fixes the vacuum state to be $\bar{\theta} = 0$ dynamically regardless of the initial value. A simple proof of this is given by Vafa and Witten[26] who used the real positivity of the mass determinant of the quarks in the Euclidean potential energy of the axion.

Consider the standard model with one generation for simplicity. Denoting the PQ charge by y, the fermion have the following gauge hypercharge Y and y: $(1/6, -y_1)$ for $(u,d)_L$; $(2/3, y_2)$ for u_R; $(-1/3, y_3)$ for d_R; $(-1/2, -y_4)$ for $(\nu, e)_L$; and $(-1, y_5)$ for e_R. For the gauge invariance to hold for the Yukawa terms, we need two Higgs scalar doublets ϕ_+ and ϕ_- of which ϕ_+ gives the mass to u and ϕ_- to d and e. The (Y,y) values of these scalars are $(-1/2, -y_1 - y_2)$ and $(1/2, -y_1 - y_3)$, and

$$y_1 + y_3 = y_4 + y_5. \tag{41}$$

The most general Higgs potential that is gauge and $U(1)_{PQ}$ invariant and renormalizable can then be constructed out of $\phi_\pm$; it contains the quartic terms made out of $\phi_+ \epsilon \phi_-$ and both the quartic and quadratic terms made out of $(\phi_+^\dagger \phi_+)$ and $(\phi_-^\dagger \phi_-)$ of which only the first term dictates the relative y-charges. In particular, they give

$$2y_1 + y_2 + y_3 = 0. \tag{42}$$

There are then three automatic $U(1)$ symmetries in principle among the five global charges. But in order for any of the three $U(1)$s to rotate away the θ-term, the corresponding current must have the color anomaly so that under $q_i \to \exp(i\alpha y_i)q_i$, the θ-value should be modified by

$$\theta \to \theta - 2\sum_i y_i = \theta - 2(2y_1 + y_2 + y_3). \tag{43}$$

In general the color anomaly is given by

$$l = 2\text{Tr}(yT^2), \tag{44}$$

where $\text{Tr}(T^2)$ is the second index of the fermion representation of $SU(3)$ with the normalization $1/2$ for the fundamental representation. Thus from (42) and (43) it is clear that none of the three automatic $U(1)$s can be used as $U(1)_{PQ}$ and it is necessary to suppress some of the quartic terms such as $(\phi_+ \epsilon \phi_-)^2$ and allow them only in the form $(\phi_+ \epsilon \phi_-)^\dagger(\phi_+ \epsilon \phi_-)$. There will be then four $U(1)$s, two of which can be identified by the well-known global symmetries B, L. The third can be combined with the hypercharge and we are still left with another $U(1)$ which is color anomalous, i.e., the needed $U(1)_{PQ}$ symmetry, albeit nonautomatic. In the original work of Peccei and Quinn, the potential is chosen to be

$$V = -\mu_+^2(\phi_+^\dagger \phi_+) - \mu_-^2(\phi_-^\dagger \phi_-) + \sum_{i,j=+,-} a_{ij}(\phi_i^\dagger \phi_i)(\phi_j^\dagger \phi_j) + \sum_{i,j=+,-} b_{i,j}(\phi_i^\dagger \tilde\phi_j)(\phi_i^\dagger \tilde\phi_j)^\dagger, \tag{45}$$

neglecting the term $\sum_{i \neq j} c_{ij}(\phi_i^\dagger \tilde\phi_j)^2 + \text{h.c.}$ where $\tilde\phi_j = i\sigma_2 \phi_j^*$ with real symmetric constants a_{ij} and b_{ij} and hermitian constants c_{ij}. The gauge and PQ invariance is then broken spontaneously by the Higgs scalars acquiring the VEV's. There will then be four Goldstone bosons associated with the four phase fields of the two doublets, three of which will be absorbed by $W^\pm$ and Z through the Higgs mechanism. The remaining one is the axion. Let us write the Higgs doublets as

$$\phi_+ = \begin{pmatrix} \frac{1}{\sqrt{2}}(v_+ + \rho_+)e^{i\xi_+/v_+} \\ \phi_1^- \end{pmatrix} \quad \phi_- = \begin{pmatrix} \phi_2^+ \\ \frac{1}{\sqrt{2}}(v_- + \rho_-)e^{i\xi_-/v_-} \end{pmatrix}, \tag{46}$$

where ρ's and ξ's are real fields and $\phi^\pm$ are complex fields. The scalars and fermions transform under $U(1)_{PQ}$ as

$$\begin{aligned}
\phi_\pm &\to e^{iy_\pm\alpha}\phi_\pm \\
q_L &\to e^{iy_1\alpha}q_L, \quad u_L^c \to e^{iy_2\alpha}u_L^c, \quad d_L^c \to e^{iy_3\alpha}d_L^c \\
l_L &\to e^{iy_4\alpha}l_L, \quad e_L^c \to e^{iy_5\alpha}e_L^c
\end{aligned} \tag{47}$$

with $y_+ = -y_1 - y_2$, $y_- = -y_1 - y_3 = -y_4 - y_5$. Since ξ's belong to the neutral components, they are related to the hypercharge and the PQ symmetry only. Then it is clear that the axion a and the orthogonal state b to be absorbed by Z must be made of these ξ's. Let us parametrize them as

$$\begin{pmatrix} a \\ b \end{pmatrix} = \begin{pmatrix} \cos\omega & \sin\omega \\ -\sin\omega & \cos\omega \end{pmatrix} \begin{pmatrix} \xi_+ \\ \xi_- \end{pmatrix} \tag{48}$$

and demand that under (47) the axion transforms as

$$a \to a + \lambda, \tag{49}$$

λ being a constant. Then we get from (46)–(49) that

$$\cos\omega = v_+ y_+/A, \qquad \sin\omega = v_- y_-/A, \tag{50}$$

where $A^2 = (v_+ y_+)^2 + (v_- y_-)^2$. It is clear that this procedure can be generalized to the case of any number of neutral Higgs phases so that the axion field is in general given by

$$a = \frac{1}{A}\sum_i (v_i y_i)\xi_i \tag{51}$$

where $A^2 = \sum_i (v_i y_i)^2$, where v_i's are the VEV's. To determine the ratio y_+/y_-, we demand that under the hypercharge rotation, $b \to b + \text{const}$. This guarantees that b can be absorbed by Z. Since the hypercharges of $\phi_\pm$ are $\mp 1/2$, it then follows that

$$\sin\omega = v_+/v, \qquad \cos\omega = v_-/v, \tag{52}$$

where $v^2 = v_+^2 + v_-^2$, so that from (50) and (52) we get

$$\tan\omega = \frac{v_- y_-}{v_+ y_+} = \frac{v_+}{v_-}, \tag{53}$$

which gives

$$y_+/y_- = (v_-/v_+)^2 \equiv x^2. \tag{54}$$

Then a- and b- fields can be expressed as

$$a = \frac{1}{v}(v_-\xi_+ + v_+\xi_-), \qquad b = \frac{1}{v}(-v_+\xi_+ + v_-\xi_-). \tag{55}$$

In addition, there are in this model four charged fields, of which two are eaten up by W and the other two remain as massive Higgs fields, and two massive neutral Higgs fields. The charged fields that are gauged away are

$$\phi'^{\pm} = (v_+ \phi_1^{\pm} + v_2 \phi_2^{\pm})/v \tag{56}$$

and the Higgs fields are

$$\phi^{\pm} = (v_- \phi_1^{\pm} - v_+ \phi_2^{\pm})/v$$
$$\rho'_{\pm} = (v_- \rho_{\pm} \pm v_+ \rho_{\mp})/v. \tag{57}$$

From the Yukawa interaction sector of the Lagrangian, one can identify the axion interaction term

$$\mathcal{L}_{\text{axion}} = \frac{1}{2}(\partial_\mu a)^2 - i\sum_j \frac{a}{v}[x m_{u_j} \bar{u}_j \gamma_5 u_j + \frac{1}{x} m_{d_j} \bar{d}_j \gamma_5 d_j + \frac{1}{x} m_{e_j} \bar{e}_j \gamma_5 e_j], \tag{58}$$

where $v^2 = (\sqrt{2} G_F)^{-1} = (247 \text{ GeV})^2$. The axial current associated with the PQ symmetry is then given by

$$j_\mu^{PQ} = -v \partial_\mu a + \frac{1}{2} \sum_j [x \bar{u}_j \gamma_\mu \gamma_5 u_j + \frac{1}{x} \bar{d}_j \gamma_\mu \gamma_5 d_j + \frac{1}{x} \bar{e}_j \gamma_\mu \gamma_5 e_j], \tag{59}$$

which has the color anomaly so that

$$\partial^\mu j_\mu^{PQ} = N_f \left(x + \frac{1}{x}\right) \left(\frac{g^2}{16\pi^2} \text{Tr} F_{\mu\nu} \tilde{F}^{\mu\nu}\right) + \text{mass terms}, \tag{60}$$

where the mass term vanishes in the limit of the quark masses $\to 0$. Because of the anomaly, the naive charge of the $U(1)_{PQ}$ current cannot be used as the chiral transformation generator. As we said before, the same anomaly breaks also the chiral $U(1)_A$ symmetry associated with the $SU(3)$ singlet meson. The latter is broken in addition by the presence of a small quark mass term as described in Section 2, but below the scale of QCD chiral symmetry where the nonperturbative effects dominate, the instanton effect is more important than the mass term and generates the mixing of the axion with π^0, η etc. As in (33), we have to modify the axion current such that the resulting charge generator is a constant of motion and its divergence can be interpreted as the pseudoscalar field to apply the standard current algebra technique. The result is that the axion current is a linear combination of the PQ current and the QCD $U(1)_A$ current and it manifests π^0–η–a mixing so that we need proper diagonalization of the mass matrix in the π^0, η and

a basis. This gives the axion current to be a linear combination of the $U(1)_{PQ}$, QCD $U(1)_A$ and the anomaly-free π^0 currents. Let us define the anomaly-free axion current by[28]

$$j_\mu^a = j_\mu^{PQ} - \frac{N_f}{2}\left(x + \frac{1}{x}\right)(\bar{u}\gamma_\mu\gamma_5 u + \bar{d}\gamma_\mu\gamma_5 d) \\ - \frac{N_f}{2}c\left(x + \frac{1}{x}\right)(\bar{u}\gamma_\mu\gamma_5 u - \bar{d}\gamma_\mu\gamma_5 d), \quad (61)$$

where N_f is the number of flavor generations and c is to be determined by the condition that a does not mix with π^0,

$$\langle 0|[\partial^\mu j_\mu^a, Q_5^3]|0\rangle = 0, \quad (62)$$

Q_5^3 being the axial isospin current coupled to π^0,

$$j_\mu^3 = \frac{1}{2}(\bar{u}\gamma_\mu\gamma_5 u - \bar{d}\gamma_\mu\gamma_5 d). \quad (63)$$

Then (61) can be written as

$$j_\mu^a = j_\mu^{PQ} - \frac{N_f(x + 1/x)}{2(1+Z)}(\bar{u}\gamma_\mu\gamma_5 u + Z\bar{d}\gamma_\mu\gamma_5 d), \quad (64)$$

where

$$Z = (1 - 2c)/(1 + 2c) = m_u\langle \bar{u}u\rangle/m_d\langle \bar{d}d\rangle, \quad (65)$$

which is 0.56 for the $SU(3)$-flavor symmetric limit $<\bar{u}u> = <\bar{d}d> = <\bar{s}s>$. The divergence of j_μ^a is

$$\partial^\mu j_\mu^a = i\frac{N_f}{2}\left(x + \frac{1}{x}\right)\frac{m_u m_d}{m_u + m_d}(\bar{u}\gamma_5 u + \bar{d}\gamma_5 d), \quad (66)$$

to be compared with

$$\partial^\mu j_\mu^3 = \frac{i}{2}(m_u\bar{u}\gamma_5 u - m_d\bar{d}\gamma_5 d). \quad (67)$$

Then it follows from the soft pion (axion) theorem that

$$m_\pi^2 f_\pi^2 = \langle 0|[iQ_5^3, \partial^\mu j_\mu^3]|0\rangle = m_u\langle\bar{u}u\rangle + m_d\langle\bar{d}d\rangle \\ m_a^2 f_a^2 = m_a^2 v^2 = \langle 0|[iQ_5^a, \partial^\mu j_\mu^a]|0\rangle = N_f^2\left(x + \frac{1}{x}\right)^2 \frac{Z}{(1+Z)^2}m_\pi^2 f_\pi^2 \quad (68)$$

so that

$$m_a = N_f\left(x + \frac{1}{x}\right)\frac{\sqrt{Z}}{1+Z}\frac{f_\pi}{f_a}m_\pi \quad (69)$$

with $f_\pi = 93$ MeV. We obtain for three generations and the $SU(3)$-flavor symmetric vacuum that

$$m_a = 73.6 \left(x + \frac{1}{x}\right) \text{ keV}. \tag{70}$$

To calculate the axion lifetime compared to the pion, we note that j_μ^3 has the electromagnetic anomaly

$$\partial^\mu j_\mu^3 = \text{Tr}(T_3 Q_{\text{em}}^2) \frac{e^2}{16\pi^2} F_{\text{em}} \tilde{F}_{\text{em}} \tag{71}$$

through the triangle diagram, whereas $a \to 2\gamma$ gives

$$\partial^\mu j_\mu^a = \text{Tr}(Y_{PQ} Q_{\text{em}}^2) \frac{e^2}{16\pi^2} F_{\text{em}} \tilde{F}_{\text{em}},$$

so that the corresponding effective Lagrangians are

$$\begin{aligned}\mathcal{L}_{\pi^0 \to 2\gamma} &= \frac{\alpha}{2\pi} \text{Tr}(T_3 Q_{\text{em}}^2) \frac{\pi^0}{f_\pi} F_{\text{em}} \tilde{F}_{\text{em}} \\ \mathcal{L}_{a \to 2\gamma} &= \frac{\alpha}{2\pi} \text{Tr}(Y_{PQ} Q_{\text{em}}^2) \frac{a}{f_a} F_{\text{em}} \tilde{F}_{\text{em}}.\end{aligned} \tag{72}$$

Since

$$\text{Tr}(T_3 Q_{\text{em}}^2) = \frac{1}{2} \quad \text{and}$$
$$\text{Tr}(Y_{PQ} Q_{\text{em}}^2) = \frac{N_f}{2}\left(\frac{4}{3}x + \frac{1}{3x} + y_e\right) - \frac{N_f}{2} \frac{x + 1/x}{1+Z}\left(\frac{4}{3} + \frac{Z}{3}\right), \tag{73}$$

we obtain

$$\tau(a \to 2\gamma) = \tau(\pi^0 \to 2\gamma) Z^{-1} (m_\pi/m_a)^5 \simeq 0.75 \left(\frac{100 \text{ keV}}{m_a}\right)^5 \text{ s}, \tag{74}$$

where $y_e = 1/x$ and (69) are used. This is true as long as $m_a < 2m_e$. If m_a is larger than $2m_e$ but less than $2m_\mu$ then the axion will decay predominantly to e^+e^- and

$$\tau(a \to e^+e^-) = \frac{8\pi v^2}{m_e^2 m_a} x^2 \left(1 - 4\frac{m_e^2}{m_a^2}\right)^{-\frac{1}{2}} = 4 \times 10^{-9} \left(\frac{\text{MeV}}{m_a}\right)\left(1 - 4\frac{m_e^2}{m_a^2}\right)^{-\frac{1}{2}} \text{ s}. \tag{75}$$

The axion couplings to the nucleons can be obtained by PCAC for the axion,

$$\lim_{p \to 0} f_a \langle N', a(p) | \mathcal{L}_{\text{eff}} | N \rangle = i \langle N' | [Q_5^a, \mathcal{L}_{\text{eff}}] | N \rangle \tag{76}$$

where
$$\mathcal{L}_{\text{eff}} = i\bar{N}\gamma_5 \left(g^{(0)} + \tau_3 g^{(3)}\right) Na, \tag{77}$$

N being the neucleon doublet. By reexpressing (64) in terms of the axial isovector and isoscalar currents, one finds the corresponding aNN coupling constants to be

$$\begin{aligned} g^{(0)} &= \frac{m_N}{2f_a}(1-N_f)\left(x+\frac{1}{x}\right) g_A^{(0)} \\ g^{(3)} &= \frac{m_N}{2f_a}\left[\left(x-\frac{1}{x}\right) - N_f\left(x+\frac{1}{x}\right)\frac{1-Z}{1+Z}\right] g_A^{(3)}, \end{aligned} \tag{78}$$

where $g_A^{(i)}$'s are the axial form factors evaluated at $q^2 = 0$ and $g_A^{(3)}(0) = 1.254$ from the β-decay experiments. $g_A^{(0)}$ can be related to $g_A^{(3)}$ by the $SU(3)$-flavor symmetry in the limit the strange quark contribution is neglected. We note that the effective pion-nucleon couplings can be worked out similarly, which gives at the low-momentum transfer $g_{aNN} \sim g_{\pi NN}(f_\pi/f_a)$. In particular using the pionic current component in (64), the isovector aNN coupling constant is given by

$$g_{aNN} = g_{\pi NN}\frac{f_\pi}{2f_a}\left[\left(x-\frac{1}{x}\right) - N_f\left(x+\frac{1}{x}\right)\frac{1-Z}{1+Z}\right], \tag{79}$$

which agrees with $g^{(3)}$ of (78) if the Goldberg-Treiman relation is used for the πNN coupling, i.e., $g_{\pi NN} = m_N g_A^{(3)}/f_\pi$. In particular for $x = 1$ and $N_f = 3$, (79) gives

$$g_{aNN}^2 \cong g_{\pi NN}^2(f_\pi/f_a)^2 \simeq (1.5 \times 10^{-7})g_{\pi NN}^2 \tag{80}$$

for $f_\pi = 93$ MeV and $f_a = 247$ GeV.

The standard axion has been searched for experimentally[29] but we will give only two results which will sufficiently disfavor its existence. Firstly, the decay $K^+ \to \pi^+ a$ gives the branching ratio to be $\text{BR}(K^+ \to \pi^+ a) \simeq 3 \times 10^{-6}(x + \frac{1}{x})^2 \geq 1.3 \times 10^{-5}$, which is to be compared to the experimental upper bound[30] for not observing the strangeness changing K^+ decay, 3.8×10^{-8}. For the heavy quarkonium decays, one gets

$$\begin{aligned} \text{BR}(J/\psi \to \gamma a) &\simeq (5.4 \pm 0.7) \times 10^{-5} x^2 \\ \text{BR}(\Upsilon \to \gamma a) &\simeq (2.4 \pm 0.2) \times 10^{-4} x^{-2}, \end{aligned} \tag{81}$$

to be compared to the experimental upper bounds[31] 1.4×10^{-5} and $(3 \sim 9) \times 10^{-4}$ respectively. Then one gets from J/ψ decay $x < 0.5$ while from the Υ-decay $x > 0.5$–0.9, which are contradictory to each other.

Finally we mention that the narrow states produced in heavy ion collisions[7] at GSI in Darmstadt stimulated to invent the so-called *variant axion* models[6], which vary $\phi_\pm$ coupling to e^- and c-quark. But there are difficulties in understanding the mass and the lifetime simultaneously with this sort of axion interpretation of the 1.8 MeV peak.

To sum up this section, the standard axion that emerges in the Peccei-Quinn-Weinberg-Wilczek model seems to be lacking experimental support. It is for this reason, we move on to the subject of the invisible axion[12].

4. Invisible Axions

From (69) and (79), we note that both the mass and coupling of the axion are inversely proportional to f_a. Furthermore from (51) it is clear that if there are more than just the two Higgs doublets that can get the VEVs, f_a will then be controlled by the largest VEV of those multiplied by its PQ charge y. Nonexistence of the standard axion may then imply that the PQ symmetry needed to resolve the strong CP problem must be broken at a scale V_{PQ} which is much higher than the electroweak scale $v = 246$ GeV so that the axion has an ultralight mass and superweak couplings with matter and escapes detection. This is the origin of the so-called invisible axion[12].

It is then clear that the PQ model should be extended to allow more Higgs scalars and the axion field should get contributions from the phases of the neutral components of these scalars. The PQ symmetry $U(1)_{PQ}$ should retain the desired color-anomalousness which is the source of the tiny mass of the axion. There are then options for the quark sector, i.e., either to modify with hitherto unknown quarks or not. The invisible axion model invented by Kim belongs to the former while the model of Dine, Fischler and Srednicki (DFS) is the latter type. In Kim's model, there is a heavy quark Q having $\underline{3}$ of $SU(3)_c$ and a complex scalar σ which is a singlet under $SU(2) \times U(1)$. These additional fields are the only ones transforming nontrivially under $U(1)_{PQ}$ and the ordinary quarks and leptons as well as the Higgs doublet in the standard model are required to carry no PQ charges so that they do not couple to the axion at the tree level.

The DFS model introduces just an additional complex singlet scalar σ in the PQ model and considers the possibility of its coupling to the ordinary quarks through the ordinary doublet scalars $\phi_\pm$. Every field in the model transforms nontrivially under the $U(1)_{PQ}$ and in particular the Higgs scalars are constrained to undergo

$$\phi_+ \to e^{i\alpha y_+}\phi_+, \quad \phi_- \to e^{i\alpha y_-}\phi_-, \quad \sigma \to e^{i\alpha y_3}\sigma \tag{82}$$

with the condition
$$y_+ + y_- = -2y_s = 1. \tag{83}$$

Since the DFS model is the most economical extension of the PQ model, we will discuss this as the representative of the invisible axion models. The Lagrangian is exactly the same as in the PQ model except for a straightforward modification in the potential $V(\phi_+, \phi_-, \sigma)$, which contains the term $(\phi_+^i \epsilon_{ij} \phi_-^j \sigma^2 + \text{h.c.})$ in addition to all harmless terms containing $|\phi_\pm|^2$, $|\phi_+ \epsilon \phi_-|^2$, and $|\sigma|^2$. To identify the axion, we concentrate on the neutral components and their phase fields in particular,

$$\phi_+ = \begin{pmatrix} v_+/\sqrt{2} \\ 0 \end{pmatrix} e^{i\xi_+/v_+}, \quad \phi_- = \begin{pmatrix} 0 \\ v_-/\sqrt{2} \end{pmatrix} e^{i\xi_-/v_-}, \quad \sigma = \frac{V}{\sqrt{2}} e^{i\xi_s/V}, \tag{84}$$

where $<\sigma> = V/\sqrt{2} \gg v = (v_+^2 + v_-^2)^{1/2}$. There are three linearly independent combinations of the phases ξ's but since the Goldstone boson b to be absorbed by the Z boson is independent of the $SU(2) \times U(1)$ singlet, we can introduce the three fields, a, b and c by the following orthogonal combinations:

$$\begin{pmatrix} a \\ b \\ c \end{pmatrix} = \begin{pmatrix} \cos\omega \cos\phi & \sin\omega \cos\phi & -\sin\phi \\ -\sin\omega & \cos\omega & 0 \\ \cos\omega \sin\phi & \sin\omega \sin\phi & \cos\phi \end{pmatrix} \begin{pmatrix} \xi_+ \\ \xi_- \\ \xi_s \end{pmatrix}. \tag{85}$$

As before, we demand $a \to a + \text{const.}$ under $U(1)_{PQ}$. This gives (51) with $A \simeq |y_s V| = V/2 \equiv f_a$. Noting again the hypercharges of $\phi_\pm$ to be $\mp 1/2$, the condition $b \to b + \text{const.}$ under $U(1)_Y$ determines ω to be as in (52) so that

$$y_\pm = (v_\mp/v)^2, \quad \tan\omega = v_+/v_-, \quad \tan\phi = -y_s \frac{vV}{v_+ v_-}. \tag{86}$$

Thus it follows that

$$\begin{aligned} a &= \frac{1}{A} \left[\frac{v_+ v_-}{v^2} (\xi_+ v_- + \xi_- v_+) + V y_s \xi_s \right] \\ b &= \xi_- \frac{v_-}{v} - \xi_+ \frac{v_+}{v} \\ c &= \frac{1}{vA} [v_+ v_- \xi_s - y_s V(\xi_+ v_- + \xi_- v_+)]. \end{aligned} \tag{87}$$

Note that in the limit $V \gg v$, the DFS axion field is mostly the phase field of σ and the massive Higgs field c is smoothly continued to the standard axion field. The mass of the c field arises from the new term $\phi_+ \epsilon \phi_- \sigma^2$ in V. The axion part of the PQ current is given by (59) in which the replacement $x \to y_+$ and $1/x \to y_-$ are made. Since the current is color-anomalous, we repeat the steps of (61) and (62) to obtain the anomaly-free axion current in the mass eigenstates of a, π^0, η

etc. The result is given by (67) in which $x \to y_+$, $1/x \to y_-$ and $y_+ + y_- = 1$ are used. Repeating again the current-algebra steps, we find

$$m_a = \frac{N_f \sqrt{Z}}{1+Z} \frac{f_\pi}{f_a} m_\pi \simeq 0.6 \times 10^7 \left(\frac{N_f \text{ GeV}}{f_a} \right) \quad \text{eV}. \tag{88}$$

The axion couplings to the neucleon can also be carried out from (77) and the isoscalar as well as the isovector coupling constants corresponding to (78) turn out to be

$$\begin{aligned} g^{(0)} &= \frac{m_N}{2f_a}(1 - N_f) g_A^{(0)} \\ g^{(3)} &= \frac{m_N}{2f_a}\left[y_+ - y_- - N_f \frac{1-Z}{1+Z} \right] g_A^{(3)}. \end{aligned} \tag{89}$$

This type of the invisible axion resolution to the strong CP problem is by far the most attractive proposal. But the mass and coupling strength of the invisible axion is so small that it escapes detection in laboratory experiments, although there are experimental proposals[32] to detect them based on the property of the axion-photon conversion in a static background magnetic field in a microwave cavity or in a strong conversion through the axion Compton-type scattering. These proposals may eventually lead to the detection of the invisible axion if the parameters are guessed right because of the enormous axion number density. But at the moment we have to turn to the astrophysical and cosmological arguments to constrain the invisible axion scale f_a. Essentially the energy emission rates[33] in stellar objects have been used to obtain the upper bound of the axion mass or equivalently the lower bound of the invisible axion scale f_a. There are comprehensive examinations of the emission rates in different regions of the electron or neutron gas, i.e., depending on the relativistic or nonrelativistic gases and degenerate and nondegenerate gases for typical H-burning stars, He-burning stars, carbon- and oxygen-burning stars, supernovae and neutron stars. There are seven types of reactions in stellar objects that contribute to the axion production, i.e., (1) the annihilation $e^+e^- \to \gamma a$, (2) the Primakoff reaction $(Z,e) + \gamma \to (Z,e) + a$, (3) the Compton-like $e + \gamma \to e + a$, (4) the plasma decay $\gamma_{pl} \to \gamma + a$, (5,6) the axion bremsstrahlung of the Compton and Primakoff type $e+(Z,A) \to e+(Z,A)+a$, and (7) the axion bremsstrahlung of the neutron-neutron collisions $n + n \to n + n + a$. Summary of the bounds are given in several places which can be expressed in general by $f_a > 10^8$ GeV. On the other hand, the invisible axion interacts so weakly that the energy density stored in the coherent oscillation of the classical axion field cannot dissipate fast enough and in fact it was shown by several groups[34] that the axion energy density would exceed the critical density to close the universe if $f_a > 10^{12}$ GeV. Thus the axion cannot be too invisible and the invisible axion should lie

$$10^8 \text{ GeV } < f_a < 10^{12} \text{ GeV}, \tag{90}$$

which is several orders of magnitude smaller than the grand unification theory (GUT) scale.

In the context of GUTs one can then imagine that the PQ symmetry is broken at an intermediate scale between the electroweak- and GUT-scales. It will then be esthetically appealing if the PQ symmetry is an automatic one as a consequence of constructing the most general renormalizable potential out of all Higgs scalars that are allowed by the gauge invariance in the Yukawa sector of the Lagrangian. In general, this is too ambitious and one has to ignore certain congruency classes of the Higgs representations or sometimes to be even more selective in the given congruency class. Another problem associated with the PQ symmetry breaking is that if the PQ current has the color anomaly l as given by (44) there are l degenerate vacuum states leading to the formation of domain walls[35] in the evolution of the early universe, which is in contradiction to the standard cosmology. The most elegant resolution to the domain-wall problem is to embed[36] the unbroken discrete subgroup Z_l in the discrete center group of the gauge group. Since such embedding is not always possible, this will constrain the GUT models further. In fact, the automatic invisible axion without any domain wall difficulty can be used together with other required guidelines of the GUT model building[14] to search for the realistic GUT model. The list of models surviving all such requirements is very exclusive. There are only two representations in $SU(9)$ GUT that possess at least three generations of the light particle families at low energy.

5. A Model for Invisible Axions and Majorons

The problem of the neutrino mass and oscillation has recently gained renewed interest. This is due to the discovery of the resonant amplification mechanism[17] of the neutrino oscillation in matter that can explain the observed depletion[16] of the electron neutrino flux in the Cl detector of the solar neutrinos. This solution is indeed very attractive because the mechanism does not require large mixing angles nor the postulation of any new particles. Several detailed calculations[37] show that a tiny mass difference of the order of 10^{-2} eV or less with a small mixing angle can account for the observed depletion. On the other hand, the most natural scheme to generate a small neutrino mass is the seesaw mechanism[38] that makes use of a large Majorana mass term of the right-handed neutrino which violates the lepton number. It was suggested some time ago[4] that the spontaneously broken global lepton number symmetry could be used to generate the Majorana mass term at the seesaw scale. Then there would be a Goldstone boson known as the majoron. But the scale at which the seesaw mechanism becomes operative is essentially an open question.

In a series of investigations[39] on the neutrino masses and their mixings in a class of realistic $SO(10)$ GUT models that contain three family generations of fermions and obey the observed KM mixing structure, it was realised that the scale of the Majorana mass should be bounded from above by the invisible axion scale of the PQ symmetry. The solutions obtained for the neutrino masses and mixing are in fact quite consistent with the ν_e–ν_μ mixing of the resonance amplification solution of the solar neutrino problem alluded to above. It appears that there are connections[40] between the two seemingly independent scales and therefore between the invisible axion and the majoron.

Recently, we have modified[15] the DFS invisible axion model by including the global lepton number symmetry $U(1)_M$ of majoron with the assumption that the Majorana mass scale is the same as the invisible axion scale. The model has an additional fermion, the righthanded neutrino η_R which has both the PQ and lepton charges and an additional complex Higgs scalar σ_M of the $SU(2) \times U(1)$ singlet that carries the lepton number 2 and the PQ charge y_M. The original doublets and σ do not have lepton numbers. The singlet σ_M gets the VEV so that

$$\sigma_M = \frac{1}{\sqrt{2}} V e^{i\xi_M/V} \qquad (91)$$

and the other neutral scalars are parametrized as in (84). For convenience, (83) is also required for the scalars under $U(1)_{PQ}$. The Majorana mass term then arises from the term $G_M \eta_L^c \sigma_M \eta_R$ + h.c. when σ_M acquires the VEV, while the Dirac mass terms appear when $\phi_\pm$ get the VEV's $v_\pm/\sqrt{2}$. Then we get the physical neutrino states

$$\nu_L^D \cong \nu_L - \frac{m}{M}\eta_L^c, \qquad \nu_R^M \cong \eta_R + \frac{m}{M}\nu_R^c \qquad (92)$$

corresponding to the mass eigenstates $m_\nu^D = m^2/M$ and $m_\nu^M \simeq M$ where $M = G_M V/\sqrt{2}$ and $m = G_\nu v_+/2\sqrt{2}$. There are four phase fields and the Goldstone bosons are then given by the four linear independent combinations of them. Anticipating that the one to be eaten up by Z^0 will only know $\xi_\pm$ and that the one to remain as a massive Higgs scalar will be independent of σ_M, the 4×4 orthogonal real matrix can be parametrized with only three angles ω, ϕ and, say ξ, which can be determined by the condition that the Goldstone fields change only by a constant under each global $U(1)$. In particular, the Goldstone boson to be absorbed by Z^0 is given by the same b in (84) while the combination to become a massive

Higgs boson is exactly the field c in (87). The axion field is given by a of (87) in which $V y_s \xi_s$ is replaced by $V(y_s \xi_s + y_M \xi_M)$ and $(y_s V)^2 \to V^2(y_s^2 + g_M^2)$ in A. Finally, there is the new massless Goldstone boson χ which becomes in the $V \gg v$ limit

$$\chi \simeq (y_M \xi_s - y_s \xi_M)/(y_s^2 + y_M^2)^{1/2}, \tag{93}$$

which is the majoron naively speaking. However, both axial currents coupled to a and χ turn out to be color-anomalous, so that further rotation in a–χ is needed to give the fields corresponding to the physical mass eigenstates. In the absence of the color-anomaly both the axion and majoron are massless but the physical axion a_p is to be identified as the physical majoron χ_p. In particular, from the interaction Lagrangian of a and χ with the quarks and leptons, one finds that

$$\begin{pmatrix} a_p \\ \chi_p \end{pmatrix} = \frac{1}{\sqrt{1 + 4y_M^2}} \begin{pmatrix} 1 & -2y_M \\ 2y_M & 1 \end{pmatrix} \begin{pmatrix} a \\ \chi \end{pmatrix} \tag{94}$$

and

$$\mathcal{L}_{\text{int}}^{a_p} = \frac{i a_p}{v_p'} [y_+ m_u \bar{u}\gamma_5 u + y_- m_d \bar{d}\gamma_5 d + y_- m_e \bar{e}\gamma_5 e \\ + m y_+ (\bar{\eta}\gamma_5 \nu + \bar{\nu}\gamma_5 \eta)] \tag{95}$$

and

$$\mathcal{L}_{\text{int}}^{\chi_p} = \frac{i \chi_p}{2 v_p} [m_\nu^M \bar{\nu}^M \gamma_5 \nu^M + m_\nu^D \bar{\nu}^D \gamma_5 \nu^D - m(\bar{\nu}^D \gamma_5 \nu^M - \bar{\nu}^M \gamma_5 \nu^D)] \tag{96}$$

where $v_p' = V\sqrt{y_s^2 + y_M^2}$. Note that (95) gives the axial current to be the same in the DFS model except for the additional neutrino terms. Thus the mass and coupling of the physical invisible axion are given by the same as in the DFS model. On the other hand the physical majoron χ_p, (96), remains massless, i.e., a pure Goldstone boson, and couples only to the neutrinos as in the original majoron model CMP[4]. The rotation (94) from a–χ to the physical a_p–χ_p states is similar to $U(1) \times U(1) \to U(1)_A \times U(1)_V$ where the $U(1)_V$-current is exactly conserved. Also the c field is smoothly connected to the old (standard) axion field in the $V \gg v$ limit. The physical majoron field is orthogonal to the physical invisible axion field when the currents are expressed in the mass eigenstates of axion, π^0, η etc. Note that the physical particles are independent of the choice of y_M.

Finally we mention that the global L symmetry could have been broken explicitly[41] by a small term $K \sigma_M \phi_+^\dagger \phi_-$ instead of using two singlets σ and σ_M. In that case, the invisible axion is smoothly connected to the majoron as $K \to 0$, whereas the massive Higgs boson extrapolates back to the old (standard) axion.

One question yet to be answered in all axion models is the origin of the PQ symmetry-breaking scale. It will certainly be desirable to understand this well-known global symmetry from a truly fundamental level.

I would like to thank J.-M. Richard for the invitation to the Les Houches Workshop, S. Hadjitheodoridis for his help in writing up this lecture note, R. Peccei for the helpful discussions on axions and majorons, and R. Vinh Mau for the warm hospitality extended to me at IPN and LPTPE. This work is supported in part by the U.S. Department of Energy under the Contract No. DE-AC02-76ER03130.A20-Task A.

References

1. See the recent review by J.E. Kim, SNU 86/09 to appear in Physics Reports.
2. Y. Nambu, Phys. Rev. Lett. **4**, 380(1960); Y. Nambu and G. Jona-Lasinio, Phys. Rev. **122**, 345(1961); **124**, 246(1961); J. Goldstone, Nuovo Cim. **19**, 154(1961); J. Goldstone, A. Salam and S. Weinberg, Phys. Rev. **127**, 965(1962).
3. S. Weinberg, Phys. Rev. Lett. **40**, 223(1978); F. Wilczek, *ibid* **40**, 279(1978).
4. Y. Chikashige, R.N. Mohapatra and R.D. Peccei, Phys. Lett. **98B**, 265(1981); G.B. Gelmini and M. Roncadelli, *ibid* **99B**, 411(1981).
5. F. Wilczek, Phys. Rev. Lett. **49**, 1549(1982); D.B. Reiss, Phys. Lett. **115B**, 217(1982).
6. L.M. Krauss and F. Wilczek, Phys. Lett. **173B**, 189(1986); R.D. Peccei, T.T. Wu and T. Yanagida, *ibid* **172B**, 435(1986); W.A. Bardeen, R.D. Peccei and T. Yanagida, Nucl. Phys. **B279**, 401(1987).
7. J. Schwepp *et al.*, Phys. Rev. **51**, 2261(1983); M. Clemente *et al.*, Phys. Lett. **137B**, 41(1984); T. Cowan *et al.*, Phys. Rev. Lett. **54**, 761(1985); **56**, 446(1985).
8. G. t' Hooft, Phys. Rev. Lett. **37**, 8(1976); R. Jackiw and C. Rebbi, *ibid* **37**, 172(1976); C.G. Callan, R. Dashen and D. Gross, Phys. Lett. **63B**, 334(1976).
9. R.D. Peccei and H.R. Quinn, Phys. Rev. Lett. **38**, 1440(1977); Phys. Rev. **D16**, 1791(1977).
10. S. Adler, Phys. Rev. **177**, 2426(1969); J.S. Bell and R. Jackiw, Nuovo Cim. **60A**, 49(1960).
11. R. Crewther, Phys. Lett. **70B**, 349(1977); see also G. t' Hooft, Utrecht preprint (1986), "How Instantons Solve the $U(1)$ Problem".

12. J.E. Kim, Phys. Rev. Lett. **43**, 103(1979); M. Dine, W. Fischler and M. Srednicki, Phys. Lett. **104B**, 199(1981).
13. F.T. Avignone *et al.*, in the A.P.S. Metting of the DPF, Salt Lake City, January 1987; D.O. Caldwell in the Proc. of the Moriond Workshop, Les Arcs, France, January 1987.
14. K. Kang, C.K. Kim, J.K. Kim, I.-G. Koh and H.-W. Lee, Phys. Lett. **133B**, 79(1983); K. Kang, C.K. Kim and J.K. Kim, Phys. Rev. **D33**, 260(1986); K. Kang, in the Proc. of the 14th Int. Coll. of Group Theor. Methods in Phys., Y.M. Cho, ed. (World Scientific Pub. Co., 1985).
15. K. Kang and A. Pantziris, Brown HET-605/IPNO/TH 87-12(1987).
16. J.K. Rowley, B.T. Cleveland and R. Davis, AIP Conference Proceedings **126** (1985).
17. S.P. Mikheyev and A. Yu. Smirnov, Nuovo Cim. **9C**, 17(1986); L. Wolfenstein, Phys. Rev. **D17**, 2369(1978).
18. M. Gell-Mann and M. Lévy, Nuovo Cim. **16**, 705(1960).
19. See for example H. Georgi, "Weak Interactions and Modern Particle Theory", (Benjamin-Cummings, 1985).
20. A. Manohar and H. Georgi, Nucl. Phys. **B234**, 189(1984).
21. P.W. Higgs, Phys. Rev. Lett. **13**, 508(1964); Phys. Lett. **12**, 132(1965); Phys. Rev. **145**, 1156(1966).
22. A.A. Belavin, A. Polyakov, A. Schwartz and Y. Tyupkin, Phys. Lett. **59B**, 85(1974).
23. G. t' Hooft, Orbis Scientia, 1977 Coral Gables Conference, ed. A. Perlmutter.
24. J.M. Pendleburg *et al.*, Phys. Lett. **136B**, 327(1984); I.S. Altarev *et al.*, *ibid* **102B**, 13(1981).
25. V. Baluni, Phys. Rev. **D19**, 227(1979); R. Crewther *et al.*, Phys. Lett. **88B**, 123(1979); see also H.Y. Cheng, IUHET-125 for a summary.
26. C. Vafa and E. Witten, Nucl. Phys. **B234**, 173(1984); Phys. Rev. Lett. **53**, 535(1984).
27. See the simple proof given below Eq.(41), which is taken from K. Kang, I.-G. Koh and S. Ouvry, Phys. Lett. **119B**, 361(1982).
28. W.A. Bardeen and S.-H.H. Tye, Phys. Lett. **74B**, 229(1978).
29. See the summaries in Kim (Ref.1), Cheng (Ref.25) and W.A. Bardeen, MPI-PAE/PTH 56/86.
30. Y. Asano *et al.*, Phys. Lett. **107B**, 159(1981).
31. C. Edwards *et al.*, Phys. Rev. Lett. **48**, 903(1982); M. Sivertz *et al.*, Phys. Rev. **D26**, 717(1982).

32. P. Sikivie, Phys. Rev. Lett. **51**, 1415(1983); Phys. Rev. **D32**, 2988(1985); L. Krauss, J. Moody, F. Wilczek and D.E. Morris, Santa Barbara preprint NSF-ITP-85-08(1985).
33. M. Fukugita, S. Watamura and M. Yoshimura, Phys. Rev. **D26**, 1840(1982); N. Iwamoto, Phys. Rev. Lett. **53**, 1198(1984); A. Pantziris and K. Kang, Phys. Rev. **D33**, 3509(1986).
34. J. Preskill, M. Wise and F. Wilczek, Phys. Lett. **120B**, 127(1983); L. Abbott and P. Sikivie, *ibid* **120B**, 133(1983); M. Dine and W. Fischler, *ibid* **120B**, 137(1983).
35. P. Sikivie, Phys. Rev. Lett. **48**, 1156(1982).
36. G. Lazarides and Q. Shafi, Phys. Lett. **115B**, 21(1982).
37. S.P. Mikheyev and A. Yu. Smirnov, INR preprint Moscow, (1985); H.A. Bethe, Phys. Rev. Lett. **56**, 1305(1986); S.P. Rosen and J.M. Gelb, Phys. Rev. **D34**,969(1986); P. Langacker *et al.*, Nucl. Phys. **B282**, 589(1987).
38. M. Gell-Mann, P. Ramond and R. Slansky, in Supergravity (North-Holland, Amsterdam, 1979); T. Yanagida, in Proceedings of the Workshop on Unified Theory and Baryon Number of the Universe, KEK, Japan, 1979.
39. K. Kang and M. Shin, Brown HET-577 contributed to the Int. Conf. on High Energy Phys., Berkeley, CA. July 16–23, 1986; Mod. Phys. Lett. **A1**, 585(1986); Phys. Lett. **185B**, 163(1987); Nucl. Phys. **B**, (1987); Proc. of the Moriond Workshop, January 24–31, 1987, Les Arcs, France.
40. The same conclusion is reached by P. Langacker, R.D. Peccei and T. Yanagita, Mod. Phys. Lett. **A1**, 541(1986) based on phenomenological arguments.
41. See Langacker *et al.* (Ref.40) and R.D. Peccei in Proc. of the Moriond Workshop, January 24–31, 1987, Les Arcs, France.

Part VII

Round Table on Future Medium Energy Accelerators

A LEAR-like Option for Brookhaven[*]

G.A. Smith

Laboratory for Elementary Particle Science, Department of Physics,
The Pennsylvania State University, University Park, PA 16802, USA

1. Introduction

Recently a working group at Brookhaven developed ideas for intense antiproton beams from very low energies (20 KeV) up to several GeV/c [1]. One of these schemes is similar to LEAR in its concept (but without cooling in its initial applications), and it is this scheme that I wish to discuss very briefly [2,3].

2. Concept

The plan is illustrated in Fig. 1. Three RF bunches (leaving 9 for other physics) are extracted onto a $\bar{p}$ target, lithium lens arrangement. Four GeV/c $\bar{p}$'s (angular acceptance = 40 mrad, $\Delta p/p$ = 2%) of maximum intensity 4×10^7 $\bar{p}$'s/10^{13} protons on target are transported to the booster accelerator presently under construction. After injection in the reverse direction, they can be decelerated to 650 MeV/c if desired and extracted to an experimental area (previously the 80-inch bubble chamber hall). The booster acts as a stretcher ring, from which pure $\bar{p}$'s are extracted every supercycle ($\sim$ 3 sec) of the AGS. Momentum compression to $\Delta p/p = 10^{-4}$ can be achieved, resulting in the extracted intensities given in Fig. 2. Stochastic cooling in the booster could be introduced to recover beam losses during deceleration. A further option allows one to inject $\bar{p}$'s into the linac and decelerate to 750 KeV, and then further decelerate to 20 KeV in the RFQ. Cost estimates have been made, assuming the availability of existing or planned (g-2 experiment) facilities, at $\sim$ 3.6M\$.

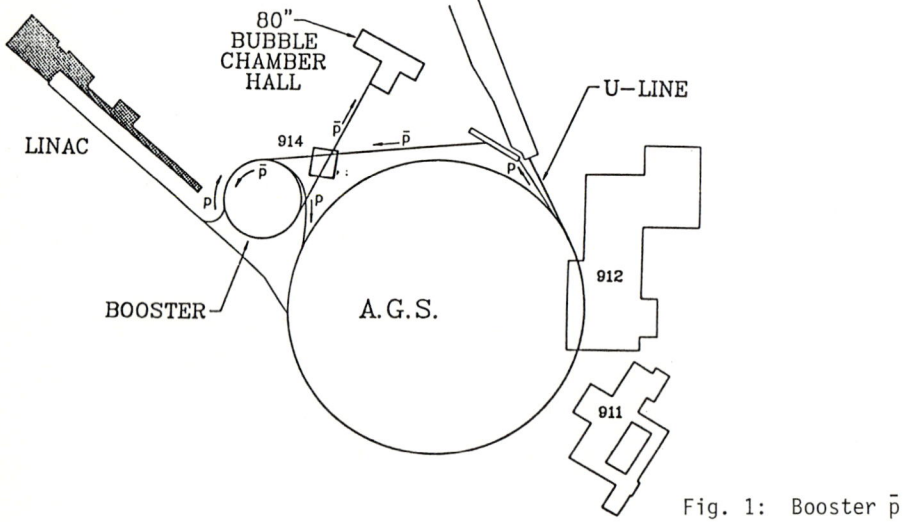

Fig. 1: Booster $\bar{p}$ system

---

[*] Work supported in part by the U.S. National Science Foundation.

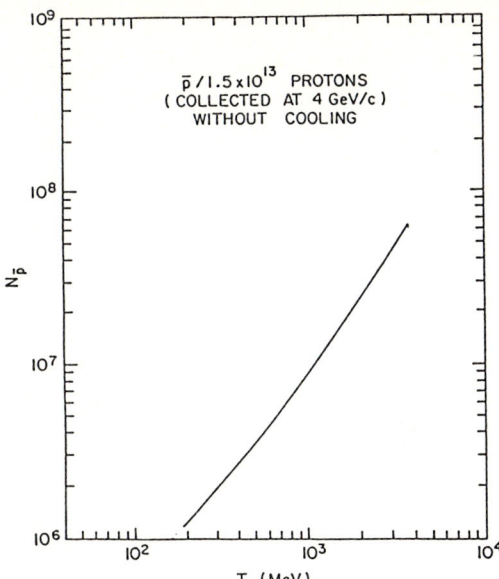

Fig. 2: Intensity of decelerated antiprotons

3. Physics

I only mention here a few ideas for experiments. With beams in the 2-4 GeV/c region, one could embark on a systematic program of studying hyperon-antihyperon final states as a test of QCD, including a measurement of spin correlations, as discussed by H. GENZ in this workshop. Of course, with careful consideration of systematic errors in mind, one could search for CP violation in $\Lambda\bar{\Lambda}$ production as recently suggested by J. DONOGHUE [4]. Here the intense and pure $\bar{p}$ beam would be extremely valuable. Finally, the ultra-low energy option could lead to experiments on trapping of antiprotons, tests of gravitational invariance, precise $\bar{p}$-p mass difference measurements, etc., as are now just getting underway at LEAR [4].

References

1. D. Peaslee et al., Summary of 1986 Workshop on Antiproton Beams, draft report, March, 1987.
2. Y.Y. Lee, BNL Accelerator Division Technical Note #266, Oct. 17, 1986.
3. Y.Y. Lee and D.L. Lowenstein, BNL Accelerator Division Technical Note #269, Dec. 3, 1987.
4. Intersections Between Particle and Nuclear Physics, ed. D.F. Geesaman, AIP Conf. Proc. 150, 1986.

The Scientific Program of a Multi-GeV cw Electron Accelerator

A. Gérard

DPhN/HE, C.E.N. Saclay,
F-91191 Gif-sur-Yvette Cedex, France

NEW FRONTIERS IN NUCLEAR PHYSICS

With present day accelerators, nuclear matter can be explored with spatial resolution *comparable to the nucleon size* (1 fm). This has allowed emergence of a description of nuclear phenomena in terms of *composite particles* (nucleons + isobars) interacting by *exchange of mesons*.

The goal of modern nuclear physics is to deepen this description, determine its domain of validity when one goes to shorter distances, and insert it in a more fundamental framework.

QCD seems to emerge as the fundamental theory of the strong interaction. Elementary constituents (quarks, gluons) are seen in *high-energy experiments*. The quark-quark force is weak at short distances, which allows the use of perturbative methods.

Nuclear physics is concerned with low energy, non-perturbative aspects of QCD, not directly accessible from first principles. Its aim is to understand the structure and dynamics of **extended and complex** objects, the nature of the **effective forces** and **collective phenomena** which are manifested in these systems, to determine the **relevant degrees of fredom** to describe them.

This situation presents some analogies to condensed matter physics but also some specific features:

- the **confinement of colour** at a scale of about 1 fermi.
- the **high density** of nuclear matter. The distance between moving nucleons fluctuates from a few fermis to short distances at which they overlap. This allows us to study baryon-baryon forces at all distances and how confinement forces act in a baryon-rich environment.
- nuclei are **finite systems** and a large number of species are available in nature, from two- and three-nucleon systems, accessible to microscopic calculations, to heavy nuclei which provide an approximation of infinite nuclear matter.

Some key-topics should be addressed by future programs:
- The **free nucleon** structure and dynamics. Quark vs meson description.
- The relevance of the concept of nucleon in nuclear matter, especially in **deep-lying shells**.

- The nature of the **two-nucleon subsystem**. Origin of short range correlations ? Many-body forces? Quark clusters? Glueballs? Colour degrees of freedom in nuclei? Percolation?

- The nucleon resonances. Understand the nucleon radial excitations. The role of N^* in nuclear dynamics. The NN^* interaction.

- **The short distance part** of nuclear wave functions. **High momentum components**.

- **New flavors in nuclei**. The strangeness degree of freedom. How do Λ and Σ hyperons behave in nuclei?

- The momentum distribution of quarks in nuclei, especially the large x part.

WHY ELECTRONS ?

1. The electron is a pointlike particle. The observed structures in electron scattering are related to the target and not to the projectile.
2. The elementary interaction is well under control from QED.
3. The interaction is weak. This allows the **entire nuclear volume** to be probed.
4. Virtual photons allow three parameters to be varied **independently**:

   - Momentum transfer = **spatial resolution**.
   - Energy transfer = **time resolution** = observation of **highly virtual phenomena** = excitation of new degrees of freedom = creation of $\pi, \rho, \omega, K, \Phi, a_1$, etc.
   - Polarisation of the photon. This allows different aspects of the dynamics to be probed: charge vs magnetic properties and mesonic currents.

WHICH MACHINE ?

Energy

The 2-3 GeV range opens π, η and K meson production and provides good conditions for (e,e'x) reactions allowing T/L separations.

The 3-5 GeV range allows (e,e'NN) and (e,e'Nπ) experiments to be performed in forward kinematics with fast outgoing particles and reasonable counting rates.

Duty cycle and luminosity

Study of very exclusive channels requires **multicoïncidence** experiments. A **continuous beam** is a necessary condition to perform such experiments.

Study of **rare phenomena** requires a high luminosity ($> 10^{37}$ cm^2 s^{-1}).

Energy resolution

Hypernuclear spectroscopy requires better than 10^{-4}.

Secondary beams and possible options

- A tagged bremsstrahlung real photon beam
- Low momentum K beams
- Storage ring in parasitic mode with internal targets
- Polarized electrons

European Proposals for a B-Factory

J. Duclos

DPhN/HE, C.E.N. Saclay,
F-91191 Gif-sur-Yvette Cedex, France

Two proposals have been made recently for a high luminosity electron-positron collider. One is a traditional storage ring collider, presented by R. Eichler et al. /1/ as a project of the Swiss Institute for Nuclear Research of Zurich (SIN). The other one is a superconducting linear collider presented by Y. Amaldi (CERN) and G. Coignet (Annecy) /2/.

The motivation for such proposals is to perform a quantitative study of a number of basic phenomena accessible by e^+e^- colliders operating in the Υ region or below, with an emphasis on the CP violation in the B-$\bar{\text{B}}$ systems.

The present machines (DORIS-II at Desy, CESR at Cornell) have a peak luminosity smaller than 3.10^{31} cm^{-2}s^{-1} and a yearly production of about 150 events/pb. It is proposed to increase the peak luminosity by more than an order of magnitude and to improve the running efficiency in order to get a yearly production of 10,000 events/pb. In addition the parameters would be optimized for maximum luminosity at $E_{CM} \sim 10$ GeV.

The basic idea is indeed to operate the machine on the Υ (4S) resonance just above the B-$\bar{\text{B}}$ threshold. If the energy resolution is smaller than the resonance width (25 MeV), an almost pure source of $1.5 \; 10^7$ B-mesons can be obtained per year. It is also possible to operate the machine in the continuum at higher energy ($\sim$ 15 GeV) and reconstruct the B($\bar{\text{B}}$) meson by tagging the associated $\bar{\text{B}}$(B). In that case the energy resolution is no longer critical.

The mixing of neutral B-mesons has been seen by the CERN experiment UA1 and the observation of the B$^\circ$-$\bar{\text{B}}^\circ$ and B$_s$-$\bar{\text{B}}_s$ oscillations could be done in about one year run. The CP violation could be observed through partial decay widths : B$^+ \to$ K$^+\rho^\circ$ and B$^- \to$ k$^-\rho^\circ$ in about the same time. The B$^\circ$-$\bar{\text{B}}^\circ$ oscillations depend on the CKM matrix elements of the standard model and the measurement of the CP violation would provide a decisive information concerning the nature of the phenomenon.

In addition to the B-meson physics, such high luminosity colliders would allow many other measurements concerning hadron spectroscopy of c and b quark systems, leptonic τ decays ... (see ref. 1). Then an extensive program would be open in the field of non-perturbative QCD and weak interactions, providing an opportunity to complete the measurement of the standard model parameters and to look for possible deviations.

THE STORAGE RING COLLIDER

Several e^+e^- storage rings have been in operation for many years and a lot of experience has been collected. The existing machines store electrons and positrons in a single ring, which limits the number of bunches and consequently the luminosity. The collider design study is based on a multibunch double storage ring with two interaction regions of zero degrees beam crossing angle.

A standard linac structure is used to accelerate electrons up to about 200 MeV. These electrons may be ejected or may be shot on a tungsten target to produce a sufficient number of positrons accelerated in a second linac.

The two beams are accelerated in a booster synchrotron up to 7 GeV, and injected in the double storage ring. The circumference of the ring is 480 m.

The maximum luminosity is expected to be 5.10^{32} cm^{-2}s^{-1} at 5.3 GeV (10.6 GeV total) with a resolution of about 10 MeV, for maximum currents of 485 mA and 10 bunches. At energies above 5.5 GeV the maximum beam currents are limited by the rf-power and consequently the luminosity drops down.

THE LINEAR COLLIDER PROJECT

The linear collider project uses superconducting cavities in a two-step linear accelerator. The first step goes up to 2.5 GeV and the second one up to 5 or 10 GeV. The cavities, working at a frequency of 350 MHz, are assumed to perform an accelerating gradient of G =7 MV/m. To minimize the cost a recirculation scheme is used for both steps. Damping rings are necessary between the two steps to adjust the emittance.

The positrons and electrons are circulating in opposite directions in the same accelerator sections. Positron production is a serious problem, because of the low energy of the collider. For example with a $\sim$ 150 MeV electron injector and a $\sim$ 30 mA current the power on the target would be 5 MW. This problem is not solved.

The machine is planned for two operating regimes. The maximum energies are respectively 10 and 20 GeV, the luminosities 10^{33} and 10^{34} cm^{-2}s^{-1} and the energy spreads 10 and 300 MeV. The bunch frequency is 12 kHz with 5 to 8.10^{10} particles per bunch. The bunch radius at the intersection point is 1.1 to 0.6 µm and the bunch length 1.3 to 0.4 mm.

The project is presented as a multipurpose machine. A fraction of the first stage, combined with a wiggler system, could be used as a free electron laser. The complete first stage would allow nuclear studies with a continuous electron beam of energy 2.5 GeV.

REFERENCES

1. Motivation and Design Study for a B-Meson Factory with High Luminosity. R. Eichler, T. Nakada, K.R. Schubert, S. Weseler, and K. Wille - SIN PR-86-13.
2. Conceptual Design of a Multipurpose Beauty Factory Based on Superconducting Cavities - U. Amaldi and G. Coignet - CERN-EP/86-211.

Physics at Laboratoire National Saturne with MIMAS

M. Roy-Stéphan

Division de Physique Théorique, IPN,
Université Paris-Sud, F-91406, Orsay Cedex, France and
L.N.S., CEN Saclay, F-91191 Gif-sur-Yvette Cedex, France

LABORATOIRE NATIONAL SATURNE has been operating the SATURNE II synchrotron from 1978. New performances will be achieved with MIMAS, the new injector, which has started running on May 1987.
SATURNE provides beams of :

- p, d, ^{3}He, ^{4}He,
- Heavy ions,
- polarized $\vec{p}$ and $\vec{d}$, (monokinetic polarized $\vec{n}$ beams are obtained by $\vec{d}$ break-up).

The energy range is :
- from 200 MeV to 2.9 GeV for protons
- from 200 MeV to 2.3 GeV for deuterons
- up to 1.15 GeV for neutrons.
- Heavy ion energy per nucleon, from 100 MeV to 1.15 GeV.

The proton polarization is about 90 % below 1 GeV and around 75 % at high energy. The deuterons may be either vector or tensor polarized. Vector polarization is about 100 % and tensor polarization about 85 % of its maximum theoretical value.
Up to now, the particles were injected in SATURNE through a Linac. The intensities which have been reached up to 1986 are (in particle per burst) :
1.7×10^{12} protons, 2×10^{10} $\vec{p}$, 4×10^{10} $\vec{d}$, 10^8 ^{12}C, 3×10^7 ^{20}Ne, (and $5 \; 10^5$ $\vec{n}$ for the neutron secondary beam). A new injector, the MIMAS synchrotron [1], has been designed to obtain an injection efficiency of 100 %. This transparency should remain close to 100 % for particles with a ratio charge over mass Z/A much less than 1/2. New beams will become available, especially much heavier ions. Moreover, MIMAS will work as an accumulator for heavy ions.
During MIMAS construction a new heavy ion source, DIONE [2] has been built. It is an improved version of the previous one, CRYEBIS. It is a cryogenic electron beam ion source, delivering fully stripped or nearly fully stripped ions. The repetition rate of the new source is much higher than the repetition rate of the previous source, thus several pulses from the source will be stored in MIMAS. In 1987 ions up to Ar will be accelerated and hopefully in 1988 up to Kr.
The intensities expected in 1987 with MIMAS are the following :

- 10^{11} to 2×10^{11} $\vec{p}$ or $\vec{d}$. 10^{11} $\vec{d}$ have actually been accelerated in May 1987. The $\vec{n}$ beam intensity should be multiplied by 10. Improvements of the polarized source are in progress [3].

- Concerning heavy ions 3×10^9 ^{12}C, 4×10^8 ^{20}Ne, 6×10^7 ^{40}Ar, per burst are planned in 1987. A factor of 3 to 10 more is expected for these ions in 1988.
- The intensities for unpolarizd light particles will not change.

Among experimental equipment at L. N. S., three spectrometers may be used. [4] SPES I is the very high resolution spectrometer : resolution

($\Delta p/p$) = 6.10^{-5}. The energy of the scattered particle is limited to 1.2 GeV. A solenoïd on the incident line can rotate the spin of incoming $\vec{p}$. SPES III has a very large momentum acceptance ($\Delta p/p$) = $\pm 40\%$ (from 600 to 1400 MeV/c) and a large solid angle 10^{-2} sr. Its detection set-up has been designed for multiparticle operation. SPES IV is the high energy spectrometer. The momentum of the scattered particle may be analysed up to 4 GeV/c. It is well suited for 0° measurements. A time-of-flight basis 17 meters long provides a very precise particle mass identification opportunity. Recoil polarimeters will soon be in operation both at SPES I and SPES IV.

Furthermore, there is a dedicated nucleon-nucleon area designed to measure $\vec{p}+\vec{p}$ and $\vec{n}+\vec{p}$ elastic amplitudes [5]. The set-up originally built by the nucleon-nucleon collaboration includes a big volume frozen spin target. The spin of the incident particle and the spin of the target proton may be oriented in three perpendicular directions. The spin of the recoil proton is analysed, through a recoil polarimeter. The momentum of the scattered proton (in $\vec{p}+\vec{p}$ system) is analysed. A neutron detector is operated for the $\vec{n}+\vec{p}$ program. A complete set of observables may be measured.

Other beam lines are devoted to specific apparatus. In particular the 4π detector DIOGENE [6] has been designed to study high multiplicity nucleus-nucleus collisions. It is used to work on nucleon nucleon inelastic channels as well and to detect decay products of resonances produced in nuclei.

A physics program taking advantage of these capabilities has been recently debated [7].

Among present and future experimental topics let us refer to nucleon-nucleon elastic and inelastic channels, with a special emphasis on narrow dibaryon "hunting". - π and heavier meson production especially in few body systems. -Spin isovector and isoscalar response of nuclei. - Δ and N* excitation in nuclei - Relativistic effects, for example Dirac phenomenology. For all these items it is essential to measure spin observables. Thus $\vec{p}$ and $\vec{d}$ beams will be intensively used. Experiments involving strangeness production are going on or planned, as well as experiments to test isospin conservation.

With heavy ions, central collisions are studied to determine nuclear equation of state at high density, research on multifragmentation will also be carried on. Fragmentation, exotic nuclei creation and Δ excitation in nuclei will be studied in peripheral collisions.

This program is very diversified, covering nuclear structure, mesonic degrees of freedom and hopefully quark degrees of freedom.

REFERENCES

1 Nouvelles de Saturne n° 9B
  and J.-L. LACLARE : 3e Journées d'Etudes Saturne - Fontevraud, avril 1983, page 433.
2 J. FAURE : Rapport LNS-SD 87/128 (1987)
  and B. GASTINEAU, J. FAURE, A. COURTOIS : Nucl. Instrum. Methods in Physics Research B9 (1985) 538.
3 R. VIENET : communication privée
  and J.-L. LEMAIRE, R. VIENET, P.-Y. BEAUVAIS : Helvetica physica acta, vol.59 (1986) 573.
4 Ensemble de détection magnétique du Laboratoire National Saturne - LNS 1980.
5 J. BYSTRICKY et al. Nucl. Phys. B262 (1985) p 715 and references included.
6 J.-P. ALARD et al. Rapport DPhN Saclay 2420 (1987) and Nucl. Instrum. Methods to be published.
7 Proceedings of 4e Journées d'Etudes Saturne - La physique avec MIMAS (novembre 1986).

A Hadron Facility for Europe

F. Bradamante

Università di Trieste, Via A. Valeria 2,
I-34100 Trieste, Italy

1. The European Hadron Facility

The European Hadron Facility is a new research facility for basic research in Nuclear and Particle Physics which is proposed for the 1990s and beyond. It consists of a complex of accelerators to produce a high-intensity (100μA) proton beam of 30 GeV kinetic energy, and is meant to provide a broad range of intense, high-quality secondary beams of neutrinos, muons, pions, kaons and antiprotons. A distinctive feature of this machine is that it has been designed to accelerate polarized proton beams to full energy.

The design intensity of 100μA primary protons is the prerequisite for the envisaged research program, and it represents an improvement of two orders of magnitude with respect to existing accelerators of similar energy. The design energy of 30 GeV is determined mainly by the desire to produce copious beams of antiprotons.

The main components of the EHF are a high-energy LINAC, accelerating a H⁻ beam to 1.2 GeV, and two fast cycling synchrotrons, a 9 GeV booster ring and a 30 GeV main ring, with radii and repetition rates of ratios 1:2 and 2:1 respectively. The repetition rates of the LINAC and of the booster are the same, 25 Hz. Two more rings complement the system, a 9 GeV holder ring, with the same radius as the booster, where the booster pulses are stored before being transferred to the main ring, and a 30 GeV stretcher ring, having the same circumference as the main ring synchrotron, where the fast extracted 30 GeV beam from the main ring is stored and then slowly extracted to produce 100% duty factor secondary beams.

Two experimental areas are foreseen, a fast extraction hall for neutrino and pulsed muon physics and a slow extraction hall with nine secondary beam lines for counter experiments. Provision is also made for laboratory buildings, common services and utilities. Figure 1 shows a schematic view of the facility.

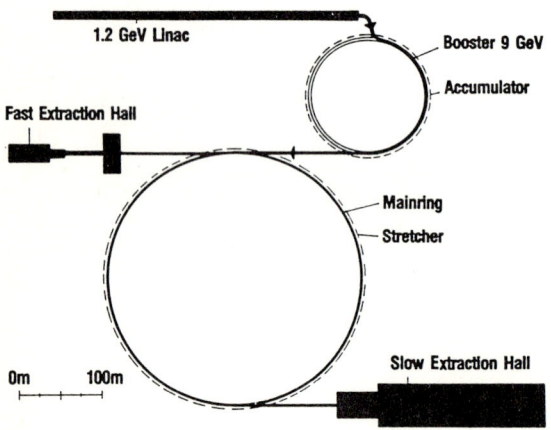

Fig. 1 - Schematic layout of the European Hadron Facility

2. The Physics Program

The availability of intense secondary beams of different particles will guarantee a rich and deversified experimental program in a field which has had a long and successful tradition in Europe. Although our description of the subatomic world has undergone spectacular improvement over the last decades, many fundamental problems are still unresolved and require dedicated experiments at low energy. Broadly speaking:

1) The dynamics of quarks and gluons is still not described by a complete and self-consistent theory. Quantum chromodynamics is surely an important step, but so many problems are either unresolved or incalculable that strong interactions are still an open problem.
2) The existence of several families of quarks and leptons is a puzzle, and the search for the presence of high masses in rare processes is probably the best way to get signals of new physics.

EHF is the accelerator which will guarantee a long term research program in these fields, and will be complementary to the high energy colliders which are already being planned.

If realized as a new laboratory, and an Italian option at present seems the most realistic, it will cost 870 Million Deutsche Marks, it could be realized in six years, and will need some 500-600 staff when running. Clearly, it could be constructed at CERN, where important savings on the investment cost could be realized.

Bibiography

The EHF Study Group, "Letter of intent for a European Hadron Facility", August 1986, edited by F. Scheck, Mainz.

Proceedings of the "International Conference on a European Hadron Facility," Mainz, Germany, March 10-14, 1986, edited by Th.Walcher, North Holland Publishing Co., reprinted from Nucl. Phys. B279 (1987) Nos. 1,2.

The EHF Design Group "Feasibility Study for a European Hadron Facility", EHF-86-33, June 1986.

"Proposal for a European Hadron Facility", EHF-87-18, May 1987.

Physics at Super-LEAR

P. Dalpiaz[2], R. Klapisch[1], P. Lefevre[1], M. Macri[4], L. Montanet[1], D. Möhl[1], A. Martin[1], J.M. Richard[3], H.J. Pirner[1], and L. Tecchio[2]

[1]CERN, European Organization for Nuclear Research,
 CH-1211 Geneva, Switzerland
[2]Istituto di Fisica, Univ. di Torino, I-Torino, Italy
[3]Laboratoire de Physique Théorique des Particules Elémentaires,
 Université Pierre et Marie Curie, F-75252 Paris, France
[4]Istituto Nazionale di Fisica Nucleare, I-Genova, Italy
[5]Istituto Nazionlae di Fisica Nucleare, I-Ferrara, Italy

1. INTRODUCTION

At the third LEAR Workshop, Tignes 1985 [1], possible future developments of "low energy" antiproton rings were discussed.

In particular, the realisation and characteristics of a superconducting antiproton ring (Super-LEAR) were presented by D. Möhl et al. with the possibilities for high energy electron cooling (L. Tecchio). The experimentation at such a new machine was discussed by several speakers, in particular by M. Poulet et al. and P. Dalpiaz while the theoretical background was reviewed by S. Narison, A. Martin, W. Buchmüller and M.G. Olsson.

After the workshop, Robert Klapisch convened a working group (P. Dalpiaz, P. Lefevre, M. Macri, A. Martin, D. Möhl, L. Montanet, H.J. Pirner, J.M. Richard, and L. Tecchio) to clarify our views on the physics arguments for proceeding further with a design study and, eventually a proposal for the construction of a Super-LEAR at CERN.

One of the main guidelines in writing this report was provided by the R704 Experiment [1] performed at the ISR. This experiment has produced very interesting results on the properties of χ_1 and χ_2 ($^3P_1$, $^3P_2$ $c\bar{c}$) [2] and has even given some hints on the longstanding question of the ^1P state [3], although the luminosity was limited to 3×10^{30} cm^{-2}s^{-1} and the running time to 5 days! (integrated luminosity: 700 nb^{-1}). From this experiment, we learn that charmonium physics is accessible through $p\bar{p}$ annihilation, provided that the following conditions are satisfied:

(a)   A high luminosity is available.

(b)   A very narrow beam momentum distribution is achievable. This is to maximize the rate of resonance formation and to allow a precise measurement of the width of the charmonium states.

(c)   The use of a thin target preserving the intrinsic $\bar{p}$ beam momentum definition given by the cooling techniques.

These conditions are met in the interaction of a coasting $\bar{p}$ beam on an internal H_2 gas jet target of a suitable density. In the case of R704 the $\bar{p}$ beam was coasting in ring 2 of the ISR. The beam momentum distribution was narrowed with the use of a stochastic cooling system to $\Delta p/p \sim 10^{-3}$.

The internal H_2 gas "jet" target was built using the technique of a clustered beam. Its thickness was chosen such that the stochastic cooling system acting on the $\bar{p}$ beam was able to counteract the blow-up of the beam dimensions and the degradation of momentum due to the energy loss in the multiple traversal of the target.

In this way the p̄p interaction probability was almost 1 (p̄ were lost only by inelastic nucleon interactions) and the beam momentum spread was kept to the initial value ($\leq 10^{-3}$) by the stochastic momentum cooling system. The largest initial luminosity with an injected p̄ beam of 10^{11} p and a target thickness of 10^{14} atoms/cm was $\sim 3 \cdot 10^{30}$ cm^{-2}s^{-1}.

A similar experiment has been approved at the Fermilab p̄ accumulator.

As we shall see, the luminosity at Super-LEAR could reach 10^{32} cm^{-2}s^{-1} with a $\Delta p/p$ of 10^{-3} to 10^{-4}. In addition, the dimensions of the source could be smaller than at the ISR, a useful characteristic for the design of the detector and for data analysis.

Two additional conditions have to be satisfied to perform the charmonium study:

(a)   knowledge of the absolute momentum value;

(b)   reproducibility of the momentum setting to a precision better than dp_{lab}/p_{lab}.

In R704, the determination of the resonance parameters was done via the measurement of the excitation curve of the resonance.

This is performed by determining the number of candidates for the resonance decay at several momentum values around the one giving a √s equal to the mass of the resonance under study. The role of the final state detector is to tag in an unambiguous way the events, but the resolution on the resonance parameters depends essentially on the knowledge of the beam momentum distribution and not on the final state detector resolution.

The resonances χ_1 and χ_2 were detected through the radiative decays to the J/ψ and the J/ψ was identified by its decay into e^+e^-. The background level was measured in a control region in between the two resonances.

The value of the background cross section was less than 5 pbarn. R704 had also the following features:

(a)   an efficient signature of $e^{\pm}$;

(b)   a high rejection power of background events from $\pi^{\pm}$, π^0 simulating electrons through δ rays production, γ conversion or Dalitz decays.

For R704, the statistics were the limiting factor on the width resolution measurements. ($\Gamma(\chi_1) < 1.25$ MeV at 90% CL and $\Gamma(\chi_2) = 2.9^{+1.8}_{-1.1}$ MeV). At Super-LEAR, the dominant limiting factor could be the p̄ beam momentum distribution which could be such that $\Delta M/M \leq 10^{-4}$ could be reached (giving a relative error of 2% on the total widths).

In sect. 2, we shall summarize the characteristics of Super-LEAR. This machine should permit precision measurement in a domain of centre-of-mass energies which is not currently accessible with p̄p at CERN (fig. 1).

In sect. 3, some considerations will be made on electron cooling at intermediate energies.

Three versions of Super-LEAR will be assumed in sect. 4, where we shall discuss the physics arguments:

SL1   One antiproton ring, from 2 to 10 GeV/c. Hydrogen (or nuclear) jet target.
       L = 10^{32} cm^{-2}s^{-1} $\Delta p/p = 10^{-3}$ to 10^{-4}.

SL2   Proton-antiproton collider up to 10 × 10 GeV/c one ring.
       L = 3×10^{30} cm^{-2}s^{-1}.

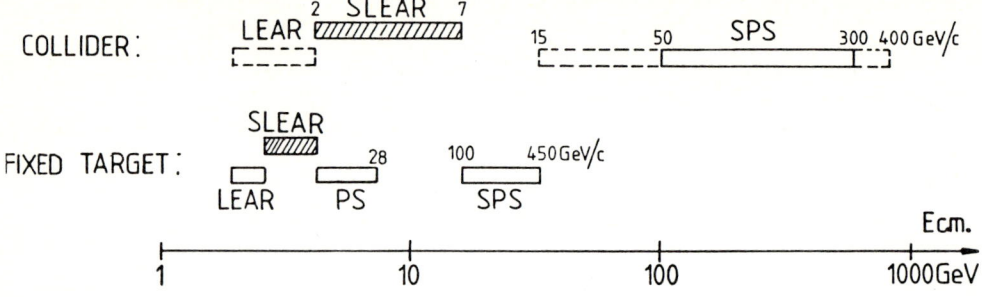

Fig. 1 Energy domain covered by the various CERN accelerators

SL3   Proton-antiproton collider (2 separated rings) with some (factor 3 to 5) gain in luminosity as compared to SL2.

2. SUPER-LEAR - SOME CHARACTERISTICS

2.1 Introduction

In this section we summarize some tentative machine aspects. The design is based on several new technologies including the use of superconducting magnets. They permit 10 GeV/c within a small circumference ("1.5 × LEAR") and hence cost effectiveness and fast revolution. This permits very economic use of antiprotons and high luminosity.

2.2 Layout

Basic parameters assumed are compiled in table 1.

TABLE 1

Momentum range	2-10 GeV/c
Injection momentum	3.5 GeV/c
Circumference	120 m
Intensity	$\leq 10^{12}$ $\bar{p}$
Possible extension of momentum range (with superconducting quads and 9 Tesla instead of 4 Tesla bending magnets)	2-15 GeV/c

The tentative layout, which uses 8 superconducting 45° bending magnets and 16 normal conducting quadrupole doublets is sketched in fig. 2. The doublet structure has been chosen as it allows us to push transition energy above the working range.

Table 2 summarizes the optical properties in the straight sections and equipment to be installed. Note that the medium straight sections have small beta (small beam size - large angular acceptance) and zero dispersion for the installation of the collision region and/or relatively thick targets. The short straights have large beta and large dispersion for targets or other applications requiring parallel beam and orbit separation by momentum. This lattice is only an example and more work is needed first to establish the final design criteria and then to work out an optimised design.

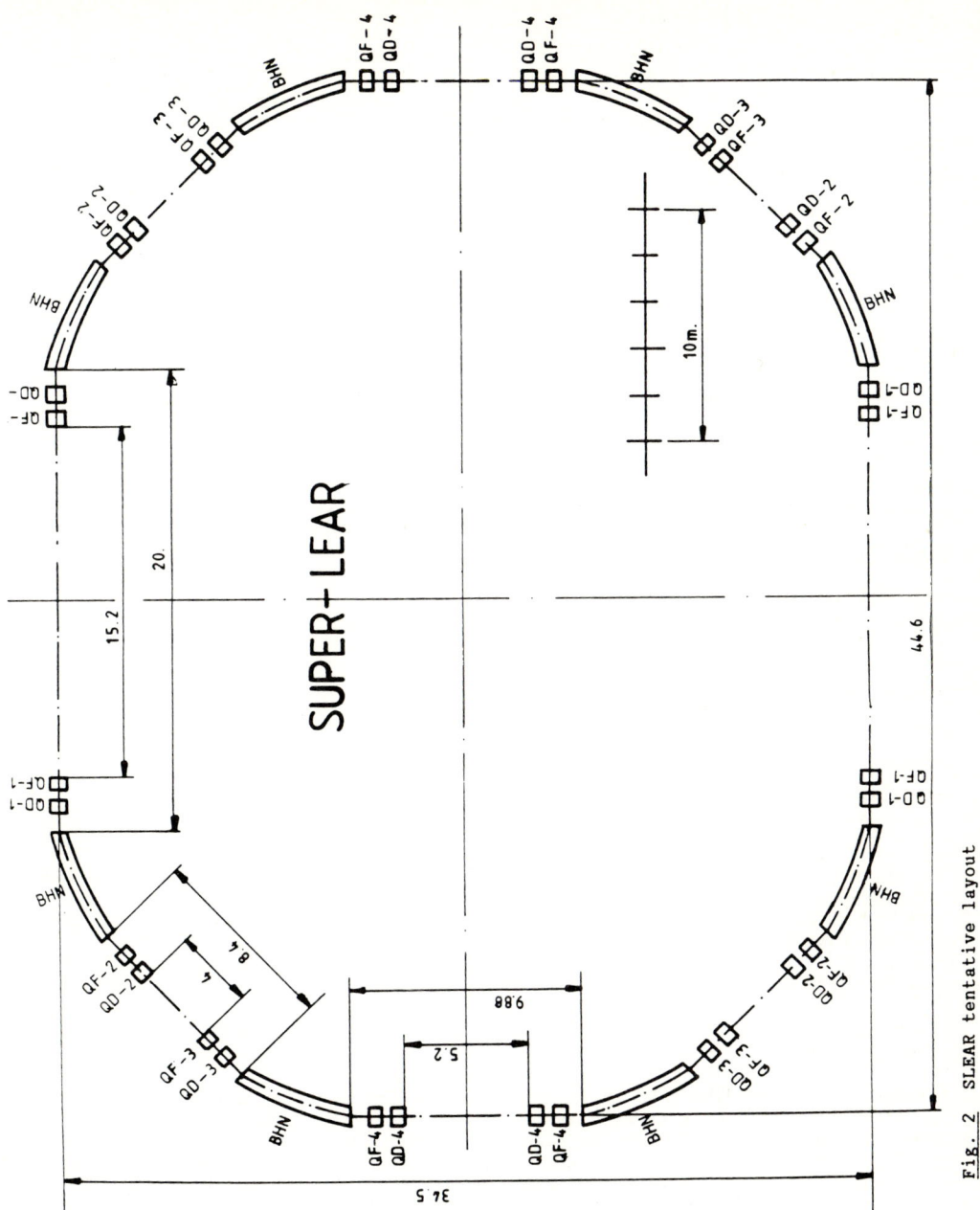

Fig. 2 SLEAR tentative layout

Some characteristics of the bending magnets are given in table 3. As an example, a 5 m long version (I) requiring 5 Tesla at 10 GeV/c and a shorter 4 m, 6.5 Tesla version (II) are considered. These fields are within the realm of present technology. With new types of superconductors now under development, fields of, say, 9 Tesla corresponding to 15 GeV/c in SLEAR could be aimed at. To work up to this momentum in a future extension, the quadrupoles also would have to be converted to a superconducting design (gradients of 30 Tesla/m instead of 15 Tesla/m assumed in the "normal" layout). Ramping of the fields from injection (3.5 GeV/c) to the final momentum can be done very slowly (dB/dt < 1 Tesla/min), so that essentially "d.c." magnets can be used.

TABLE 2

Type		Long	Medium	Short
Number		2	2	4
Length (m)		15	5	4
Typic. Beta(h) (m)		10	1.5	10
Beta(v) (m)		12	0.8	5
D (m)		-1.3	0	4.5
Equipment		Injection	Collisions	Cooling
		diagnostic	targets	diagnostic
		corrections	RF	targets

TABLE 3

Superconducting magnets		Version I	Version II
Length (cent. orbit)	(m)	5.0	4.0
Bending angle	(deg)	45	45
Bending radius	(m)	6.37	5.09
Sagitta	(m)	0.5	0.4
Field at 10 GeV/c	(T)	5.3	6.6
Bore diameter	(mm)	100	100
Gradient (nominal)		0	0
Tolerances DB/B		1E-4	1E-4

Injection uses a septum in one of the long straight sectors and a fast kicker in the subsequent short straight sector. Antiprotons are directly transferred from ACOL-AA at 3.5 GeV/c. Proton injection (from the PS) for the collider mode can be done in the same long section or in the second one, depending on the location chosen for SLEAR. To save cost, location in an existing hall (West Hall?, ISR service building 181?, East Hall?) should be envisaged.

A vacuum of a few 10^{-12} Torr N_2 would give sufficient beam life (~ 100 h) even at lowest energy.

A modest RF system (about 15 kV at 2.5 MMHz, 10% frequency swing) is sufficient in the internal target mode to accelerate or decelerate the beam but very powerful systems are required to compress the beam into a very short bunch for head-on collisions; see table 4 where a two and a three-stage version are considered. The high voltage bunch compression cavity could possibly be superconducting.

Phase-space cooling both stochastic (with time constants of several hours at 10^{12} $\bar{p}$) and electron cooling are desirable to prepare the beam and to keep it in shape. This is to be discussed in the next section.

This concludes our overview of the layout considered. Work is needed to arrive at a more detailed design. Possible modifications to be considered are structures with more (e.g. 10 or 12 instead of 8 straight sections) and the possibility of adding a second intersecting ring at a later stage so that protons and antiprotons have their own ring. But the considerations presented should allow us to make performance estimates. This will be the subject of the rest of this section.

2.3 Performances

In any interaction, the luminosity attainable is ultimately limited by the flux of particles available. Matching the consumption

$$dN/dt = L \cdot \sigma_{loss} \tag{1}$$

to the production rate (ϕ = dN/dt produced, 10^7/s in our case) and taking a total cross section of $\sigma \sim 100$ mb (= 10^{-25} cm^2) to approximate p-$\bar{p}$ interactions in the 2-10 GeV/c range, one obtains

$$L \leq \phi/\sigma \sim 10^{32} \text{ cm}^{-2} \text{ s}^{-1} . \tag{2}$$

For the internal target mode, conditions close to this performance limit can (in principle) be reached by a judicious choice of target thickness and filling cycle. As an example, let us assume a filling cycle ($1/t_{++}$) of once per 10^5 s ($\sim$ 24 h) with a transfer of 10^{12} $\bar{p}$, as a result of a production rate of 10^7/s. A matched target uses up this beam (to say 1/e) in 10^5 s. It has a density

$$nd = (f_{rev} \sigma t_f)^{-1} = 4 \times 10^{13} \text{ (H atoms/cm}^2) = 0.7 \times 10^{-10} \text{ g/cm}^2 . \tag{3}$$

This is about half the density of the ISR gas target [4] and thus is perfectly feasible. In fact, it may be preferable to work with a, say, 5-times thicker target ($\rho d \sim 3 \times 10^{-10}$ g/cm^2) needing transfers of $\sim 2 \times 10^{11}$ $\bar{p}$ once every 5 h. In both cases, $L \rightarrow 10^{32}$ cm^{-2}s^{-1}.

For the colliding beams, various intensity and beam density limitations enter into play which make it difficult to reach the limit of $L \rightarrow 10^{32}$ cm^{-2}s^{-1}. Probably most stringent is the beam-beam effect and only this effect will be discussed here: current understanding is [5] that the non-linear space-charge field of beam 1 experienced by beam 2 in the interaction region leads to a degradation of beam 2 (and vice versa for the effect of beam 2 on beam 1). The linear part (ΔQ) of the tune shift is used as a measure of the beam-beam effect and rapid beam degradation is expected in hadron colliders when $\Delta Q > 5 \times 10^{-3}$.

The tune shift of the antiprotons can be approximated by [5]

$$\Delta Q_{\bar{p}} = r_o \frac{N_p \beta_v^*}{A} \left(\frac{1 + \beta^{-2}}{4 \gamma} \right) , \tag{4}$$

where

r_o = 1.5×10^{-18} m.

β_v^* = focusing function of storage ring in interaction point (1 m assumed).

$A \sim \pi(\sigma_h \times \sigma_v)$ = effective transverse beam area in the interaction region: a horizontal beam size $\sigma_h \gg \sigma_v$ much larger than the vertical one is assumed.

This expression (4) as well as the corresponding relation for the proton tune shift depends in much the same way on beam density N/A as does the luminosity

$$L = \frac{N_{\bar{p}} N_p f_{rev}}{4 A} . \tag{5}$$

Combining (4) and (5) we find a luminosity limit

$$L = \frac{N_{\bar{p}} f_{rev} \Delta Q}{r_o \beta_v^*} \left(\frac{\gamma}{1 + \beta^{-2}} \right) . \tag{6}$$

419

For a given $N_{\bar{p}}$ (equal to N_p in an optimised design) and given revolution frequency f_{rev} (large for our small ring!), the only "free" parameter is the focusing strength ($1/\beta^*$) of the storage ring at the interaction point. This dependence on β^* expresses the fact that stronger space-charge forces are acceptable at points where the focusing is strong, i.e. where the beta function is small. However, it is well known [5] that the focusing strength decreases rapidly with the distance (s) from a low beta point ($\beta = \beta^* + s^2/\beta^*$) and, to avoid interaction outside the small beta region, the beams have to be well separated at a distance $\Delta s = \pm \beta^*$ from the centre of the interaction region. In the head-on collision scheme discussed here, this is realised having a bunch length $\ell \leq \beta^*$. Since the RF voltage to make short bunches becomes very high (table 4) and since the longitudinal and transverse stability of a bunch of 10^{12} p becomes very critical for bunches shorter than 1 m, we choose $\ell = 1$ m and adjust $\beta^* = \ell = 1$ m.

TABLE 4

	System	Harmonic	Frequency	Amplitude	Bunch length
Two stage system	RF1	1	2.5 MHz	15 kV	100-10 m
	RF2	10	25.	100 kV	10--1 m
Three stage alternative	RF1	1	2.5 MHz	15 kV	100-10 m
	RF2	10	25. MHz	20 kV	10--3 m
	RF3	30	75. MHz	40 kV	3--1 m

We then arrive at a set of parameters given in table 5 which yields a luminosity of 3×10^{30} cm^{-2}s^{-1} at 7 GeV/c. This is a factor of ~ 30 below the "flux limit" (2) and to come closer to this limit, ways to beat the beam-beam effect or to further decrease β^* are desired.

TABLE 5

Number of particles per beam	$N_{\bar{p}} = N_p = 10^{12}$
Beam-beam tune shift	$\Delta Q = 5 \times 10^{-3}$
Lattice function at interaction point	$\beta^* = 1$ m
Bunch length	$\ell_b = 1$ m
Number of bunches per beam	$n_b = 1$
Revolution frequency	$f_{rev} = 2.5$ MHz
Resulting luminosity at 7 GeV/c	$L = 3 \times 10^{30}$ cm^{-2}s^{-1}

In fig. 3, we reproduce results of ISR measurements [6] which show the very steep dependence of beam life (τ_b) on the tune shift discussed. One may hope to stabilize the effect with a cooling system of sufficient strength ($\tau_{cooling} \ll \tau_b$). This agrees with results from electron storage rings which - for radiation cooling times of the order of 0.1 to 1 s - manage to work with $\Delta Q = 0.025$. In SLEAR, assuming very powerful electron cooling at 7 GeV with time constants of the order of a minute, we may hope to go to $\Delta Q = 0.015 - 0.02$ and thus gain a factor 3-4 in luminosity.

Any further improvement has probably to come from a (still) lower beta. In the head-on scheme, this may pose difficulties for the detector - in addition to the storage ring problems sketched above - as the short bunch tends to introduce momentum spread and strong "modulation" of the interaction rate. The solution then could be to have two rings with unbunched beams crossing at an angle. This

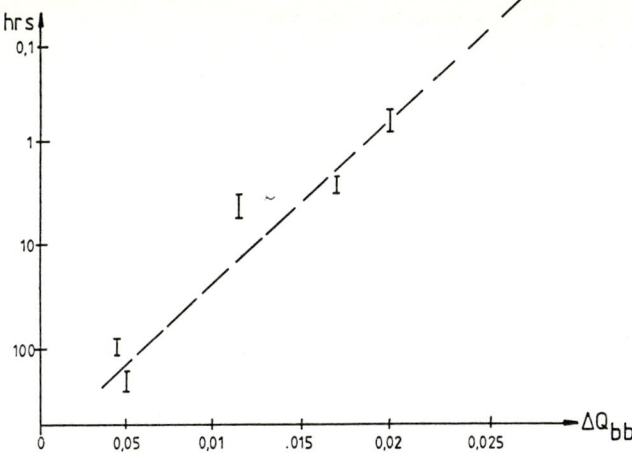

Fig. 3 Beam lifetime (without cooling) versus beam-beam tune shift

could be a future extension to aim at luminosities beyond 10^{31} cm^{-2}s^{-1}. We envisage a staged design where the second ring can be added at a later stage.

To summarize the machine aspects one can say that a compact high luminosity antiproton ring working in the 2 to 10 GeV/c range looks (so far) quite feasible. Small circumference and hence efficient use of antiprotons become possible by use of superconducting magnets. Ultimate luminosities of 10^{32} cm^{-2}s^{-1} in the internal target mode and 3×10^{30} cm^{-2}s^{-1} with p-$\bar{\text{p}}$ colliding beams at 7 GeV/c can be expected for a single ring scheme with head-on collision. Strong phase-space cooling, a powerful RF system with superconducting cavities to make very short bunches and later on maybe a two ring scheme with coasting beams may allow us also in the collider mode to aim at $L \to 10^{32}$ cm^{-2}s^{-1} which is the best obtainable with an antiproton production rate of 10^7/s.

Apart from being a valuable tool for particle physics, such a machine could be an attractive "test bench" of new accelerator technology.

3. ELECTRON COOLING AT INTERMEDIATE ENERGIES

3.1 Cooling time evaluation

In order to achieve (anti)proton cooling, we take into account the possibility of installing the electron cooling devices in two short straight sections of Super-LEAR. An alternative possibility could be to have the two cooling devices in one long straight section.

Electron cooling can be first used at injection energy. After this operation (precooling), both beams can be accelerated/decelerated to the working energy, practically without changing their temperature. In order to compensate beam broadening, due to beam-beam interaction, electron cooling must be kept working. We assume the electron beam parameters shown in table 6.

If at injection energy the beams are first cooled stochastically, we can then assume that the condition $\theta_{p,\bar{p}} < \theta_e$ is verified. Then the cooling time is given by

$$\tau_e = \frac{1}{2\pi} \frac{e\beta\gamma^2}{r_e r_p n j_e L} \left(\frac{T_e}{m_e c^2}\right)^{3/2} = 40.9 \, \beta q^2 \qquad (7)$$

where $\eta = 1.8 \times 10^{-2}$ is the radio of cooling length to circumference and $L = 20$ is the Coulomb logarithm.

Table 7 summarizes the cooling times at different momenta.

TABLE 6

Electron energy	0.7-5 MeV
Electron current	3 A d.c.
e-beam diameter	1.5 cm
Current density	1.7×10^4 A/m^2
$\Delta p/p$	$< 10^{-3}$
e-beam temperature	5 eV
Cooling region length	2 m

TABLE 7

$p_{\bar{p}p}$ (GeV/c)	E_e (MeV)	τ_e (s)
2	0.7	204 (3.4')
3.5	1.46	587 (9.8')
5	2.25	1176 (19.2')
6	2.8	1693 (28.2')
7	3.3	2262 (37.7')
10	5.0	4672 (77.8')

Cooling times can be shorter if a strong magnetic field confinement can be applied to the electron beam (fast cooling).

3.2 Jet target operation

Electron cooling is an efficient method to increase the beam density, to improve the momentum resolution and to counteract the multiple scattering of the target. In conjunction with a thin internal target this leads to an excellent definition of momentum and interaction point and to an increased beam lifetime. The improvements in beam size that we can obtain applying electron cooling are shown in table 8.

TABLE 8

$p_{\bar{p},p}$ (GeV/c)	$\Delta Q_{Laslett}$	σ (mm)
2	1×10^{-2}	4
3.5	10^{-2}	2
7	2.8×10^{-3}	1
10	10^{-3}	1

3.3 $\bar{p}p$ collider operation

In the collider mode the electron cooling can be used to improve the resolution and the luminosity.

The gain in luminosity is obtained by the decrease of the beam area (eq. (5)). The concurrent beam-beam tune shift (eq. 4) can be tolerated as long as the cooling compensates the beam-beam blow up (fig. 3) as discussed before.

Unfortunately, this gain factor is limited because in the head on collision scheme, the Laslett tune shift of the bunched beam is large.

To improve the luminosity, it could be appropriate to consider a machine consisting of two separated rings [7], operating with coasting beams that cross at an angle.

Without cooling the luminosity of hadron storage rings is calculated assuming a permissible beam-beam tune shift of 5×10^{-3} as discussed above. In the present case, assuming that the cooling time just compensates the beam decay constant we can deduce from fig. 3 and the cooling time given in table 3, that the allowed $\Delta\nu_{\bar{p}}$ is ranging from 0.020 and 0.028. This corresponds to an increase by about a factor 5 of the luminosity compared with the example without cooling, where $\Delta\nu_{\bar{p}} = 5 \times 10^{-3}$.

4. PHYSICS AT SUPER-LEAR

4.1 Introduction

When discussing "elementary particle physics" today, the arguments used fall on two fundamental questions:

- What is the origin of the standard model?

- How can we explain the dynamics of confinement in QCD?

Related to the first question are problems like the origin of the weak symmetry breakdown, the source of quark and lepton masses or the origin of CP violation. Relevant to the dynamics of confinement in QCD are the problems raised by the QCD (gluonic, hybrids) and relativistic degrees of freedom, the understanding of the hadronic spectrum, the hadron fragmentation.

We would like to show that Super-LEAR could give fundamental and original opportunities of testing nature and answer, at least partially, some of the questions mentioned above. In particular, we shall discuss:

- the possibility of performing new tests on CP and CPT;

- the possibility of studying heavy quarkonia ($c\bar{c}$ and $b\bar{b}$) inaccessible to e^+e^- colliders and of measuring their widths and decay angular distributions;

- the possibility of studying the production and decay properties of hadrons in a new quark environment (heavy flavours - charmed hyperons);

- the possibility of studying QCD dynamics in $\bar{p}p$;

- the possibility of searching for exotic states predicted by QCD (glueballs-hybrids).

We shall not include in this report questions like spin physics, $\bar{p}$ nuclear or atomic physics. These very interesting topics belong for the time being to a field reserved to LEAR, but could also benefit from the higher energies accessible with Super-LEAR.

4.2 CP and CPT

So far CP violation has only been studied in the neutral kaon decays of K_S^0 and K_L^0. CP violation in this system is characterized by one parameter, ε, with a magnitude of 2.3×10^{-3}. An experiment will be carried out at LEAR to study CP violating phenomena, taking advantage of the increase of intensity obtained with ACOL to produce tagged $\bar{K}^0$ and K^0 in the reaction

$$\bar{p}p \to K^0 K^{\pm} \pi^{\mp}$$

with the necessary statistics (Pavlopoulos [1]).

With SL2 (Super-LEAR in $\bar{p}p$ collider mode), one could envisage studying CP and CPT violation in a new system, namely for the Λ^0-$\bar{\Lambda}^0$ produced in the exclusive reaction

$$\bar{p}p \to \Lambda^0 \bar{\Lambda}^0 .$$

The Λ^0 and $\bar{\Lambda}^0$ being produced in a symmetric way in the laboratory the systematic errors could be reduced below the precision required to measure effects of the order of 10^{-4}. Optimal incident momentum for the detection of Λ^0 and $\bar{\Lambda}^0$ decays could be in the 2 to 3 GeV/c range.

With $L = 10^{30}$ cm^{-2}s^{-1}, 10^6 $\Lambda\bar{\Lambda}$ could be produced within one day.

(a) Lifetimes of Λ and $\bar{\Lambda}$ could be compared (test of CPT). For the time being, our knowledge is limited to:

$$\frac{\tau_\Lambda - \tau_{\bar{\Lambda}}}{\tau_\Lambda + \tau_{\bar{\Lambda}}} = (4.4 \pm 8.5) \times 10^{-2} .$$

(b) Branching ratios: Ling Lee Chau [8] has shown that CP violation can be studied by comparing the branching ratios

$$\Lambda \to p\pi^- , \quad \bar{\Lambda} \to \bar{p}\pi^+ .$$

Using the quark diagrams (for Λ)

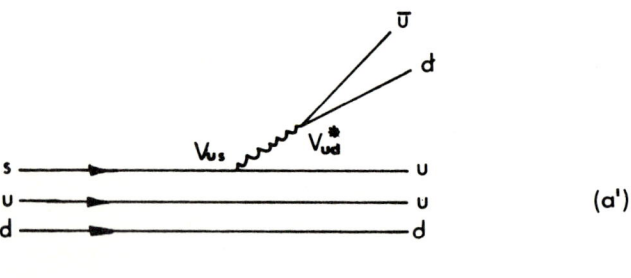

(a')

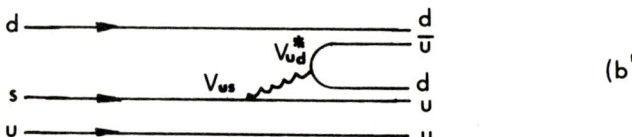

(b')

424

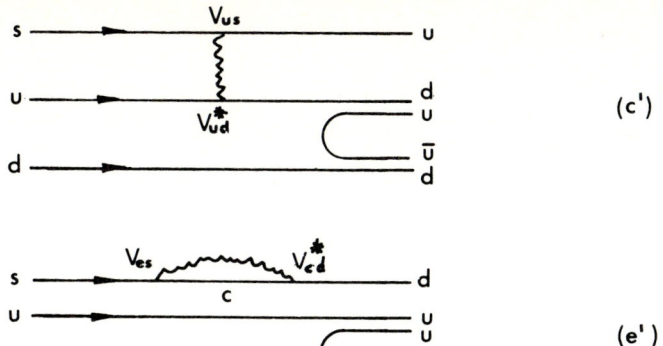

(c')

(e')

and the same diagrams for $\bar{\Lambda}$, replacing u,d,s,c by $\bar{u},\bar{d},\bar{s},\bar{c}$, $V^* \to V$ and $V \to V^*$, one gets

$$A(\Lambda \to \pi^- p) = V_{us} V^*_{ud} A'_1 + V_{cs} V^*_{cd} A'_2 ,$$

$$\bar{A}(\bar{\Lambda} \to \pi^+ \bar{p}) = V^*_{us} V_{ud} A'_1 + V^*_{cs} V_{cd} A'_2 ,$$

where

$$A'_1 = a' + b' + c' + e'$$

$$A'_2 = e' .$$

The difference in partial rates becomes

$$\frac{Br(\Lambda \to \pi^- p) - Br(\bar{\Lambda} \to \pi^+ \bar{p})}{Br(\Lambda \to \pi^- p) - Br(\bar{\Lambda} \to \pi^+ \bar{p})} = \frac{4(s_2 s_3 s_\delta c_1 c_2 c_3)\, Im(A'_1 A'^*_2)}{(|A'|^2 + |\bar{A}'|^2)\, s_1^{-2}}$$

c_1, c_2, c_3, s_2, s_3, s_δ, are the usual Kobayashi-Maskawa parameters ($c_1 = \cos\theta, \ldots$). The term $(s_2 s_3 s_\delta c_1 c_2 c_3)$ is of the order of 10^{-3} to 10^{-4} depending on the t-quark mass.

$$\frac{Im(A'_1 A'^*_2)}{(|A'|^2 + |\bar{A}'|^2)\, s_1^{-2}} \sim \frac{Im(a_{1/2} a_{3/2})}{|a_{1/2}|^2}$$

is suppressed by $\Delta I = 1/2$ dominance. One can therefore expect a small difference for the branching ratios, which could be however measured at SL2 with a significant result.

Present measurements give

$$Br(\Lambda \to \pi^- p) = (64.2 \pm 0.5)\%$$

$Br(\bar{\Lambda} \to \pi^+ \bar{p})$ is unknown.

(c) Decay asymmetry: J. Donoghue has underlined [9] that any asymmetry in the decay parameters of the Λ versus $\bar{\Lambda}$ would be indicative of CP violation effects. If we write the lab momentum of the incident $\bar{p}$, decay products p, π^-, $\bar{p}$, π^+ of the Λ, $\bar{\Lambda}$: $\vec{k}$, $\vec{p}$, $\vec{q}$, $\vec{\bar{p}}$ and $\vec{\bar{q}}$, respectively,

$$A = \vec{k} \cdot (\vec{p} \times \vec{q} - \vec{\bar{p}} \times \vec{\bar{q}})$$

$\langle A \rangle \neq 0 \rightarrow$ CP violation.

The transition matrix for hyperon decay may be written

$$M = s + p\, (\vec{\sigma} \cdot \vec{q})$$

where s and p are the parity changing and parity conserving amplitudes, respectively, σ is the Pauli spin operator and $\vec{q}$ a unit vector along the decay baryon in the hyperon rest frame.

The asymmetry parameters are defined by

$$\alpha = \frac{2\,\mathrm{Re}\,(s^*p)}{|s|^2 + |p|^2}$$

$$\beta = \frac{2\,\mathrm{Im}\,(s^*p)}{|s|^2 + |p|^2}$$

$$\gamma = \frac{|s|^2 - |p|^2}{|s|^2 + |p|^2}$$

and the angular distribution of the baryon in the hyperon rest frame is of the form

$$W(\theta) = 1 + \alpha\, \vec{P}_y \cdot \vec{q}$$

($\vec{P}_y$: hyperon polarization).

Introducing now a CP violating phase ϕ_I, we have, for S and P amplitudes ($I = 1/2, 3/2$, δ_I: final state interaction phase)

$$S = S_1\, e^{i(\delta_1^S + \phi_1^S)} + S_3\, e^{i(\delta_3^S + \phi_3^S)},$$

$$P = P_1\, e^{i(\delta_1^P + \phi_1^P)} + P_3\, e^{i(\delta_3^P + \phi_3^P)},$$

$$\bar{S} = -S_1\, e^{i(\delta_1^S + \phi_1^S)} + S_3\, e^{i(\delta_3^S + \phi_3^S)},$$

$$\bar{P} = P_1\, e^{i(\delta_1^P + \phi_1^P)} + P_3\, e^{i(\delta_3^P + \phi_3^P)}.$$

With the dominance of $\Delta I = 1/2$, $A_3/A_1 \sim 1/20$,

$$\delta_1^S = 6.0°, \quad \delta_3^S = -3.8°,$$

$$\delta_1^P = 1.1°, \quad \delta_3^P = -0.7°\ .$$

Measuring α, $\bar{\alpha}$, β, $\bar{\beta}$, one gets

$$A = \frac{\alpha + \bar{\alpha}}{\alpha - \bar{\alpha}} = \sin(\phi_1^S - \phi_1^P)\,\mathrm{sm}\,(\delta_1^P - \delta_1^S)$$

and

$$B = \frac{\beta + \bar{\beta}}{\alpha - \bar{\alpha}} = \sin(\phi_1^S - \phi_1^P) \ .$$

Some model predictions for B/ϵ show striking differences:

KM: $B/\epsilon \sim 1/10$,

Higgs: $B/\epsilon \sim 1$,

Superweak: $B = 0$,

left-right: $B/\epsilon \sim 1/10$.

The magnitude of the effect may be within the range of the sensitivity of experiments at Super-LEAR and could discriminate between models.

Present measurements give: $\alpha_\Lambda/\alpha_{\bar{\Lambda}} = -1.04 \pm 0.29$.

Similar effects could be expected for $\Sigma\bar{\Sigma}$, etc.

4.3 Heavy quarkonia ($c\bar{c}$, $b\bar{b}$)

As mentioned in the introduction, one can refer here to the ISR Experiment R704 which obtained interesting results on the χ spectroscopy within a few days of run at the ISR (Macri [1]).

In a $p\bar{p}$ experiment (SL1) one could:

(a) Discover $c\bar{c}$ states not accessible to e^+e^- colliders: $^1P_1$, $^1D_2$, $^3D_2$ (although $^1D_2$ and $^3D_2$ are predicted with masses above $D\bar{D}$ threshold, they cannot decay into $D\bar{D}$ by J^{PC} conservation and are therefore narrow).

They would end up somehow with a J/ψ in the final state which could be used to trigger (e^+e^-, $\mu^+\mu^-$ modes) the detector (R704 technique):

$$\bar{p}p \to {}^1P_1 \to J/\psi\,\pi^0 \to (e^+e^-)\,\pi^0 \text{ (isospin violating)}, J/\psi\,\pi\pi \to (e^+e^-)\pi\pi$$

$$\bar{p}p \to {}^3D_2 \to \chi\gamma_{E1} \to J/\psi\gamma_{E1}\gamma_{E1}$$
$$\to (e^+e^-)\gamma\gamma$$

$$\bar{p}p \to {}^1D_2 \to J/\psi\,(\pi^+\pi^-)^{J=1} \to (e^+e^-)\pi^+\pi^-.$$

In a formation experiment the cross section can be expressed in terms of the branching ratios of the resonance under study to the initial and final state as

$$\sigma(E) = \pi\lambda^2(2J+1)\Gamma_i\Gamma_f/\{(2S_1+1)(2S_2+1)[(E-E_0)^2+\Gamma^2/4]\}$$

where $\lambda^2 = (\hbar/p)^2$ with p being the c.o.m. momentum and $\hbar = 0.197$ f.m. GeV/c; S_1, S_2 and I the spin of the incident particle, target and resonance respectively; Γ, Γ_i, Γ_f are the total width, the partial width to initial and to the final state, respectively.

If one takes into account the incident $\bar{p}$ beam momentum distribution one can calculate the effective cross section at energy E from the convolution of the resonance Breit-Wigner and the $\bar{p}$ beam momentum distribution:

$$\sigma(\bar{E})) = \int_{\Delta M} L(E)\sigma(E)dE.$$

The interval ΔM reflects the size of the beam momentum distribution dp_{lab} and is

$$\Delta M = \beta \, dp_{lab}/[2(1 + \gamma)]^{1/2}.$$

The dp_{lab} results in a reduced ΔM due to the Lorentz boost. For a very narrow beam momentum distribution ($\Delta M \ll 1$ MeV), which corresponds to a $dp_{lab}/p_{lab} \sim 10^{-4}$ for charmonium states, we obtain for resonances with $\Gamma \gtrsim 1$ MeV the maximum cross section at the $\sqrt{s}$ equal to the mass of the resonance

$$\sigma_p = \pi \lambda^2 (2J + 1) BR\,(R \to \bar{p}p) BR(R \to \text{final}).$$

If the beam momentum distribution is wider, the effective cross section is lower. It can be expressed as

$$\sigma_{eff} = \sigma_p \, F(\frac{\Delta M}{\Gamma}).$$

The function $F(\Delta M/\Gamma)$ is shown in fig. 4.

Assuming SL1 is running for 15 days in a search for narrow resonances ($\Gamma \sim 1$ MeV) in a 100 MeV mass range, the presence of a resonance X would be revealed (at 95% CL), provided that $(2J + 1) Br(\bar{p}p \to X) Br(X \to J/\psi \ldots) > 7 \times 10^{-9}$.

For comparison, this factor is

5×10^{-4} for $\bar{p}p \to J/\psi \to e^+e^-$

5×10^{-6} for $\bar{p}p \to x_2 \to J/\psi \to e^+e^-$

2×10^{-6} for $\bar{p}p \to x_1 \to J/\psi \to e^+e^-$

0.5×10^{-6} for $\bar{p}p \to \eta_c \to \gamma\gamma$

i.e., all these factors (resulting from experimental observations – R704) are much larger than the limit 7×10^{-9}. The discovery of $^1P_1$, $^3D_2$ and $^1D_2$ should be easy.

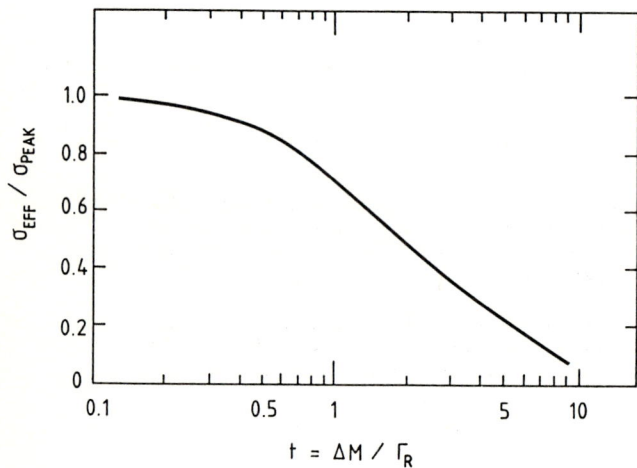

Fig. 4 $F(\Delta M/\Gamma)$

(b) Get reliable data on not well established $c\bar{c}$ states ($^1P_1$, η_c').

The interest of these discoveries is underlined by the relations which link the masses of the $c\bar{c}$ states to the spin-spin, spin-orbit, tensor and relativistic terms in the charmonium potential (Martin [1]). The fine and hyperfine splittings provide direct tests of QCD [10]. The spin-dependent components of the potential, indeed, do not result anymore from ad-hoc assumptions, but are calculated directly on lattices [11].

In particular, the short-range character of the spin-spin forces is tested by the vanishing of the singlet-to-triplet splitting of P states

$$^1P_1 - \frac{1}{9}(\chi_0 + 3\chi_1 + 5\chi_2).$$

The ratio $(\psi'-\eta_c')/(J/\psi-\eta_c)$ turns out also to be very sensitive to the range of the hyperfine interaction.

(c) Measure accurately the widths of the $c\bar{c}$ states, which are calculable in QCD. The unique feature of SL1 is its mass resolution: $\Delta M \sim 50$ keV.

With adequate factorization of the non-perturbative and perturbative part, the hadronic widths are calculable, typically of the type (Buchmüller [1], Remiddi [12])

$$\Gamma(c\bar{c} \to \text{had}) \sim |\phi(0)|^2 \alpha^n(M^2),$$

n being the number of gluons involved in the decay

(n = 2 for 0^{-+}, 0^{++}, 2^{-+}, 2^{--}, 2^{-+} ...)

(n = 3 for 1^{--}, 1^{++}, 1^{+-} ...).

Most of these widths are not yet measured or measured accurately. The agreement with QCD is fair. Some large relativistic corrections have to be taken into account (for E1 transitions).

To accumulate $\sim 10^3$ events on a $c\bar{c}$ state, with 15 days of SL1, one must have

$$(2J+1)\text{Br}(\bar{p}p \to X)\text{Br}(X \to J/\psi ...) > 1.4 \times 10^{-7},$$

a condition which is easily satisfied for the known states (see above).

The systematics of J/ψ decay branching ratios is very instructive and has been continuously studied since 1974 (presently with MARK 3 and DM 2). Super-LEAR offers the possibility of studying the decay of other narrow $c\bar{c}$ states.

(d) Measure the production and decay angular distributions (multipole analysis).

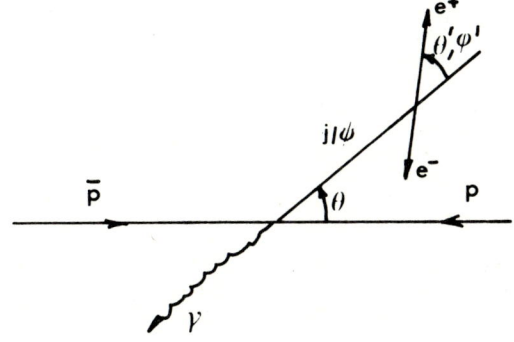

Consider: $\bar{p}p \to \chi_J \to J/\psi\gamma \to (e^+e^-)\gamma$.

The joint angular distribution $W(\theta, \theta', \phi')$ can be expanded in terms of the electric dipole (E1), magnetic quadrupole (M2) and electric octopole (E3) transitions (for J = 2).

In the single quark radiation model one gets predictions for these transitions. Comparing to experimental observations gives access to the anomalous magnetic momentum of the c quark. This is therefore a test for the constituent structure (Olsson [1]).

(e) Repeat all this charmonium physics on the bottomium with SL3. However, even with SL3, the feasibility of these experiments may become marginal for the $b\bar{b}$ system.

Straightforward QCD predictions (which may be wrong by 2 orders of magnitude) give (Buchmüller [1]):

for "allowed" decays (1^{--}, 1^{++}, 2^{++})

$$\frac{(b\bar{b})}{(c\bar{c})} \sim \left(\frac{m_c}{m_b}\right)^8 \sim 10^{-4},$$

for "forbidden" decays (0^{-+}, 0^{++}, 1^{+-})

$$\left(\frac{b\bar{b}}{c\bar{c}}\right) \sim 10^{-5}.$$

With SL3 running for 4 months, we may get

$0^{-+}(\eta_b)$ $s_{\bar{p}p} \sim$ 2pb $\to 10^3$ events.

$1^{--}(\Upsilon)$ $s_{\bar{p}p} \sim$ 100pb $\to 5 \times 10^4$

$1^{++}, 2^{++}(\chi_b)$ $s_{\bar{p}p} \sim$ 10pb $\to 5 \times 10^3$ events.

One must however remember that these predictions may be wrong, that they do not take into account the fact that $\bar{p}p$ are much more efficient than pp at producing $Q\bar{Q}$ at low energy (Omega beam dump experiment) and that we have already had some surprises in this field (R704).

One must also consider that the direct formation in $\bar{p}p$ annihilation is the only laboratory for a clear study of OZI forbidden diagrams in the Υ energy region.

4.4 Super-LEAR as a heavy flavour hadron factory

Only one charmed baryon is "known" (i.e. its mass and lifetime), the Λ_c. The decay properties of the other multiplet members ($J^{PC} = 1/2^+, 3/2^+$) with cud, cuu, cdd, csd, csu ... quarks would shed light on the behaviour of quarks and diquarks in the environment of a heavy quark.

Little is known about hadrons with open charm. For mesons, we know only the ground state with various spin and flavour configurations (D, D^*, D_s, D_s^*) and there is only one candidate for a radial or orbital excitation, reported at the Bari Conference. For baryons the mass and lifetime are available only for the Λ_c; there is some indication for a Σ_c, but nothing for the Σ_c^*. Charmed strange baryons (csu) and (css) have also been reported.

The spectroscopy of these states is extremely interesting since in the environment of a heavy quark the light quark experiences more relativistic corrections than in any other configuration.

The decay properties of the charmed hadron should be investigated. The observed lifetime ratio $\tau(D^+)/\tau(D^0) \sim 2.3$ has already shown that several types of diagram contribute.

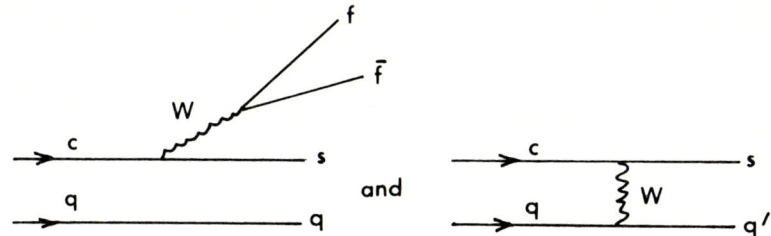

These dynamics will be usefully tested with baryons.

One could set up $e^{\pm}$, $\mu^{\pm}$, $e^{\pm}K^{\mp}$ triggers (as in the pioneer ISR experiments).

QCD predictions for exclusive cross sections are

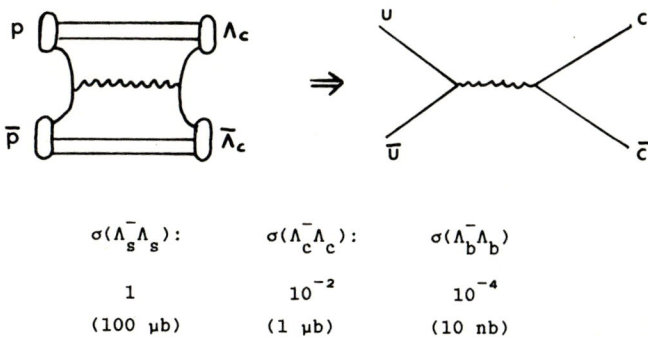

$\sigma(\Lambda_s \bar{\Lambda}_s)$: $\sigma(\Lambda_c \bar{\Lambda}_c)$: $\sigma(\Lambda_b \bar{\Lambda}_b)$

1   10^{-2} 10^{-4}

(100 μb)   (1 μb)   (10 nb)

This order of magnitude seems reasonable when compared to our present experimental knowledge

NA27	($\sqrt{s} = 27$ GeV)pp	$\sigma_{c\bar{c}} = 30$ μb
Serpukhov	($\sqrt{s} = 10$ GeV)n Be	$\sigma(\Lambda_c)_{incl} = 50$ μb
Hyperon beam	($\sqrt{s} = 20$ GeV) Σ^- Be	$\sigma(\Lambda_1) \sim 5$ μb

One may safely assume that $\sigma(\bar{p}p \to c\bar{c}) \gg \sigma(pp \to c\bar{c})$ at $\sqrt{s} = 20$ GeV and therefore expect inclusive cross sections of the order of ~ 10 μb for Λ_c and of order 10 nb for Λ_b, the collider operation making also easier the detection of leptons at large angle for heavy particle production.

4.5 QCD dynamics in $\bar{p}p$ interactions

(a) What is the class of diagrams relevant to the binding of 3 quarks in a baryon? This question could be investigated by studying $\bar{p}p$ annihilations as a function of $\sqrt{s}$. G. Veneziano has shown [13] that three Regge trajectories may play a role in $\bar{p}p$ annihilations, with slopes 1, 1/2, 1/3 and corresponding to one, two or three $q\bar{q}$ annihilation pairs:

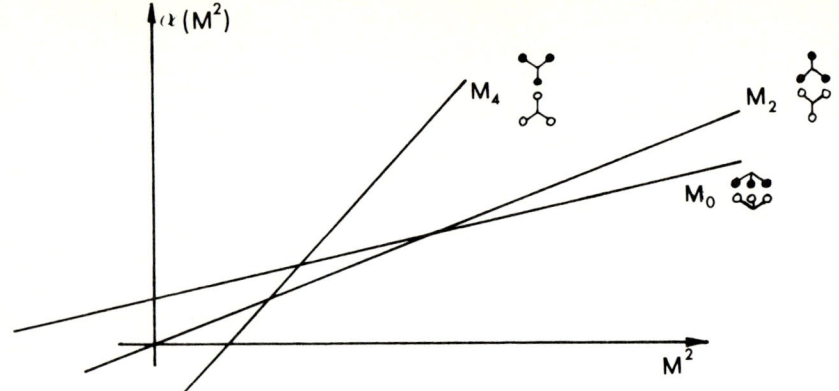

(b) One can test perturbative QCD or, at least, to which extent perturbative QCD works already at a few GeV and test the momentum dependence of the hadron wave-function [14] by studying the scaling reactions

$$\bar{p}p \to \gamma\gamma \qquad \frac{d\sigma}{d\Omega} \sim \frac{\alpha^2}{p_T^{10}} \, f(\theta)$$

$$\bar{p}p \to \gamma M \qquad \frac{d\sigma}{d\Omega} \sim \frac{\alpha}{p_T^{12}} \, f(\theta)$$

$$\bar{p}p \to MM \qquad \frac{d\sigma}{d\Omega} \sim \frac{1}{p_T^{14}} \, f(\theta)$$

and check the crossing behaviour

$$\bar{p}p \to \gamma\gamma: \qquad \gamma p \to \gamma p: \qquad \gamma\gamma \to \bar{p}p$$

One expects:

$$\sigma(\bar{p}p \to \gamma\gamma) \sim \frac{4\pi}{(\frac{s}{5})^2} \, \text{nb}$$

with SL3, 15 days → > 10^4 $\bar{p}p \to \gamma\gamma$.

4.6 Glueballs-hybrids-exotics

Why $\bar{p}p$? Because $\bar{p}p$ is expected to be a good source of glue (in particular to produce states with C = -1 since the other "good" source of glue, $J/\psi \to \gamma X$, can only produce states with C = +1).

(a) The study of complete $\bar{p}p$ annihilations should constitute an ideal "hard" glue factory.

The basic diagram would lead to two or three gluons depending on the quantum numbers of the final state.

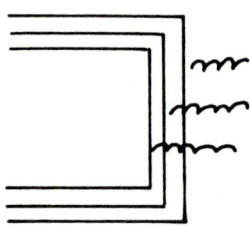

To isolate these processes the most important issue is the choice of the final state.

Only final states with no quark content present in the initial states should be the flag of the presence of "hard" gluons in the intermediate state.

For example consider $\bar{p}p \to D\bar{D}$, we may imagine two diagrams

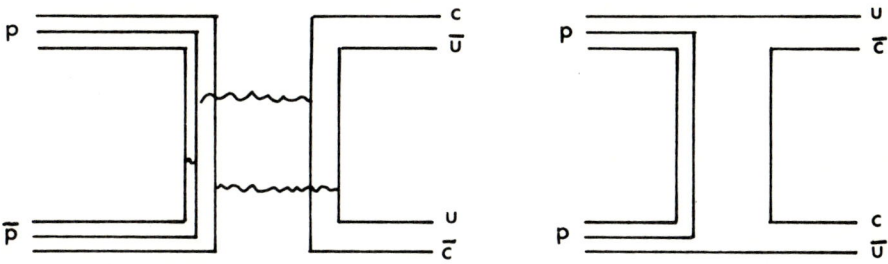

but for example in the case of $\bar{p}p \to \bar{D}_s D_s$:

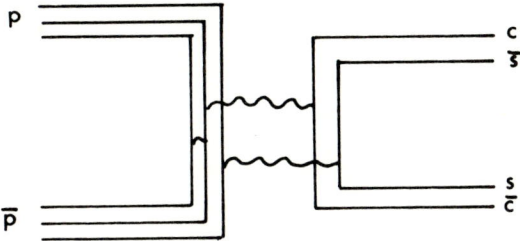

So all diagrams which have final states completely decoupled from initial states would give very valuable information on the intermediate gluon states.

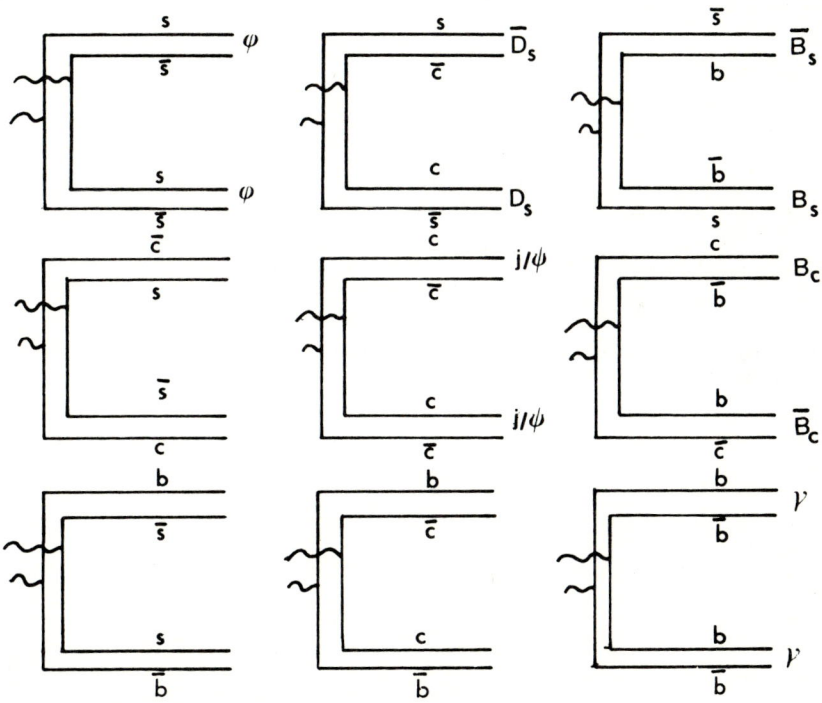

The same arguments apply to baryons with b, c, s quarks as constituents.

(b) One may also expect that annihilation leads frequently to a situation where the formation of a hybrid state ($q\bar{q}g$) together with a $q\bar{q}$ state is favoured [15]

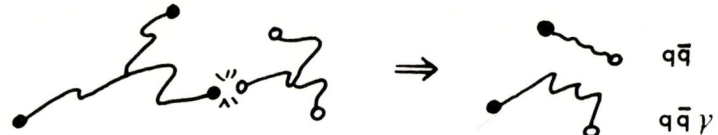

(c) $\bar{p}p$ may lead to the production $\bar{p}p \to X\pi$ of states X with exotic quantum numbers.

(d) In $\bar{p}p$ annihilation, there is no baryon in the final state. This helps in partial wave analysis.

(e) With SL1 one could cover the 3-4 GeV mass region where $q\bar{q}g$ are expected to be produced and the 4-10 GeV mass region where $Q\bar{Q}g$ are expected. These predictions correspond to a consensus between several approaches:

- bag model;
- QCD sum rules;
- flux tube;
- valence g;
- lattice calculations.

(f) The predictions for hybrids are particularly interesting since they involve several <u>octets</u>, some of them having exotic quantum numbers:

$$0^{+-}, \ 0^{--} \ 1^{-+}, \ 2^{+-}$$

with mass relations of the type

$$M_{q\bar{q}g}(\omega) \sim M_{q\bar{q}g}(\rho) + 100 \text{ MeV}$$

and decays which may be characterized by the presence of a $q\bar{q}$ P-wave state in the decay products [15]

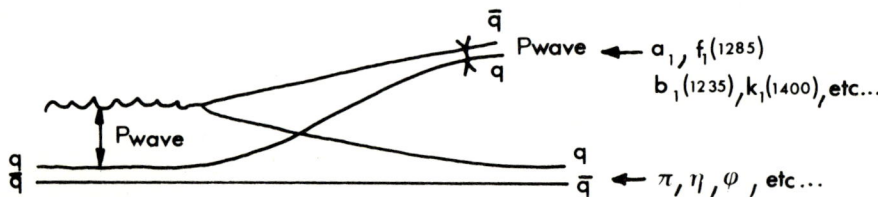

For example, a search for a 1^{-+} hybrid with $1.5 < M < 3.5$ GeV could be undertaken in annihilations of the type

$$\bar{p}p \to X^+\pi^-$$
$$X^+ \to D^\circ \pi^+$$
$$D^\circ \to K^\circ K^\pm \pi^\mp$$

with a trigger: $K^+ \times K^\circ \times M(K^\circ K^+\pi^-) = 1285 \pm 10$ MeV one could collect $10^6(D\pi)$ in 5 days.

(g) $\bar{p}p$ annihilations have already been the source of "exotic" mesons: the E(1420) was first discovered in $\bar{p}p$ annihilations at rest ($J^{PC} = 0^{-+}$) and several curious effects have been observed which may be investigated in a good environment in $\bar{p}p$ annihilations:

the	3.1 GeV	$\Lambda\bar{p}\pi^+\pi^-$	resonance [16]
the	1.5 GeV	$\phi\pi$	resonance [17]
the	2.3 GeV	$\phi\phi$	resonance [18]
the	1.59 GeV	$\eta\eta'$	resonance [19]

the exotic ($p\pi^+\pi^+$) bumps observed in 12 GeV $\bar{p}p$ annihilations into $\bar{p}p\pi^+\pi^+\pi^-\pi^-$ [20].

4.7 $\bar{p}$ nucleus interaction

The antiproton-nucleus interaction has been accurately measured at LEAR (see various contributions in [1]). At low energy, the antiproton interacts mostly at the surface of the nucleus. At higher energy, the antiproton will penetrate more deeply inside the nucleus, and interesting cascades will occur, with possibly the formation of hot nuclei [21].

The physics of hypernuclei has provided us with unexpected phenomena, which are at present being further investigated. Charmed hypernuclei could reveal even more interesting features [22]. When attached to a nucleus, a charmed hyperon should fall deeply inside the nuclear medium and appreciably disturb the structure of the inner nucleon shells.

4.8 Conclusions

Super-LEAR would open a new window on particle physics with the energy range and the mass resolution it would cover. For important aspects of the charmonium spectroscopy it could offer experimental conditions which are unsurpassable elsewhere. The $\bar{p}p$ annihilation channel, analysed with high statistics and good trigger, may reserve pleasant surprises for all the questions relative to confinement.

It could also offer the possibility of studying the fundamental symmetries in a new context (like CP violation in $\Lambda\bar{\Lambda}$).

With a mass resolution which is at least a 100 times better than with the present accelerators, it is also possible that "unexpected" results will introduce fundamental changes in our present understanding of particle physics.

REFERENCES

1. Physics with antiprotons at LEAR in the ACOL era, Third LEAR Workshop, Tignes 1985, (ed. Frontières).

2. C. Baglin et al., Phys. Lett. B172 (1986) 455.

3. C. Baglin et al., Phys. Lett. B171 (1986) 135.

4. M. Macri, A clustered H_2 beam, Proc. of the 2nd LEAR Workshop, Erice (1982) 691 (Plenum Press, New York and London 1984, ed U. Gastaldi and R. Klapisch).

5. See e.g. M. Sands, The physics of electron storage ring, SLAC report 121 (1970);

E. Keil, Beam-beam interaction in p-p storage rings, in: Theoretical aspects of the behaviour of beams, CERN/EP 77-13 (1977) 314.

S. Kheifets, Experimental observations and theoretical models for beam-beam phenomena, SLAC report 2700 (1981).

6. B. Zotter, Experimental investigation of the beam-beam limit of proton beams, in Proc. Xth Int. Conf. on High Energy Accel., Protvino, USSR (1977) 23.

7. U. Bizzarri et al., LEAR, Double-LEAR, Super-LEAR as colliders, Proc. 2nd LEAR Workshop, Erice (1982) 729 (Plenum Press, New York and London 1984, ed. U. Gastaldi and R. Klapisch).

8. Ling Lee Chau, Phys. Rep. C95 (1983) 62.

9. J. Donoghue, Phys. Lett. B179 (1986) 319.

10. R. Barbieri, R. Gatto and E. Remiddi, Phys. Lett. 106B (1981) 497.

11. C. Michael, Phys. Rev. Lett. 56 (1986) 1219;
    M. Campostrini et al., Phys. Rev. Lett. 57 (1986) 44.

12. E. Remiddi, Proc. 2nd LEAR Workshop, loc. cit.

13. G. Veneziano, Phys. Rep. C63 (1980) 149.

14. S. Brodsky, Fermilab Workshop on low energy $\bar{p}$ physics, April 1986.

15. N. Isgur, Fermilab Workshop on low energy $\bar{p}$ physics, April 1986.

16. M. Bourquin et al., Phys. Lett. B172 (1986) 113.

17. R. Landsberg et al., Phys. Lett. 188B (1987) 383.

18. D. Bisello et al., Phys. Lett. B179 (1986) 294.

19. D. Alde et al., Phys. Lett. B177 (1986) 120.

20. Helsinki-Amsterdam-Liverpool-Stockholm Collaboration, private communication by A. Grégorian).

21. W.R. Gibbs, Fermilab Workshop on low energy $\bar{p}$ physics, April 1986.

22. C.B. Dover and S.H. Kahana, Phys. Rev. Lett. 39 (1977) 1506;
    R. Gatto and F. Paccanoni, Nuovo Cimento 46A (1978) 313.

Round Table Discussion on Future Accelerators: RHIC

L. McLerran

Fermi National Laboratory, P.O. Box 500, Batavia, IL 60510, USA

RHIC is an ultra-relativistic nuclear collider envisioned for Brookhaven National Laboratory. In this accelerator, beams of nuclei as heavy as gold would collide with center of mass energy of 100 GeV per nucleon (in each beam). This machine would use the AGS as an injector for heavy ions, which would then be accelerated in the tunnel originally built for the now defunct CBA project. It would use the existing experimental halls, support buildings, and refrigeration system.

Thoeretical considerations suggest that this energy is sufficient to produce and study a quark gluon plasma. These energies are unique since they allow a clean separation of the fragmentation regions of the colliding nuclei. At about this energy, it is expected that the achieved local thermal energy density becomes a slowly varying function of beam energy, and for many purposes has saturated. The need for large A of the beams has been shown to be extremely important at Bevelac energies where A greater than 100 seems to be required for measuring collective effects.

It is anticipated that there will be a suite of detectors which will study global event characteristics, strangeness and charm production, photons and dileptons and perhaps jets. Possible experiments involve a calorimeter with multiple small aperture spectrometers, a dimuon detector, electron detectors, a large magnetic spectrometer, and possibly an experiment to probe the fragmentation region.

The community participating in such experiments, judging by the experience at CERN, will no doubt be quite diverse, and involve an almost equal mixture of high energy and nuclear experimentalists. There was a workshop two years ago at BNL, and this year there will be one at LBL to discuss experimental plans.[1] The number of experimentalists participating in a single experiment is anticipated to be quite large, of the order of 100. The number participating in the workshop was about 100. The number of people currently involved in the programs at AGS and at CERN is greater than 400.

It is expected that construction money will be available for this project in FY 1989. The approximate construction time will be 5 years. In FY 1986 dollars, the cost of the machine is 178M and detectors an additional 65M. In FY 1988 dollars, the cost including detectors is estimated to be 275M.[2]

References

[1] Proceedings of RHIC Workshop, April 15-19, 1985, Edited by P. Haustein and C. L. Woody, Published by Brookhaven National Lab. BNL 51921.
[2] T. Ludlam: Private communication

Part VIII

Astrophysics

Supernova Theory and 1987a (Shelton)[1]

S.H. Kahana

Physics Department, Brookhaven National Laboratory,
Upton, NY 11973, USA

The fortunate occurrence of supernova (Shelton) 1987a has provided a very good excuse for describing supernova theory to the general public and also, of course, for discussing neutrino properties. A brief statement on present facts about 1987a and some reference to the theory of the prompt mechanism is given.

1. INTRODUCTION

I generally begin talks on supernova with some justification for studying such a subject. This now hardly seems necessary. The rather spectacular optical and neutrino event in the large Magellanic Clouds at some 160,000 light years from us has been very much in the news, both popular and scientific. There was at the outset some controversy about the classification and progenitor for this supernova [1]. The B3 (blue giant) identified [2] initially as a possible progenitor appeared to still be there emitting in the ultraviolet, but by now may finally have disappeared [3]. Some observers (not the Canadians) suggested a type I supernova at the outset, i.e. no hydrogen lines in the spectrum, but Balmer lines were very quickly found. The light curve rose dramatically, but after a few days levelled out at a luminosity some 1.5 orders of magnitude below that for a normal type II. One was left with something intermediate between a type II (SNII) and a type Ib (SNIb) supernova; the latter suspected in the past to arise from a type II progenitor which had lost its hydrogen envelope, probably through its presence in a binary.

In actual fact, the light curve continued to rise slowly, and at about 30 days showed a dramatic upturn, perhaps due to radioactive decay (see fig. 2 [4,13]).

The small time differential between the prompt neutrino bursts of IMB [5] or Kamiokande [6] and the first optical observation argued for a very small progenitor. Taking the shock velocity on its way out from the collapsed core optimistically as $\varepsilon/20$ and allowing perhaps an hour or so for bolometric shift from high energy photons into the visible leads to a radius certainly less than 12 light minutes (4×10^{12} m), i.e. somewhat outside the earth's orbit if the star were centered at the sun. A small star, considerably smaller than this reasonable upper limit was postulated from the start to explain the low luminosity. ARNETT [7] had pointed out that the SNII luminosity is proportional to the envelope radius and to shock velocity. Given the observed 16,000 km/sec Doppler shifts, one knows the velocities are not small and concludes the radius is. Some calculations [8] indeed took very little hydrogen in the envelope, not necessarily consistent with the strong hydrogen lines seen, and a radius as small as two light minutes. These authors [8] had concluded they were dealing with a hydrogen-poor super red giant, and predicted an early drop in the light curve, perhaps after one month or so. Again, this has not turned out as imagined.

More recent observations [3] have demanded a return to the B3 star SK-69202 as the likely progenitor, and have turned the atmospheric supernova theorists in the opposite direction. Blue giants are not expected under normal circumstances to explode, representing too early a phase in the evolution towards super reds. However, BRUNISH and TRURAN [9] had pointed out that low metallicity can induce premature carbon ignition. WOOSLEY, PINTO, and ENSMAN [4], and no doubt others [10], have used this low metallicity, actually a feature of the LMC, as a mechanism for accelerating evolution to the point of carbon ignition and thence forward to collapse and perhaps explosion. All of these presupernova and postcollapse people seem to be requiring a large explosive energy which pleases the prompt shock mechanism proponents [11].

The observation of neutrino bursts, seemingly simultaneous in the IMB and Kamiokande detectors, labels SN1987a as definitely due to gravitational collapse. Shortcutting an enumeration of the rather complicated divisions into types, we will simply differentiate our supernova by supposed mechanisms. True type I's involve less massive stars, $M \lesssim 8\,M_\odot$, and a disruption of the entire core through thermonuclear detonation, perhaps C-O detonation. Such an end to the star's life would not lead to neutrinos, so we can concentrate for the rest of this discussion on type II. SNII's involve more massive progenitors, perhaps $M \gtrsim 12\,M_\odot$, and occur directly as a consequence of gravitational collapse.

In the following section I will briefly refer to: the observational aspects of supernova as we know these; to the existing theoretical mechanisms producing type II's, including their underlying nuclear physics; and finally will discuss in some detail the lessons learned from neutrino emission from 1987a.

2. HISTORY AND OBSERVATIONS

There have been some seven historical supernova events, i.e. optical sightings in our own galaxy [12]. The most recent of these are the Crab (1054), Tycho (1572), Kepler (1604), and Cas A (1660 ? possibly seen by Flemsteed). Most of our knowledge then comes from extragalactic observations in the present century, beginning with ZWICKY, BAADE, HUBBLE, and others, and extending to the recently inaugurated automatic searches. Figure 1 from the review aricle by Virginia TRIMBLE displays a histogram of the many hundreds of such events. Figure 2, [4,13], shows the light curve from 1987a, figure 3 [14] similar composite curves from type II events seen in other galaxies, and figure 4 [15] a time-evolving spectrum also for type II's. The brightest of SNI's are some 10^{10} more luminous than the sun ($L_\odot$) approaching an entire galaxy in intensity. "Normal" SNII's are perhaps only 1/3 or 1/4 of this, but clearly in view of SHELTON [1] with an early luminosity $L \sim 3-4 \times 10^7\,L_\odot$ one must have a large variation in brightness for this category.

The closest giants to us, Antares (blue, and unlikely to explode soon) and Betelgeuse (red) at 520 light years would of course provide unbelieveably spectacular displays should they go (or have gone) in the appropriate time interval. These neighbours would lead to optical luminosities in excess of 10^5 times that of 1987a, and to similar increases in neutrino emission. Given that only part of our galaxy is electromagnetically visible, the use of present proton-decay detectors, or newer larger dedicated neutrino detectors, would with luck greatly increase our knowledge of the frequency and character of SNII's, even with sources nowhere as close as Betelgeuse.

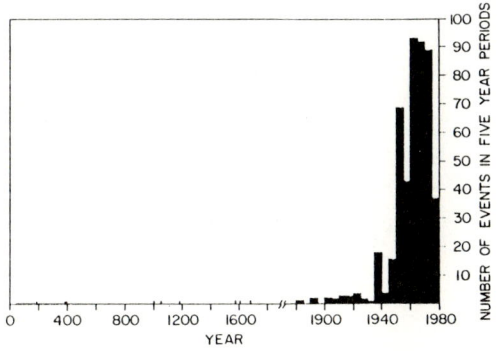

Figure 1

Histogram for discovery of supernovae, including the six historical events. From V. Trimble [12]

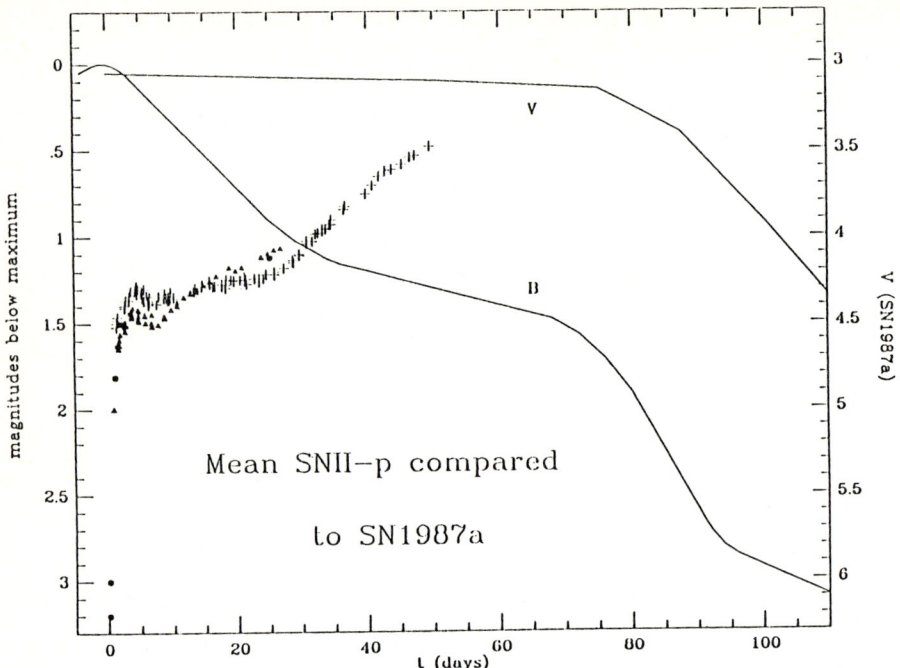

Figure 2 Visible 1987a light curve with comparisons to the Blue (B) and Visible (V) light curves for average SNII's. From [13]

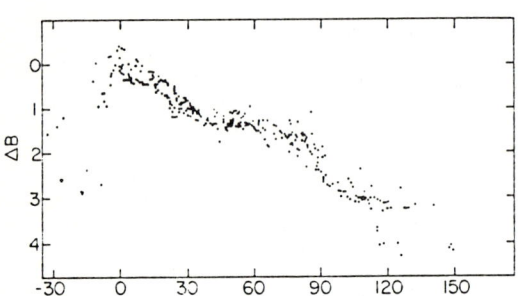

Figure 3 Composite blue light curves of type II supernovae showing 15 events. Time (t) is in days and luminosity in blue magnitude normalized at the peak. From R. Barbon et al.[14]

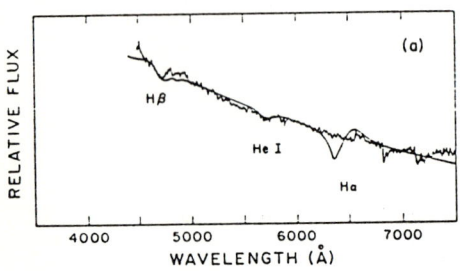

Figure 4 (caption see opposite page)

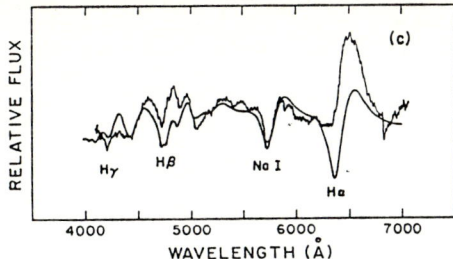

Figure 4

Observed spectrum (uneven line) and fit for SN1979c, a type II (a) at maximum light (b) six days later, and (c) 36 days later. The appearance of hydrogen lines and then later of other elements is clear.
From D. Branch et al.[15]

I end this all too brief exposition with a reminder that most of our fascination with supernova surely centers securely on their intrinsic but catastrophic beauty.

3. MECHANISM AND THEORY

There can be no doubt that the long quasi-static presupernova evolution of massive stars $M \gtrsim 10\ M_\odot$ is driven by gravity and leads eventually to the complete exhaustion of the nuclear fuels generated in the central regions of the star. The recent scurrying [4,8,10] to rewrite the possible evolutions for the progenitor of 1987a is an example of the uncertainties inherent in theoretical modelling of presupernova evolution. The broad outlines of the scenario accepted by most begin with the various stages of quasi-static collapse which leads to helium, carbon-oxygen, neon, and eventually silicon burning. The primordial star is, say, 75% hydrogen, 25% helium, with the total main sequence mass M in units of the solar mass $M_\odot$ as the only "free" parameter in the eventual development.

The evolution little concerns the hydrogen component and often the helium core mass can be used as an equivalent label. A SNII progenitor in a binary will evolve somewhat differently because of mass loss.

The end point of the early development, in a sufficiently large star ($\gtrsim 12\ M_\odot$) is an onion-skinned structure with an inner core of iron-like elements. The mass of this "Fe" core is near the Chandresakhar limit $M_{ch} \sim 5.8\ Y_e^2$ with $Y_e \approx Z/A$ the electron ratio per baryon. The temperature at the center of the core is $\lesssim 0.5$ MeV. The core is supported by the fermi pressure of the highly degenerate electron gas, $E_F \approx 8$ MeV, and its highest density is near 10^{10} gm/cm^3, some four orders of magnitude below that of normal nuclear matter. The central entropy, and indeed the entropy up to the silicon-burning shell, is low [16], which in turn keeps Y_e low and the mass low. This core mass had better be sufficiently small or there will be no explosion. This mass is by far the most important single initial parameter. Historically, in calculations it has evolved downward from $\lesssim 1.5\ M_\odot$ [17] to $1.36\ M_\odot$ [18] and now but perhaps not finally to $1.2\ M_\odot$ [19] for a main sequence mass $13\ M_\odot$.

What follows next is the collapse started by photodisintegration and then by the pressure drop due to rapid β-capture. The speed of collapse increases to an appreciable fraction of free fall, densities rise, nuclei melt into nuclear matter, which, because of the electron capture, contains some two neutrons for every proton. Given that nuclear matter is sufficiently stiff, else the core plunges directly into a black hole, the collapse is halted. The clash of incoming matter with outgoing matter at bounce generates an outward moving shock.

The two mechanisms competing for attention diverge at this point or slightly later. The shock will, if sufficiently energetic and if it doesn't have to plow its way through too much material, blow off the mantle and the envelope i.e. yield a visible SNII. This would be a prompt explosion. The first computationally adequate hydrodynamical calculations [20], with of course the initially believed $1.5\ M_\odot$ core mass [17], led to stalled shocks which in time would accrete to black holes.

Given a stalled shock, WILSON [21] and collaborators [22] found that the energetic neutrinos emitted from the core can deposit sufficient energy to revitalize the shock. This is the so-called delayed mechanism, which relies on a quite compressed core, and of course the absence of prompt explosions, and leads to rather weak explosions [22].

Following a study of the equation of state [23,24], altered phenomenologically in the direction of softness, and with the introduction of full general relativity BARON, COOPERSTEIN, and KAHANA [11] (BCK) were able to generate prompt explosions in the 12 $M_\odot$, 15 $M_\odot$ models of WOOSLEY and WEAVER [18] and also in the more recent and improved 13 $M_\odot$ model of Hashimoto and Nomoto [19]. Both calculations, prompt or delayed, obtain a shock energy as a delicate balance between internal (positive) and gravitational (negative) energy; the total shock energy is only a few percent of, say, gravitational. However, the prompt explosions of BCK yield $1-3 \times 10^{51}$ ergs plus a small amount from burning of outer regions by the shock, while the delayed explosion produces only perhaps as much as 1×10^{31} ergs. Much hinges on the eventual energy in 1987a, and at this moment calculations suggest an energy in excess of 2×10^{31} ergs and increasing as the practitioners include more ejected mass in their calculations [4,8].

I have not said much about the high density equation of state (EOS) for nuclear matter, at the densities one to four times saturation density important to supernovae, which was used in the BCK calculations. The pressure of cold nuclear matter is given by

$$P(n) = \frac{K_0(Z/A)}{9\gamma} \left[\left(\frac{n}{n_0(Z/A)}\right) - 1 \right], \qquad (1)$$

where the saturation compressibility $K_0(Z/A)$ and density $n_0(Z/A)$ are given as functions of the baryon charge-to-mass ratio. The thermal energy and pressure of a degenerate fermi gas are of course included. The softening of matter necessary to permit more gravitational compression and an increased initial shock energy is:

(1) consistent with the compressibility $K_0(1/2) \approx 180-210$ MeV extracted from the breathing mode in heavy nuclei;

(2) also produced naturally by the neutron-rich nature of dense supernuclear matter, i.e. $Z/A = 1/3$ rather than $1/2$.

One has [11,23,24]

$$K_0(Z/A) \approx K_0(1/2) \left[1 - 3.0 (Z/A - 1/2)^2\right]. \qquad (2)$$

The symmetry energy in fact plays a double role in the theory. We have identified two important criteria for successful explosions, a favourable equation of state and a low initial electron fraction $Y_e = Z/A$. We see above that a high symmetry energy results in soft neutron-rich matter, while in more recent work [25] it now appears that a high symmetry energy suppresses initial stage, β-capture, and thus keeps Y_e up.

4. NEUTRINO EMISSION IN GRAVITATIONAL COLLAPSE

The number of papers written on neutrinos from 1987a is by now approaching astronomical proportions. Two and perhaps three actual observations were made by the IMB [5], Kamiokande [6], and Mont Blanc [26] proton-decay groups. Table 1 and 2 list the IMB and Kamiokande events in time and electron energy.

In principle, the very long time-of-flight path from the LMC permits one to say something about electron-antineutrino ($\bar{\nu}_e$) masses [27,28], and the number, energy, and time structure of the events gives us a fix on the total binding energy of the compact remnant of the explosion. Many fanciful conclusions have been drawn, spanning the gamut from definite (not just limiting) neutrino masses and very large magnetic moments to ν-emission influenced by the frequency of the created pulsar.

Table 1. Neutrino Events. From IMB (reference 5)

Event No.[a]	Time (UT)	No. of PMT's	Energy[b] (MeV)	Angular distribution[c] (degrees)
33162	7:35:41.37	47	38	74
33164	7:35:41.79	61	37	52
33167	7:35:42.02	49	40	56
33168	7:35:42.52	60	35	63
33170	7:35:42.94	52	29	40
33173	7:35:44.06	61	37	52
33179	7:35:46.38	44	20	39
33184	7:35:46.96	45	24	102

[a]The event numbers are not sequential. Interspersed with the contained neutrino events are fifteen entering cosmic-ray muons.
[b]Error in energy determination is ±25% (systematic plus statistical).
[c]Individual track reconstruction uncertainty is 15°. Note that this angular distribution will be systematically biased toward the source because of the location of the inoperative PMT's.

Table 2. Neutrino events. From Kamiokande (reference 6)

Event number	Event time (sec)	Number of PMT's (N_{hit})	Electron energy (MeV)	Electron angle (degrees)
1	0	58	20.0 ± 2.9	18 ± 18
2	0.107	36	13.5 ± 3.2	15 ± 27
3	0.303	25	7.5 ± 2.0	108 ± 32
4	0.324	26	9.2 ± 2.7	70 ± 30
5	0.507	39	12.8 ± 2.9	135 ± 23
6	0.686	16	6.3 ± 1.7	68 ± 77
7	1.541	83	35.4 ± 8.0	32 ± 16
8	1.728	54	21.0 ± 4.2	30 ± 18
9	1.915	51	19.8 ± 3.2	38 ± 22
10	9.219	21	8.6 ± 2.7	122 ± 30
11	10.433	37	13.0 ± 2.6	49 ± 26
12	12.439	24	8.9 ± 1.9	91 ± 39

It is easy to understand a limit being drawn on neutrino masses from the observed tight pulses. Two $\bar{\nu}_e$'s with energies E_1, E_2, and mass emitted simultaneously arrive after a distance L with a time separation

$$\Delta t = \frac{L}{2}\left(\frac{m^2}{E_1^2} - \frac{m^2}{E_2^2}\right). \qquad (3)$$

Examining IMB neutrinos and being somewhat sanguine about the energy limits one could extract a mass limit using $E_2 \approx 40$ MeV for any of the first five events, and $E_1 \approx 20$ MeV from event #7 at 5.0 sec. One finds

$$m_{\bar{\nu}_e} \lesssim \left[\frac{10}{L}\right]^{1/2} \cdot \frac{40 \cdot 20}{[1600-400]^{1/2}} \approx 32 \text{ eV} . \qquad (4)$$

Presumably Kamioka data with its lower energies and up and down energy structure can be used to do better. There are problems here also if you take the authors at their word, since the first two events are supposed to be ν_e's, which can arrive earlier, and not $\bar{\nu}_e$'s, while #6 is to be excluded. The other events up to #7 have reasonably similar energies if one takes an error a bit larger than stated. Nevertheless, ignoring these quibbles, one can without sophisticated analysis take, say, #1 and #2, subtract 2 seconds (the entire initial pulse length) from the time for the lower energy of these and then deduce

$$m_{\nu} \lesssim \left[\frac{4}{L}\right]^{1/2} \cdot \frac{(20)(10)}{[400-100]^{1/2}} \approx 10 \text{ eV} . \qquad (5)$$

If one were to take seriously the time structure dictated by a cooling model for the supposed neutron star remnant, then a better limit might follow from the Kamiokande data. However, one must be sure temperatures, say, consistently fall with time and not rise and then fall, or else one must really have confidence in the model. Given the very low statistics, it is difficult, even with sophisticated statistical analysis, to see how one can do better than the limit in [5], for such a low statistical sample with other inconsistencies or difficulties in interpretation.

Other neutrino issues:

(1) Clearly they do not decay in a way to help the solar neutrino problem [6].

(2) The possibility of matter induced mutation of species is ever present in the supernova, the core, mantle and envelope providing a broad range of densities. However, one does see mostly $\bar{\nu}_e$'s in the detectors, and so the solar problem and the supernova are somewhat at odds. One point to be made is that if oscillations occur in the core interior before or during collapse, i.e. if there is a species with a reasonably high mass, then one has a serious problem. Division of the neutrino number into more than one species drops the pressure in the core and implies a smaller homologously collapsing core mass. The shock wave then forms at too low a mass point and will never propagate out of the core. One can write for the effective mass shift of an electron neutrino in matter

$$\Delta E \approx 1.41 \rho(\text{gm/cm}^3) Y_e 10^{-12} \text{ eV} .$$

Thus, if we assume all neutrinos trapped within a density of, say, 10^{14} gm/cm^3 or higher pass through this density during their diffusive passage around the collapsing core, we obtain an <u>upper</u> limit for a stable heavy neutrino of any species $X \neq e$,

$$m_X \approx 45 \text{ eV} .$$

A lower limit might result from more detailed calculation within the actual hydrodynamics.

(3) One can, however, attack one point more seriously: the total neutrino energy emitted in the cooling phase. I have indicated neutrinos are trapped at nuclear densities above 10^{12} gm/cc in the infalling core, at least for the dynamic times, which are of the order of milliseconds. When the shockwave reaches this density point, i.e. the neutrinosphere, on its outward progress, a burst of ν_e's is emitted in the order of milliseconds followed by perhaps a 10-20 millisecond tail of the same species. Total deleptonization of the compact core by β-capture yields only ~10^{57} neutrinos of rather high energy. The main group seen by the detectors comes

shortly after this when the core star acquires a large gravitational binding energy ~ 7.0% to 10% of its rest mass. This binding energy cooling by neutrinos, again only neutrinos (or any other sufficiently weakly interacting particle) can transport out of the dense core, favours ν_e's slightly, but is essentially from processes like

$$e^+ + e^- \rightarrow \nu_x + \bar{\nu}_x$$

at or near the neutrinosphere, and hence is spread evenly over all species.

COOPERSTEIN, BARON, and myself [28], amongst many others [27], are doing simple model calculations of the cooling and emission. BURROWS and LATTIMER [29] had done a more serious calculation prior to 1987a, and we rely on this for guidance, but a truly hydrodynamic evolution through bounce, explosion, and cooling, for a prompt explosion, awaits completion.

In a cooling model with two parameters, the initial temperature and the cooling rate, we find that central values for the binding energy are near 1.5×10^{53} ergs. This implies a gravitational mass for the initial core remnant, dependent on the softness of the EOS of between 1.2 and 1.35 solar masses [30].

The analysis of average energy and total number of events for IMB and Kamioka separately yield this binding energy. One finds a common initial temperature near $T_0 \approx 5$ MeV could work for both. We have, however, some trouble with the later events, certainly in Kamiokande, but also at IMB where the measured and displayed background is one event per ten days. We suggest then as other explanations for truly late events aside from the low probability of statistical fluctuations, background or thermal:

(1) a high threshold for the "base" cooling temperature reached at the end of the initial burst, a temperature maintained for some time;
(2) a mass of 32 eV (highly unlikely);
(3) a later falling back of material, raining through the shock;
(4) a delayed phase transition to, say, quark matter.

Kamioka's later events especially, if taken seriously, provide some support to the non-exponential cooling. We tend to take (3) most seriously as a possibility, which would imply a larger final mass for the neutron star of perhaps 1.5 M_o, although we also intend to look into (4).

Given our rather low value for the total binding energy, inferred from the total neutrino detection, it is difficult to place a strong constraint on the number of neutrino species. Raising the number used in our analyses from six to eight would lead to proportionally higher total binding energies, to near 2×10^{53} ergs, which implies a still higher cutoff mass in the initial phase of creating the neutron star. This point and others will be discussed further in work on the "long" time hydrodynamical calculation my colleagues Ed Baron and Jerry Cooperstein are performing [28].

One last comment: we all await the uncovering and observation of the neutron star (black hole) left behind 1987a. Estimates of the time taken for the ejected material to become transparent to x-rays and γ-rays from the surface of the neutron stars depend on (a) a knowledge of the total mass ejected and (b) the shock velocity. Any time from six months to a year is possible, so we don't have long to wait.

ACKNOWLEDGEMENTS

I am especially grateful to my colleagues Jerry Cooperstein and Ed Baron for allowing me to quote some of our results in advance of publication, and to G. E. Brown for continual discussion and encouragement. I was also very pleased to receive guidance and comments from M. Goldhaber. This manuscript has been authored under Contract No. DE-AC02-76CH00016 with the U. S. Department of Energy.

REFERENCES

1. I. Shelton (Las Campanas Observatory, Chile): communicated by W. Kunkel and B. Madore, International Astronomical Circular No. 4316, 24 February 1987
2. Private communication from University of Toronto, Feb. 1987
   See also, G.L. White and D.F. Malin: IAU Circular no. 4330 (1987);
   R.M. West, A. Lauberts, H.E. Jorgensen, and H.E. Schuster: Astron. Astrophys., submitted (1987)
3. R. Kirshner: APS Annual Meeting, Washington DC, April 1987. The indications are that SK-69202 (Sanduleak) is the B3 that may have exploded.
4. S.E. Woosley, P.A. Pinto, and L. Ensman: ApJ, submitted April 1987
5. R.M. Bionta et al (IMB): Phys. Rev. Lett. $\underline{58}$, 494 (1987)
6. K. Hirata et al.(Kamiokande): Phys. Rev. Lett. $\underline{58}$, 1490 (1987)
7. W.D. Arnett: ApJ Letters $\underline{230}$, L37 (1979)
8. T. Shigeyama, K. Nomoto, M. Hashimoto, and D. Sugimoto: Nature, submitted April 1987
9. W.M. Brunish and J.W. Truran: ApJ Suppl. $\underline{49}$, 447 (1982)
10. W. Hillebrandt, P. Höflich, J.W. Truran, and A. Weiss: Max Planck Institut preprint, MPA-286, submitted to Nature
11. E. Baron, J. Cooperstein, and S. Kahana: Phys. Rev. Lett. $\underline{55}$, 126 (1984)
12. V. Trimble: Rev. Mod. Phys. $\underline{54}$, 1183 (1982)
13. N. Gehrels and B. Teergarten: ApJ, submitted April 1987 (see ref. [4])
14. R. Barbon, R. Ciutti, and L. Rosino: Astron. Astrophys $\underline{72}$, 287 (1979)
15. D. Branch, S.W. Falk, M.L. McCall, P. Rybski, A.K. Uomoto, and B.J. Willis: ApJ $\underline{244}$, 780 (1981)
16. H.A. Bethe, G.E. Brown, J. Applegate, and J. Lattimer: Nucl. Phys. $\underline{A234}$, 487 (1979)
17. T.A. Weaver, B. Zimmerman, and S.E. Woosley: ApJ $\underline{225}$ (1978)
18. S.E. Woosley and T.A. Weaver: Bull. Am. Astr. Soc. $\underline{16}$, 971 (1984);
    S.E. Woosley: private communication (1984)
19. K. Nomoto and M. Hashimoto: BNL preprint and private communication
20. T. Mazurek, J. Cooperstein, and S. Kahana: In <u>DUMAND '80</u>, ed. V.J.Stenger (Hawaii Dumand Center, Honolulu, 1981; and In <u>Supernovae: A Survey of Current Research</u>, ed. M.Rees and R.J.Stoneham (Reidel, Dordrecht, 1982)
21. J.R. Wilson: In <u>Numerical Astrophysics</u>, eds J. Centrella, J. Leblanc, and R. Bowes (Jones and Bartlett, Boston 1985)
22. R.W. Mayle: PhD thesis, University of California at Berkeley, unpublished (1985)
23. E. Baron, J. Cooperstein, and S.H. Kahana: Nucl. Phys. $\underline{A440}$, 744 (1985)
24. S.H. Kahana: Prog. in Part. & Nucl.Phys. $\underline{17}$, 231 (1987);
    S.H. Kahana: In Conf. Proc. Weak and Electromagnetic Interactions in Nuclei, Heidelberg, FRG, July 1986
25. E. Baron, J. Cooperstein, L. van den Horn, and S. Kahana: work in progress
26. Mont Blanc Proton Decay Group, unpublished
27. J.N. Bahcall and S.L. Glashow: Nature $\underline{326}$, 476 (1987);
    J.N. Bahcall, T. Piran, W.H. Press, and D.N. Spergel: Nature, submitted, April 1987;
    A. Burrows and J. Lattimer: "Neutrinos from SN1987a" preprint April 1987
    J. Lattimer: Invited talk at XIth Particle and Nuclei Conference, Kyoto, Japan, April 1987
28. S.H. Kahana, J. Cooperstein, and E. Baron: in preparation
29. A. Burrows and J. Lattimer: ApJ $\underline{307}$, 178 (1986)
30. J. Cooperstein: private communication (1987)

Particle Physics and Astrophysics

E. Schatzman

Nice Observatory, BP 139, F-06003 Nice Cedex, France

1. The Solar Neutrino Problem

It is now well known that the predicted solar neutrino flux (Davis experiment) from the ^{7}Be$(p,\gamma)^8$B$(e^+,\nu)^8$Be(2α) reactions is overestimated by at least a factor of 3 (see for example Schatzman and Ribes 1987; Lebreton and Maeder 1986, 1987; Cahen 1986 and references therein). Various explanations have been proposed (Schatzmann and Ribes 1987):

a)   the nuclear cross section of the beryllium reaction is overestimated;
b)   the nuclear cross section of the $p(p,e^+\nu)D$ reaction is underestimated;
c)   the radiative transfer is wrong (opacities);
d)   mixing in the solar core lowers the central temperature;
e)   cosmions in the solar core lower the central temperature;
f)   interaction between massive neutrinos and solar matter produces at the surface a deficiency in electron neutrinos.

Compatibility with heliosismology excludes c). Furthermore, there are still some difficulties with solar models, e.g. the abundance of helium (see for example Lebreton and Maeder 1986).

Cosmions give an easy solution (for example, Faulkner and Gilliland 1985) but with a large arbitrariness in the parameters. Consistency with astrophysical data (Renzini 1987) or underground experiments (Gaisser et al. 1987) limit the range of possibilities.

Assumption f) (Mikheyev, Smirnov, Wolfenstein mechanism) will eventually be proved by other neutrino experiments (Gallex. See Kirsten 1986).

2. The Missing Mass

The virial theorem says that

$$\frac{d^2I}{dt^2} = 2T + \Omega \quad ,$$

where I is the momentum of inertia, T the kinetic energy, Ω the potential energy of a Newtonian system. For a stationary system,

$$2T + \Omega = 0 \quad .$$

Roughly speaking, $\Omega \approx -(GM^2/R)$, $T \approx 1/2M\langle V^2\rangle$. Measuring the mean square velocity and the length scale of a system gives its mass, $M \approx (R\langle V^2\rangle/G)$. This is the principle on which mass determination is based for large systems. On the other hand, the determination of the mass of galaxies from the rotation curve provides the mass to luminosity ratio (Faber and Gallagher 1979) of the order of 10 in solar units. When comparing systems of galaxies, the (M/L) ratio seems to increase with the size of the system, growing to 250 for large systems. The dark matter (Blumenthal et al. 1984) seems to represent 90% of the mass of the universe, which eventually is flat ($\Omega_0 = 8\pi G\varrho/H^2 = 1$) where ϱ is the average density of the universe and H the Hubble constant, G the constant of gravitation. It should be noticed that $\Omega_0 = 0.1$ today means $\Omega_0 = 1 - 10^{-61}$ in the early universe (at the Planck time). The presence of cold dark matter makes it easier to explain the formation of galaxies (Blumenthal et al. 1984). However, the consistency of the rotation curve of galaxies (Bahcall and Caserato 1985) suggests that the dark matter in galaxies is *baryonic*. The production of shells around galaxies (Dupraz 1987) can be well explained by collisions of galaxies without any amount of cold dark matter; there are several examples showing that the assumption of stationarity in applying the virial theorem is not valid (Binggali et al. 1987); the value of Ω_0 as derived from primordial nucleosynthesis of light elements is in the interval 0.014–0.14 (Yang et al. 1984) and the same order of magnitude is obtained by Kirshner et al. (1983), Aaronson et al. (1982), and Press and Davis (1982) from dynamical arguments.

The question of galaxy formation does not improve the problem. Cold dark matter makes things easier (Blumenthal et al. 1984, and others) especially if one considers that galaxies are formed before clusters (Binggali et al. 1987). However, the early production of very active galaxies can heat up the pregalactic plasma to temperatures where gravothermal instabilities can take place (Ostriker et al. 1981; Vishniac et al. 1985; see Schwarz et al. 1975). This would explain the production of voids of a large size (about 10 megaparsecs) on the boundary of which the galaxy formation would have taken place.

3. Conclusion

One cannot at present decide exactly what the nature of the dark matter is. The dark matter is there, as shown by its gravitational effects, but one can hardly exclude the possibility that it is entirely baryonic.

Our aim in this contribution is to provide an introduction to the physical problems which are raised by the question of new particles in astrophysics. It is impossible to review the whole bibliography of the subject. I apologize for the large number of missing references.

Bibliography

Aaronson M., Huchra J., Mould J., Schechter P.L., Tully R.B. (1982): Ap. J. **258**, 64
Bahcall J.N., Caserato S. (1985): Ap. J. **293**, L7
Binggali et al. (1987): preprint ESO
Blumenthal G.R., Faber S.M., Primack J.R., Rees M.J. (1984): Nature **311**, 517; **313**, 72
Cahen S. (1986): Recontres de Moriond, January 1986
Dupraz C. (1987): preprint ENS Radioastronomie Paris
Faber S.M., Gallagher J.S. (1979): ARAA **17**, 135
Faulkner J., Gilliland R. (1985): Ap. J. **299**, 994
Gaisser T.K., Steigman G., Tilav S. (1987): preprint, Bartol Research Foundation (University of Delaware)
Kirshner et al. (1983): A.J. **88**, 1285

Kirsten T. (1986): Rencontres de Moriond, January 1986
Lebreton Y., Maeder A. (1986): AA **161**, 119
Lebreton Y., Maeder A. (1987): AA **175**, 99
Ostriker J.P. et al. (1981): Ap. J. **243**, L127
Press W.A., Davis M. (1982): Ap. J. **259**, 449
Renzini A. (1987): AA **171**, 121
Schatzman E., Ribes E. (1987): in Recontres de Moriond, January 1987
Schwarz et al. (1975): Ap. J. **202**, 1
Vishniac et al. (1985): Ap. J. **291**, 399
Yang J.Y., Turner M.S., Steigman G., Schramm D.N., Olive K.A. (1984): Ap. J. **281**, 493

Part IX

Conclusions

Theoretical Perspective

H.B. Nielsen

The Niels Bohr Institute, University of Copenhagen,
Blegdamsvej 17, DK-2100 Copenhagen Ø, Denmark

1. Introduction

First of all, I have to apologize for all the omissions I will commit in this summary. In fact, I can and will only cover a rather small part of the work presented by the contributors at this conference. So, in the following, I shall mainly talk about strong interaction physics although there were also talks on important and fashionable subjects too.

So for example S. Kahana (note that I will not give explicit reference to the material found in the proceedings of this conference) told us about an event that had come out of good luck for science: the recent supernova explosion 1987A (or rather the fact that the light from this explosion recently reached the earth). Another interesting topic that does not really fit in the line of this summary is the mysterious result of the GSI experiments (as reported by Stiebing). They perhaps tell us that something might be missing in our understanding of atomic physics. But really nobody knows (yet) the explanation of the strange peak observed in the e^+e^--spectrum. Also a talk like Kungsik Kang's about axions and majorons (starting from the problem of strong interaction CP-violation due to the topological Θ-term) falls a bit outside the main subject. Finally, as the title suggests, pure experimental topics will not be included.

The situation of intermediate and low energy hadron physics today is the following: We have *a theory*, QCD. It is, however, extremely difficult to make use of this theory in practice (in this energy regime). Therefore, typically one replaces QCD by a more applicable model which approximates QCD in the low energy regime. Different intermediate models approximating QCD may be used under different conditions, in, say, the cold hadron and in the hot quark-gluon plasma phases. An attempt to illustrate the situation is made in Fig.1. Here, we have three different physical conditions or regimes listed:

1) The high transverse energy jet regime of hard quark and gluon collisions. In this regime one can apply perturbative QCD because of asymptotic freedom. One can expect that a perturbative calculation can give information about, for example, the angular distribution of jets representing quark and gluons. In fact, such jet cross sections, nicely measured for instance in the SPS (Super Proton Synchrotron), constitute a beautiful test of QCD and represent a strong reason for the belief that QCD is the correct theory. However, the subject of the present conference

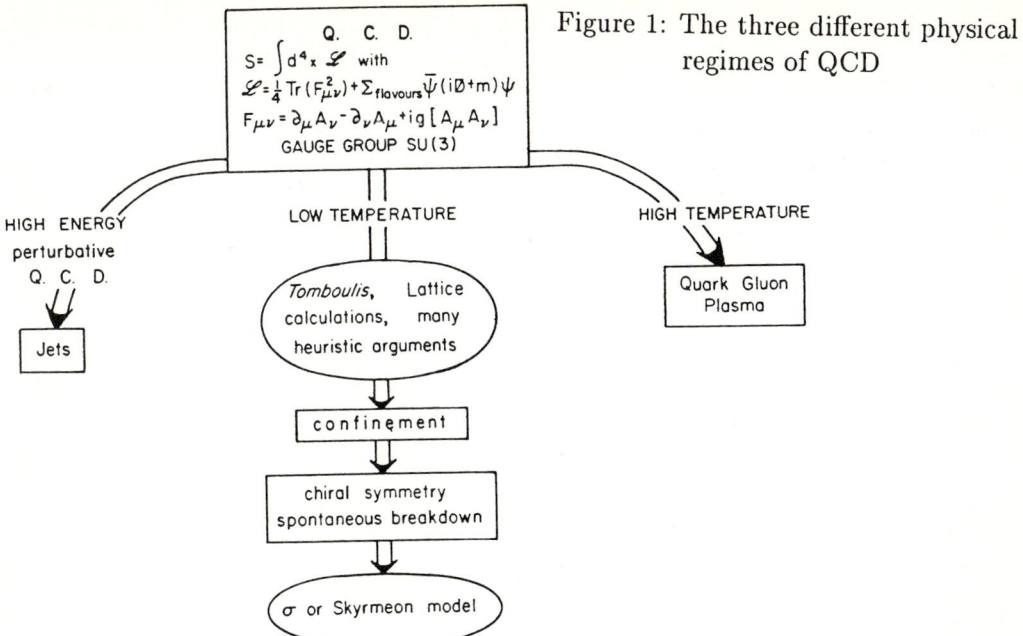

Figure 1: The three different physical regimes of QCD

was not this region, but rather the more difficult and therefore more interesting region of lower energy where QCD has "infrared slavery" and is not accessible to any (almost trivial) direct approach.

2) The low energy and cold (or normal) phase of QCD (or strong interaction physics): This is the region of greatest importance for two reasons. Firstly, it is the phase which is realized in nature under normal conditions. Its complexity is enormous since it includes nearly all nuclear physics and astrophysics as special cases. Secondly, since one has not yet reached the quark-gluon plasma experimentally, all practical experimental achievements in the low and intermediate energy regime are concerned with QCD matter which is in the cold phase.

3) There is the quark-gluon plasma phase which may become the "playground" of the experimentalist of the future – or even the near future. However, studying the quark-gluon plasma might not be as simple as one might naively think. From the experimental point of view, one has to find an answer to the problem of clear characteristic signatures of this phase in comparison to the normal one (phase 2). There are also growing theoretical indications that this phase will be complicated by nonperturbative effects [1].

2. Interpretation of QCD

In the low energy phase, QCD strongly modifies itself in comparison to its naive content: First, there is the modification by the so-called confinement. It is in such a phase that none of the fundamental particles of QCD, the quarks and gluons, are seen as true particles. That it is really true that QCD has this (frequently speculated) feature, was argued by Tomboulis [2] by using blockspinning (i.e. lattice renormalization group) arguments on lattice QCD theory. It is well established from computer Monte Carlo calculations too. In fact, Dr. Pendleton showed in his lecture a figure demonstrating that there is, within 20%, scaling for the string tension. This means that the expectation value of Wilson loops

$$\langle \mathbf{Tr}\, \mathbf{P}\,(\exp{(i \int dx^\mu\, A^a{}_\mu \frac{\lambda^a}{2})})\rangle \tag{1}$$

calculated in lattice Monte Carlo $SU(3)$ Yang-Mills theory behaves as a function of the coupling constant as expected in a phase where the force between a couple $Q\bar{Q}$ of heavy quarks-antiquarks grows linearly with the distance (at large distance of course). The "string tension" σ, i.e. the proportionality constant of the increase of the potential energy of a $Q\bar{Q}$-pair versus the increase in distance between the Q and the $\bar{Q}$ can be calculated in terms of the lattice coupling constant in such a way that it depends only on the QCD scale Λ_{QCD} of the $SU(3)$ Yang-Mills theory (to an accuracy of 20%). We consider this as strong evidence that QCD does indeed predict such (for large distances) a linearly rising potential which will confine the quarks (at least the heavy ones), since it will cost infinite energy to remove the quark Q infinitely far from the antiquark $\bar{Q}$.

On top of this departure from a naive interpretation of the QCD quantum field theory model, there is the phenomenon of spontaneous chiral symmetry breakdown. In this conference, Dr. Pene told us how one might imagine that QCD leads to a vacuum state extremely analogous to the state of a superconductor. Just as in the superconductor, we have the formation of Cooper pairs, i.e. in the case of a superconductor, of two (quasi) electrons, in the case of the QCD vacuum, of quark and antiquark pairs of different chirality forming zero angular momentum, zero momentum bound states. These Cooper pairs build up a condensate in the sense that there is a macroscopic – but quantum-mechanically (somewhat) uncertain – number of them present in a piece of vacuum. This makes it possible for the vacuum to exchange a quark of one chirality (i.e. one γ^5-eigenvalue) with one of the other chirality. Then it becomes possible to confine massless quarks too (or rather, previously or a priori massless quarks). The point is namely that in the presence of the Cooper pair condensate the quarks acquire (an extra contribution to their) masses and a quark can now be stopped and returned without invoking necessarily a spin-flip. In fact, when the quark has lost enough energy, it can add itself to the vacuum by being part of a Cooper pair provided that another quark with opposite chirality but same spin angular momentum is emitted. This quark can now run in the opposite direction and so confinement is realizable. Without breaking either chirality conservation or angular momentum conservation, it seems, however, not possible to confine a chiral quark. At least, the corresponding amount of angular momentum or chirality has to be transferred away when the (chiral, i.e. massless) quark turns back (towards the center of the hadron in which it is confined). Dr. Pene told us that in his toy model confinement ensured also that the ground state (vacuum) of his model would be the one with spontaneous symmetry breakdown of chiral symmetry. That is, it would be a vacuum which is not chirally invariant.

Both the confinement which essentially gets rid of the a priori particles – gluons and quarks – and the spontaneous symmetry breakdown of chiral symmetry, are crucial for model building as well as for the physics of low or intermediate energy.

In this "theoretical perspective" I would like to stress the (in some sense) suspicious encounter of these most important reinterpretations or "explaining-aways" of the first guess about the content of QCD. At first, one would have guessed to find quarks and gluons as physical particles if one had stared for a moment at the supposedly fundamental Lagrangian of QCD

$$\mathcal{L} = -\frac{1}{4}F_{\mu\nu}^a F^{a\mu\nu} + i\sum_f \bar{\Psi}_f \partial_\mu \gamma^\mu \Psi_f + \sum_f m_f \bar{\Psi}_f \Psi_f \quad . \tag{2}$$

Furthermore, if the quark masses were negligible one would have expected to find at least particles representing the chiral symmetry group $U(N_f) \otimes U(N_f)$. Actually, however, all this has to be explained away: Quarks and gluons are not seen and chiral symmetry is spontaneously broken so that there are "massless" pseudoscalar Nambu-Goldstone bosons instead. It is suspicious that one must let QCD dress itself up to be almost unrecognizable, in order that it fits in with Nature.

3. Intermediate Effective Models – How far away from QCD?

Let us as an example look at the effective model presented by W. Weise which is an extension of the work of Bando et al. [3]. The essential idea in this model is the identification or "finding" of (composite) vector meson (ρ and ω) degrees of freedom by some notational shift in the nonlinear σ-model. Then Weise et al.[4] add a kinetic term for the vector particles (by hand) in order to give a dynamical meaning to the vector mesons. Furthermore, they arrange for vector meson dominance, when a minimal gauge interaction is introduced. This is done so as to obtain the KSFR-relation [5].

In this model, one has several of the lowest mass nonstrange mesons represented by explicit fields π, ρ, ω. The baryons can be obtained as solitons. Now, one may ask, though, how much of this model is coming from QCD and how much is phenomenological input and how much is free phantasy. Table 1 is an attempt to roughly answer this question. It is presumably not surprising that most of the features of a modern model for hadrons have been discovered phenomenologically at first rather than being derived from QCD. The simple reason is of course that the phenomenology of hadron physics goes much further back in history than the discovery of QCD. Still, one may consider it to be embarrassing for QCD that so few of the main features – at least historically – are derived from it. However, at least logically, both confinement and chiral symmetry breakdown are – as we heard – consequences of QCD – and then quite a lot follows: the σ-model – we may say – is a natural way to incorporate the above constraints at low energies into an effective language, and then, one is able to find vector mesons by rewriting the σ-model [3,4] (Perhaps it is more honest though to say that vector meson dominance is still put in). Thus logically (rather than historically), the status of QCD even at low energies is better than one would naively guess.

Table 1. Origin of the effective model suggested by Weise et al.[4].

Feature	From QCD	From phenomenology	Output of model	Pure phantasy
Chiral symmetry	Nicely but needs $m_q \approx 0$.	Historically yes.	Not really, it is built in.	No.
Action	Via chiral symmmetry and using confinement. (Large N_c helps.)	To some extent: Known particles ρ and ω are "found" hidden in σ-model.	Much derived from σ-model, but additional term inserted (also using vector dom.).	A bit, but mostly it can be argued for.
Vector dom. KSFR relation	No, not that I know.	Yes.	No, the parameter a was fitted to V.D.	No.
Baryons	(Yes) They are implied by QCD to exist.	(Yes) But baryons are not put in!	Skyrmeon idea: Yes, they come out of the model.	No, not at all.
Isospin symmetry	Yes, nicely when just $m_u, m_d \ll \Lambda_{QCD}$.	Historically, yes.	Really rather input.	No, it is well established.
Hadron sizes	No.	Rather from vector dom.	Well for $a = 2$.	No, no.
Vector meson mass relation		Rather from vector dom. (size of the coeff. of the kin. term).	Yes.	No.

Let us discuss some other ways of making intermediate models which simulate QCD. One can – naturally – look at the old nonrelativistic quark model (see H. Lipkin's talk or, for example, [6]) as such an intermediate model. These as well as potential models are first of all applicable to heavy quarkonium states where the nonrelativistic approximation may be justified.

Yet another intermediate model is the bag model. These days one may regard the bag model combined with the σ-model as the hybrid chiral bag model (sometimes also called "little Brown bag model")[7]. In my talk, I elaborated that such a bag model can be reinterpreted in a "Cheshire Cat" way (this phrase was coined by K. Johnson). M. Rho reported on this subject too. This means that one looks at the bag model as an intermediate model simulating QCD, but only in part of space-time – the outside region of the bag. In the rest of space-time, one keeps

QCD a priori exactly. Of course, in order to do practical calculations, one must make approximations also in the inside – but this might be justified since one can now work with rather small bags.

One can say, the resonating group method is a way of handling effective models. However, the scattering tests on the scattering length of the K^+ proton for instance do not seem to agree well with experiment when the RGM is taken as it stands. On the other hand, a phenomenologically broadened one (provided with a linear confining potential) works fine. So in the end, the RGM also had to be improved phenomenologically or speculatively in order to function in agreement with experiment.

The purpose of some intermediate models is mainly to illustrate important physical features. A most interesting phenomenon deserving such illustration is quark confinement. Historically, the bag model was made as such an illustration. Here, we have heard E.J. Moniz presenting some simple quantum mechanical models made to illustrate confinement under hadronic scattering.

All these models are to some extent based on QCD, but far from completely derivable if one takes, for example, the quark model. The potential between a heavy quark-antiquark pair can be computed in lattice Monte Carlo QCD. The bag model has been speculated from various QCD vacuum pictures - instantons, merons, Copenhagen vacuum or the like, but from the Cheshire Cat point of view, one should rather say that to derive the little Brown bag from QCD means to derive the σ-model. Then we only make use of that derivation in part of spacetime. In a similar "derivation" of the cloudy bag model, one should first derive another intermediate model to be used inside the bag – namely one with σ-model fields ($\vec{\pi}$ at least) and with quark fields. The latter model is the one favoured by S. Kahana and G. Ripka at this conference.

In practice, results tend to depend on the calculational method because one has to understand phenomena in a sufficiently simplified way to be able to connect with experimental data at all. So, the question to be settled may turn out to be which diagram "wins" or dominates. For instance, Tony Green (A.M. Green) sought to convince us that nucleon-antinucleon annihilation is explained by lowest-order diagrams of the $^3P_0$-model (often called the pair-creation model). Somewhat more general diagrammatic considerations on the same subject were shown by Genz.

In addition, in order to study annihilation processes, one could use the so-called tube model. This is an example of how one makes models in successive steps based on QCD ideas: the tubes are "derived" in a bag model which again is "derived" from QCD with some confinement picture put in.

Stassard described baryon decays in a QCD inspired model in which quark pairs are produced. The agreement with data is quite fair.

4. More Direct Approaches to QCD

If one intends to test QCD more closely, there are several possibilities:

1) The most direct – in some sense – is to use the Monte Carlo technique for the QCD lattice formulation. Pendleton told us about ongoing progress in the method of putting quarks (fermions) in a hybrid technique on the lattice. Another piece of progress was more like a step back: Previously, it looked as if the string tension calculated for weaker and weaker QCD coupling constant had reached its region of scaling in the Monte Carlo calculations already performed, but now it seems still not to have fallen into place. There is scaling only to 20% accuracy for it.

2) Another approach rather close to QCD is the type of model put forward by H. Pirner. He uses QCD as the source of inspiration and takes – one may say – his starting point from there, seeking an effective theory calculated to approximate QCD. A major point is to allow for further effective fields describing dielectric properties of the "vacuum". H. Pirner uses lattice calculations to determine the parameters of the effective theory, so this approach is indeed very strongly connected to QCD.

3) The Russian sum rule [8] – as presented by H. Rubinstein in this conference – can be considered as a real test of QCD where the extra input to QCD is introduced in a rather hidden – or only "technical" – way. Here, the basic idea is a type of duality principle suggesting that the (imaginary part of) Green's functions for some quark-currents are roughly equal whether estimated from diagrams with quarks (and gluons) or from hadron state information. This "rough equality" is studied via the sum rules, i.e. by smoothing out integrals which are identified in both languages rather than the amplitudes themselves. Maybe there is also in this approach an appreciable amount of extra input or assumption, but now some of this input is replaced by judgements as to which moments we shall agree to trust and by how much one allows some extra fields in the vacuum to be included explicitly. It seems to me to be a method that is very close to QCD in the sense that the "small helps" are performed in a more automatic way which does not involve any obvious strong assumptions. Or in other words, one sits and looks at the numbers and curves and decides what to trust, and there is less making of glorious big extra assumptions in addition to QCD itself and its identification to hadron physics. A little allowance for an $\langle F_{\mu\nu}^2 \rangle \neq 0$, but that is not much. Really the assumption is that it is not necessary to take a lot of other field fluctuations into account.

4) Werner et al.[9] told us about a mean field calculation in QCD in the Fock-Schwinger gauge $x^\mu A_\mu^a = 0$. This model might be very promising as a QCD attached approach. They got, for example, a lower bound of 1 GeV for the glueball mass(es). A different technique – namely, use of many-body theory in QCD – was presented by G. Schütte.

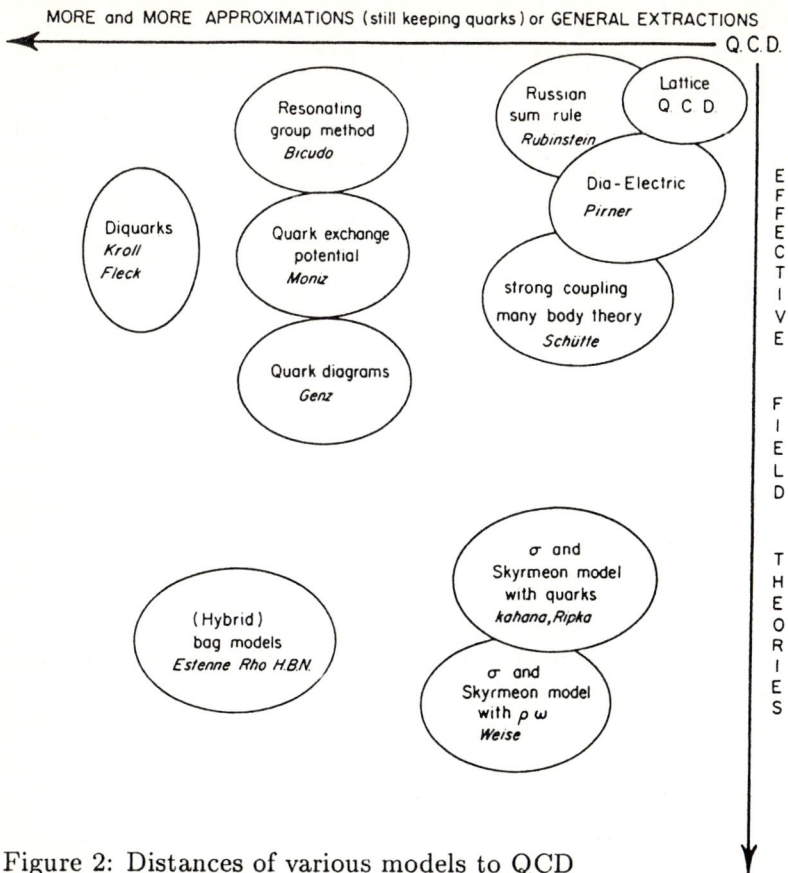

Figure 2: Distances of various models to QCD

Figure 2 crudely illustrates the distances of several approaches (presented at this conference) from QCD.

5. The Quark-Gluon Plasma – The QCD Phase of the Future?

Another approach to get a bit closer to QCD is to search for a phase, a state, of matter in which the naive QCD picture makes more sense. Such a phase would be the quark-gluon plasma in which one has a finite temperature blackbody radiation of quarks and gluons. So in first approximation, one should have freely moving particles (the ones one naively gets from the QCD action). That there might be some caveats (i.e. nonperturbative effects could be still present in this phase) was pointed out by several authors [1]. Such plasmas are, however, not yet reached experimentally.

L. McLerran showed us how the phases of the quark-gluon plasma and usual low temperature confinement are found to have an – albeit rather narrow – gap connecting them in a quark mass m versus temperature T diagram. Even if there is such a connection, the two phases are separate phases for zero quark mass m. So there should indeed be a new phase to be found. The place to find it is of course in high energy collisions of heavy ions. However, there is the problem of how one would be able to tell that one had actually produced a quark-gluon plasma in a given experiment. Here, the production of strangeness – strange quarks in equilibrium with the other quarks – plays an important role. Especially the ratio of Ξ hyperons to anti-Ξ hyperons would be sensitive to the plasma. Without such a quark-gluon plasma it seems to be difficult to get more than one $\bar{\Xi}$ for every 100 Ξ's. With a fully developed high temperature plasma, equally many Ξ's and $\bar{\Xi}$'s are expected. The baryon number density no longer plays any role because of enough pure thermally produced matter.

Maurice Jacob told us how the measurement of the $\bar{\Lambda}$ longitudinal polarization in nucleus-nucleus collisions at SPS energies (200 GeV per nucleon) could provide us with a direct and unambiguous test of the quark-gluon plasma. If one can measure a 15% effect of such longitudinal polarization one can thereby determine the relative abundance of anti-cascade particles $\bar{\Xi}$ in the $\bar{\Lambda}$ sample. So one would notice the quark-gluon plasma via its cascade hyperons.

6. Diquarks and Exotics – Which Are More Exotic?

It is possible – although far from being conclusive – to add to QCD the idea that couples of quarks (or couples of antiquarks) could be somewhat strongly bound so as to form the so-called "diquarks" (or anti-diquarks).

There is some indication that quarks might hang together as diquarks. In fact, Kroll showed us how one can successfully use diquarks as constituents in the Brodsky-Farrar-Lepage power law rule for exclusive reactions [10]. Actually, there is some phenomenological evidence for diquarks. In a model involving diquarks – "the end point model" – they (Kroll et al.) even explain the polarization for the process $\pi^- p \to \rho^- p$.

S. Fleck investigated instead under which conditions – inside the hadrons – the diquark states are of importance. For light quark baryons, they are suggested to be there for high (orbital) angular momentum ℓ in order to explain that the Regge trajectories are linear for the baryons (at least at high angular momentum) quite in analogy to the $\bar{q}$-q systems forming the mesons. So, the diquarks are important for the light quarks for large angular momentum ℓ. For heavy quarks, however, the opposite holds: The diquark built of a couple of heavy quarks is tightest bound at low angular momentum, and dissolves at higher angular momenta.

At first, diquarks were considered with suspicion in a similar way as the exotics. The suspicion of the latter is due to the lack of empirical findings. The diquark idea is based on phenomenological input and fundamental support is lacking, whereas in the case of exotics it is just the other way around. In fact, the idea that diquarks should actually exist would be a bit of a surprise, if one had just guessed blindly. On the other hand, I suppose that a theoretical first guess about exotics would be the one that there should be no reason that the so-called exotic states should not exist – at least as resonances. Of course, states with exotic quantum numbers might have somewhat higher masses than normal hadrons. Even with confinement, there are ways to have exotics. We have heard here serious estimates showing on theoretical grounds that exotics should exist.

Actually, J.M. Richard (Gignou and Silvester-Brac) told us about stable multiquark states. In fact, if one composes an exotic state of $QQ\bar{q}\bar{q}$ it becomes stable when the mass M (or mass ratio M/m) of the heavy quark Q becomes sufficiently large since then the masses obey the mass inequality: $QQ\bar{q}\bar{q} < Q\bar{q} + Q\bar{q}$. The same qualitative conclusion was reached by Lipkin [11], and by Heller and Tjon [12]. Also Gignou and independently Lipkin found that heavy exotic baryons $\bar{Q}uuds$ should have the same chromomagnetic binding with respect to the channel $[\bar{Q}q] + [qqq]$ as the H dibaryon with respect to the channel $\Lambda\Lambda$. So states like $\bar{Q}uuds$ should have a chance of being stable for a very heavy Q-quark.

It seems that exotics are now taken very seriously in spite of the lack of empirical evidence in favor of them. The fact is that at low energies (where resonance studies are feasible) no conclusive indication for exotics has been found until now.

7. Strangeness – A Gift to Science?

It may happen that Nature provides us with some tool that helps us to study physics (of hadrons, say) in a way that would otherwise not be possible. It is for instance yet not understood why Nature has "ordered" a quark type - the strange one s - with a current algebra mass of the same order of magnitude as the scale Λ_{QCD} which determines when confinement sets in. The other flavors are either very heavy (t, b, c) or very light (u and d) compared to Λ_{QCD}. We should presumably consider the strange quark s – or you may say "strangeness" (as such) – a gift from Nature to science.

For instance, we heard from Zituon about the suppression of the production of strange particles K versus nonstrange ones π in a hadronic process. He explained this suppression in terms of the "suppression factor" λ defined to mean

$$\lambda = \frac{K\text{rate}}{\pi\text{rate}}\Big|_{\text{genuine}} = \frac{K^\star\text{rate}}{\rho\text{rate}}, \qquad (3)$$

which in turn may be well understood in a string or bag tube breaking picture. The phenomenological value $\lambda = 0.3$ is essentially calculable from such a picture (i.e. $\lambda = 0.31$ theoretically). A more important application (reported by Rafelski) is to use strange particle production as a diagnostics for the quark-gluon plasma (see also Sect.5). By seeing an excess of strange particles – an increased λ – one should have the signal that a quark-gluon plasma was formed.

Strangeness may also help science by putting effective models to the test. Let us take the Skyrme model as an example. M. Praszalowicz is one that took up this challenge by investigating the inclusion of strangeness into this model. It does, however, not seem easy to get good phenomenological results out (Nowak).

8. Influence of Nuclear Environment

An important general problem in addition to testing QCD and adapting it to more accessible intermediate energy hadron physics is the further question: how are the properties of the hadrons – first of all the nucleons of course – changed inside nuclear matter. Especially, we heard of two questions of this nature and their possible interdependence:

1) Nucleons in iron swell in comparison to their free partners.

2) There was much discussion about the famous EMC effect.

Hong Jung and G. Miller told us that there are N_c (with $N_c \to \infty$) theories explaining why one does not find – as expected – the same deep inelastic structure function for nucleons inside iron, say, as for the ones inside deuterium. However, experimentally deviations showed up [13]. They reported on the growing understanding that the models called "nucleon pionic excess" are really models in which the important and best understood effects are nuclear binding prescribed by the standard nuclear shell model and its extensions, see also [14]. Their philosophy is that one shall take into account the effects that must be there before trying to explain all by new physics. A point explaining the surprisingly large effect is the somewhat accidental quick variation of the nucleon structure function.

9. Conclusion of Summary

In spite of the belief that we have in principle *a theory* – QCD – for strong interaction physics and therefore in principle even for nuclear physics, the reduction of all hadron interaction physics to this model still involves several helping assumptions and intermediate modelling. But, that is the way reductionism usually works: It is almost never performed in a complete way. We illustrate in Fig.3 how QCD – symbolized by a naked man – gets dressed by various features which are a priori non-obvious: Confinement and then – to a certain degree as a consequence – spontaneous chiral symmetry breaking, and then – although of less importance – still further features are added: vector meson dominance, KSFR, the so-called constituent quarks etc. and even diquarks. This latter freedom of adding more and more things to QCD must be – presumably – looked upon as a weakness in the testing of QCD via these phenomena since they are indeed more speculated into QCD (of course based on phenomenological studies of Nature) than truly derived from QCD.

Some of the most direct ways to QCD were the lattice Monte Carlo calculations, the Russian sum rules, H. Pirner's attempt to get more fields in a QCD inspired way, and mean fields in the Fock-Schwinger gauge, say. As an ideal picture, we should have a series of intermediate models which are successively based on each other. First, at short distances, we have of course QCD. Then at a little longer distance where the coupling constant becomes stronger the Pirner et al. type of constructing effective fields would be the appropriate approach. Next, at

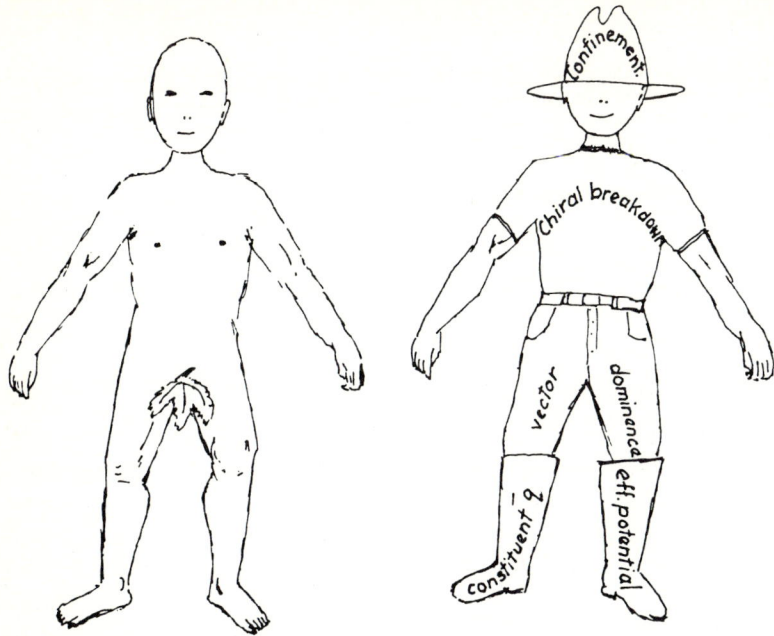

Figure 3: QCD - symbolized by a naked man - gets dressed by various features

even longer distances, we should find a σ-type model with vector mesons etc. Perhaps some glueball fields have still to be included too. At even longer distances, we may treat the topological solitons of the σ-model as pointlike baryons which interact via boson exchanges (with cutoffs due to, for example, the finite size of the solitons). Finally, at an even longer distance scale, one reduces the boson exchange to effective forces as they are used in nuclear physics.

An important thing is that QCD is supposedly applicable under various conditions: QCD plasma, high momentum transfer events, the low energy regime with confinement – both for few-particle processes and in nuclear matter – and under cosmological big bang high temperature conditions. There might be possibilities of learning especially about the QCD quark gluon plasma in future experiments.

The fact that there are so many facets and developments in the "antireduction" from "QCD" to real physics – as accessible by experiments – is really a charming feature that makes theoretical work needed. So, we should be happy for that.

Acknowledgement
It is a pleasure to thank Andreas Wirzba (NORDITA) for helping very much in working out the present "Theoretical Perspective" text. Richard McGough is thanked for helping with the English language. Thanks are also due to the editors for their patience in waiting for the manuscript.

References

[1] S. Nadkarni, *Phys. Rev. D* **33**, 3738 (1986), *Phys. Rev. D* **34**, 3904 (1986).

[2] E.T. Tomboulis, *Phys. Rev. Lett.* **50**, 885 (1983).

[3] M. Bando et al., *Phys. Rev. Lett.* **54**, 1215 (1984).

[4] U.-G. Meissner, N. Kaiser, A. Wirzba and W. Weise, *Phys. Rev. Lett.* **57**, 1676 (1986); U.-G. Meissner, N. Kaiser and W. Weise, *Nucl. Phys. A* **466**, 685 (1987).

[5] K. Kawarabayashi and M. Suzuki, *Phys. Rev. Lett.* **16**, 255 (1966); Riazuddin and Fayyazudin, *Phys. Rev.* **147**, 1071 (1966).

[6] N. Isgur and G. Karl, *Phys. Rev. D* **19**, 2653 (1979), *Phys. Rev. D* **20**, 1191 (1979).

[7] G.E. Brown and M. Rho, *Phys. Lett.* **28 B**, 177 (1979).

[8] M.A. Shiffman et al., *Nucl. Phys. B* **147**, 385,448 (1979).

[9] M. Schaden et al., "Hartree Approximation to QCD in Fock-Schwinger Gauge", Regensburg Preprint, TPR-86-30, (1986).

[10] S.J. Brodsky and G.R. Farrar, *Phys. Rev. Lett.* **31**, 1153 (1973), *Phys. Rev. D* **11**, 1309 (1975); S.J. Brodsky and G.P. Lepage, *Phys. Rev. Lett.* **43**, 545 and 1625(E) (1979).

[11] H. Lipkin, *Phys. Lett.* **172 B**, 242 (1986).

[12] L. Heller and J.A. Tjon, *Phys. Rev. D* **35**, 969 (1987).

[13] J.J. Aubert et al., *Phys. Lett.* **123 B**, 275 (1983).

[14] S.V. Akulinichev et al., *Phys. Rev. Lett.* **55**, 2239 (1985).

Index of Contributors

Arvieux, J. 205

Barnes, P.D. 292
Bertini, R. 205,272
Bicudo, P. 51
Boschitz, E. 205
Bradamante, F. 412

Carbonell, J. 311
Catz, H. 205
Chaumeaux, A. 205
Cugnon, J. 211

D'Agostini, G. 235
Dalpiaz, P. 414
Descroix, E. 205
Duclos, J. 408
Durand, J.M. 205

Fabbro, B. 205
Faivre, J.-C. 205
Fanet, H. 205
Fleck, S. 148
Fried, H.M. 164

Gérard, A. 226,406
Gignoux, C. 42,148,311
Glazek, St. 155
Green, A.M. 190
Grossiord, J.Y. 205
Guichard, A. 205
Gyles, W. 205

Harountunian, R. 205
Heller, L. 35

Jacob, M. 333
Jung, Hong 259

Kahana, S.H. 125,440
Kang, Kyungsik 378
Klupisch, R. 414
Konter, J. 205
Kroll, P. 138

Lavelle, M. 155
Le Yaouanc, A. 115
Lefevre, P. 414
Lipkin, H.J. 24
London, G.W. 352,361

Macri, M. 414
Mango, S. 205
Martin, A. 414
McLerran, L. 320,437
Miller, G.A. 259
Möhl, D. 414
Moniz, E.J. 56
Montanet, L. 414
Moussallam, B. 160

Nielsen, H.B. 72,454

Oliver, L. 115
Ottermann, C.R. 205

Pène, O. 115
Pain, J. 205
Pendleton, B.J. 2
Perrot, F. 205
Pirner, H.J. 14,414
Praszalowicz, M. 133

Rafelski, J. 340
Raynal, J.-C. 115
Richard, J.M. 42,148,414
Rith, K. 245
Roy-Stéphan, M. 410

Schatzman, E. 449
Schütte, D. 168
Silvestre-Brac, B. 42,148,311
Smith, G.A. 197,202,219,404
Soffer, J. 173
Stancu, Fl. 46
Stassart, P. 46
Stiebing, K.E. 372

Tacik, T. 205
Taxil, P. 184
Tecchio, L. 414

Van den Brandt, B. 205
Vercellin, E. 205

Weise, W. 101
Werner, E. 155
Wirzba, A. 72

Yonnet, J. 205

RAYMOND H. FOGLER LIBRARY
DATE DUE

RECALL AFTER
DEC 16 1988